MW00512203

Data Integrity and Data Governance

Practical Implementation in Regulated Laboratori

Data Integrity and Data Governance

Practical Implementation in Regulated Laboratories

By

R. D. McDowall
R.D. McDowall Ltd, UK
Email: rdmcdowall@btconnect.com

Print ISBN: 978-1-78801-281-2
PDF ISBN: 978-1-78801-327-7
EPUB ISBN: 978-1-78801-662-9

A catalogue record for this book is available from the British Library

Reprinted 2020

The Royal Society of Chemistry is a charity, registered in England and Wales, Number 207890, and a company incorporated in England by Royal Charter (Registered No. RC000524), registered office: Burlington House, Piccadilly, London W1J 0BA, UK, Telephone: +44 (0) 20 7437 8656.

Visit our website at www.rsc.org/books

Printed in the United Kingdom by CPI Group (UK) Ltd, Croydon, CR0 4YY, UK

Preface

Data integrity and data governance are the hottest topics within the pharmaceutical industry now and will continue to be for a long time. However, data integrity has a history of over twenty-five years, initially limited to a few major cases but has gathered pace as the regulatory authorities have seen the extent and depth of either poor record data practices or deliberate falsification of records over the past ten to fifteen years. Data integrity is not limited to a single company, country or continent but is a world-wide issue as the pharmaceutical supply chain is now globalised.

In response to these data integrity non-compliances, the regulatory authorities have issued data integrity guidance after guidance document. These guidances call for data governance and data integrity controls within companies and provide a definition of data governance as *the arrangements to ensure that data, irrespective of the format in which they are generated, are recorded, processed, retained and used to ensure the record throughout the data lifecycle*. However, "*the arrangements*" provide scant detail of what to do or indeed how they should be implemented. Industry guidance documents attempt to go further but the guidance tends to be broad brush to provide information on what should be done but again little detail on the how, especially for regulated laboratories. In addition, qualification of analytical instruments and validation of analytical procedures are both omitted from these all regulatory and industry guidance documents but these are essential components of both data integrity but also data quality in any regulated laboratory.

Therefore, this book has been written to address the lack of specific guidance in GXP regulated analytical laboratories for data integrity from sampling to reporting the final results. However, on its own this is insufficient as the subject of data governance is also a mandatory requirement and provides the overarching umbrella for data integrity. Senior management are

Data Integrity and Data Governance: Practical Implementation in Regulated Laboratories
By R. D. McDowall

Published by the Royal Society of Chemistry, www.rsc.org

responsible for leadership developing data integrity policies, resourcing the data integrity programme and for creating the environment for an open culture and addressing breaches of corporate data integrity policies.

This book is practical and is intended to take data integrity and data governance from the boardroom to the laboratory bench and provides details for effective implementation. It looks in more detail about the integrity of existing paper based processes and computerised workflows to highlight the issues that face many regulated analytical laboratories working in or for the pharmaceutical industry. This book also proposes ways forward for the laboratories to ensure the integrity of these processes and the data generated by them. However, in many pharmaceutical laboratories, management are wanting quick fixes based on procedural controls that will result in a waste of money and resources.

Solving data integrity problems should result in the implementation of solutions that make laboratories more effective and efficient, as there is little point spending money to paper over cracks in a process without providing significant business benefit. Unfortunately, minimalist remediation is often all that some companies are prepared to consider. However, short remediation is acceptable but long-term solutions are much more beneficial and cost effective and are, in the opinion of the author, the only way forward to ensure data integrity issues are resolved in the laboratory. Short term remediation such as procedural controls will result in inefficient and error prone processes that will delay analytical work. However, management do not think through the consequences of this short-sighted approach.

Quality oversight is an important facet within a comprehensive data governance and data integrity framework. This involves raising data integrity concerns, conducting data integrity audits as well as investigations into violations of integrity policies and procedures. Finally, how outsourcing work should be handled in today's data integrity environment is discussed from assessment of potential supplier's data governance and data integrity approaches to audits of the work carried out on a company's behalf.

The book is intended to be a comprehensive approach to data governance and data integrity in regulated laboratories.

Bob McDowall
Bromley, UK

Acknowledgements

I am grateful for the help, discussions and advice from Chris Burgess, Mark Newton and Paul Smith during the writing of this book. Sections of Chapters 11 and 24 have been co-written by Chris Burgess and myself, Paul Smith and I have collaborated on ideas and input to Chapter 14. Mark Newton and I have worked on Chapter 19 on quality metrics and as well as input to Chapter 16 on Performing an Analysis.

Data Integrity and Data Governance: Practical Implementation in Regulated Laboratories
By R. D. McDowall

Published by the Royal Society of Chemistry, www.rsc.org

Glossary, Abbreviations and Data Integrity Terms

Data integrity terms presented here are quoted from several sources that can be found in the references at the end of this section. In some cases, a term may have several definitions from different sources to illustrate the differences between regulatory authorities. Also, an authority may have published two or three versions of guidance over time and the same term may have different definitions from the same authority that are included here, this is deliberate so that readers can see how the definition of a single term has evolved over time. Note that detailed references are not given for all terms in this glossary as the focus is on data integrity terms.

Glossary

Term	Meaning
Acceptance Criteria	The criteria a system must meet to satisfy a test or other requirement.[1]
ALCOA	A commonly used acronym for "attributable, legible, contemporaneous, original and accurate".[2]
ALCOA+	A commonly used acronym for "attributable, legible, contemporaneous, original and accurate", which puts additional emphasis on the attributes of being complete, consistent, enduring and available – implicit basic ALCOA principles.[2]

(continued)

Data Integrity and Data Governance: Practical Implementation in Regulated Laboratories
By R. D. McDowall

Published by the Royal Society of Chemistry, www.rsc.org

Glossary (*continued*)

Term	Meaning
ALCOA-plus	The guidance refers to the acronym ALCOA rather than "ALCOA+". ALCOA being Attributable, Legible, Contemporaneous, Original, and Accurate and the "+" referring to Complete, Consistent, Enduring, and Available. ALCOA was historically regarded as defining the attributes of data quality that are suitable for regulatory purposes. The "+" has been subsequently added to emphasise the requirements. There is no difference in expectations regardless of which acronym is used since data governance measures should ensure that data is complete, consistent, enduring and available throughout the data lifecycle.[3]
Analytical Instrument Qualification (AIQ)	AIQ is the collection of documented evidence that an instrument performs suitably for its intended purpose. Use of a qualified instrument in analyses contributes to confidence in the validity of generated data.[4]
Application	Software installed on a defined platform/hardware providing specific functionality.[5]
Archival	Archiving is the process of protecting records from the possibility of being further altered or deleted, and storing these records under the control of independent data management personnel throughout the required retention period. Archived records should include, for example, associated metadata and electronic signatures.[2]
Archive	Long term, permanent retention of completed data and relevant metadata in its final form for the purposes of reconstruction of the process or activity.[6] A designated secure area or facility (*e.g.* cabinet, room, building or computerised system) for the long term, permanent retention of complete data and relevant metadata in its final form for the purposes of reconstruction of the process or activity.[7] A designated secure area or facility (*e.g.* cabinet, room, building or computerised system) for the long term, retention of data and metadata for the purposes of verification of the process or activity.[3]
Archivist	An independent individual designated in good laboratory practice (GLP) who has been authorised by management to be responsible for the management of the archive, *i.e.* for the operations and procedures for archiving.[2]

Glossary (*continued*)

Term	Meaning
Audit	An audit is a formal, independent, disciplined and objective review activity designed to assess the performance of a process or system with regards to established regulations and procedures. There are internal audits, second party audits (between two companies) and third-party audits (using an independent auditor).
Audit Trail	GMP/GDP audit trails are metadata that are a record of GMP/GDP critical information (for example, the change or deletion of GMP/GDP relevant data), which permit the reconstruction of GMP/GDP activities.[6] A secure, computer-generated, time-stamped electronic record that allows for reconstruction of the course of events relating to the creation, modification, or deletion of an electronic record. An audit trail is a chronology of the "who, what, when, and why" of a record. Electronic audit trails include those that track creation, modification, or deletion of data (such as processing parameters and results) and those that track actions at the record or system level (such as attempts to access the system or rename or delete a file).[8] Audit trail means a secure, computer-generated, time-stamped electronic record that allows for reconstruction of the course of events relating to the creation, modification, or deletion of an electronic record. An audit trail is a chronology of the "who, what, when, and why" of a record.[2] Audit trails are metadata that are a record of critical information (for example, the change or deletion of relevant data) that permit the reconstruction of activities.[7] The audit trail is a form of metadata containing information associated with actions that relate to the creation, modification or deletion of GXP records. An audit trail provides for secure recording of life-cycle details such as creation, additions, deletions or alterations of information in a record, either paper or electronic, without obscuring or overwriting the original record. An audit trail facilitates the reconstruction of the history of such events relating to the record regardless of its medium, including the "who, what, when and why" of the action.[3]

(*continued*)

Glossary *(continued)*

Term	Meaning
Back-up	A copy of current (editable) data, metadata and system configuration settings (*e.g.* variable settings that relate to an analytical run) maintained for the purpose of disaster recovery.[6] A true copy of the original data that is maintained securely throughout the records retention period (for example, § 211.180). The backup file should contain the data (which includes associated metadata) and should be in the original format or in a format compatible with the original format.[8] A copy of one or more electronic files created as an alternative in case the original data or system are lost or become unusable (*e.g.* in the event of a system crash or corruption of a disk). It is important to note that backup differs from archival in that back-up copies of electronic records are typically only temporarily stored for the purposes of disaster recovery and may be periodically overwritten. Such temporary back-up copies should not be relied upon as an archival mechanism.[2] A copy of current (editable) data, metadata and system configuration settings (variable settings that relate to a record or analytical run) maintained for the purpose of disaster recovery.[7] A copy of current (editable) data, metadata and system configuration settings maintained for recovery including disaster recovery.[3] FDA uses the term backup in § 211.68(b) to refer to a true copy of the original data that is maintained securely throughout the records retention period (for example, § 211.180). The backup file should contain the data (which includes associated metadata) and should be in the original format or in a format compatible with the original format. This should not be confused with backup copies that may be created during normal computer use and temporarily maintained for disaster recovery (*e.g.*, in case of a computer crash or other interruption).[8]
Bespoke/Customised Computerised System	A computerised system individually designed to suit a specific business process.[9]
Boundary Value	A minimum or maximum input, output or internal data value, applicable to a system.[1]
Branch Testing	Execution of tests to assure that every branch alternative has been exercised at least once.

Glossary (*continued*)

Term	Meaning
Calibration	The demonstration that a particular instrument or device results within specified limits by comparison with those produced by a reference or traceable standard over an appropriate range of measurements.[10] An operation that, under specified conditions, in a first step, establishes a relation between the quantity values, with measurement uncertainties provided by measurement standards, and corresponding indications with associated measurement uncertainties and, in a second step, uses this information to establish a relation for obtaining a measurement result from an indication. Note that: 1. A calibration may be expressed by a statement, calibration function, calibration diagram, calibration curve, or calibration table. In some cases, it may consist of an additive or multiplicative correction of the indication with associated measurement uncertainty. 2. Calibration should not be confused with adjustment of a measuring system, often mistakenly called "self-calibration", or with verification of calibration. 3. Often, the first step alone in the above definition is perceived as being calibration.[4]
Change	The addition, modification or removal of approved, supported or baselined hardware, network, software, application, environment, system, desktop build or associated documentation (ITIL).
Change Advisory Board	A group of people who can give expert advice to change management on the implementation of changes. This board is likely to be made up of representatives from all areas within IT and representatives from business units (ITIL).
Change Control	A formal monitoring system by which qualified representatives of appropriate disciplines review proposed or actual changes that might affect a validated status to determine the need for corrective action that would assure that the system retains its validated state.[11]
Change History	Auditable information that records, for example, what was done, when it was done by who and why (ITIL).
Change Management	Process of controlling Changes to the system or any aspect of services, in a controlled manner, enabling approved Changes with minimum disruption (ITIL).

(*continued*)

Glossary *(continued)*

Term	Meaning
Code of Federal Regulations (CFR)	The codification of the general rules of the United States of America published in the Federal Register by the executive departments and agencies of the Federal Government. Divided into 50 titles that represent the areas regulated by the US Government.
Commercial off the shelf software	Software commercially available, whose fitness for use is demonstrated by a broad spectrum of users.[9]
	Author note: this term can be abused and can be used to confuse. GAMP software categories should be used instead.
Commissioning	The setting up, adjustment and testing of equipment or a system to ensure that it meets all the requirements, specified in the user requirements specification, and capacities specified by the designer or developer. Commissioning is carried out before qualification and validation (WHO GMP).
Computer Hardware	Various hardware components in the computer system, including the central processing unit, printer, screen and other related apparatus.
Computer System	Computer hardware components assembled to perform in conjunction with a set of programs, which are collectively designed to perform a specific function or group of functions.
	A group of hardware components and associated software designed and assembled to perform a specific function or group of functions.[9]
Computerised System	A system including the input of data, electronic processing and the output of information to be used either for reporting or automatic control.[5]
	People, machines, and methods organised to accomplish a set of specific functions. Computer or related systems can refer to computer hardware, software, peripheral devices, networks, cloud infrastructure, operators, and associated documents (*e.g.*, user manuals and standard operating procedures).[12]
	A computerised system collectively controls the performance of one or more automated processes and/or functions. It includes computer hardware, software, peripheral devices, networks and documentation, *e.g.* manuals and standard operating procedures, as well as the personnel interfacing with the hardware and software, *e.g.* users and information technology support personnel.[2]
Computerised System Specification	A document or set of documents that describe how a computerised system will satisfy the system requirements of the computer related system.

Glossary (*continued*)

Term	Meaning
Computer System Transactions	A computerised system transaction is a single operation or sequence of operations performed as a single logical "unit of work". The operation(s) that makes a transaction may not be saved as a permanent record on durable storage until the user commits the transaction through a deliberate act (*e.g.* pressing a save button), or until the system forces the saving of data.[3]
Computerised System Validation	Establishing documented evidence that provides a high degree of assurance that a specific computer related system will consistently operate in accordance with predetermined specifications.[1] Confirmation by examination and provision of objective evidence that software specifications conform to user needs and intended uses, and that the particular requirements implemented through software can be consistently fulfilled.[13]
Control Strategy	A planned set of controls, derived from current protocol, test article or product and process understanding, which assures protocol compliance, process performance, product quality and data reliability, as applicable. The controls should include appropriate parameters and quality attributes related to study subjects, test systems, product materials and components, technologies and equipment, facilities, operating conditions, specifications and the associated methods and frequency of monitoring and control.[2]
Configuration – 1	The arrangement of a computer system or component as defined by the number, nature, and interconnections of its constituent parts (IEEE).
Configuration – 2	Changing the business process automated by an application by modifying the parameters within the software provided by the supplier (also known as parameterisation).
Configuration Baseline	Configuration of a system established at a specific point in time, which captures the structure and details of the system and enables that system to be rebuilt at a later date (ITIL).
Configuration Item (CI)	Component of an infrastructure or system or an item such as Request for Change that is under the control of Configuration Management. Configuration Items may vary widely in complexity, size and type – from an entire system (including all hardware, software and documentation) to a single module or a minor hardware component (ITIL).

(*continued*)

Glossary (*continued*)

Term	Meaning
Configuration Management	A system for identifying the configuration of hardware, software, firmware or documentation of a computerised system at discrete points in time with the purpose of systematically controlling changes to the configuration and maintaining the integrity and traceability of the configuration throughout the system life cycle.
Corrective Action and Preventative Action (CAPA)	System for implementing corrective actions and preventative actions resulting from the investigation of non-conformances, deviations, audits, regulatory inspections and findings.[14]
	Actions taken to improve an organisation's processes and to eliminate causes of non-conformities or other undesirable situations. CAPA is a concept common across the GXPs (good laboratory practices, good clinical practices and good manufacturing practices), and numerous International Organization for Standardization business standards.[2]
COTS Software	Commercial off the shelf software application is used as is without altering the basic program.
	Configurable off the shelf software applications that can be configured to specific user applications by "filling in the blanks" without altering the basic program.
	Author Note: This term should not be used in a validation project as the term is confusing. A much better approach is to use the GAMP Software category.
Controlled Function	A process and any related equipment controlled by a computer system.
Customisation	Changing the business process automated by an application by writing software modules to add to an existing commercial application.
	Note enhancement of an application *via* a supplier language is custom code.
Data	Facts, figures and statistics collected together for reference or analysis.[6]
	Data means all original records and true copies of original records, including source data and metadata and all subsequent transformations and reports of these data, which are generated or recorded at the time of the GXP activity and allow full and complete reconstruction and evaluation of the GXP activity.[6]
	Facts and statistics collected together for reference or analysis. Data governance measures should also ensure that data are compete, consistent and enduring throughout the lifecycle.[7]

Glossary (*continued*)

Term	Meaning
Data Governance	The sum total of arrangements to ensure that data, irrespective of the format in which it is generated, is recorded, processed, retained and used to ensure a complete, consistent and accurate record throughout the data lifecycle.[6] The totality of arrangements to ensure that data, irrespective of the format in which they are generated, are recorded, processed, retained and used to ensure a complete, consistent and accurate record throughout the data life cycle.[7] The sum total of arrangements to ensure that data, irrespective of the format in which it is generated, is recorded, processed, retained and used to ensure a complete, consistent and accurate record throughout the data lifecycle.[7] The arrangements to ensure that data, irrespective of the format in which they are generated, are recorded, processed, retained and used to ensure the record throughout the data lifecycle.[3]
Data Integrity	The extent to which all data are complete, consistent and accurate throughout the data lifecycle.[6] Data integrity refers to the completeness, consistency, and accuracy of data. Complete, consistent, and accurate data should be attributable, legible, contemporaneously recorded, original or a true copy, and accurate (ALCOA).[8] Data integrity is the degree to which data are complete, consistent, accurate, trustworthy and reliable and that these characteristics of the data are maintained throughout the data life cycle. The data should be collected and maintained in a secure manner, such that they are attributable, legible, contemporaneously recorded, original or a true copy and accurate. Assuring data integrity requires appropriate quality and risk management systems, including adherence to sound scientific principles and good documentation practices.[2] The extent to which all data are complete, consistent and accurate throughout the data lifecycle.[15] Data integrity is the degree to which data are complete, consistent, accurate, trustworthy, reliable and that these characteristics of the data are maintained throughout the data life cycle. The data should be collected and maintained in a secure manner, so that they are attributable, legible, contemporaneously recorded, original (or a true copy) and accurate. Assuring data integrity requires appropriate quality and risk management systems, including adherence to sound scientific principles and good documentation practices.[3]

(*continued*)

Glossary (*continued*)

Term	Meaning
Data Lifecycle	All phases in the life of the data (including raw data) from initial generation and recording through processing (including transformation or migration), use, data retention, archive/retrieval and destruction.[6] All phases of the process by which data are created, recorded, processed, reviewed, analysed and reported, transferred, stored and retrieved and monitored until retirement and disposal. There should be a planned approach to assessing, monitoring and managing the data and the risks to those data in a manner commensurate with potential impact on patient safety, product quality and/or the reliability of the decisions made throughout all phases of the data life cycle.[2] All phases in the life of the data (including raw data) from initial generation and recording through processing (including analysis, transformation or migration), use, data retention, archive/retrieval and destruction.[15] All phases in the life of the data from generation and recording through processing (including analysis, transformation or migration), use, data retention, archive/retrieval and destruction.[3]
Data Processing	A sequence of operations performed on data in order to extract, present or obtain information in a defined format. Examples might include: statistical analysis of individual patient data to present trends or conversion of a raw electronic signal to a chromatogram and subsequently a calculated numerical result.[3]
Data Quality	The assurance that data produced is exactly what was intended to be produced and fit for its intended purpose. This incorporates ALCOA.[3]
Data Retention	Data retention may be classified as either archive (protected data for long term storage) or backup (dynamic data for the purposes of disaster recovery).[7]
Data Review	There should be a procedure that describes the process for the review and approval of data. Data review should also include a review of relevant metadata, including audit trails.[7]
Data Transfer/Migration	Data transfer is the process of transferring data and metadata between storage media types or computer systems. Data migration changes the format of data to make it usable or visible on an alternative computerised system. Data transfer/migration should be designed and validated to ensure that data integrity principles are maintained.[7]

Glossary (*continued*)

Term	Meaning
Dynamic Data/Record	Dynamic data means that the record format allows interaction between the user and the record content, *e.g.* interpretation of a chromatography data file for integration of the peaks of interest.[16] Records in dynamic format, such as electronic records, that allow for an interactive relationship between the user and the record content. For example, electronic records in database formats allow the user to track, trend and query data; chromatography records maintained as electronic records allow the user (with proper access permissions) to reprocess the data and expand the baseline to view the integration more clearly.[2] Information that is originally captured in a dynamic state should remain available in that state.[3]
Design Qualification (DQ)	The documented verification that the proposed design of the facilities, systems and equipment is suitable for the intended purpose.[11] DQ is the documented collection of activities that define the functional and operational specifications of the instrument, including the criteria for selection of the supplier, based on the intended purpose of the instrument. DQ states what the laboratory wants the instrument to do and shows that the selected one is suitable.[17] DQ is the documented collection of activities that define the functional and operational specifications and intended purpose of the instrument. DQ states what the laboratory wants the instrument to do and shows that the selected one is suitable.[4]
Developer	The company, group or individuals responsible for developing a system or some portion of a software application.
Electronic Signature	A signature in digital form (bio-metric or non-biometric) that represents the signatory. This should be equivalent in legal terms to the handwritten signature of the signatory.[3]
Excluding Data	Data may only be excluded where it can be demonstrated through sound science that the data is anomalous or non-representative. In all cases, this justification should be documented and considered during data review and reporting. All data (even if excluded) should be retained with the original data set, and be available for review in a format that allows the validity of the decision to exclude the data to be confirmed.[7]
Exception Report	A validated search tool that identifies and documents predetermined "abnormal" data or actions, which requires further attention or investigation by the data reviewer.[6]
Flat File	A "flat file" is an individual record that may not carry any additional metadata with it, other than that which is included in the file itself.[6] A "flat file" is an individual record that may not carry any additional metadata with it, other than that included in the file itself.[7]

(*continued*)

Glossary *(continued)*

Term	Meaning
Firmware	A software program permanently recorded in a hardware device such as a chip or EPROM. Typified by GAMP software category 2 (now unfortunately discontinued) or USP <1058> Group B analytical instruments.[4]
Fully Electronic Approach	This term refers to use of a computerised system in which the original electronic records are electronically signed.[2]
Functional Requirements	Statements that describe functions a computer related system must be capable of performing.
Functional Specification	Statements of how the computerised system will satisfy functional requirements of the computer related system. Typically for a CDS validation, this is replaced by a configuration specification.
Functional Testing	A process for verifying that software, a system or a system component performs its intended functions.
Good Data and Record Management Practices	The totality of organised measures that should be in place to collectively and individually ensure that data and records are secure, attributable, legible, traceable, permanent, contemporaneously recorded, original and accurate and that if not robustly implemented can impact on data reliability and completeness and undermine the robustness of decision making based upon those data records.[2]
Good Documentation Practices	The measures that collectively and individually ensure documentation, whether paper or electronic, is secure, attributable, legible, traceable, permanent, contemporaneously recorded, original and accurate.[2]
GXP	Acronym for the group of good practice guides governing the preclinical, manufacturing, testing, storage, and distribution for regulated pharmaceuticals, biologicals and medical devices, such as good laboratory practice, good clinical practice and good manufacturing practice.[2]
Harm	Physical injury or damage to the health of people, or damage to property or the environment. Note this is for a medical device; this needs to be interpreted as the consequences of a software error or malfunction of the system.[18]
Hazard	A potential source of harm.[18]
Hybrid System	The use of a computerised system in which there is a combination of original electronic records and paper records that comprise the total record set that should be reviewed and retained. The hybrid approach requires a secure link between all record types, including paper and electronic, throughout the records retention period.[2]
Inspection	The action by a regulatory authority of conducting an official review of facilities, documents, records, systems and any other resources to determine compliance with applicable regulations.

Glossary (*continued*)

Term	Meaning
Installation Qualification (IQ)	Documented verification that the facilities, systems and equipment, as installed or modified, comply with the approved design and the manufacturer's recommendations.[11] IQ is the documented collection of activities necessary to establish that an instrument is delivered as designed and specified, and is properly installed in the selected environment, and that this environment is suitable for the instrument.[17] IQ is the documented collection of activities necessary to establish that an instrument is delivered as designed and specified, is properly installed in the selected environment, and that this environment is suitable for the instrument.[4]
Instrument	Instrument includes any apparatus, equipment, instrument, or instrument system used in pharmacopoeial analyses.[4]
IT Infrastructure	The hardware and software such as networking software and operating systems, which makes it possible for the application to function.[9]
Life cycle (Computerised System)	All phases in the life of the system from initial requirements until retirement including design, specification, programming, testing, installation, operation, and maintenance.[9]
Maintenance	Actions performed to keep an analytical instrument in a state of proper function so that it continues to operate within the boundaries set during qualification or validation.[4]
Metadata	Data that describe the attributes of other data, and provide context and meaning.[6] Metadata is the contextual information required to understand data. A data value is by itself meaningless without additional information about the data. Metadata is often described as data about data. Metadata is structured information that describes, explains, or otherwise makes it easier to retrieve, use, or manage data.[8] Metadata are data about data that provide the contextual information required to understand those data. These include structural and descriptive metadata. Such data describe the structure, data elements, interrelationships and other characteristics of data. They also permit data to be attributable to an individual. Metadata necessary to evaluate the meaning of data should be securely linked to the data and subject to adequate review.[2] Metadata are data that describe the attributes of other data and provide context and meaning. Typically, these are data that describe the structure, data elements, inter-relationships and other characteristics of data, *e.g.* audit trails. Metadata also permit data to be attributable to an individual (or if automatically generated, to the original data source). Metadata forms an integral part of the original record. Without metadata, the data has no meaning.[3]

(*continued*)

Glossary *(continued)*

Term	Meaning
Ongoing Evaluation	The dynamic process employed after a system's initial validation to maintain its validated state.
Operating Environment	Those conditions and activities interfacing directly or indirectly with the system of concern, control of which can affect the system's validated state.
Operating System	A set of software programs provided with a computer that function as the interface between the hardware and the applications program.
Operational Qualification (OQ)	The documented verification that the facilities, systems and equipment, as installed or modified, perform as intended throughout the anticipated operating ranges.[11]
	Operational Qualification (OQ) is the documented collection of activities necessary to demonstrate that an instrument will function according to its operational specification testing in the selected environment. OQ demonstrates fitness of purpose for the user's ways of working, and should reflect the contents of the DQ document.[17]
	OQ is the documented collection of activities necessary to demonstrate that an instrument will function according to its operational specification testing in the selected environment. OQ demonstrates fitness for the selected use, and should reflect the contents of the DQ document.[4]
Original record	Data as the file or format in which it was originally generated, preserving the integrity (accuracy, completeness, content and meaning) of the record, *e.g.* original paper record of manual observation, or electronic raw data file from a computerised system.[7]
	The first or source capture of data or information, *e.g.* original paper record of manual observation or electronic raw data file from a computerised system, and all subsequent data required to fully reconstruct the conduct of the GXP activity. Original records can be Static or Dynamic.[3]
Out of Specification (OOS) Result	A reportable result outside of specification or acceptance criteria limits. As we are dealing with specifications, OOS results can apply to test of raw materials, starting materials, active pharmaceutical ingredients and finished products but not for in-process testing. If a system suitability test fails this will not generate an OOS result as the whole run would be invalidated, however, there needs to be an investigation of the failure.[19]
Out of Trend (OOT) Result	Not an out of specification result but does not fit with the expected distribution of results. This can include a single result outside of acceptance limits for a replicate result used to calculate a reportable result. If investigated, the same rules as OOS should be followed.

Glossary *(continued)*

Term	Meaning
Outsourced Activities	Activities conducted by a contract acceptor under a written agreement with a contract giver.[14]
Path Testing	Execution of important control flow paths throughout the program.
Performance Qualification (PQ)	The documented verification that the facilities, systems and equipment, as connected together, can perform effectively and reproducibly, based on the approved process method and product specification.[11]
	The documented verification that the integrated computerised system performs as intended in its normal operating environment, *i.e.* that computer related system performs as intended.[1]
	Performance Qualification (PQ) is the documented collection of activities necessary to demonstrate that an instrument consistently performs according to the specifications defined by the user, and is appropriate for the intended use. The PQ verifies the fitness for purpose of the instrument under actual conditions of use. After IQ and OQ have been performed, the instrument's continued suitability for its intended use is demonstrated through continued performance qualification.[4,17]
Policy	A directive usually specifying what is to be accomplished.
Process	Structured activities intended to achieve a desired outcome.
Process Owner	The person responsible for the business process.[9]
Prospective Validation	Validation carried out before routine use of the system.[11]
Qualification	Action of proving that any equipment works correctly and actually leads to the expected results. The word validation is sometimes widened to incorporate the concept of qualification.[5]
	Action of proving and documenting that equipment or ancillary systems are properly installed, work correctly and actually lead to the expected results. Qualification is part of validation, but the individual qualification steps alone do not constitute process validation.[10]
	Action of proving that any instrument works correctly and delivers the expected results; demonstration of fitness for purpose.[4]
Qualification Protocol	A prospective experimental plan that when executed is intended to produce documented evidence that a system or subsystem has been qualified properly.
Quality	The degree of which a set of properties of a product, system or process fulfils requirements.[20]
Quality Assurance	The sum total of organised arrangements made with the object of ensuring that all APIs are of the quality required for their intended use and that quality systems are maintained.[10]
Quality Control	Checking or testing that specifications are met.[10]

(continued)

Glossary (*continued*)

Term	Meaning
Quality Unit	An organisational unit independent of production that fulfils both Quality Assurance and Quality Control responsibilities. This can be in the form of separate QC and QA units or a single individual or group depending on the size of the organisation.[10]
Raw Data	Original records, retained in the format in which they were originally generated (*i.e.* paper or electronic), or as a "true copy". Raw data must be contemporaneously and accurately recorded by permanent means. The definition of "original records" currently varies across regulatory documents. By its nature, paper copies of raw data generated electronically cannot be considered as "raw data". Raw data must permit the full reconstruction of the activities resulting in the generation of the data. In the case of basic electronic equipment that does not store electronic data, or provides only a printed data output (*e.g.* balance or pH meter), the printout constitutes the raw data.[21] Raw data is defined as the original record (data) that can be described as the first-capture of information, whether recorded on paper or electronically.[3] Raw data means any laboratory worksheets, records, memoranda, notes, or exact copies thereof, that are the result of original observations and activities of a nonclinical laboratory study and are necessary for the reconstruction and evaluation of the report of that study 21 CFR 58.3(k).[22]
Recording Data	Companies should have an appropriate level of process understanding and technical knowledge of systems used for data recording, including their capabilities, limitations and vulnerabilities.[21]
Reportable Result	The term reportable result as used in this document means a final analytical result. This result is appropriately defined in the written approved test method and derived from one full execution of that method/procedure, starting from the sample. Compared with the specification to determine pass/fail of a test.[19]
Retrospective Validation	Validation of a process or system after it has become operational.[10]
Revalidation	A repeat of all or part of the validation to provide assurance that changes in the system introduced in accordance with change control procedures do not adversely affect the system operation or data integrity.
Risk	Combination of the probability of occurrence of harm and the severity of that harm.[18]
Risk analysis	The systematic use of available information to identify hazards and estimate the risk.[18]
Risk assessment	The overall process of a risk analysis and risk evaluation.[18]

Glossary (*continued*)

Term	Meaning
Risk evaluation	Judgement, on the basis of risk analysis, of whether a risk that is acceptable has been achieved in a given context.[18]
Risk management	The systematic application of management policies, procedures and practices to the tasks of analysing, evaluating and controlling risk.[18]
Self-Inspection	An internal evaluation of compliance with applicable regulation in relevant regulated areas.[2]
Senior Management	Person(s) who direct and control a company or site at the highest levels with the authority and responsibility to mobilise resources within the company or site.[2]
Severity	Measure of the possible consequences of a hazard.[18]
Software Configuration	Adaptation of software functions to a business process using tools provided within the application by the supplier of the software.[4]
Software Customisation	Changing the way software automates a business process by the addition of externally custom coded software modules using a recognised programming language or the development of macros within the application software.[4]
Standard Operating Procedure	Instructions that specify how an activity or process is to be performed
Static Data/Record	Static record format is used to indicate a fixed-data document such as a paper record or an electronic image.[8] A static record format, such as a paper or pdf record, is one that is fixed and allows little or no interaction between the user and the record content. For example, once printed or converted to static pdfs, chromatography records lose the capability of being reprocessed or enabling more detailed viewing of baselines.[2]
Structural Integrity	Software attributes reflecting the degree to which source code satisfies specified software requirements and conforms to contemporary software development.
Structural Verification	An activity intended to produce documented assurance that software has the appropriate structural integrity.
Supplier	The company or group responsible for developing, constructing and delivering a software application, computerised system or part of a system. This term is used generically and can mean the manufacturer, a vendor, a service agent, or a consultant, depending on the circumstances.[4]
System Owner	The person responsible for the availability, and maintenance of a computerised system and for the security of the data residing on that system.[9]
Third Party	Parties not directly managed by the holder of the manufacturing and/or import authorisation.[9]

(*continued*)

Glossary (*continued*)

Term	Meaning
True Copy/Certified Copy/ Verified Copy	A true copy is a copy of an original recording of data that has been verified and certified to confirm it is an exact and complete copy that preserves the entire content and meaning of the original record, including, in the case of electronic data, all essential metadata and the original record format as appropriate.[2] A copy of original information that has been verified as an exact (accurate and complete) copy having all of the same attributes and information as the original. The copy may be verified by dated signature or by a validated electronic signature. A true copy may be retained in a different electronic file format to the original record, if required, but must retain the equivalent static/dynamic nature of the original record.[7] A copy (irrespective of the type of media used) of the original record that has been verified (*i.e.* by a dated signature or by generation through a validated process) to have the same information, including data that describe the context, content, and structure, as the original.[3]
User	The company or group responsible for the operation of the system.
User access/system administrator roles	Full use should be made of access controls to ensure that people have access only to functionality that is appropriate for their job role, and that actions are attributable to a specific individual. Companies must be able to demonstrate the access levels granted to individual staff members and ensure that historical information regarding user access level is available. Controls should be applied at both the operating system and application levels. Shared logins or generic user access should not be used. Where the computerised system design supports individual user access, this function must be used. This may require the purchase of additional licences.[7]
Validated scanning process	A process whereby documents/items are scanned as a process with added controls such as location identifiers and OCR so that each page duplicated does not have to be further checked by a human.[3]
Validation – for intended purpose	Computerised systems should comply with regulatory requirements and associated guidance and be validated for their intended purpose. This requires an understanding of the computerised system's function within a process. For this reason, the acceptance of vendor-supplied validation data in isolation of system configuration and intended use is not acceptable. In isolation from the intended process or end user IT infrastructure, vendor testing is likely to be limited to functional verification only and may not fulfil the requirements for performance qualification.

Glossary (*continued*)

Term	Meaning
Validation Master Plan	A document providing information on the company's validation work programme. It should define details of and timescales for the validation work to be performed. Responsibilities relating to the plan should be stated.[23]
Validation Plan	A document that identifies all systems and subsystems involved in a specific validation effort and the approach by which they will be qualified and the total system will be validated; includes the identification of responsibilities and expectations.
Worst Case	Requirements or set of requirements that can include upper and lower limits and circumstances for a computerised system. Typically represent the limits of use within a specific laboratory.

Abbreviations

Abbreviation	Meaning
AIP	Application Integrity Policy
ALCOA	Attributable, Legible, Contemporaneous, Original, Accurate
ANSI	American National Standards Institute
API	Active Pharmaceutical Ingredient
ASTM	American Society for Testing and Materials
ATP	Analytical Target Profile
BCP	Business Continuity Plan
CAPA	Corrective Action and Preventative Action
CBER	Center for Biologics Evaluation and Research (FDA)
CDER	Center for Drug Evaluation and Research (FDA)
CDRH	Center for Devices and Radiological Health (FDA)
CDS	Chromatography Data System
CE	Capillary Electrophoresis
CFR	Code of Federal Regulations (*e.g.* 21 CFR 11)
CMC	Chemistry, Manufacturing and Controls (of a New Drug Application/Product Licence Application)
CMO	Contract Manufacturing Organisation
COA	Certificate of Analysis
COTS	Commercial/Configurable Off The Shelf (Software) This term is not recommended for use as it is confusing when not defined. Define software type using GAMP software categories.
CPG	Compliance Program Guide Compliance Policy Guide
CRO	Contract Research Organization
cGMP	current Good Manufacturing Practices
CS	Configuration Specification
CSF	Critical Success Factor
CSV	Computer System Validation
CTD	Common Technical Document
DHHS	Department of Health and Human Services (US)
DI	Data Integrity

(*continued*)

Abbreviations *(continued)*

Abbreviation	Meaning
DG	Data Governance
DQ	Design Qualification
DR	Disaster Recovery
DS	Design Specification
EC	European Community
ECD	Electron Capture Detector
EDMS	Electronic Document Management System
EIR	Establishment Inspection Report
ELN	Electronic Laboratory Notebook
EMA	European Medicines Agency
EMEA	European Agency for the Evaluation of Medical Products (now EMA)
EP	European Pharmacopoeia (Ph.Eur)
EU	European Union
FDA	Food and Drug Administration
FID	Flame Ionisation Detector
FMEA	Failure Mode and Effects Analysis
FOI (A)	Freedom of Information (Act)
FR	Federal Register
FS	Functional Specifications
GAMP	Good Automated Manufacturing Practice guidelines
GC	Gas Chromatography
GDocP	Good Documentation Practice
GLP	Good Laboratory Practice
GRDP	Good Records and Documentation Practice
GXP	Good X Practices (where X can be clinical, laboratory and/or manufacturing)
GMP	Good Manufacturing Practice
HACCP	Hazard Analysis Critical Control Point
HPLC	High Pressure/Performance Liquid Chromatography
IaaS	Infrastructure as a Service
ICH	International Conference on Harmonisation (1990–2015) International Council for Harmonisation of Technical Requirements for Pharmaceuticals for Human Use (from October 2015)
ICP	Inductively Coupled Plasma
IEEE	Institute of Electrical and Electronic Engineers, Inc.
ILT	Instructor Lead Training
IND	Investigational New Drug Application
IQ	Installation Qualification
IP	Internet Protocol
IR	Infra-Red
ISO	International Organization for Standardization
ISPE	International Society of Pharmaceutical Engineers
IT	Information Technology
ITIL	IT Infrastructure Library
ITT	Invitation to Tender
KPI	Key Performance Indicator
LAN	Local Area Network
LES	Laboratory Execution System
LIMS	Laboratory Information Management System
MBA	Minus Brain Activity

Abbreviations (*continued*)

Abbreviation	Meaning
MHLW	Ministry of Health, Labour and Welfare (Japan)
MHRA	Medicines and Healthcare products Regulatory Agency (UK)
MOU	Memorandum of Understanding
MRA	Mutual Recognition Agreement
MS	Mass Spectrometer
NCE	New Chemical Entity
NDA	New Drug Application
NIR	Near Infra-Red
NIST	National Institute of Standards and Technology (Gaithersville, Maryland, USA)
NMR	Nuclear Magnetic Resonance
OECD	Organization for Economic Cooperation and Development
OOE	Out Of Expectation
OOS	Out Of Specification
OOT	Out Of Trend
OQ	Operational Qualification
ORA	Office of Regulatory Affairs (FDA)
OTS	Off The Shelf (refers to software – use is not recommended)
PaaS	Platform as a Service
PAI	Pre-Approval Inspection
PDA	Parenteral Drug Association
PDF	Portable Document Format
Ph.Eur.	European Pharmacopoeia
PIC	Pharmaceutical Inspection Convention
PIC/S	Pharmaceutical Inspection Convention/Scheme
PKI	Public Key Infrastructure
PPQ	Procedure Performance Qualification
PPV	Procedure Performance Verification
PQ	Performance Qualification
PQS	Pharmaceutical Quality System
QA	Quality Assurance
QAU	Quality Assurance Unit
QC	Quality Control
QMS	Quality Management System
QP	Qualified Person
QRM	Quality Risk Management
R&D	Research and Development
RAID	Redundant Array of Inexpensive Disks
RFC	Request for Change
RFID	Radio Frequency Identity
RFP	Request for Proposal
RSD	Relative Standard Deviation
SaaS	Software as a Service
SAN	Storage Area Network
SDLC	System Development Life Cycle
SDMS	Scientific Data Management System
SILC	System Implementation Life Cycle
SLA	Service Level Agreement
SOP	Standard Operating Procedure
SPC	Statistical Process Control
SQA	Society for Quality Assurance

(*continued*)

Abbreviations *(continued)*

Abbreviation	Meaning
SRS	System Requirements Specification (equivalent to URS)
SST	System Suitability Test
TAT	Turn Around Time
TMU	Total Measurement Uncertainty
UAT	User Acceptance Testing
UPS	Uninterruptible Power Supply
URS	User Requirement Specifications
USB	Universal Serial Bus
USP	United States Pharmacopoeia
UV	Ultra Violet
VMP	Validation Master Plan
VSR	Validation Summary Report
WAN	Wide Area Network
WI	Work Instruction
WHO	World Health Organisation

References

1. *Technical Report 18: Validation of Computer-related Systems*, Parenteral Drug Association (PDA), Bethesda, MD, 1995.
2. *WHO Technical Report Series No. 996 Annex 5 Guidance on Good Data and Records Management Practices*, World Health Organisation, Geneva, 2016.
3. *MHRA GXP Data Integrity Guidance and Definitions*, Medicines and Healthcare products Regulatory Agency, London, 2018.
4. *USP 41 General Chapter <1058> Analytical Instrument Qualification*, United States Pharmacopoeia Convention, Rockville, MD, 2018.
5. *EudraLex – Volume 4 Good Manufacturing Practice (GMP) Guidelines, Glossary*, European Commission, Brussels, 2004.
6. *PIC/S PI-041 Draft Good Practices for Data Management and Integrity in Regulated GMP/GDP Environments*, Pharmaceutical Inspection Convention/Pharmaceutical Inspection Co-Operation Scheme, Geneva, 2016.
7. *MHRA GXP Data Integrity Definitions and Guidance for Industry, Draft Version for Consultation July 2016*, Medicines and Healthcare products and Regulatory Agency, London, 2016.
8. *FDA Draft Guidance for Industry Data Integrity and Compliance with cGMP*, Silver Spring, MD, USA, 2016.
9. *EudraLex – Volume 4 Good Manufacturing Practice (GMP) Guidelines, Annex 11 Computerised Systems*, European Commission, Brussels, 2011.
10. *EudraLex – Volume 4 Good Manufacturing Practice (GMP) Guidelines, Part 2-Basic Requirements for Active Substances Used as Starting Materials*, European Commission, Brussels, 2014.
11. *EudraLex – Volume 4 Good Manufacturing Practice (GMP) Guidelines, Annex 15 Qualification and Validation*, European Commission, Brussels, 2015.

12. *FDA Glossary of Computerized System and Software Development Terminology*, 1995, Available from: http://www.fda.gov/iceci/inspections/inspectionguides/ucm074875.htm#_top.
13. *FDA Guidance for Industry General Principles of Software Validation*, Food and Drug Administration, Rockville, MD, 2002.
14. *ICH Q10 Pharmaceutical Quality Systems*, International Conference on Harmonisation, Geneva, 2008.
15. *MHRA GMP Data Integrity Definitions and Guidance for Industry 2nd Edition*, Medicines and Healthcare products Regulatory Agency, London, 2015.
16. *MHRA Expectation Regarding Self Inspection and Data Integrity*, 2013 [cited 2013 01 Jan 2016], Available from: http://webarchive.nationalarchives.gov.uk/20141205150130/http://www.mhra.gov.uk/Howweregulate/Medicines/Inspectionandstandards/GoodManufacturingPractice/News/CON355490.
17. *USP 31 General Chapter <1058> Analytical Instrument Qualification*, United States Pharmacopoeial Convention, Rockville, MD, 2008.
18. *ISO 14971: Risk Management for Medical Devices*, International Standards Organisation, Geneva, 2012.
19. *FDA Guidance for Industry Out of Specification Results*, Food and Drug Administration, Rockville, MD, 2006.
20. *ICH Q9 Quality Risk Management*, International Conference on Harmonisation, Geneva, 2005.
21. *MHRA GMP Data Integrity Definitions and Guidance for Industry 1st Edition*, Medicines and Healthcare products Regulatory Agency, London, 2015.
22. *21 CFR 58 Good Laboratory Practice for Non-clinical Laboratory Studies*, Food and Drug Administration, Washington, DC, 1978.
23. *PIC/S Recommendations on Validation Master Plan, Installation and Operational Qualification, Non-sterlie Process Validation and Cleaning Validation (PI-006-3)*, Pharmaceutical Inspection Convention/Pharmaceutical Inspection Co-Operation Scheme (PIC/S), Geneva, 2007.

Contents

Data Integrity and Data Governance: Practical Implementation in Regulated Laboratories
By R. D. McDowall

Published by the Royal Society of Chemistry, www.rsc.org

CHAPTER 1

How to Use This Book and an Introduction to Data Integrity

Data Integrity is the hottest topic in the pharmaceutical industry now and will continue to be in the future. Regulatory authorities around the world have issued a number of guidance documents on the data integrity and data governance since the start of 2015 as well as two industry guidance documents. However, all documents are vague and most do not contain detailed examples or advice to help regulated laboratories to implement policies, procedures and processes to ensure integrity: they outline what needs to be done but there is often insufficient detail for effective implementation. From an analytical perspective there has not been a detailed focus on data integrity in a regulated analytical laboratory. Hence, the rationale for writing this book.

1.1 Aims and Objectives

The aim of this book is to provide practical and detailed advice on how to implement data integrity and data governance for regulated analytical laboratories working in the pharmaceutical and allied industries. Although the main thrust of the book is for chemical laboratories some microbiological analysis will also be discussed.

This book is written for analytical scientists, laboratory managers and supervisors and quality assurance personnel working in regulated laboratories in and for the pharmaceutical industry and who are involved with data integrity and data governance programmes. Where networked systems are discussed IT professionals may also find useful information here.

Data Integrity and Data Governance: Practical Implementation in Regulated Laboratories
By R. D. McDowall

Published by the Royal Society of Chemistry, www.rsc.org

1.2 Structure of This Book

This book is comprised of 24 chapters that are divided into five sections of this book, as follows:

1. How to Use this Book is covered in this chapter.
2. The Regulatory Environment is discussed in Chapters 2 and 3.
3. Data Governance is presented and explained in Chapters 4 to 10 as well as Chapter 19.
4. Data Integrity is covered in Chapters 11 to 18.
5. Quality Assurance Oversight and Outsourcing are discussed in Chapters 20–24.

The detailed structure of the book and the constituent chapters are shown in Figure 1.1 and discussed in more detail in the following sections of this chapter. It is important to understand that data integrity and data governance is a complex subject that is not just focused on the accuracy, completeness and correctness of numbers generated in a regulated laboratory.

1.2.1 Chapter Structure

The majority of chapters in this book are structured and written in the same way:

- A chapter starts with a brief overview why the chapter is important within the overall context of data integrity and data governance.
- This is followed by a section on regulatory requirements or regulatory guidance that are relevant to the chapter; thereby positioning the regulatory rationale for the topic of the chapter.
- Where appropriate, there is also the business rationale for the tasks contained in the chapter.
- Then there is a discussion of how to achieve the objective of each chapter. For example, if you are assessing a process or a computerised system the chapter discusses how this can be achieved and how to avoid some of the common pitfalls.
- Each chapter finishes with a list of the references used.

1.2.2 You Do Not Read the Regulations!

As many people working in the pharmaceutical industry do not read the applicable regulations or applicable guidance documents, the intention of this approach is to put the regulatory and business rationale for performing a task at the reader's fingertips. It also allows an individual chapter to stand alone if a quick reference to a specific data integrity or data governance topic is all that is required. Overall, the aim is to give any reader the practical basis and confidence to implement or perform any of the topics covered by this book.

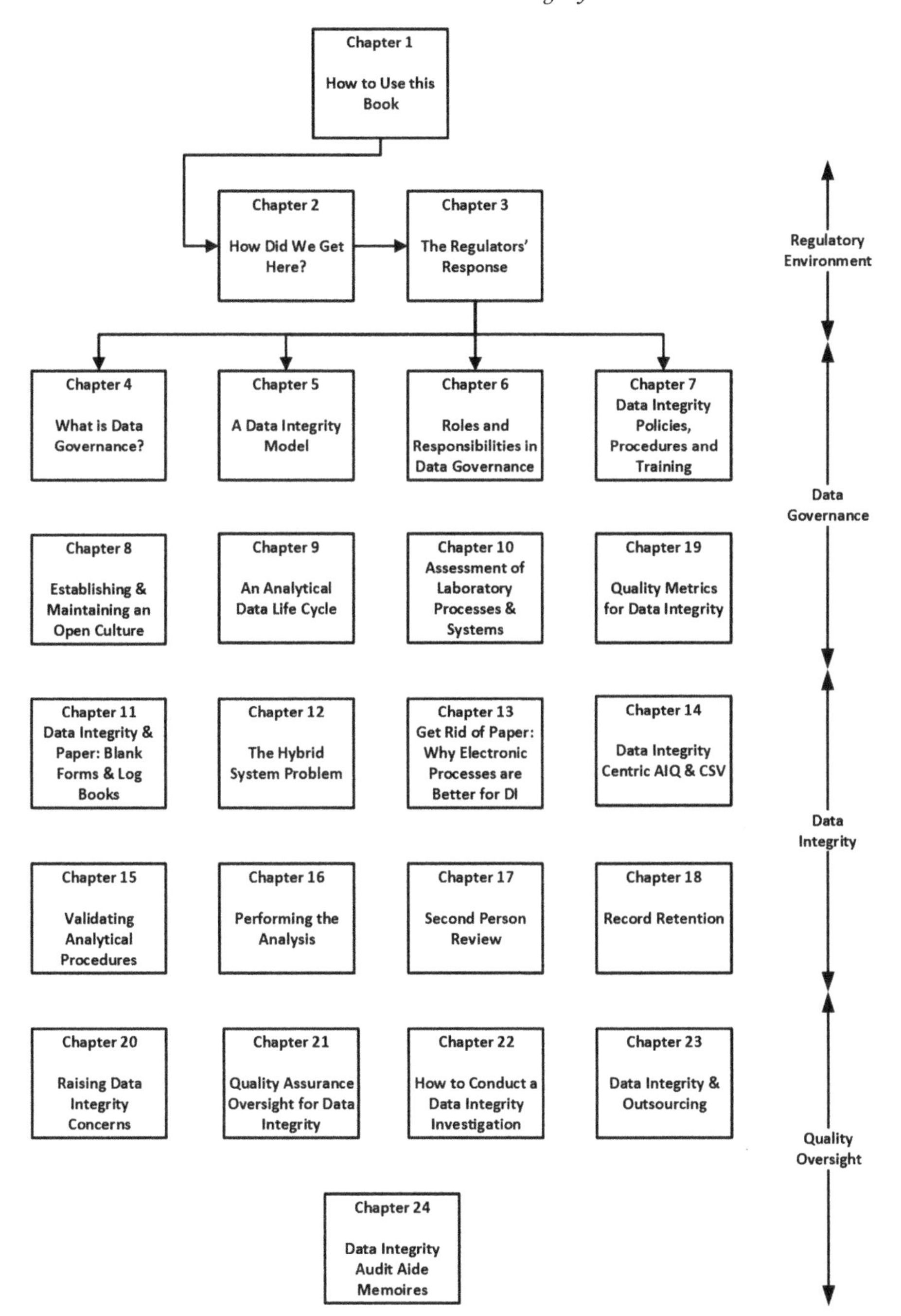

Figure 1.1 Outline structure of this book.

1.2.3 The Regulatory Environment

This topic is covered in two introductory chapters.

- Chapter 2, entitled *How Did We Get Here?*, provides the background to data integrity in regulated laboratories of the pharmaceutical industry over the past quarter of a century. The story starts with the Barr Laboratories court case and ruling in 1993[1–3] and then moves on to the Able Laboratories fraud case in 2005.[4] The latter case triggered regulatory authorities to change their inspection focus from paper to electronic records and consequently they discovered many more cases of data falsification and poor data management practices in companies throughout the world. The key compliance issues, derived mainly from FDA warning letters for paper records and computerised systems, are used to highlight areas of regulatory concern for laboratory operations.
- Chapter 3 is *The Regulators' Responses* and outlines the response to the increased occurrence of data falsification and poor data integrity practices by the various regulatory agencies by issuing guidance documents and where necessary updating regulations. This chapter looks at the various guidance documents issued. The first was the FDA's Inspection of Pharmaceutical Quality Control Laboratories issued in 1993 following the Barr Laboratories case.[5] Since 2015 there have been a many data integrity guidance documents issued from regulatory agencies and industry bodies.[6–12]

1.2.4 Data Governance

Data governance is a relatively new concept to the pharmaceutical industry and the term comprises several interlinked topics that are discussed over the course of eight chapters.

- Chapter 4, entitled *What is Data Governance?*, sets out the strands of data governance that are discussed in more detail in the rest of the book. Data governance is not new and has been used outside the pharmaceutical industry for at least 15 years. We will compare and contrast DG definitions from inside and outside the industry. From the data governance definitions and input from the various regulatory and industry guidance documents we will draw out the key features of a programme of work: roles and responsibilities, DI policy and training, culture and ethics, open and honest approach to work as examples of a corporate DI policy. Are there areas in an organisation, *e.g.* research, where data integrity should not apply?
- Chapter 5 presents a *Data Integrity Model*[13] consisting of four layers that describes data integrity within a pharmaceutical company from a GMP perspective: production, quality control and quality assurance,

however, this book will focus on an analytical laboratory with quality assurance oversight. The four layers of the Data Integrity Model comprise:

- Foundation: data governance and setting right culture and ethos for data integrity.
- Level 1: Ensuring the right system and right instrument for the job.
- Level 2: Developing and validating the right analytical procedure for the job.
- Level 3: Allying all lower layers of the model to ensure the right approach for the right reportable result.

The Data Integrity Model will be used throughout this book to demonstrate how the layers of the model interact together.

In addition, there is a comparison of the Data Management Maturity (DMM) Model from CMMI Institute[14] and Data Integrity Maturity Model from GAMP Guide on Records and Data Integrity.[15]

- Chapter 6 focuses on all those involved with data governance and outlines the key *Roles and Responsibilities of a Data Governance Programme*. Here, the various roles, from the boardroom to the laboratory bench, are presented and the responsibilities of each one discussed. The roles of process owner and system owner presented in EU GMP Annex 11[16] will be mapped to the data governance roles so that the roles are aligned with the regulations. We will also see how a corporate data integrity programme impacts a regulated laboratory.
- Chapter 7 discusses *Data Integrity Policies, Procedures and Training*. Data integrity policies and associated procedures must be present throughout an organisation. These will vary from an overall data integrity policy, good documentation practices for paper and computerised systems, interpretation of laboratory data and second person review. Coupled with the policies and procedures there must be effective staff training and where necessary evidence of the effectiveness of training. Typical poor documentation practices and record keeping failures are highlighted.
- Chapter 8 covers the necessity for *Establishing and Maintaining an Open Culture for Data Integrity*. It will explore what is open culture as well as the role of senior and laboratory management in setting expectations and maintaining a culture to ensure data integrity. This will include ways of balancing the pressure of work that could result in staff cutting corners due to management pressure that could compromise data integrity. This is the most difficult part of the Data Integrity Model as it requires management leadership and will require constant nurturing.
- Chapter 9 presents *An Analytical Data Life Cycle* from acquisition through processing, retention and finally destruction at the end of the retention period. The currently published data life cycles are

inadequate as they do not cover analytical procedures in any detail as they are generic. Presented here is a flexible analytical data life cycle from sampling to reporting that can adapt to any analysis from simple to complex. There are potential problems as some analyses do not generate objective evidence that is essential to demonstrate that an activity took place and for second person review. Key phases of the life cycle are sampling, sample preparation, acquisition and interpretation and these are where poor practices and falsification can occur, often without detection. Reiterating, an analytical data life cycle model needs to be flexible as not all analytical procedures require all phases.

- Chapter 10 concerns the *Assessment and Remediation of Laboratory Processes and Systems*. Management needs to have inventories of computerised systems and paper processes available and have them assessed for data integrity risks. The priority of assessment should be based on the criticality of the process and the records produced in it. Data process mapping is one methodology for assessing a process or system that will highlight the data generated and the record vulnerabilities for paper process and hybrid and electronic systems. From this will come short term quick wins for immediate implementation and also the development of long-term solutions options offering business benefit.
- Chapter 19 focuses on *Quality Metrics for Data Integrity*. Although placed towards the back of the book, this is a data governance topic as the metrics are a major input into the management review of the data governance of an organisation. It is important to understand the data integrity issues before developing metrics to monitor the programmes. There are two main areas for quality metrics, the first is to assess the progress of the overall data integrity assessment and remediation programme as well as the metrics for day to day routine analysis to identify areas of concern for monitoring, setting key performance indicators (KPIs) that may trigger further investigation of a process or system.

1.2.5 Data Integrity

Moving from the data governance umbrella, we come down into the detail of data integrity with eight chapters looking at the three top layers of the data integrity model.

- Chapter 11 looks at the *Data Integrity and Paper: Blank Forms and Instrument Log Books*. The use of blank forms for documenting analytical records will be used. An analysis based on observation will be discussed and the main data integrity issue with such a process is there objective evidence available for second person review? Many manual processes in a laboratory such as sample preparation lack evidence of activities other than a trained analyst following a procedure. In addition, there are now stringent regulatory control requirements for master templates and blank forms that will be outlined to demonstrate data integrity. Instrument log books are key records for data integrity but are they completed correctly and if using an instrument data system why cannot the data system generate the log automatically?

- Chapter 12 discusses *The Hybrid System Problem*. A hybrid computerised system is one that creates electronic records with signed paper printouts with the problem that two incompatible record formats must be managed and synchronised. What can software suppliers do to ensure data integrity? Many software applications used in the laboratory were designed before data integrity issues became a major issue. We also discuss the most common hybrid system: spreadsheets and present what to do to ensure data integrity. The regulatory guidance from WHO is that "use of hybrid systems is discouraged"[9] will be presented as an introduction to Chapter 13.
- Chapter 13 presents a contentious topic: *Get Rid of Paper: Why Electronic Processes are Better for Data Integrity* and looks at why electronic processes are better. This is a goal for long term solution to existing systems and paper based processes. If resources are used to update and validate new systems to improve data integrity, there must be substantial business benefits. This is a business opportunity that will be compared with approaches taken with two other major programmes of work for the pharmaceutical industry: Year 2000 and Part 11 remediation projects, as discussed in Section 1.9 of this chapter..
- Chapter 14 focuses on *Data Integrity Centric Computer System Validation and Analytical Instrument Qualification*. It compares the traditional approach to top down computer validation: improving the process and then configuring the application to match the new process. However, this can leave the records generated by the system vulnerable especially if access *via* to data files *via* the operating system is available. Therefore, a combination of top-down validation coupled with a bottom-up approach identifying the record vulnerabilities and mitigating them is essential. The integrated approach of analytical instrument qualification and computerised system validation in the 2018[17] version of USP <1058> will be discussed. The data integrity issues of allowing a service provider to have access to your computerised systems will be debated.
- Chapter 15 presents *Validation of Analytical Procedures* that discusses the new approach of a life cycle approach that has been proposed by the publication of a draft version of USP <1220> by the United States Pharmacopoeia (USP).[18] The omission of ICH Q2(R1)[19] is that method development is not mentioned. In contrast, the draft USP <1220> is a three-stage process looking at defining the aims of the method with an Analytical Target Profile (ATP) and the method design is based on this. When developed, the method is

validated and then deployed for use with continuous monitoring of the performance.

- Chapter 16 discusses *Performing the Analysis*. When an analysis is performed, the lower levels of the data integrity model (Foundation, Level 1 and Level 2) must work effectively so that the work is carried out correctly at Level 3 to ensure data integrity. We will look at analytical procedures requiring observation alone, sample preparation followed by instrumental analysis, instrumental analysis with data interpretation and finally a method involving sample preparation, instrumental analysis with data interpretation and calculation of the reportable result. Data integrity requires that out of specification results are documented and will be a lead-in to the next chapter.
- Chapter 17 entitled *Second Person Review*. The second person review is important for ensuring data integrity and is the second line of defence (the first being the performer who does the work). A reviewer needs to ensure that all elements in the procedure have been followed and to check for falsification or hidden records. Review of paper, hybrid and electronic records will be discussed as well as risk based audit trail review including review by exception.
- Chapter 18 looks at *Record Retention*. It presents and discusses options for retention of records: migration, museum, virtualisation or read only database. Discussion of data standards: history and current efforts as a way forward for the future. Some data integrity guidance documents require that dynamic data is not converted to static data.[8,20]

1.2.6 Quality Oversight for Data Integrity

The final five chapters present and discuss quality oversight within a regulated laboratory and also when outsourcing analysis to a Contract Manufacturing Organisation (CMO) or Contract Research Laboratory (CRO):

- Chapter 20 is entitled *Raising Data Integrity Concerns* and discusses how data integrity concerns can be raised by an employee. This is different from the open culture discussed in Chapter 8 and is focused on how a concern can be raised and investigated by a company that protects the confidentiality of all concerned when handling such issues. This policy includes non-retaliation against individuals either raising or being subject to the data integrity concern.
- Chapter 21 discusses *Quality Assurance Oversight for Data Integrity* and the role of the Quality Assurance department in self-inspections and risk based data integrity audits. The chapter will discuss the potential overlap between data integrity audits and computer system periodic reviews. What do I audit in a data integrity? This chapter is linked with Chapter 24 where several data integrity audit aide memoires are presented based on the data integrity model presented in Chapter 5.

- Chapter 22 is focused on *How to Conduct a Data Integrity Investigation.* It begins with the regulations and guidance covering data integrity investigations from various regulatory authorities including the FDA. There is a spectrum of issues from mistakes through to deliberate falsification that will be discussed before focusing on falsification of data. Using a case study, the reader is taken through how to conduct a data integrity investigation including writing the report and the corrective and preventative actions to be taken. One important aspect is understanding the material impact of the data violation and do I report the violation to the regulators?
- Chapter 23 looks at the *Data Integrity and Outsourcing.* The pharmaceutical supply chain is globally outsourced with Active Pharmaceutical Ingredient (API) manufacturers, excipient suppliers, contract manufacturers and contract laboratories being used extensively. This chapter takes the salient points from the preceding chapters and applies then to assessing and monitoring any supplier of a chemical, product or a service.
- Chapter 24 entitled *Data Integrity Audit Aide Memoires* provides several aide memoires based on the Data Integrity Model described in Chapter 5. Many of these aide memoires are unusual as they are based on figures rather than a checklist of questions that will constrain an auditor. As such, they are intended for use by experienced auditors and referenced during an audit.

1.3 Mapping This Book to the Data Integrity Model

Although appearing premature, it is appropriate to introduce here the Data Integrity Model that will be discussed in more detail in Chapter 5. The model consists of a foundation and three levels that can describe data integrity and data governance in a single diagram. The model also includes production, analytical development/quality control and quality assurance. Shown in Figure 1.2 is the portion of the Data Integrity Model covered in this book (Analytical Development/Quality Control and Quality Assurance) with the chapters of the book mapped against it. You will see that one of the main areas of focus is on the Foundation level as it is winning hearts and minds that allows an organisation to change its culture.

1.4 Pharmaceutical Quality System and Data Integrity

1.4.1 Integration Within the Pharmaceutical Quality System

All the data governance and data integrity topics covered in this book must not exist in isolation but be integrated within the Pharmaceutical Quality System (PQS) of an organisation. For example, in Chapter 19 data integrity

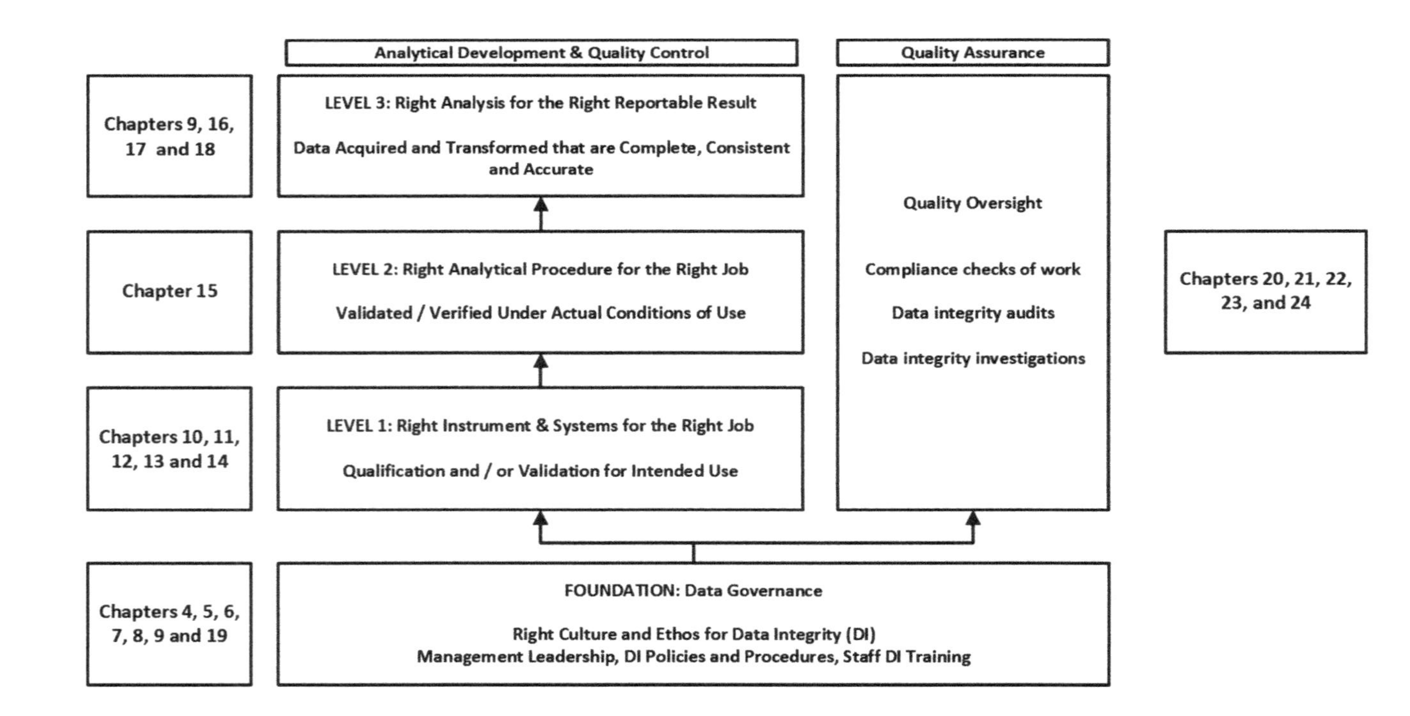

Figure 1.2 Mapping this book to the Data Integrity Model.

quality metrics are discussed that will be fed into a management review of not just data integrity but of the whole PQS. Similarly, the policies and procedures for data integrity together with the associated training will be integrated into the normal staff training covering both the scientific and compliance aspects of their position descriptions.

By taking this approach, data integrity coupled with the development of an open culture described in Chapter 8, becomes integrated into the normal operation of a regulated laboratory.

1.4.2 No Chapter on Risk Management

There is not a specific chapter on risk management in this book. This is a deliberate approach to risk management as it is integrated into the chapters throughout the book either implicitly or explicitly. For example, in an analytical data life cycle we look at the criticality of data generated through different mechanisms. The risks posed with using hybrid systems and standalone workstations are discussed in Chapters 12 and 10, respectively, and data process mapping in Chapter 9 on the analytical data life cycle is used to identify risks and vulnerabilities of records produced during an analytical procedure. In summary, we should not think of risk management as a separate subject but it must be integrated into normal work and should be second nature to all involved.

1.4.3 Back to the Future 1: Understanding Current in cGMP

The FDA GMP regulations, 21 CFR 211, are entitled Current Good Manufacturing Practice Regulations for Finished Pharmaceuticals.[21] Typically, this is shortened to GMP or sometimes cGMP. What does the word "current" mean in the context of the US GMP regulations? To understand the intent of the regulation this is where we need to go back to understand the intent of the regulation when it was first promulgated in 1978.[21] This will be discussed in more detail in Chapter 13 where the advantages of electronic working over paper are discussed.

1.4.4 The European Equivalent of cGMP

When you look at EudraLex volume 4 to see the various parts, chapters and annexes of EU GMP regulations there is nothing in them that is the equivalent to the "current" of cGMP described in the previous section. However, buried in the overarching and incredibly dry European Union directive 2001/83/EC on the Community code for medicinal products for human use there is Article 23 that states[22]:

1. After a marketing authorisation has been granted, the marketing authorisation holder shall, in respect of the methods of manufacture and control provided for in Article 8(3)(d) and (h), take account of

scientific and technical progress and introduce any changes that may be required to enable the medicinal product to be manufactured and checked by means of generally accepted scientific methods.

Note the phrase "take account of scientific and technical progress and introduce any changes... and checked by generally accepted scientific methods". This is the European equivalent of cGMP. In the 2015 MHRA GMP data integrity guidance document[7] notes that:

> Manufacturers and analytical laboratories should be aware that reverting from automated/computerised to manual/paper-based systems will not in itself remove the need for data integrity controls.

Therefore, ignoring computerised systems and going back to paper records alone may be a failure to comply with Article 23 [22] and hence a GMP non-compliance.

1.5 What Is Data Integrity?

1.5.1 How Many Definitions Would You Like?

The starting point in a discussion of data integrity must be to define what we mean by the term. There are several definitions from regulatory agencies as well as government bodies.

Table 1.1 shows six definitions of either integrity or data integrity, including two from the FDA. I have deliberately listed all these definitions in Table 1.1 to illustrate that different organisations or even different divisions of the same regulatory organisation can have different definitions for the same term.

1.5.2 What Do These Definitions Mean?

From these various definitions of integrity and data integrity is there a common ground to explain what is data integrity? Therefore, let us reconcile and combine these different definitions into a single approach for data integrity:

- Data must be complete, consistent and accurate.
- Data have a life cycle.
- Data must not have been improperly modified.
- If using a computerised system, then software should prevent unauthorised modification of data.
- The software should not be improperly modified (such as modifying the system configuration to turn the audit trail off and then on again to make unauthorised changes to data or deletions).
- To have integrity a record needs to meet the ALCOA criteria (we will discuss what ALCOA means in Section 1.5.3).

Table 1.1 Data integrity and integrity definitions.

Source	Term	Definition
FDA	Data integrity	The degree to which a collection of data are complete, consistent and accurate.[20]
FDA	Data integrity	The completeness, consistency, and accuracy of data. Complete, consistent, and accurate data should be attributable, legible, contemporaneously recorded, original or a true copy, and accurate (ALCOA).[23]
IEEE	Integrity	The degree to which a system or component prevents unauthorized access to, or modification of, computer programs or data.[24]
MHRA	Data integrity	The extent to which all data are complete, consistent and accurate throughout the data lifecycle.[7]
NIST	Data integrity	The property that data has not been altered in an unauthorized manner. Data integrity covers data in storage, during processing, and while in transit.[25]
WHO	Data integrity	The degree to which a collection of data are complete, consistent and accurate throughout the data life cycle. The collected data should be attributable, legible, contemporaneously recorded, original or a true copy and accurate.[9]

The first three and sixth bullet points hold for both manual processes as well as computerised systems and the fourth and fifth points are necessary for both hybrid and electronic computerised systems.

Although the focus of these definitions is on the data, metadata and records created and updated during an analytical analysis, however, data integrity is more than just data and numbers. It involves data governance that will be discussed in Chapters 4–10 and 19.

1.5.3 ALCOA+ Criteria for Integrity of Laboratory Data

Any record that is generated during regulated laboratory analysis needs to have its data integrity assured as discussed above. This now raises the question how does an analytical scientist assess data integrity of regulatory records? In Table 1.2 the term ALCOA is mentioned as a means of assessing the integrity of data.

As a quick introduction to the basic criteria of laboratory data integrity we need to explain the acronym ALCOA mentioned in the previous section. This is a term developed in the 1980's by an FDA GLP inspector as a means of teaching his colleagues about the integrity of data and records generated during non-clinical toxicology studies. The five ALCOA terms are:

- Attributable;
- Legible;
- Contemporaneous;
- Original;
- Accurate.

Table 1.2 ALCOA+ criteria for data integrity.

Criterion	Meaning
Attributable	• Attributable means information is captured in the record so that it is uniquely identified as executed by the originator of the data (*e.g.* a person or a computer system).[9] • Attributable to the person generating the data.[7] • Who acquired the data originally or performed an action subsequently to it and when?
Legible	• The terms legible and traceable and permanent refer to the requirements that data are readable, understandable, and allow a clear picture of the sequencing of steps or events in the record so that all GXP activities conducted can be fully reconstructed by the people reviewing these records at any point during the records retention period set by the applicable GXP.[9] See also consistent and enduring. • Legible.[7] • Can you read the data together with any metadata or all written entries on paper? • Can you read and understand all audit trail entries?
Contemporaneous	• Contemporaneous data are data recorded at the time they are generated or observed.[9] • Documented (on paper or electronically) at the time of an activity.
Original	• Original record: Data as the file or format in which it was originally generated, preserving the integrity (accuracy, completeness, content and meaning) of the record, *e.g.* original paper record of manual observation, or electronic raw data file from a computerised system.[10] • True copy: An exact verified copy of an original record.[7] • Original data include the first or source capture of data or information and all subsequent data required to fully reconstruct the conduct of the GXP activity. The GXP requirements for original data include the following: original data should be reviewed; original data and/or true and verified copies that preserve the content and meaning of the original data should be retained; as such, original records should be complete, enduring and readily retrievable and readable throughout the records retention period. • Written observation or printout or a certified copy thereof. • Electronic record including metadata of an activity.
Accurate	• Accurate.[7] • The term "accurate" means data are correct, truthful, complete, valid and reliable.[9] • No errors in the original observation(s). • No editing without documented amendments/audit trail entries by authorised personnel.
Complete	• All data from an analysis including any data generated before a problem is observed, data generated after repeating part or all of the work or reanalysis performed on the sample. • For hybrid systems, the paper output must be linked to the underlying electronic records used to produce it.
Consistent	• All elements of the analysis such as the sequence of events follow on and data files are date (all processes) and time (when using a hybrid or electronic systems) stamped in the expected order.

Table 1.2 (*continued*)

Criterion	Meaning
Enduring	• Recorded on authorised media, *e.g.* laboratory notebooks, numbered worksheets for which there is accountability or electronic media. • Not recorded on the backs of envelopes, laboratory coat sleeves, cigarette packets or Post-it notes.
Available	• The complete collection of records can be accessed or retrieved for review and audit or inspection over the lifetime of the record.

Implicit in the ALCOA requirements are that the records should be complete, consistent, enduring and available according to the WHO Guidance on Good Records Management.[9]

In addition, there are four more terms that were added by an EMA document on clinical trial electronic source data[26]:

- Complete;
- Consistent;
- Enduring;
- Available.

These nine terms are collectively called ALCOA+ or sometimes ALCOA-plus and are detailed in Table 1.2. Although many of the regulatory guidance documents talk about ALCOA and claim that the four additional terms can be derived from the original five, in this book we will use ALCOA+ criteria for data integrity as there are differences.

Analytical scientists need to understand these criteria and apply them in their respective analytical methods. The WHO guidance provides the best understanding of the ALCOA principles where Appendix 1 gives definitions of each of the five ALCOA terms with side by side tables of expectations for paper and electronic records, followed by special risk considerations for each of the five terms.[9] This is a very useful source of information about ALCOA requirements for data integrity of records and to help understand the issues associated with each term.

1.6 Data Quality and Data Integrity

1.6.1 From Sample to Reportable Result

Although this book is focused on data governance and data integrity, it is important to understand the difference between data integrity and data quality. Data integrity definitions have been presented in Table 1.2 and discussed in Sections 1.5.1 and 1.5.2. Data quality is defined by MHRA as:

> The assurance that data produced is exactly what was intended to be produced and fit for its intended purpose. This incorporates ALCOA.[8]

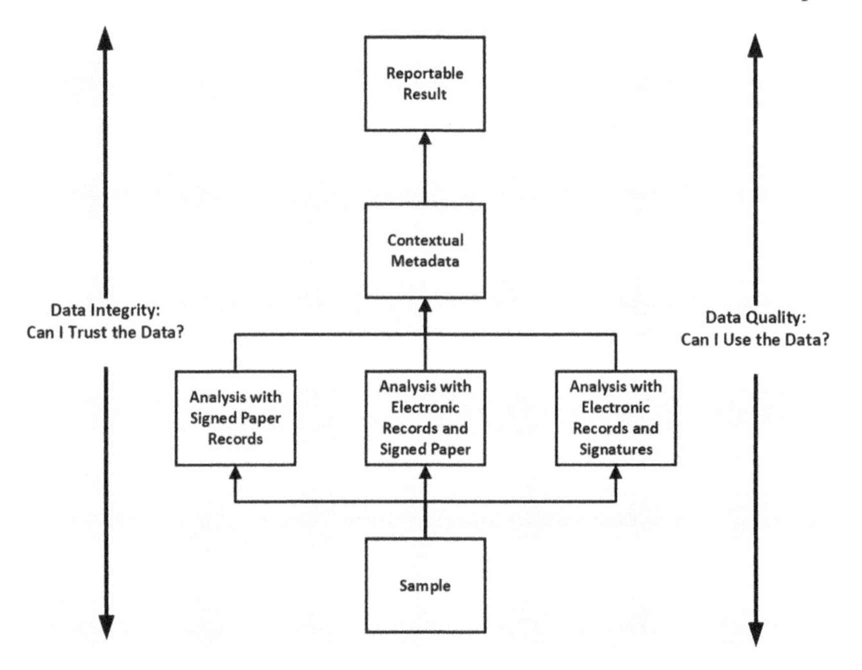

Figure 1.3 Difference between Data Quality and Data Integrity.

In summary, data quality can be considered as "can I use the data" and data integrity as "can I trust the numbers". Therefore, data integrity is an important component of data quality. This is shown diagrammatically in Figure 1.3. In the centre of the figure is the analytical process from the sample to the reportable result. An analysis can be conducted using one of the following options:

- Paper records that are signed by the analyst and the reviewer.
- Analysis by an analytical instrument controlled by a computer application that creates electronic records, may perform calculations and generates a reportable result that is printed out and the paper signed (a hybrid computer system or simply a hybrid).
- Analysis by an analytical instrument controlled by a computer application that creates electronic workflow, may perform calculations and generates a reportable result and the report is electronically signed.

Regardless of the option used for a test, contextual metadata is generated that is essential for interpretation of the reportable result and puts the value in context, as we shall see in the next section.

1.6.2 Contextual Metadata and a Reportable Result

As a sample is analysed, there will be additional data gathered, these are called metadata. The term was originally derived from using computerised systems and has been sometimes defined as data about data, which is not the most helpful definition. Although many data integrity guidances use the term metadata, which is defined in the glossary at the front of this book, I will use the term contextual metadata as it reminds us that it refers to data that put the reportable result or equivalent into context.

To illustrate this, Table 1.3 shows the reportable result for a chromatographic analysis of a drug substance in a finished product or even the

Table 1.3 A reportable result and the contextual metadata for a chromatographic analysis.

Reportable result	Phase of analysis	Data and contextual metadata
98.3	Sample management	• Sampling plan and sampling information • Sample transport and sample management, as appropriate • Sample number and identity • Batch/study number of sample(s) • Initials/signatures of the analyst(s) and reviewer
	Overall analysis	• Start and end dates and times of the analysis • Dates and times of any reanalysis
	Sample preparation	• Sample weights on balance printout • Analytical balance used and calibration records. Balance log book completed • Sample preparation records including pipettes and glassware used • Initials/signatures of the analyst(s) and reviewer
	Instrumental analysis	• Instrument identity and current qualification status • Column identity • Instrument control method and any changes made • Data acquisition method and any changes made • Sequence file of the order of injections, *e.g.* system suitability samples, blanks, quality controls, standards, samples, injection volumes, purities, dilutions, correction factors • Integration method: Automatic integration of peak areas and any manual changes (if allowed) • System Suitability Test results *versus* acceptance criteria • Calculation of individual results and the reportable result • Result units (percent or concentration) • Completed instrument log book • Initials/signatures of the analyst(s) and reviewer
	Reporting	• Printouts with date and time of printout linked to underlying electronic records and with handwritten signatures and dates (OR) Electronic report electronically signed with date and time of generation and electronic signature signing • Audit trail entries of all activities involved with the analysis • Initials/signatures of the analyst(s) and reviewer

plasma concentration of a drug from a clinical study and some of the contextual metadata associated with the separation. As can be seen from the table a single result requires a large volume of contextual metadata that is generated throughout the analytical process including the chromatographic run. Figure 1.4 shows the same information graphically. In addition, the figure also indicates the scope of second person review that needs to be carried out that will be discussed in Chapter 17 and the scope of raw data and complete data that is discussed in Chapter 7.

Note that Table 1.3 only refers to the chromatographic analysis and it does not consider the following:

- sampling process;
- sample transport to the laboratory;
- sample management in the laboratory;
- reagent and standard preparation used in the analysis;
- preparation of the samples for analysis;
- identification of any instruments (analytical balance, electronic pipettes), glassware (volumetric flasks, pipettes) and equipment (sample thief) used in the above processes;
- log books for instruments and equipment used.

These activities will create additional data and contextual metadata as part of the overall analytical procedure that will need to be considered for data retention and assessment of data integrity. All together they support the overall quality of the reportable result.

1.6.3 Data Integrity – Can I Trust the Data?

Data integrity provides assurance that the analytical work in the laboratory from the sampling to the calculation of the reportable result has been carried out correctly. Selecting three items from the definitions from Section 1.5.1 that the analysis is:

- Complete.
- Consistent.
- Accurate.

In short, can I trust the analysis and the reportable result(s)? If you cannot trust the analysis, then you cannot take a decision based on poor, incomplete or falsified data. Therefore, data integrity comes before data quality; you cannot make a quality decision without the integrity of the underlying analysis. It is the role of the analytical scientist and the second person reviewer to ensure data integrity but that also requires that the instruments, computer systems, analytical procedures, laboratory environment and organisation within which they work are set up to ensure that they are not pressured to cut corners or falsify data.

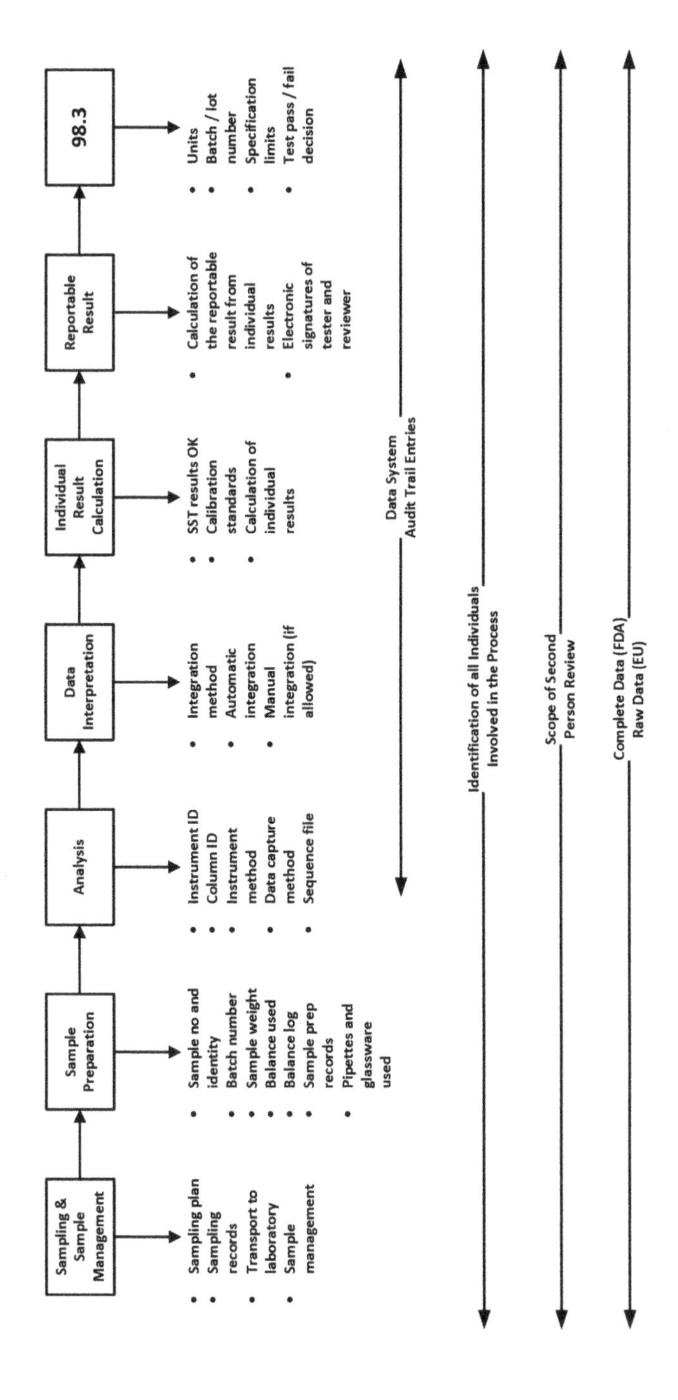

Figure 1.4 Reportable result and the supporting contextual metadata.

The focus of this book is on data integrity and the overall data governance in the context of a regulated analytical laboratory, however, in focusing on data integrity we must not lose sight of data quality.

1.6.4 Data Quality – Can I Use the Data?

Once you have assured the integrity of data, we turn to data quality. In part, this focuses on the laboratory's ability to deliver the results to the sample submitter and meet their requirements for the analysis such as:

- precision and accuracy or measurement uncertainty required;
- reportable value format;
- supporting information for the analysis;
- speed of analysis (*e.g.* releasing a production batch or turnaround of analysis from a clinical trial).

Data quality is a term that means can I use the data to make a decision, *e.g.* for batch release?

1.6.5 The Proposed FDA GLP Quality System

In 2016, the FDA issued a proposed update to the Good Laboratory Practice regulations under the title GLP Quality System.[27] One of the proposed changes was a new section 58.180 on Data Quality and Integrity:

(a) All data generated during the conduct of a nonclinical laboratory study must be accurate, legible, contemporaneous, original, and attributable (ALCOA). Also, data must be credible, internally consistent, and corroborated.
(b) All data must be recorded indelibly, directly, and promptly to a permanent medium at the time of observation and must identify unambiguously the person entering the data. Any change to any entry must be made so as not to obscure the original entry, must indicate the reason for such change, must indicate when the change was made, and must identify who made the change. When data are either captured or maintained, or both captured and maintained electronically, these requirements are fulfilled by the use of an electronic records system fully compliant with applicable regulations.
(c) All data accrued as required in paragraphs (a) and (b) of this section must be included in the final study report.

If these proposals are unchanged and become law, then for the first time there is the unambiguous requirement for the integrity of data, changes to data are transparent and that all data generated during a study must be reported without any omissions.

1.6.6 Continual *Versus* Continuous Improvement

A data integrity programme of work will require changes within an organisation, to processes, to systems and attitudes of staff. Some can be achieved without involving regulatory authorities but it is important when change must be made to understand the difference between continual and continuous improvement. The ISO 9001 standard referred to continual improvement until the 2015 version[28] and this has now changed to continual improvement in Section 10.3. ISO 9001 now aligns with EU GMP Chapter 1 [29] – derived from ICH Q10 [30] that also uses the phrase continual improvement in Section 1.4 [29]:

> (xi) Continual improvement is facilitated through the implementation of quality improvements appropriate to the current level of process and product knowledge.

The question arises do continuous and continual have the same meaning? This is important for some readers whose first language is not English there is sometimes no word for continual, only continuous and therefore it may appear that there is no difference between the words. There is some overlap in meaning between the two but they are not completely synonymous. Both can mean roughly "without interruption", but this is where continuous is used much more. Continual, by contrast, usually means "happening frequently, with intervals between".

This difference is very important in the context of the pharmaceutical industry as it is regulated and change cannot simply occur, especially if a change impacts a marketing authorisation. Change control, impact analysis, risk assessment and informing the regulatory authorities are the reasons for continual change being enshrined in the regulations.

1.7 Static *Versus* Dynamic Data

The FDA, MHRA and WHO data integrity guidance documents[8,9,20] refer to data as either static and dynamic and the definitions from these two publications are presented in Table 1.4.

Static records and dynamic records are very common in regulated laboratories. For example, an analytical balance with a printer will only produce a static record *via* the printer. However, if the balance is connected to an instrument data system, ELN or LIMS and the weight and other information collected electronically, then the weight becomes a dynamic record as it is capable of being manipulated and transformed later in the analytical process. Most instrument data systems used in laboratories produce dynamic records, *e.g.* chromatography data systems, spectrometry data systems, *etc.*

In the US GMP regulations there is the requirement for production information in 21 CFR 211.188 and complete data in 21 CFR 211.194(a).[31] Most production records are static in that they are typically temperatures, pressures, rotation speeds and relative humidity measurements and do not need

Table 1.4 Definitions of static and dynamic records from the FDA and WHO guidances.

Record type	Definition
Static record	• A static record format, such as a paper or pdf record, is one that is fixed and allows little or no interaction between the user and the record content. For example, once printed or converted to static pdfs, chromatography records lose the capability of being reprocessed or enabling more detailed viewing of baselines.[9] • A static (record) is used to indicate a fixed-data document such as a paper record or an electronic image.[20]
Dynamic record	• Records in dynamic format, such as electronic records, that allow for an interactive relationship between the user and the record content. For example, electronic records in database formats allow the user to track, trend and query data; chromatography records maintained as electronic records allow the user (with proper access permissions) to reprocess the data and expand the baseline to view the integration more clearly.[8,9] • Dynamic record means that the record format allows interaction between the user and the record content. For example, a dynamic chromatographic record may allow the user to change the baseline and reprocess chromatographic data so that the resulting peaks may appear smaller or larger. It also may allow the user to modify formulas or entries in a spreadsheet used to compute test results or other information such as calculated yield.[20] • Information that is originally captured in a dynamic state should remain available in that state.[8]

to be interpreted. Although there are static records in the laboratory as discussed above such as weights, pH values and colours, most laboratory records are dynamic as they must be interpreted to obtain a reportable result. Hence, the requirement for production information and laboratory data: they are typically static and dynamic records respectively.

1.8 Important Data Integrity Concepts

As you read this book it should become apparent that there are several important data integrity concepts that I will discuss now.

1.8.1 Data Integrity Is More than Just Numbers

The first concept is that in just reading this chapter and looking at the structure in Figure 1.1 is that data integrity is more than just numbers generated in a regulated laboratory. Although numbers and data including the associated integrity are important there is much, much more to consider. A holistic approach to data integrity and data governance:

- Starts at the top and works throughout an organisation: from the boardroom to the laboratory bench.
- Management lead, the rest follow.

- Integrity is integral within the overall pharmaceutical management system or quality management system.
- Organisation culture must be open and honest so that mistakes can be admitted and turned to an advantage of continual improvement.
- Instruments must be qualified and computerised systems validated.
- Analytical procedures must be validated.

Have we mentioned numbers yet? No! All the above items, and more, need to be in place to ensure that when analytical work is carried out the numbers generated have the integrity and quality required.

1.8.2 Quality Does Not Own Quality Anymore

The second concept is quality is not owned by the Quality Assurance Department as quality is now everybody's job. All staff when they work should ensure that they are adequately trained and follow the applicable procedures. If they make a mistake the work should stop and the issue discussed with a supervisor to determine what to do to resolve the issue. It may be that a procedure is too complex and needs to be revised or that the analyst needs more training. Regardless of the outcome it is the analytical scientists working in regulated environments that must own quality. It is their job and their responsibility. This approach must be supported by management behaviours to ensure that there is not pressure to get work out of the door as this could lead to short cuts being taken.

Quality starts in the laboratory. Quality Assurance staff are not there to identify errors as that is a laboratory function of the two most important people in any analysis: the performer of a test and the reviewer of the work undertaken. Data integrity and data quality are their responsibility. Quality Assurance provides the advice, ensures that work is compliant with regulations and the quality oversight *via* audits and data integrity investigations.

1.8.3 Data Integrity Is Not Just 21 CFR 11 or Annex 11 Compliance

In some people's minds data integrity is a second attempt to get compliance with 21 CFR 11 or Annex 11 right. It must be understood that data integrity covers much, much more than Part 11/Annex 11 compliance. It includes:

- Management attitudes and leadership.
- Definition and allocation of roles and responsibilities for data integrity.
- Development of an open culture.
- Establishment and maintenance of data integrity policies and procedures.
- Assessment and remediation of manual, hybrid and electronic processes.

In contrast, 21 CFR 11 would just be focused on computerised systems.

1.8.4 Data Integrity Is an IT Problem

Some attendees at data integrity training courses I have participated in have stated that management thinks that data integrity is a computer problem and left it to IT to resolve. This approach indicates that senior management has no clue. The issue is that it impacts both paper processes as well as computerised ones. Indeed, as we shall see in later in this book control of blank paper forms without a computer in sight is a key regulatory issue.

1.8.5 Data Integrity Is a Laboratory Problem

Equally concerning is the perception that data integrity is a laboratory problem. This is not even close! As we shall see in Chapters 4 and 5, EVERYBODY in an organisation from the Boardroom to the laboratory bench is involved in data integrity.

1.8.6 We Are Research – Data Integrity Does Not Impact Us

Whoever thinks this is naïve, stupid or both. If data are submitted to the FDA in a licencing application (*e.g.* IND or NDA) then 21 CFR 11 regulations apply to that data as noted under the Scope of the regulation in 11.1(b)[32]:

> This part applies to records in electronic form that are created, modified, maintained, archived, retrieved, or transmitted, under any records requirements set forth in agency regulations.
> This part also applies to electronic records submitted to the agency under requirements of the Federal Food, Drug, and Cosmetic Act and the Public Health Service Act, even if such records are not specifically identified in agency regulations.

If research data are contained in a New Drug Application, then they can be subject to inspection during a Pre-Approval Inspection under CPG 7346.832[33] that is discussed in Chapter 3.

1.9 It's Déjà vu all Over Again!

For those with short memories, data integrity is the third major IT systems improvement programme that has faced the pharmaceutical industry over the past 20 years, the other two being Year 2000 (Y2K) and electronic records/signatures (Part 11) assessment and remediation. Is the pharmaceutical industry going to make the same mistakes again? Let us explore this question.

- The Y2K programme was simply replacing applications and operating systems that could handle dates past 31st December 1999. Typically, it was a case of updating, rather than process improvement, to complete the work before the deadline; this was a technical project with a fixed due date.

- In contrast, a 21 CFR 11 assessment and remediation programme was an opportunity to upgrade and provide substantial business benefit by changing the business process to use electronic signatures and eliminate paper. However, not many laboratories took advantage of this approach and simply replaced non-compliant with technically compliant software and continued to use hybrid systems.
- Data integrity and data governance is a programme not a project of work. It covers management, culture, policies, procedures, training, instrument, computerised systems, analytical procedures and analysis with quality oversight. From this list it should be very apparent that the programme work is much greater that just computer remediation and assessment as with Y2K and 21 CFR 11 projects.

Is the industry going to repeat the Part 11 behaviour?

Read the various guidance documents[7–9,11,12,20] and you will see the storm clouds on the horizon: tight and bureaucratic control of blank forms, discouraging use of hybrid systems. The message is clear – get rid of paper or control it rigorously. The cost of electronic management is steady or declining – BUT the cost of using paper records is rising. Consider not just the physical storage cost but also the time to access reports and trend data and realize that paper's cost is considerable and labour intensive. Unfortunately, the management of many organisations are short sighted and only focus on fixing immediate problems and ignore the big picture.

Working electronically is the only long term viable option for the pharmaceutical industry.

An alternative view for data integrity remediation is seeing it as a second chance to get Part 11 right by looking at the intent rather than the letter of the regulation. Seen in this way, the industry can both align with regulators and position themselves for productivity—and data integrity—improvements in their processes.

However, many organisations complain that that this will cost money. Yes, but what is the effect on the organisation's cash flow if every batch can be released a few days earlier? Do the sums and then put in your project proposals.

References

1. *Barr Laboratories: Court Decision Strengthens FDA's Regulatory Power*, 1993, Available from: https://www.fda.gov/Drugs/DevelopmentApprovalProcess/Manufacturing/ucm212214.htm.
2. C. L. Burgess, Issues related to United States versus Barr Laboratories Inc., in *Development and Validation of Analytical Methods*, ed. C. L. Riley and T. W. Rosanske, Pergamon Press, Oxford, 1996, p. 352.
3. R. J. Davis, Judge Wolin's interpretations of current good manufacturing practice issues contained in the Court's ruling in the United States versus Barr Laboratories, in *Development and Validation of Analytical Methods*, ed. C. L. Riley and T. W. Rosanske, Pergamon Press, Oxford, 1996, p. 352.

4. *Able Laboratories Form 483 Observations*, 2005 [1 Jan 2016], Available from: http://www.fda.gov/downloads/aboutfda/centersoffices/officeofglobalregulatoryoperationsandpolicy/ora/oraelectronicreadingroom/ucm061818.pdf.
5. *Inspection of Pharmaceutical Quality Control Laboratories*, Food and Drug Administration, Rockville, MD, 1993.
6. *MHRA GMP Data Integrity Definitions and Guidance for Industry 1st Edition*, Medicines and Healthcare Products Regulatory Agency, London, 2015.
7. *MHRA GMP Data Integrity Definitions and Guidance for Industry 2nd Edition*, Medicines and Healthcare Products Regulatory Agency, London, 2015.
8. *MHRA GXP Data Integrity Guidance and Definitions*, Medicines and Healthcare Products Regulatory Agency, London, 2018.
9. *WHO Technical Report Series No. 996 Annex 5 Guidance on Good Data and Records Management Practices*, World Health Organisation, Geneva, 2016.
10. *WHO Handbook Good Laboratory Practices (GLP) Quality Practices for Regulated Non-Clinical Research and Development Second Edition*, World Health Organisation, Geneva, 2009.
11. *PIC/S PI-041 Draft Good Practices for Data Management and Integrity in Regulated GMP/GDP Environments*, Pharmaceutical Inspection Convention/Pharmaceutical Inspection Co-Operation Scheme, Geneva, 2016.
12. *EMA Questions and Answers: Good Manufacturing Practice: Data Integrity*, 2016, Available from: http://www.ema.europa.eu/ema/index.jsp?curl=pages/regulation/general/gmp_q_a.jsp&mid=WC0b01ac058006e06c#section9.
13. R. D. McDowall, *Validation of Chromatography Data Systems: Ensuring Data Integrity, Meeting Business and Regulatory Requirements Second Edition*, Royal Society of Chemistry, Cambridge, 2017.
14. M. Mecca, R. Young and J. Halcomb, *Data Management Maturity (DMM) Model*, CMMI Institute, Pittsburgh, PA, 2014.
15. *GAMP Guide Records and Data Integrity*, International Society for Pharmaceutical Engineering, Tampa, FL, 2017.
16. *EudraLex – Volume 4 Good Manufacturing Practice (GMP) Guidelines, Annex 11 Computerised Systems*, European Commission, Brussels, 2011.
17. *USP 41 General Chapter <1058> Analytical Instrument Qualification*, United States Pharmacopoeia Convention, Rockville, MD, 2018.
18. G. P. Martin, *et al.*, Stimuli to the revision process: proposed new USP general chapter: the analytical procedure lifecycle <1220>, *Pharmacopeial Forum*, 2017, **43**(1).
19. *ICH Q2(R1) Validation of Analytical Procedures: Text and Methodology*, International Conference on Harmonisation, Geneva, 2005.
20. *FDA Draft Guidance for Industry Data Integrity and Compliance with CGMP*, Silver Spring, MD, USA, 2016.
21. Part 211 – Current Good Manufacturing Practice for Finished Pharmaceuticals, *Fed. Regist.*, 1978, **43**(190), 45014–45089.

22. Directive 2001/83/EC of the European Parliament and of the Council of 6 November 2001 on the Community code relating to medicinal products for human use, *OJ L 311*, 28 Nov 2001, **311**, 67.
23. *FDA Glossary of Computerized System and Software Development Terminology*, 1995, Available from: http://www.fda.gov/iceci/inspections/inspectionguides/ucm074875.htm#_top.
24. *IEEE Standard 610.12-1990 – Glossary of Software Engineering Terminology (Replaced by ISO 24765: 2010)*, Institute of Electrical and Electronic Engineers, Piscataway, NJ, 1990.
25. *SP 800-33: Underlying Technical Models for Information Technology Security*, National Institute for Standards and Technology, Gaithersburg, MD, 2001.
26. *Reflection Paper on Expectations for Electronic Source Data and Data Transcribed to Electronic Data Collection Tools in Clinical Trials*, European Medicines Agency, London, 2010.
27. 21 CFR Parts 16 and 58 Good Laboratory Practice for Nonclinical Laboratory Studies; Proposed Rule, *Fed. Regist.*, 2016, **81**(164), 58342–58380.
28. *ISO 9001: 2015 Quality Management Systems – Requirements*, International Standards Organisation, Geneva, 2015.
29. *EudraLex – Volume 4 Good Manufacturing Practice (GMP) Guidelines, Chapter 1 Pharmaceutical Quality System*, European Commission, Brussels, 2013.
30. *ICH Q10 Pharmaceutical Quality Systems*, International Conference on Harmonisation, Geneva, 2008.
31. *21 CFR 211 Current Good Manufacturing Practice for Finished Pharmaceutical Products*, Food and Drug Administration, Sliver Spring, MD, 2008.
32. *21 CFR 11 Electronic Records; Electronic Signatures, Final Rule*, in *Title 21*, Food and Drug Administration, Washington, DC, 1997.
33. *FDA Compliance Program Guide CPG 7346.832 Pre-Approval Inspections*, Food and Drug Administration, Silver Springs MD, 2010.

CHAPTER 2

How Did We Get Here?

Data integrity is not a new subject as the requirements are explicit and implicit in Good Manufacturing Practice, Good Laboratory Practice and Good Clinical Practice regulations, regardless of which regulation is used by an organisation. However, data integrity is the hottest topic now in the pharmaceutical and allied industries. It is a global issue that is not focused on any one country or region. Most of the issues are caused by poor data management practices that have changed little in the past 20–30 years in many organisations and regulated laboratories. Most people though tend to focus on the cases of falsification and fraud that are highlights of reading numerous FDA warning letters over the past 15 years. Reading these you may be forgiven for thinking that data integrity is a recent topic, it is not. This chapter will provide the historical background to the current regulatory emphasis on data integrity and data governance.

Although there are cases of data falsification outside of the pharmaceutical industry, *e.g.* the Schön scandal where apparent breakthroughs in semiconductor research were eventually shown to be false.[1] The focus in this chapter will be on GXP regulated laboratories working in and for the pharmaceutical industry.

With some of the regulatory issues highlighted, it can be strongly argued that there is a breakdown of the pharmaceutical quality system along with the trust that the regulators place with the industry.

2.1 Barr Laboratories 1993: You Cannot Test into Compliance

Although there are cases of laboratory data falsification going back 40 or 50 years, this chapter will only look at the recent pharmaceutical industry history starting with the Barr Laboratories case from 1993.

Data Integrity and Data Governance: Practical Implementation in Regulated Laboratories
By R. D. McDowall

Published by the Royal Society of Chemistry, www.rsc.org

2.1.1 Background to the Court Case

The recent history of data integrity in the GXP laboratory can be traced to Barr Laboratories court judgement in 1993. The company was a US based generic drug manufacturer that had a history of disagreements with the FDA over some of their unconventional working practices in manufacturing and quality control testing. In the previous four years there were increasingly widespread manufacturing deficiencies reported by FDA investigators when reviewing Barr's practices and products during inspections. The reported problems include misplaced records, test data recorded on scrap paper, failure to control manufacturing steps such as those governing products' physical properties, the release of products not meeting their specifications, inadequate investigation of failed products, and failure to validate test methods and manufacturing processes, including cleaning processes.

For example, Barr had a QC practice that if a test was out of specification then they conducted two more tests and took the best two of the three to determine batch release. FDA was less than impressed. Barr sued the FDA and the Agency reciprocated by suing Barr. The two law suits resulted in a single court case presided over by Judge Alfred Wolin, who was a layman from the perspective of the science and statistics presented in the case. However, the judge was remarkably prescient in his judgement of the case.

2.1.2 Key Laboratory Findings from the Judgement

From the judgement there are a number of findings that directly impact any regulated laboratory. Reading the judgement, you can see an impartial and reasoned interpretation of the GMP regulations.[2–4]

1. Any out of specification (OOS) result requires a failure investigation to determine an assignable cause. The extent of the investigation depends on the nature and location of the error, *e.g.* if the error is in the laboratory *versus* if there is a production problem. The judge also rejected Barr's two out of three testing approach as unscientific and at the same time also rejected the unreasonable FDA request that one OOS result should result in rejection of the whole product batch.
2. Good science and judgement are needed for reasonable interpretation of GMPs. This is especially true in regulated laboratories. Industry practice cannot be relied on as the sole interpretation of GMP, guidance from literature, from seminars and pharmaceutical firms, textbooks and reference books, and FDA letters to manufacturers are additional sources of help. However, any interpretation must be "*reasonable and consistent with the spirit and intent of the cGMP regulations*".

2.1.3 Regulatory Response

Following the resulting court case the judge ruled that outliers could not be rejected unless allowed by the United States Pharmacopoeia (USP).[5] The FDA also responded by issuing a guide on Inspection of Pharmaceutical Quality Control Laboratories in 1993.[6] This guidance still is relevant as many processes in regulated laboratories are still paper based or use hybrid systems and should be read as it still contains good advice on how regulators will conduct an inspection of a Quality Control laboratory. A further outcome of the Barr case was the eventual issue of an FDA guidance for industry on out of specification (OOS) results.[7]

2.2 Able Laboratories 2005: You Cannot Falsify into Compliance

2.2.1 Background to the Inspection

There has been an increasing trend when chromatography data systems (CDS) operating in GXP regulated laboratories have been inspected by the United States, Canadian, WHO and the European regulatory agencies to discover that falsification and fraud have occurred. The inspection focus has changed. Instead of wading through reams of paper printouts, the inspection has gone to review the electronic records in the CDS. The reason for the change in focus started with the Able Laboratories fraud case in 2005.[8]

Here, the company had at least seven FDA inspections with no non-compliances until a whistle blower called the local Agency field office to raise concerns about the working practices that were not exactly compliant with the regulations.

2.2.2 483 Observations

Some of the innovative analytical techniques employed were a combination of:

- Copy and pasting chromatograms from passing batches to failing ones.
- Extensive reintegration of chromatograms to ensure passing results.
- Adjustments of weights, purity factors, calculations to ensure acceptable results.

This was how an original result of 29% that failed a dissolution test specification of >85% was falsified to a passing result of 91%[8] At the heart of the fraud was a Chromatography Data System (CDS), which when investigated by the FDA had an audit trail that identified the individuals responsible for the data falsification. Identification of these problems in the laboratory led to the recall of 3184 batches of product, suspension of 7 manufacturing licences and the closing of the company in 2005 followed by the criminal prosecution of four former employees in 2007.

2.2.3 Regulatory Response

The Able Laboratories fraud case led to a review of the FDA's inspection approach, the Agency was very concerned that throughout seven inspections the falsification was not detected. The main reason is that the inspectors focused on paper printouts from the CDS rather than looking at the electronic records and audit trail entries. The outcome of Able resulted in the rewrite of Compliance Program Guide (CPG) 7346.832 for Pre-Approval Inspections (PAI).[9] There are three objectives for any PAI:

- Objective 1: Readiness for Commercial Manufacturing.

  Determine whether the establishment(s) has a quality system that is designed to achieve sufficient control over the facility and commercial manufacturing operations.

- Objective 2: Conformance to Application.

  Verify that the formulation, manufacturing or processing methods, and analytical (or examination) methods are consistent with descriptions contained in the CMC section of the application for the biobatch (and other pivotal clinical batches, when applicable), the proposed commercial scale batch, and the API(s).

- Objective 3: Data Integrity Audit.

  Audit the raw data, hardcopy or electronic, to authenticate the data submitted in the CMC section of the application. Verify that all relevant data (*e.g.*, stability, biobatch data) were submitted in the CMC section such that CDER product reviewers can rely on the submitted data as complete and accurate.

At first sight the focus would be on objective 3 and the laboratory, but it is important to realise that data integrity and laboratory data runs through all three objectives. Before this became effective in May 2012, most of the FDA's inspectors were given training in understanding and detecting data integrity issues. The training focused on computer systems and the records they contain rather than the paper printouts. This focus on electronic records in regulated GXP laboratories has seen the increased number of warning letters containing data integrity citations in recent years that we will discuss in the next section.

Both Barr Laboratories and Able Laboratories were US companies. However, over the past 20 years to reduce overall manufacturing costs the pharmaceutical supply chain has been globalised; now 80% of active pharmaceutical ingredients are produced in China and India where we shall now look.

2.3 Ranbaxy Warning Letters and Consent Decrees

2.3.1 Background to the Regulatory Action

Following on from the Able Laboratories fraud case, Ranbaxy was the next company with major falsification and fraud issues to be highlighted. From 2006 onwards several Ranbaxy Laboratories in India and a US subsidiary, Ohm Laboratories, were the subject of several regulatory actions from the FDA as follows:

- Warning letters for the following sites:
  Paonta Sahib, India (2006)
  Dewas, India (2008)
  Batamandi (Unit II), in Paonta Sahib, India (2008)
  Ohm Laboratories, Inc. (2009).
- Import Alert for Dewas and Paonta Sahib, India, Facilities (2008).
- Application Integrity Policy Letter and Action for Paonta Sahib, India, Facility (2009).
- Consent Decree of Permanent Injunction for Paonta Sahib, Batamandi, and Dewas, India, Facilities and Ohm Laboratories in Gloversville, NY (2012).
- Import Alert and Consent Decree for Mohali, India, Facility (2013).
- Department of Justice Action Against Ranbaxy (2013): Pleads Guilty and Agrees to Pay $500 Million to Resolve False Claims Allegations, CGMP Violations and False Statements to the FDA.
- Consent Decree for Toansa, India, Facility (2014).

This is summarised on a single FDA web page with links to the 483 observations, warning letters as well as other regulatory and legal documents.[10]

2.3.2 Details of the 2012 Consent Decree

It is important to remember that if a Consent Decree is agreed with the FDA and the document is signed under a court order that the full title is "Consent Decree of Permanent Injection". The key word here is "permanent" meaning for indefinite, perpetual, for ever, everlasting and eternal. Once signed you cannot go back.

There was a specific requirement for all Ranbaxy sites[11]:

- Identify an individual who shall be authorized and responsible for establishing, implementing, and maintaining a comprehensive written QA and QC program to ensure that all drugs meet specifications.

The main thrust of the consent decree was the provisions for a comprehensive data integrity management programme:

- Establish an Office of Data Reliability, responsible for conducting pre-submission audits of all ANDA applications from all facilities. All new applications must be signed off by this group and certified to contain no misrepresentations of data before submitting to the FDA.
- A Chief Data Reliability Officer, reporting directly to the Managing Director of Ranbaxy Laboratories, Ltd., heads up the Office of Data Reliability. This individual has the authority to recommend that individual employees be disciplined or terminated and that any of Ranbaxy applications be withdrawn from FDA consideration for approval.
- A disclosure program must be established including a worldwide toll-free compliance phone line and a system to receive and maintain written submissions from individuals who wish to disclose to the Chief Data Reliability Officer any issues or questions associated with Ranbaxy's data integrity violation. The system will have a non-retribution, non-retaliation policy, and shall facilitate anonymous communications for which appropriate confidentiality shall be maintained. A log of issues raised together with CAPA plans will be maintained.
- Independent data integrity experts were hired to review existing applications for evidence of falsified data.
- Specific sites could not begin US manufacture until they were compliant with GMP and were independently certified by a GMP expert before FDA would consider any further inspections. This includes GMP basics such as having SOPs that were updated and reflected what work was being done.
- FDA halted review of all ANDA applications submitted and the data reviewed by an independent data integrity expert and if falsification found, it was withdrawn and could not be resubmitted.
- Current and former staff were to be interviewed to identify managers who encouraged or condoned the falsification of data and to determine if they were still able to influence GMP.
- The FDA has also calculated the cost of a lie in any new ANDA application at $3 million up to a maximum of $30 million per application.

As you can see, the cost of this consent decree is huge and this is before factoring in the impact of the second consent decree issued in 2014 and the Department of Justice case resulting in a $500 million fine. The total cost to the company is well over $1 billion considering fines, legal fees, external expert professional service fees, lost sales and corrective actions.

2.4 Court Case for GLP Data Falsification

A UK scientist, Steven Eaton, was convicted under the GLP regulations[12] of falsifying data in bioanalytical studies. Eaton was the only scientist in the laboratory to get a difficult bioanalytical procedure to work consistently, when his work was scrutinised it was found to be falsified. His employer informed the UK regulator, the MHRA, and following a further investigation it was found that he was responsible for data falsification and selective reporting going back to 2003. In a court case in 2013 he was jailed for three months for the crime.[13]

2.5 Semler Research Data Falsification

Data falsification is not limited to GMP laboratories, in 2015 a Contract Research Facility in Bangalore was inspected by the FDA. A spreadsheet was found on a server detailing how subject samples were substituted for several bioanalytical studies. In addition, discarded analytical records were found thrown into waste containers, instrument audit trails were not turned on for a five-month period when bioequivalence studies were analysed, analysts had the ability to delete electronic files plus raw data did not permit reconstitution of the study. In April 2016 an untitled letter was issued to the company[14] and at the same time the FDA posted the following notice on their web site[15]:

> FDA is notifying sponsors of New Drug Applications (NDAs) and Abbreviated New Drug Applications (ANDAs) that clinical and bioanalytical studies conducted by Semler Research Private Limited (Semler) located in Bangalore, India, are not acceptable as a result of data integrity concerns, and need to be repeated. Semler is a contract research organization that conducts bioequivalence and bioavailability studies for a number of pharmaceutical companies. ......
> FDA is taking this action as a result of an inspection of Semler's bioanalytical facility in Bangalore, India, conducted between September 29, 2015, and October 9, 2015. The inspection found significant instances of misconduct and violations of federal regulations, including the substitution and manipulation of study subject samples. ......
> FDA is sending letters to sponsors of applications that are currently under review that include data from studies conducted at Semler. The letter informs the sponsor that it must repeat the bioequivalence/bioavailability studies using an entity other than Semler at an acceptable alternate study site.

The Semler Research case highlights the risks associated when any regulated work is outsourced. Companies placing work with the organisation did not carry out sufficient due diligence in assessing if the company had good data integrity practices and conducted an audit to verify that the answers were truthful or not.

Before the Semler case, a similar situation with Cetero Research, a US based CRO, was cited in 2009 with an untitled letter and pharmaceutical sponsors were required to repeat bioequivalence studies. The Cetero case will be discussed in Chapter 23 of this book when we discuss outsourcing laboratory work where approaches for due diligence will be discussed in more detail.

2.6 The Cost of Data Integrity Non-compliance

The Able Laboratories and Ranbaxy Consent regulatory actions illustrate well the maxim that the cost of non-compliance always exceeds the cost of compliance.

- In May 2005, Able Laboratories had a stock market valuation of about \$100 million. In July, Able had seven drug applications withdrawn and had to recall 3184 batches of product from the market. The firm went bankrupt and the shell was sold to a competitor in October 2005.
- Ranbaxy has had since 2006 4 warning letters, 3 import alerts, 2 consent decrees, 1 application integrity policy letter and a \$500 million fine. This list does not include any costs for remediation of the inspection findings.

2.6.1 Relative Costs of Compliance *Versus* Non-compliance

To put this into context, we need to look at the cost of non-compliance *versus* the cost of compliance, as shown in Figure 2.1. The horizontal axis is the percentage of compliance from 0 to 100%. The only fixed points are at the ends of the scale where 0% is where little or no attention is paid to compliance and data integrity and/or falsification is rife and 100% where anything that can be done is done regarding data integrity and compliance. In between is the relative scale of compliance.

The left-hand vertical axis is the cost of non-compliance and the right-hand axis the cost of compliance. You will note the cost of compliance axis is smaller than the cost of non-compliance axis. This is one of the balances you need to consider, the right-hand side shows the cost of doing it right first time and the left-hand side is essentially the cost of getting caught.

Fixing a regulatory problem that has been identified in an inspection is always more expensive than doing the right job or finding a problem and fixing it yourself. If any reader is in doubt look at the Ranbaxy regulatory

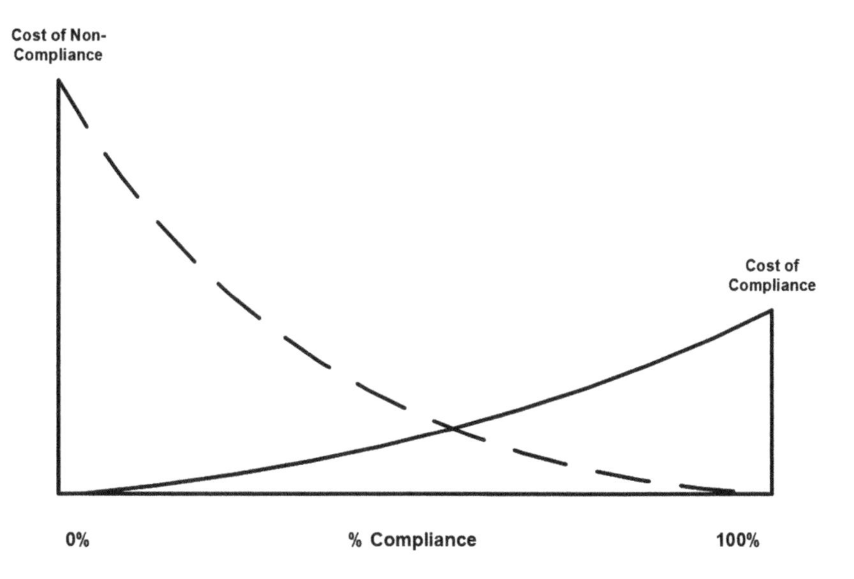

Figure 2.1 Balancing the costs of compliance and non-compliance. Reproduced from ref. 16 with permission from the Royal Society of Chemistry.

action and the consent decree discussed in Section 2.3.2 or look at some data integrity warning letters that have resulted in import alerts. The cost of non-compliance can now be quantified as up to hundreds of millions of dollars for some companies.

2.6.2 Is It Worth It?

In discussions with attendees at training and educational courses that I have participated in over the past 25 years, there has been a common theme that in some organisations management refuse to invest resources and money into systems that will improve laboratory efficiency. Often, laboratories still work with paper processes and very commonly with hybrid systems. Money is difficult to justify for improvement projects.

Perversely, if there are adverse regulatory inspection findings then money is no object to fixing the problems.

However, let us look at Figure 2.1 in a different way. If a project to improve laboratory efficiency was also able to ensure regulatory compliance, then the overall cost would fall on the cost of compliance curve. The project would be scoped and run with the timescale driven by the laboratory. Compare this with a post inspection project with the same aim but after major or critical inspection findings. Amazingly money and resources magically appear. There will be a greater urgency to complete the project quickly and this may mean bringing in expert help to support the project to be completed in a shorter time frame than above. The company wants to provide an aggressive timeframe as the regulator's hands are around the company's throat. Project risk will increase and there is a greater chance of project failure. This is the cost of non-compliance for the same project.

2.7 A Parcel of Rogues: FDA Laboratory Data Integrity Citations

In this section, the main areas of regulatory concern around data integrity discovered by the FDA are discussed and debated.

2.7.1 Why Use Only FDA Warning Letters and 483 Observations?

Although there are several sources of regulatory citations, *e.g.* FDA warning letters and 483 observations, EU GMP non-compliances, WHO notices of concern (NOC), Health Canada non-compliances. It is FDA warning letters and 483 observations that, in general, contain more detail and be far more critical of non-compliances.

Therefore, in researching the warning letters and 483 observations for poor data management and falsification citations I will focus on the laboratory for the most part rather than surrounding areas such as training, the failure to qualify or calibrate analytical instruments or validate software. It is the use, or rather misuse, of laboratory informatics software together with the operating system and utility software such as backup and the ability for paper processes to be easily falsified until the correct result is obtained that will be discussed here.

Rather than list all offending companies and go through each warning letter in excruciating detail, a summary of most of the regulatory citations involving data systems in FDA warning letters between 2013 and 2018 is presented in Figure 2.2. I will discuss each collated area in more detail, where appropriate, I will reference an individual warning letter or 483 observation where it serves to make a specific point of poor practice or falsification. Where there is no GMP regulatory citation in a warning letter I have given it the one that I think is most appropriate, however, this is my interpretation rather than the Agency's.

2.7.2 Quality Management System Failures

Please note that processes, instruments, or software on their own cannot commit fraud or falsification, so entering stage right are the intrepid users and their management. These are the main culprits either through poor data management practices or deliberate falsification. However, it should be noted that the design of many laboratory data systems makes it easier for analysts to commit falsification when the data files stored on standalone workstations reside in operating system directories rather than controlled *via* an integrated database.

It has become clear to inspectors that management and senior management are responsible for instigating falsification by a variety of means such as direct pressure on analysts to pass material regardless of the results or lax control within an organisation. The overall quality management system (QMS) has been cited in warning letters involving data integrity in the laboratory.[17,18] Unlike EU GMP Chapter 1[19] there is no direct US GMP reference to a QMS but the FDA have published ICH Q10 on pharmaceutical quality systems as a guidance for industry.[20] However, mention of the QMS raises the question of the roles and responsibilities of senior and laboratory management to prevent and detect data falsification. These people set the expectation for their staff to follow – if data falsification is found, management is responsible and often may be culpable. In the Ranbaxy consent decree[11] and in an ever-increasing number of warning letters,[21–23] FDA advises hiring a third party data integrity consultant and in some cases one of their tasks is to identify the managers who were responsible for falsification. Those identified for falsification will then be barred by the Agency from working on GMP work.

Moreover, in two 2016 warning letters there were citations of senior management's failure to act:

- Your site's senior management failed to take sufficient corrective action and prevent the recurrence of these problems. For example, an anonymous email dated August 5, 2013 notified your quality management about data falsification and manipulation in your laboratory. This email stated: "*...[t]here is no control of data in the department...Falsification is going on...Take action as early as possible...*".[22]

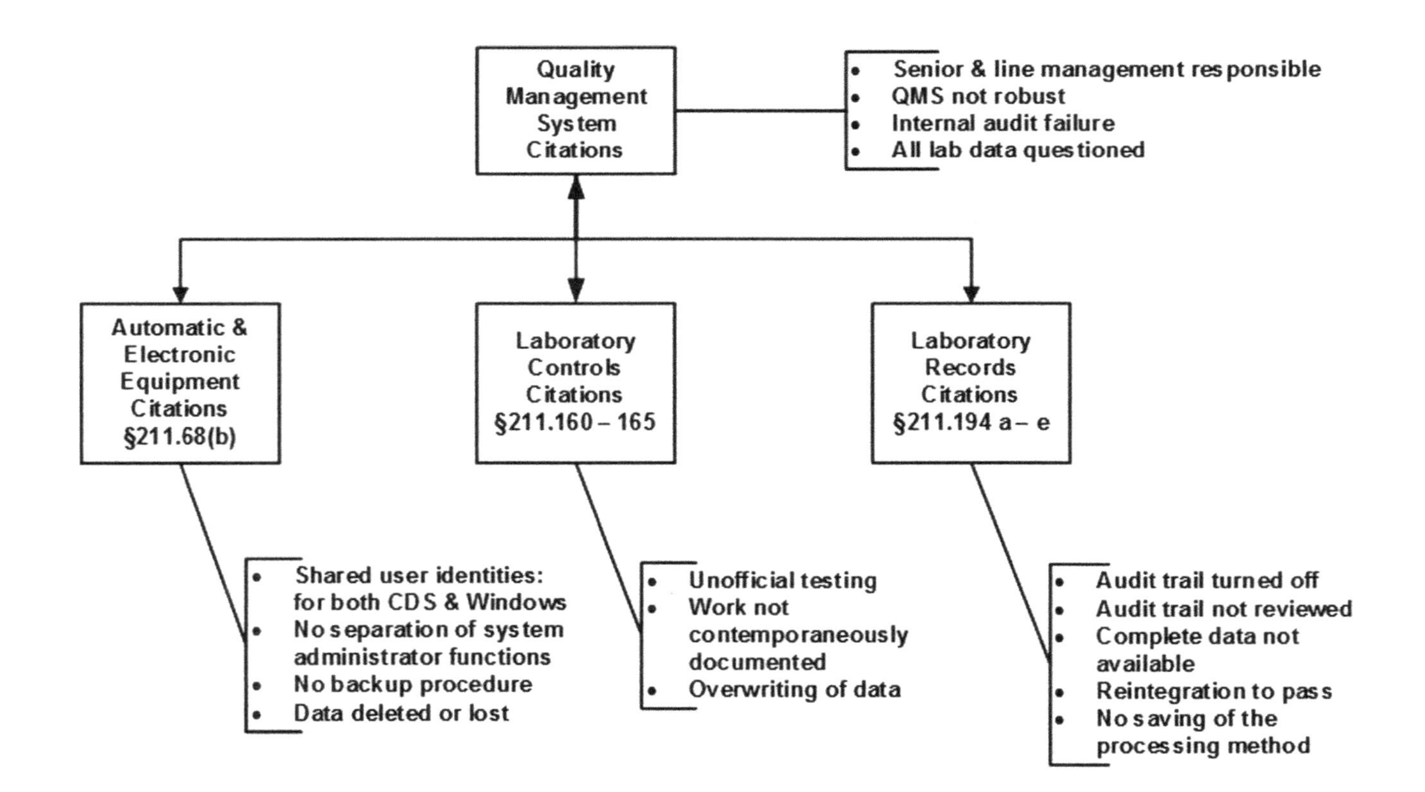

Figure 2.2 Classified citations from FDA warning letters for laboratory data systems and processes 2013–2018. Reproduced from ref. 16 with permission from the Royal Society of Chemistry.

- The examples above are serious CGMP violations demonstrating that your quality system does not adequately ensure the accuracy and integrity of the data generated at your facility to support the safety, effectiveness, and quality of the drug products you manufacture. *We observed similar issues at your facility in 2007.* At that time, we found you had improperly integrated HPLC peaks and had not identified and investigated out-of-specification test results.[23]

Therefore, even when management was aware of data integrity issues either by formal notification at the end of the inspection by the FDA or informally *via* an anonymous tip-off, but nothing was done. Hence the warning letters. There are no excuses or hiding places in these cases.

In addition, there is also the failure of internal audits to identify the data integrity problems, as highlighted in one warning letter.[18] As noted earlier, MRHA is requiring companies to perform self-inspections under EU GMP Chapter 9[24] to focus on data integrity within their own organisations and their suppliers.[25–27] Therefore, these internal audits have to be thorough and if problems are found, investigated with appropriate action taken. Although internal audit reports are confidential, the FDA can insist on seeing them if they conduct a for cause audit.[28]

Data integrity issues are not new but there is a renewed focus either due to either falsification of data or poor data management practices. Data integrity is not just confined to a single country or continent but it is a global issue. The current regulatory focus tends to be on the falsification for which the whole industry is paying the price for the sins of the (relatively) few.

2.7.3 Instrument Citations

A frequent citation in the laboratory warning letters, as shown in Figure 7.3, is §211.68(b) in the section on automatic, mechanical, and electronic equipment,[29] which requires that:

- Access is restricted to authorised individuals.
- Changes are only instituted by authorised individuals.
- The accuracy of calculations must be verified.
- Backups must be exact and complete.
- Backups must be secure from alteration, erasure or loss.

Non-compliances in this area involve:

- Sharing of user identities between two or more users, so making it impossible to identify the individual who was responsible for a particular action within the system. Therefore, ensure that there are enough user licences for each user to have one for their job. Sharing user accounts may seem to be a smart way to save money but, if you get caught, the cost of rectifying the non-compliance makes the saving pale into insignificance. A list of current and historical users is essential for compliance with both 21 CFR 11,[30] Annex 11[31] and the FDA guidances on Computerised Systems in Clinical Investigations[32] and Data Integrity and Compliance with cGMP.[33]

- Access privileges must be appropriate to a user's job function, therefore everyone cannot be a system administrator. There will need to be at least three user types or roles with corresponding access privileges that need to be documented outside of the system either in a configuration specification or an SOP, *e.g.* analyst, reviewer and administrator.
- The system administrator needs to be independent of the function of the laboratory so that configuration settings cannot be changed and the audit trail turned on and off to hide falsification activities. Therefore, the system administration activities such as the configuration of the software including controlled changes to it and user account management and access privileges need to use IT rather than laboratory staff. However, small laboratories may only have 2 or 3 users per data system, then an alternative should be considered. Each user should have two roles: the first is an administrator role with no user privileges and the second is a user role with no administration privileges. To administer the system any user would have to log out as a user and log in again as an administrator. This is not ideal from a regulatory perspective but as some guidances state that data integrity should not impose any additional burden,[34] to do otherwise is not practical.
- Limit access to the workstation operating system, as there are many citations for deletion of data in some file based instrument data systems. Failure to do this results in users being able to access the workstation clock, directories and files and the recycle bin, which can result in deletion of records, time travelling and rerunning of samples.
- Failing to backup data, incomplete backup of data, being incompetent and losing data when upgrading the application software or not having the software to interpret the data files are just some of the ways companies have been cited under this section of the regulations. The simplest way to avoid this citation is to give the job of backup to the IT professionals who will do the job for you. There are a few catches here, are the IT staff trained including GXP awareness, is there a backup SOP with evidence of actions, is recovery tested regularly and has the backup process and software been validated? This is acceptable for a networked data system but if there are standalone workstations then data may be located on the local workstation drive. This is not acceptable and, in my view, any data system must acquire data to the network where the files can be backed up professionally. At a minimum, a laboratory system should be connected to the network and an automated script developed by the IT Department used to transfer data to a secure and resilient network drive.
- Manual backup to USB flash drives or optical drives is unacceptable and should never be used as these have been cited in FDA warning letters, *e.g.* BBT Biotech[35] and Zhejiang Hisun Pharmaceutical Co.[36] The latter company was cited for an *analyst removed a USB thumb drive*

from a computer controlling an HPLC. When asked to provide the drive, the analyst instead exited the room with the thumb drive. After approximately 15 minutes, management provided our investigator with what they asserted was the USB thumb drive in question. It is impossible to know whether management provided the same USB thumb drive that the analyst had removed.[36]

- Manual backup using optical drives or USB sticks should not be attempted in a regulated laboratory under any circumstances.

Most of the citations above are where laboratories have standalone workstations rather than networked solutions.

2.7.4 Citations for Lack of Laboratory Controls

Human inventiveness knows no bounds when it comes to data falsification. One enterprising company[37] removed some of their chromatographs and workstations from site to hide data manipulation from inspectors. Other non-compliance citations include:

- Unofficial testing – which we discuss in more detail in the next section under complete data.
- Failing to document work contemporaneously. One way this can be achieved is by waiting until the chromatography has been performed, then working out the sample weight required and then falsifying the weighing result.[18]
- Overwriting data is possible with some file based systems and this was used by a number of companies that used file based instrument data system with poor controls to protect electronic records.
- Paper processes are easy to falsify as if the analysis is not right, the paper form can be discarded and another one printed and used. Control of blank forms and master templates is now a requirement in several of the data integrity guidances, as we shall see in Chapter 11.

One of the corrective actions requested by the FDA was the writing of an SOP describing a comprehensive computer life cycle to ensure that data integrity was better in the computer systems used by the organisation.[18]

2.7.5 Failure to Have Complete Laboratory Records

Here is where compliance failures become very interesting.

Audit trails in some instrument data systems were found to be turned off, which is a poor approach to compliance in a regulated environment.[17,18,36,38,39]

It is imperative that the audit trail is turned on, otherwise changes that are made to data cannot be attributed to the individual who made them and the old and new values are not recorded. Designers of CDS audit trails must embed them in the basic operation of the system so that they cannot be turned off and the only configuration item is if the laboratory wants to turn on the function to enter a reason for change.[16] When the audit trail in the system was turned on, nobody reviewed the entries (except the inspectors)[8,21] but the audit trail is part of complete data[29] that the second person reviewer needs to check.

Further non-compliance citations, outlined in Figure 2.2, are the reintegration of chromatograms to pass and not saving the integration method – here there need to be technical controls in the CDS software and well as an SOP and training on when it is permissible to reintegrate and when it is not.[16] The use of integrate inhibit during a chromatography run has also been cited as a non-compliance for hiding impurities.[40]

A common theme with many of the warning letters was the use of trial or test injections or unofficial testing in chromatographic analyses.[21,22] This practice is a test injection of samples to check if a batch is going to pass or not, furthermore, the test injections are either conveniently forgotten or, worse, deleted from the CDS as if the test never occurred. The failure to document and/or delete the test injections brings a citation under 211.194(a)[29] for not having complete data for the analysis. The worst case is the deletion of 5301 data files from a data system.[21]

In a citation for Wockhardt in the November 2013 warning letter[17] for using test injections there is the following:

> Neither the International Conference on Harmonisation of Technical Requirements for Registration of Pharmaceuticals for Human Use (ICH) document Q2R1, "Validation of Analytical Procedure: Text and Methodology," nor the United States Pharmacopoeia General Chapter <1058>, "Analytical Instrument Qualification," includes instructions for performing "trial" injections for a method that is validated.

This is an interesting citation and rationale, examining these two references in more detail we find:

- ICH Q2(R1)[41] outlines the experiments for validation of an analytical procedure. In Section 9 there is a single paragraph that outlines the use of system suitability tests (SSTs) for checking that the whole analytical system is suitable for conducting an analysis and cross references the pharmacopoeias for more information.
- United States Pharmacopoeia general chapter <1058>[42] is focused on analytical instrument qualification (AIQ). It is not surprising that it does not mention an operational detail about the test injections as it is not within the scope of the general chapter! Therefore, the citation of this reference as justification for not permitting "test" injections is plainly wrong and spurious.

In my view the Agency would be on much better ground if they cited USP <621> on chromatography[43] or even 211.160(a)[29] for scientific soundness.

2.7.6 Too Much Data – Duplicate Record Sets

In some cases companies have been found to keep duplicate records: records that failed and those that passed as can be seen in a warning letter to Jilin Shulan Synthetic Pharmaceutical Company from May 2018[44]:

> Dual sets of laboratory records and uninvestigated OOS results:
>
> Our investigator also found that you failed to document, investigate, and resolve out-of-specification (OOS) results in your laboratory. The investigator identified two sets of laboratory testing records for four (b)(4) batches and five (b)(4) batches: one set of records included OOS results; the second set included results within specifications. You could not provide evidence to support the passing results. You also failed to conduct investigations for the OOS results. Your quality department acknowledged this practice during the inspection.

2.7.7 Industrial Scale Shredding and Discarding of GMP Documents

At part of failing to have complete data, Hetero Laboratories had a site inspected in December 2016 and citation 1 of the 483 Observations form states[45]:

> The responsibilities and procedures applicable to the quality unit are not fully followed.
>
> 1. Specifically, your QA technicians and other individuals were recorded destroying and altering records pertaining to commercial batch manufacturing immediately prior to regulatory inspection. The loss of data and documents are evidenced by the following:
>
> Through a review of your firm's Closed Circuit TV we identified the following:
>
> a) A document shredder was introduced into your firm's "Documents Storage Area" on December 03, 2016 at 15.44, approximately 4 days prior to the current US FDA inspection.
>
> b) After introduction of the document shredder we observed extensive shredding of what appears to be controlled documents and extensive signing of documents by QA. These documents were of a colour consistent with batch packaging records and batch manufacturing records, among other documents. Your firm failed to maintain documentation of what had been shredded.
>
> c) On December 06, 2016, at <redacted> we observed that a contract employee with QA removed documents from the shredder and placed them in his pocket.
>
> d) On December 07, 2016 at approximately 1.13 (in the middle of the night) individuals were shredding documents. Your firm stated that event represented cleaning staff shredding documents.

In addition, evidence was found of post-dated records:

2. On December 12, 2016, we observed the scrap area behind the production area of Buildings <redacted> to contain controlled documents that had been discarded:
   a. A balance printout with drug product <redacted> dated "14-Dec-2016". After discussing this finding with your firm, you failed to explain why the balance printout was post-dated by two days, and therefore indicating an alteration to dates on balances. Your firm's VP of Operations explained that not all balances are password protected.
   d. A plethora of documents with written numbers and signatures.

These are not isolated cases, many inspections have found discarded original records as in many cases the records had been transcribed to be perfect with no errors.

2.7.8 Responses by the Regulatory Authorities

In light of these cases of data falsification and poor data management practices, regulatory agencies have responded by issuing data integrity guidance documents and these are discussed in the next chapter.

References

1. *Schon Scandal*, Available from: https://en.wikipedia.org/wiki/Sch%C3%B6n_scandal.
2. *Barr Laboratories: Court Decision Strengthens FDA's Regulatory Power*, 1993, Available from: https://www.fda.gov/Drugs/DevelopmentApprovalProcess/Manufacturing/ucm212214.htm.
3. C. L. Burgess, Issues related to United States versus Barr Laboratories Inc., in *Development and Validation of Analytical Methods*, ed. C. L. Riley and T. W. Rosanske, Pergamon Press, Oxford, 1996, p. 352.
4. R. J. Davis, Judge Wolin's interpretation of current good manufacturing practice issues contained in the Court's ruling United States versus Barr Laboratories, in *Development and Validation of Analytical Methods*, ed. C. L. Riley and T. W. Rosanske, Pergamon Press, Oxford, 1996, p. 352.
5. *USP General Chapter <1010> Outlier Testing*, United States Pharmacopoeia Convention Inc., Rockville, MD.
6. *Inspection of Pharmaceutical Quality Control Laboratories*, Food and Drug Administration, Rockville, MD, 1993.
7. *FDA Guidance for Industry Out of Specification Results*, Food and Drug Administration, Rockville, MD, 2006.
8. *Able Laboratories Form 483 Observations*, 2005 [1 Jan 2016], Available from: http://www.fda.gov/downloads/aboutfda/centersoffices/officeofglobalregulatoryoperationsandpolicy/ora/oraelectronicreadingroom/ucm061818.pdf.

9. *FDA Compliance Program Guide CPG 7346.832 Pre-Approval Inspections*, Food and Drug Administration, Silver Springs, MD, 2010.
10. *Regulatory Action Against Ranbaxy*, 2017, Available from: https://www.fda.gov/drugs/guidancecomplianceregulatoryinformation/enforcementactivitiesbyfda/ucm118411.htm.
11. *Ranbaxy Laboratories Ltd & Ranbaxy Inc: Consent Decree of Permanent Injunction*, 2012.
12. *OECD Series on Principles of Good Laboratory Practice and Compliance Monitoring Number 1, OECD Principles on Good Laboratory Practice*, Organisation for Economic Co-Operation and Development, Paris, 1998.
13. *Scientist Steven Eaton Jailed for Falsifying Drug Test Results*, 2013, Available from: http://www.bbc.co.uk/news/uk-scotland-edinburgh-east-fife-22186220.
14. *FDA Untitled Letter: Semler Research Center Private Limited*, Food and Drug Administration, Silver Spring, MD, 2016.
15. *Notification to Pharmaceutical Companies: Clinical and Bioanalytical Studies Conducted by Semler Research Are Unacceptable*, 2016, Available from: https://www.fda.gov/Drugs/DrugSafety/ucm495778.htm.
16. R. D. McDowall, *Validation of Chromatography Data Systems: Ensuring Data Integrity, Meeting Business and Regulatory Requirements Second Edition*, Royal Society of Chemistry, Cambridge, 2017.
17. *FDA Warning Letter Wockhardt Limited (WL 320-13-21)*, Food and Drug Administration, Silver Springs, MD, 2013.
18. *FDA Warning Letter, USV Limited (WL: 320-14-03)*, Food and Drug Administration, Silver Spring, MD, 2014.
19. *EudraLex – Volume 4 Good Manufacturing Practice (GMP) Guidelines, Chapter 1 Pharmaceutical Quality System*, European Commission, Brussels, 2013.
20. *ICH Q10 Pharmaceutical Quality Systems*, International Conference on Harmonisation, Geneva, 2008.
21. *FDA Warning Letter Sun Pharmaceuticals*, Food and Drug Administration, Rockville, MD, 2014.
22. *Ipca Laboratories Limited Warning Letter (WL: 320-16-07)*, Food and Drug Administration, Silver Spring, MD, 2016.
23. *Sri Krishna Pharmaceuticals Ltd. – Unit II Warning Letter (WL: 320-16-09)*, Food and Drug Administration, Silver Spring, MD, 2016.
24. *EudraLex – Volume 4 Good Manufacturing Practice (GMP) Guidelines, Chapter 9 Self Inspection*, European Commission, Brussels, 2001.
25. *MHRA Expectation Regarding Self Inspection and Data Integrity*, 2013 [cited 2013 01 Jan 2016], Available from: http://webarchive.nationalarchives.gov.uk/20141205150130/http://www.mhra.gov.uk/Howweregulate/Medicines/Inspectionandstandards/GoodManufacturingPractice/News/CON355490.
26. *MHRA GMP Data Integrity Definitions and Guidance for Industry 2nd Edition*, Medicines and Healthcare Products Regulatory Agency, London, 2015.
27. *MHRA GXP Data Integrity Guidance and Definitions*, Medicines and Healthcare Products Regulatory Agency, London, 2018.

28. *Compliance Policy Guide CPG Sec. 130.300 FDA Access to Results of Quality Assurance Program Audits and Inspections*, 2007 [21 June 2016], Available from: http://www.fda.gov/ICECI/ComplianceManuals/CompliancePolicyGuidanceManual/ucm073841.htm.
29. *21 CFR 211 Current Good Manufacturing Practice for Finished Pharmaceutical Products*, Food and Drug Administration, Sliver Spring, MD, 2008.
30. *21 CFR 11 Electronic Records; Electronic Signatures, Final Rule*, in *Title 21*, Food and Drug Administration, Washington, DC, 1997.
31. *EudraLex – Volume 4 Good Manufacturing Practice (GMP) Guidelines, Annex 11 Computerised Systems*, European Commission, Brussels, 2011.
32. *FDA Guidance for Industry Computerised Systems Used in Clinical Investigations*, Food and Drug Administration, Rockville, MD, 2007.
33. *FDA Draft Guidance for Industry Data Integrity and Compliance with cGMP*, Silver Spring, MD, USA, 2016.
34. *PIC/S PI-041 Draft Good Practices for Data Management and Integrity in Regulated GMP/GDP Environments*, Pharmaceutical Inspection Convention/Pharmaceutical Inspection Co-Operation Scheme, Geneva, 2016.
35. *FDA Warning Letter BBT Biotech Gmbh (Warning Letter 320-16-12)*, Food and Drug Administration, Sliver Spring, MD, 2016.
36. *FDA Warning Letter: Zhejiang Hisun Pharmaceutical Co., Ltd (Warning Letter: 320-16-06)*, Food and Drug Administration, Silver Spring, MD, 2015.
37. *FDA Warning Letter Fresenius Kabi Oncology (WL: 320-13-20)*, Food and Drug Administration, Silver Springs, MD, 2013.
38. *Wockhardt Limited Warning Letter 2 (WL: 320-14-01)*, Food and Drug Administration, Silver Spring, MD, 2013.
39. *FDA Warning Letter: Reine Lifescience (Warning Letter 320-18-50)*, Food and Drug Administration, Silver Spring, MD, 2018.
40. *FDA Warning Letter: Divi's Laboratories Ltd. (Unit II) (Warning Letter 320-17-34)*, Food and Drug Administration, Silver Spring, MD, 2017.
41. *ICH Q2(R1) Validation of Analytical Procedures: Text and Methodology*, International Conference on Harmonisation, Geneva, 2005.
42. *USP 31 General Chapter <1058> Analytical Instrument Qualification*, United States Pharmacopoeial Convention, Rockville, MD, 2008.
43. *USP General Chapter <621> Chromatography*, United States Pharmacopoeia Commission Inc, Rockville, MD.
44. *FDA Warning Letter: Jilin Shulan Synthetic Pharmaceutical Company (Warning Letter 320-18-51)*, Food and Drug Administration, Silver Sprint, MD, 2018.
45. *FDA 483 Observations: Hetero Labs Limited, Telangana, India*, Food and Drug Administration, Silver Spring, MD, 2016.

CHAPTER 3

The Regulators' Responses

In response to data integrity issues, regulatory authorities have issued guidance documents covering data integrity, good record keeping practices, good data management practices and data integrity and compliance with GMP regulations. There have always been requirements for data integrity within all GXP regulations but often these are implicit. As regulations are updated, then more data integrity requirements are incorporated into the regulations such as EU GMP Chapter 1 on Pharmaceutical Quality Systems and Annex 11 on Computerised Systems. Additionally, the GAMP Forum has issued a Guide entitled Records and Data Integrity that provides more information about data governance and data integrity than contained in the regulatory guidance documents. Lastly, there is a discussion of what records constitute raw data and can raw data and complete data be considered the same?

3.1 What Do the Regulators Want?

In this section the existing regulatory requirements for GXP testing in regulated laboratories are presented. Data integrity is not explicitly mentioned by name but implied in the excerpts presented here but also in the rest of the regulations.

3.1.1 EU Good Manufacturing Practice Chapter 1

Chapter 1 has in Clause 1.9 for Quality Control testing the following requirements[1]:

> (iv) Records are made, manually and/or by recording instruments, which demonstrate that all the required sampling, inspecting and testing procedures were actually carried out. Any deviations are fully recorded and investigated;

Data Integrity and Data Governance: Practical Implementation in Regulated Laboratories
By R. D. McDowall

Published by the Royal Society of Chemistry, www.rsc.org

(vi) Records are made of the results of inspection and that testing of materials, intermediate, bulk, and finished products is formally assessed against specification. Product assessment includes a review and evaluation of relevant production documentation and an assessment of deviations from specified procedures;

Any QA work performed must have taken place and the resulting records of the work are required for comparison with the specification.

3.1.2 EU GMP Chapter 4 on Documentation

In the Principle of Chapter 4 it states[2]:

> Suitable controls should be implemented to ensure the accuracy, integrity, availability and legibility of documents. Instruction documents should be free from errors and available in writing. The term ‘written’ means recorded, or documented on media from which data may be rendered in a human readable form.

Note that in the context of Chapter 4, the word *documentation* refers to any GMP record regardless of the media used for the record.

3.1.3 21 CFR 211 cGMP Regulations for Finished Pharmaceutical Goods

Under the laboratory Controls section there are the following regulatory requirements under 21 CFR 211.160[3]:

(b) Laboratory controls shall include the establishment of scientifically sound and appropriate specifications, standards, sampling plans, and test procedures designed to assure that components, drug product containers, closures, in-process materials, labeling, and drug products conform to appropriate standards of identity, strength, quality, and purity. Laboratory controls shall include:

(1) Determination of conformity to applicable written specifications for the acceptance of each lot within each shipment of components, drug product containers, closures, and labeling used in the manufacture, processing, packing, or holding of drug products. The specifications shall include a description of the sampling and testing procedures used. Samples shall be representative and adequately identified. Such procedures shall also require appropriate retesting of any component, drug product container, or closure that is subject to deterioration.

(2) Determination of conformance to written specifications and a description of sampling and testing procedures for in-process materials. Such samples shall be representative and properly identified.

(3) Determination of conformance to written descriptions of sampling procedures and appropriate specifications for drug products. Such samples shall be representative and properly identified.
(4) The calibration of instruments, apparatus, gauges, and recording devices at suitable intervals in accordance with an established written program containing specific directions, schedules, limits for accuracy and precision, and provisions for remedial action in the event accuracy and/or precision limits are not met. Instruments, apparatus, gauges, and recording devices not meeting established specifications shall not be used.

In addition, in the section on Laboratory Records there are the following requirements in 21 CFR 211.194[3]:

(a) Laboratory records shall include complete data derived from all tests necessary to assure compliance with established specifications and standards, including examinations and assays, as follows:
(1) A description of the sample received for testing with identification of source (that is, location from where sample was obtained), quantity, lot number or other distinctive code, date sample was taken, and date sample was received for testing.
(2) A statement of each method used in the testing of the sample. The statement shall indicate the location of data that establish that the methods used in the testing of the sample meet proper standards of accuracy and reliability as applied to the product tested. The suitability of all testing methods used shall be verified under actual conditions of use.
(3) A statement of the weight or measure of sample used for each test, where appropriate.
(4) A complete record of all data secured in the course of each test, including all graphs, charts, and spectra from laboratory instrumentation, properly identified to show the specific component, drug product container, closure, in-process material, or drug product, and lot tested.
(5) A record of all calculations performed in connection with the test, including units of measure, conversion factors, and equivalency factors.
(6) A statement of the results of tests and how the results compare with established standards of identity, strength, quality, and purity for the component, drug product container, closure, in-process material, or drug product tested.
(7) The initials or signature of the person who performs each test and the date(s) the tests were performed.
(8) The initials or signature of a second person showing that the original records have been reviewed for accuracy, completeness, and compliance with established standards.

3.1.4 EU GMP Annex 11 on Computerised Systems

In Annex 11 Clause 1 on Risk Management makes the following statement[4]:

> Risk management should be applied throughout the lifecycle of the computerised system taking into account patient safety, data integrity and product quality. As part of a risk management system, decisions on the extent of validation and data integrity controls should be based on a justified and documented risk assessment of the computerised system.

3.1.5 Regulatory Requirements Summary

From the above regulatory requirements, any work in a regulated laboratory must:

- Be scientifically sound.
- Use calibrated and maintained instruments.
- Be carried out using analytical procedures verified under actual conditions of use.
- Work must actually take place.
- Complete records must be gathered regardless of format of the records, *e.g.* paper, electronic, photograph, *etc.*
- Work must be reviewed by a second individual to ensure that all records are in place and that procedures have been followed.
- Results must be formally compared against specifications.
- If using a computerised system apply documented risk assessment throughout the system life cycle.

It is the failure of some regulated organisations to comply with these requirements that have forced regulatory agencies to issue data integrity guidance documents that we will discuss in the next section of this chapter.

3.2 The Proposed FDA GLP Quality System

3.2.1 Background to the Proposed Regulation

In 2016, the FDA published their proposed update for the 21 CFR 58 regulations for non-clinical laboratory studies to be known as the GLP Quality System.[5] Although not finalised, it is the first update of the GLP regulations since 1978[6] and provides some current thinking of the FDA on both data quality and data integrity for toxicology studies that could be transferred to the GMP regulations if 21 CFR 211 was updated. As the update is not finalised I have separated presentation of key sections of the other regulations as there may be changes after the review process is completed. The major change is the introduction of requirements for both data quality and data integrity plus two other consequential changes that impact data integrity and these will be discussed below.

3.2.2 New Data Quality and Integrity Requirements

The first change is the introduction of a new section, §58.180, concerned specifically with data quality and integrity[5]:

> §58.180 Data quality and integrity.
> (a) All data generated during the conduct of a nonclinical laboratory study must be accurate, legible, contemporaneous, original, and attributable (ALCOA). Also, data must be credible, internally consistent, and corroborated.

For the first time in a regulation, the term ALCOA appears with the additional requirements for data to be credible, internally consistent and corroborated. Whilst the terms credible and internally consistent are relatively self-explanatory, in the preamble to the regulation the Agency does not explain the term corroborated. As this could be interpreted in many ways from a witness for an observation to a second person review, it is important that the FDA explain what is required in the final version of the regulation.

> (b) All data must be recorded indelibly, directly, and promptly to a permanent medium at the time of observation and must identify unambiguously the person entering the data.
>
> Any change to any entry must be made so as not to obscure the original entry, must indicate the reason for such was made, and must identify who made the change.
>
> When data are either captured or maintained, or both captured and maintained electronically, these requirements are fulfilled by the use of an electronic records system fully compliant with applicable regulations.

In the section on recording data there are three main items. The first is a requirement for contemporaneously recording data on an approved medium (*e.g.* paper, photography, computer capture, *etc.*) and attributing the capture to the individual who performed the action. The second item is the current section 58.130(e) that has been moved here and is the GLP audit trail requirements, again with attribution of action. The third section is an implicit reference made to 21 CFR 11 so that the requirements for computerised systems are not in conflict between the two regulations but are aligned.

> (c) All data accrued as required in paragraphs (a) and (b) of this section must be included in the final study report.

The requirement in sub-clause (c) is to remove bias in analysing the results of any non-clinical study by selecting only those data that support the study conclusions. Here, all data are required to be reported and could be considered the GLP equivalent to complete information in the GMP regulations presented in Section 3.1.3. Such data would allow FDA reviewers to check the study conclusions when a new drug application was submitted by conducting their own statistical analysis.

3.2.3 A New Data Integrity Role for the Study Director

Under GLP regulations, the responsibility for the overall conduct and reporting of a GLP study rests with an individual called a study director. In the proposed update to 21 CFR 58 there is a new role for the study director in clause §58.33(b)(9)[5]:

> Study Director....
>
> (9) Document unforeseen circumstances that may affect the quality and integrity of the nonclinical laboratory study when they occur and the corrective action taken.

During the conduct of a non-clinical study, the study director must document any situations that could compromise either data quality or data integrity, assess the impact of the issue as well as any document and CAPA actions that arose as a result. This is not just data falsification but any issue human or technical that could impact the study data. This then leads to the final item to discuss in the proposed update of the GLP regulations.

3.2.4 The GLP Study Report

There is a new data integrity and data quality requirement for the GLP study report and this is found in §58.185 entitled reporting of nonclinical laboratory study results[5]:

> (a) A final study report ....
>
> (10) A description of all circumstances that may have affected the quality or integrity of the data, including those documented by the study director as described in § 58.33(b)(9) ......

All data quality and data integrity issues that have been documented during the course of the study must be reported and discussed in the final study report.

3.2.5 No Hiding Place for GLP Data Integrity Issues

If these sections of the proposed GLP Quality System regulations are unchanged in the final GLP rule, then any data quality and/or data integrity issues will be documented in the final study report that will be submitted to the FDA in support of a new drug application. This will mean that the Agency will have an overview of the integrity issues when they come to inspect a facility. Given the current concern that the FDA has with data integrity, I would foresee that these new sections will be largely unchanged even if there is vociferous pharmaceutical industry feedback against them.

3.3 Overview of Regulatory Guidance Documents for Data Integrity

The focus on data integrity from regulatory agencies worldwide has been on Good Manufacturing Practice as this part of the pharmaceutical industry is the closest to the patient. However, this will change in the future as regulatory authorities will expand guidance or regulations to GLP and GCP disciplines. The overall approach to data integrity has been individual regulatory authorities to issue guidance rather than change the regulations, as shown in Figure 3.1. The Chinese Food and Drug Administration (SFDA) has also published data integrity guidance but this is omitted from the figure and discussion due to the lack of an adequate English translation.

The main regulatory guidance sources on data integrity are:
Food and Drug Administration:

- Guide to the Inspection of Pharmaceutical Quality Control Laboratories.[7]
- Level 2 guidance on the FDA web site.[8] Although this has largely been superseded by the draft data integrity guidance published in April 2016, the discussion on why electronic records is better on the web site than the draft guidance.
- Compliance Policy Guide 7346.832 for Pre-Approval Inspections.[9]
- Draft Guidance for Industry on Data Integrity and cGMP Compliance.[10]

Medicines and Regulatory Healthcare products Regulatory Agency has published two versions of their data integrity guidance for industry in 2015 and a GXP guidance published in draft in 2016 and the final document issued in 2018.[11–14]

The World Health Organisation issued a guidance on Good Data and Records Management Practice in draft in 2015 with a final version issued in 2016.[15,16]

There is guidance available from two other regulatory sources:

- EMA has issued a Question and Answer web site on data integrity.[17]
- PIC/S PI-041 has issued a draft good data management guidance document for GMP/GDP environments.[18]

In addition, there are industry guidance documents from:

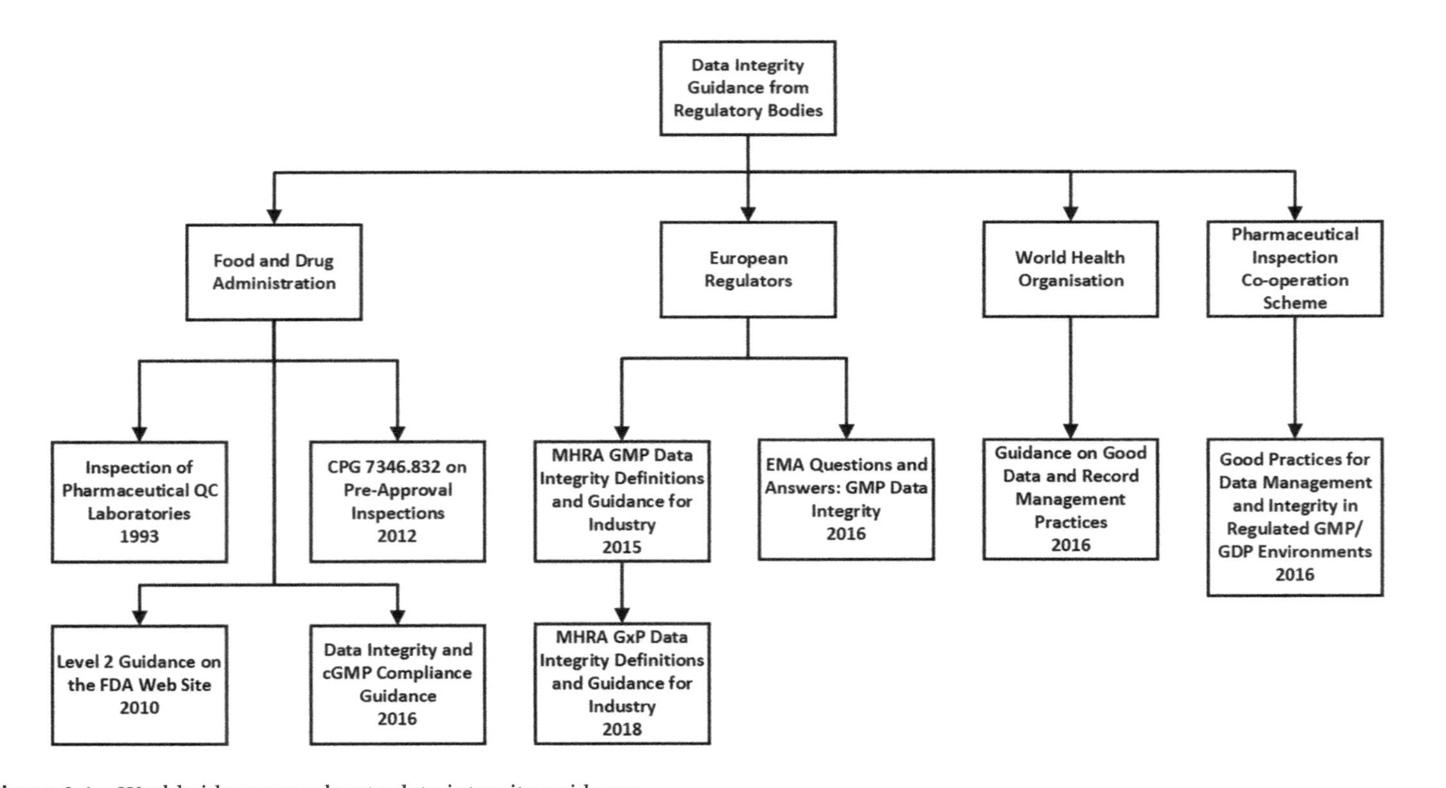

Figure 3.1 Worldwide approaches to data integrity guidance.

- ISPE have the GAMP Good Practice Guide on a Risk Based Approach to Compliant Part 11 Electronic Records and Signatures[19] have also published a companion Guide to GAMP 5 entitled Records and Data Integrity.[20] In draft are three GAMP Good Practice guides on Data Integrity.
- A Data Integrity Good Practice Guide version 2 published by the European Compliance Academy (ECA).[21]

As this area is moving rapidly, it is important to keep up with current events and publications.

3.4 Food and Drug Administration Guidance Documents

There has been a number of guidance documents on the integrity of results in regulated laboratories issued in various forms by the FDA since the Barr Laboratories court case in 1993. We will review them all in chronological order.

3.4.1 FDA Guide to Inspection of Pharmaceutical Quality Control Laboratories

Issued in 1993 following the Barr Laboratories court case, this guide outlines the approach that FDA inspectors should take when inspecting QC laboratories.[7] Why include a guide that is over a quarter of a century old and discuss the content here? The simple reason is that it is still relevant as many QC laboratories have not changed their working practices since its publication. There is a large emphasis on out of specification (OOS) results as this was the focus of the court case with Barr Laboratories[22–24] discussed in Chapter 2 but this has been replaced by the FDA Guidance on Out Of Specification (OOS) Results.[25] The more interesting points for data integrity are items such as:

Section 4 on Pre-Approval Inspections:

> Inspections should compare the results of analyses submitted with results of analysis of other batches that may have been produced. Evaluate the methods and note any exceptions to the procedures or equipment actually used from those listed in the application and confirm that it is the same method listed in the application.
> The analyst is expected to evaluate raw laboratory data for tests performed on the test batches (biobatches and clinical batches) and to compare this raw data to the data filed in the application.

This section is very consistent with the approach outlined in CPG 7346.832 for Pre-Approval Inspections[9] published in 2010 that is presented in Section 3.4.2.

Section 13 on Laboratory Records and Documentation is also very similar to readers of FDA warning letter citations[7]:

> Review personal analytical notebooks kept by the analysts in the laboratory and compare them with the worksheets and general lab notebooks and records. Be prepared to examine all records and worksheets for accuracy and authenticity and to verify that raw data are retained to support the conclusions found in laboratory results.
> Review laboratory logs for the sequence of analysis *versus* the sequence of manufacturing dates. Test dates should correspond to the dates when the sample should have been in the laboratory. If there is a computer data base, determine the protocols for making changes to the data. There should be an audit trail for changes to data.
> We expect raw laboratory data to be maintained in bound, (not loose or scrap sheets of paper), books or on analytical sheets for which there is accountability, such as prenumbered sheets. ....
> Carefully examine and evaluate laboratory logs, worksheets and other records containing the raw data such as weighings, dilutions, the condition of instruments, and calculations. Note whether raw data are missing, if records have been rewritten, or if correction fluid has been used to conceal errors. Results should not be changed without explanation. Cross reference the data that has been corrected to authenticate it. ....
> Test results should not have been transcribed without retention of the original records, nor should test results be recorded selectively. ....
> Cut charts with injections missing, deletion of files in direct data entry systems, indirect data entry without verification, and changes to computerized programs to override program features should be carefully examined. These practices raise questions about the overall quality of data.

The inspection focus can come down to detail as shown in Section 14 covering Laboratory Standard Solutions[7]:

> It is highly unlikely that a firm can "accurately and consistently weigh" to the same microgram. Therefore data showing this level of standardization or pattern is suspect and should be carefully investigated.

Hence in this guide, there is a wealth of information to help laboratories ensure the integrity of data generated as well as QA staff and auditors plan audits of laboratory data to assess the integrity of the records and data generated during analysis.

3.4.2 FDA Compliance Program Guide 7346.832 on Pre Approval Inspections

As a direct result of the Able Laboratories fraud case,[26] the FDA completely rewrote the Compliance Program Guidance (CPG) 7346.832[9] for Pre Approval Inspections that became effective in May 2012. The new version of the CPG has three objectives:

1. Readiness for Commercial Manufacturing;
2. Conformance to the Application;
3. Data Integrity Audit.

At first glance, the obvious focus for laboratory data integrity is objective 3. However, after a close reading of the document, one realises that laboratory data integrity permeates all three objectives and to focus only on objective 3 for laboratory data is unwise. In parallel with rewriting the CPG, the FDA has also trained their inspectors in data integrity, which means that there is now a focus on computerised systems and the electronic data contained therein rather than paper printouts. For hybrid and electronic systems, the inspection focus begins with the software application and the electronic records that each system has generated. Paper printouts are incidental to the main mode of inspection but the paper will be checked to see that they match the data in the underlying electronic records.

Returning to the CPG,[9] Objective 3 lists some advice for the inspectors undertaking a PAI:

- Compare raw data, hardcopy or electronic, such as chromatograms, spectrograms, laboratory analyst notebooks, and additional information from the laboratory with summary data filed in the CMC section.
- Raw data files should support a conclusion that the data/information in the application is complete....
- Lack of contextual integrity includes the failure by the applicant to scientifically justify non-submission of relevant data, such as aberrant test results or absences in a submitted chromatographic sequence.

Reiterating, the advice above, this document should be read in conjunction with the 1993 guidance on Inspection of QC Laboratories[7] to gain an overall perspective of a regulatory inspection for any regulated laboratory working to a GXP discipline.

3.4.3 FDA Level 2 Guidance

In addition, there is level 2 guidance on the FDA's web site for some aspects of data integrity such as: shared user log-ins, why paper cannot be raw data from a computerized system, and using samples as SST injections. Questions and Answers on Current Good Manufacturing Practices, Good Guidance Practices, Level 2 Guidance – Records and Reports[8]:

- Q3. How do the Part 11 regulations and predicate rule requirements (in 21 CFR Part 211) apply to the electronic records created by computerized laboratory systems and the associated printed chromatograms that are used in drug manufacturing and testing? (posted in 2010)

- Q5. Why is FDA concerned with the use of shared login accounts for computer systems? (posted in August 2014, removed in 2015)
- Q7. In warning letters to firms, why has FDA objected to the practice of using actual samples to perform system suitability testing (sometimes also referred to as "trial," "test," or "prep" runs)? (posted in August 2014, removed in 2015)

Questions 5 and 7 have been removed from the FDA web site as they are now included in the Guidance for Industry on Data Integrity and cGMP Compliance.[10] Although the guidance covers part of Q3, the better discussion on the topic is still on the FDA web site.[8]

3.4.4 Delaying, Denying, Limiting or Refusing an FDA Inspection

Changes to Food, Drug and Cosmetic (FD&C) Act in 2012 added section 501(j) that deems a drug adulterated if it

> ... has been manufactured, processed, packed, or held in any factory, warehouse, or establishment and the owner, operator, or agent of such factory, warehouse, or establishment delays, denies, or limits an inspection, or refuses to permit entry or inspection.

This tasked FDA with writing a guidance for industry on Circumstances that Constitute Delaying, Denying, Limiting, or Refusing a Drug Inspection, a draft appeared in July 2013 and the final version was issued in October 2014.[27] Delaying and limiting an FDA inspection were cited in both the Wockhardt[28] and Fresenius Kabi[29] warning letters of July 2013. This now is an important issue for ensuring speedy record retrieval during an inspection.

3.4.5 FDA Guidance on Data Integrity and Compliance with cGMP

As shown in Figure 3.1, FDA has issued a draft guidance for industry entitled Data Integrity and Compliance with CGMP.[10] It unlike the MHRA and WHO guidance documents in that it is presented in the format of 18 questions and answers as shown in Table 3.1. There is not the breadth of scope of MHRA, PIC/S and WHO guidance documents that consider data governance, the role of management and extension to an organisation's suppliers but the FDA guidance is complementary and is entirely focused on the GMP regulations and how these can be interpreted for data integrity. In question 1e, it also

Table 3.1 Questions from the FDA guidance for Industry on Data Integrity and cGMP compliance.[10]

Number	Question
1.	Please clarify the following terms as they relate to CGMP records: a. What is "data integrity"? b. What is "metadata"? c. What is an "audit trail"? d. How does FDA use the terms "static" and "dynamic" as they relate to record formats? e. How does FDA use the term "backup" in § 211.68(b)? f. What are the "systems" in "computer or related systems" in § 211.68?
2.	When is it permissible to exclude CGMP data from decision making?
3.	Does each workflow on our computer system need to be validated?
4.	How should access to CGMP computer systems be restricted?
5.	Why is FDA concerned with the use of shared login accounts for computer systems?
6.	How should blank forms be controlled?
7.	How often should audit trails be reviewed?
8.	Who should review audit trails?
9.	Can electronic copies be used as accurate reproductions of paper or electronic records?
10.	Is it acceptable to retain paper printouts or static records instead of original electronic records from stand-alone computerized laboratory instruments, such as an FT-IR instrument?
11.	Can electronic signatures be used instead of handwritten signatures for master production and control records?
12.	When does electronic data become a CGMP record?
13.	Why has the FDA cited use of actual samples during "system suitability" or test, prep, or equilibration runs in warning letters?
14.	Is it acceptable to only save the final results from reprocessed laboratory chromatography?
15.	Can an internal tip regarding a quality issue, such as potential data falsification, be handled informally outside of the documented CGMP quality system?
16.	Should personnel be trained in detecting data integrity issues as part of a routine CGMP training program?
17.	Is the FDA investigator allowed to look at my electronic records?
18.	How does FDA recommend data integrity problems identified during inspections, in warning letters, or in other regulatory actions be addressed?

illustrates the current in cGMP as backup is now interpreted by the FDA as long-term archive for records retention.

Question 1d presents the concept of static and dynamic data that was discussed in Chapter 1. To reiterate, static data are typically discrete values such as temperature and pH that cannot be interpreted although they can be averaged or trended. In contrast, dynamic data requires interpretation or processing such as UV or NIR spectra and chromatography data files and this type of data is of major concern to the FDA as inappropriate interpretation or manipulation can result in falsification of reportable results.

3.4.6 Key Points from the FDA Data Integrity Guidance

The key points from the FDA guidance on data integrity are:

- When performing GMP work all records must be retained and reviewed as *when generated to satisfy a cGMP requirement, all data become a cGMP record.*[10]
- Blank forms must be uniquely numbered and accounted for (this will be discussed in Chapter 12).
- Printouts of electronic records are not accepted as the original records as they are not exact and complete or true copies of the original.
- Access to computerised systems should be restricted to authorised individuals and their privileges must be appropriate to their role and there must be no shared user accounts.
- Audit trail review must be risk based and be performed by the operational unit where the data were generated as they comprise part of the associated records of a batch or study.
- Staff reviewing records should be trained to detect falsification of data.

These are the main points of the guidance document but the list is not exhaustive and readers are recommended to read the guidance document itself.

3.5 MHRA Data Integrity Guidance Documents

3.5.1 Initial Request to Industry December 2013

The MHRA (Medicines and Healthcare products Regulatory Agency) has been involved with data integrity since December 2013 when they announced on their web site[30] that stating from January 2014:

> *The MHRA is setting an expectation that pharmaceutical manufacturers, importers and contract laboratories, as part of their self-inspection programme must review the effectiveness of their governance systems to ensure data integrity and traceability.*

This was an extension of self-inspections (internal audits) that need to be carried out under Chapter 9 of EU GMP.[31] However, in addition to the pharmaceutical company itself it was also an expectation that the data integrity of a company's suppliers (*e.g.* API suppliers, contract manufacturing and contract laboratories, *etc.*) were included in these assessments as well.

In April 2014, MHRA as well as other European inspectors received training in data integrity from one of the consultants to the FDA on the subject.

3.5.2 MHRA GMP Data Integrity Guidance for Industry

In January 2015, MHRA released the first guidance for industry on data integrity[12] and after pharmaceutical industry feedback the Agency issued a second version in March 2015.[11] The changes were focused on relaxing the need for an independent IT function to administer standalone computerised systems that industry thought was too onerous and impractical to implement. Both versions consist of three pages of discussion about various data integrity topics followed by 13 pages of definitions with meaning and expectations. With the release of the GXP guidance for data integrity detailed discussion of the MHRA guidance will be found in the next section.

3.5.3 MHRA GXP Data Integrity Guidance for Industry

In 2016, the MHRA updated the March 2015 GMP guidance, reissuing the document as a GXP guidance by incorporating GCP and GLP elements that they are responsible for inspecting into the document but maintaining the original document format.[13] Following industry comment, the revised document was released as version 1 in March 2018.[14] In the introduction, MHRA has harmonised the guidance as much as possible and see it as a companion to PIC/S, WHO, OECD and EMA guidances and regulations. The focus is on data integrity and not data quality, the difference between the two terms was presented in Chapter 1.

The current format of the document is much improved over the GMP guidance document with numbered paragraphs and more detailed explanations of the regulatory expectations and as a result the document has expanded to 21 pages. In terms of structure, the document is consistent with the GMP guidance published in 2015 with the same sections:

- Introduction

  In the guidance it mentions that the expectations are the minimum to achieve compliance but the problem with this approach is that industry always interprets that as all they need to do. The document also advises reading this document with applicable regulations and other guidelines, which is good advice.

- The Principles of Data Integrity

  This section now consists of ten clauses covering the role of management for data integrity, open culture, expected behaviour of all staff, risk assessment of processes to identify where the vulnerabilities to records are and to put in place adequate controls to protect them. It also refers to the ALCOA and ALCOA+ principles, discussed in Chapter 1, and says that they are essentially the same.

- Establishing Data Criticality and Inherent Integrity Risk

This is a much better written section and the rather poor diagram from the 2015 GMP guidances has been removed and replaced by text. It describes the scope of data generation: paper, hybrid, electronic or other means, however it stresses that suitably configured technical controls *should be considered* (*i.e.* implemented). As the pharmaceutical industry has been deskilling, questions still arise about simple electronic systems with no record retention, *e.g.* balances and pH meters where the paper printout is the original record. However, the explicit statement of *automatic data capture or printers attached to equipment such as balances* used in the 2015 GMP guidance[11] is a better and more succinct approach to guide laboratories that has not made it to the new guidance. Risk assessment and risk management must be used to understand the vulnerability of records and implement remediation actions; management needs to be aware of these risks and prioritise their remediation.

- Designing Systems and Processes to Assure Data Integrity; Creating the "Right Environment"

  Listed here are a number of elements such as staff trained in procedures including data integrity, qualified and calibrated instruments, control of blank forms to prevent repeat testing, appropriate access rights to computerised systems, an adequate work environment with time to perform tasks and management oversight of quality metrics pertinent to data governance.

- Definition of Terms and Interpretation of Requirements

  This is the main section of the guidance where terms, listed in Table 3.2, are defined and there is an interpretation to understand it.

3.5.4 MHRA Definition of Raw Data

One of the major problems with EU GMP Chapter 4 on documentation[2] is that it mentions raw data but does not define it. The reason was the team updating the chapter in 2011 could not write a definition that covers both production and laboratory records. In contrast, US GMP regulations define production records as complete information in 211.188 and laboratory records as complete data in 211.194(a).[3] MHRA have defined raw data as[14]:

> Raw data is defined as the original record (data) which can be described as the first-capture of information, whether recorded on paper or electronically. Information that is originally captured in a dynamic state should remain available in that state.

Table 3.2 MHRA definition of terms and interpretation of data integrity requirements.[14]

MHRA definition of terms and interpretation of requirements	
• Data	• Original record and true copy
• Raw data	• Computerised system transactions
• Metadata	• Audit trail
• Data integrity	• Electronic signatures
• Data governance	• Data review and approval
• Data lifecycle	• Computerised system user access/system administrator roles
• Recording and collection of data	• Data retention: archive and backup
• Data transfer/migration	• File structure
• Data processing	• Validation – for intended purpose
• Excluding data	• IT suppliers and service providers including Cloud (*e.g.* SaaS and IaaS)

The problem is that the first sentence of the definition is not consistent with the OECD and FDA definitions of raw data. These require initial observations AND activities necessary for reconstruction of the GLP study report[6,32] and gives a limited focus for readers of the definition. What is good is that dynamic data must remain dynamic and cannot be converted to static data and this is reiterated below and is also consistent with question 10 of the FDA data integrity guidance.[10]

The MHRA definition is clarified by two sentences to help interpret it:

> Raw data must permit full reconstruction of the activities.
>
> Where this has been captured in a dynamic state and generated electronically, paper copies cannot be considered as 'raw data'.

The fact that raw data must permit full reconstruction of the activities is good but why not put this in the definition rather than the interpretation as this will lead to misinterpretation or not even read? We will return to a more detailed discussion of raw data and compare this with complete data in Section 3.9.

3.6 PIC/S Guidance Documents

PIC/S have an aide memoire for inspectors on the Inspection of Quality Control Laboratories[33] as well as their guidance on Computerised Systems in GXP environments.[34] The former publication has a section on documentation where there are small sub-sections on data traceability and computerised systems that can be used for understanding data integrity.[33] The latter has sections 23 and 24 covering inspections of computerised systems; Section 23 is a general approach to inspections and Section 24 has six checklists for systems.[34] However, there is not a specific focus on data integrity, which reflects the age of the document as it is based on GAMP 4 principles.

3.6.1 PIC/S PI-041 Good Practices for Data Management and Integrity in Regulated GMP/GDP Environments

However, as shown in Figure 3.1, PIC/S have published Good Practices for Data Management and Integrity in Regulated GMP/GDP Environments as a practical guide for GMP inspectors to use when inspecting processes and systems.[18] A first version was issued in 2016 because of the importance of publishing guidance for inspectors quickly.

The main topics in the document are:

- Data governance system covering what is data governance, management involvement including a review of the data governance system, identifying the criticality and risk of data.
- Quality culture and ethics in regulated organisations.
- Modernising the Pharmaceutical Quality System by implementing quality metrics.
- General data integrity principles and enablers.
- Specific data integrity considerations for paper-based systems, control of blank forms/templates/records, filling out records, making corrections on records and verification of records, direct print-outs from electronic systems and true copies.
- Specific data integrity considerations for computerised systems including qualification and validation, system security, data capture and processing, audit trails and security.
- Data integrity considerations for outsourced activities.
- Remediation of data integrity failures.

Of note in this document are the following:

- Mapping of EU and PIC/S GMP *versus* the ALCOA principles.
- Sections showing how paper, hybrid and electronic systems can meet the ALCOA principles.
- Mapping of EU GMP regulations to the ALCOA principles.
- Data integrity examples for some major laboratory and production systems.

3.7 WHO Guidance on Good Data and Records Management Practices

In September 2015, the World Health Organisation (WHO) issued a draft guidance document on Good Data and Records Management Practices requesting industry input before the final version was published.[15] Despite the title, this is a data integrity guidance. The document was issued as a final version in June 2016 and the main sections of this guidance are listed below and shown in Figure 3.2[16]:

- Aims and Objectives;
- Principles;
- Quality Risk Management to Ensure Good Data Management;
- Management Governance and Quality Audits;
- Contracted Organizations, Suppliers, and Service Providers;
- Training in Good Data and Record Management;
- Good Documentation Practices;
- Designing and Validating Systems to Assure Data Quality and Reliability;
- Managing Data and Records Throughout the Data Lifecycle;
- Addressing Data Reliability Issues;
- Appendix 1: Expectations and examples of special risk management considerations for the implementation of ALCOA (-plus) principles in paper-based and electronic systems.

Holistically, this guidance has a greater scope that is covered in more depth than either the FDA, MHRA or PIC/S guidance documents.[10,14,18] There is much good advice that can be used within the regulated laboratory regardless of which version of GXP is applicable.

However, from the perspective of data integrity in the laboratory, Chapter 9 on Good Documentation Practices and the associated Appendix 1 are the most applicable. In these sections the five ALCOA (Accurate, Legible, Contemporaneous. Original and Attributable) data integrity principles are each defined followed by a table listing the expectations for both paper and electronic records side by side then underneath there is a discussion of special risk factors to be considered. This is the best description of the ALCOA principles and should be read by all involved with data integrity programmes.

3.8 GAMP Guide for Records and Data Integrity

Although not a regulatory authority data integrity guidance, the GAMP Guide Records and Data Integrity was published in 2017[20] and the inclusion of this publication is relevant to the discussion on data integrity. The overall structure of the Guide is shown in Figure 3.3. The main text consists of five chapters covering regulatory focus, a framework for data governance, data life cycle and quality risk management. Underneath the main chapters are four groups of appendices for management, development, operation and general.

Three areas of interest are human factors in data integrity and a data integrity maturity model under the data governance framework and the data life cycle. The data life cycle will be discussed in Chapter 9, so no further comment is required here.

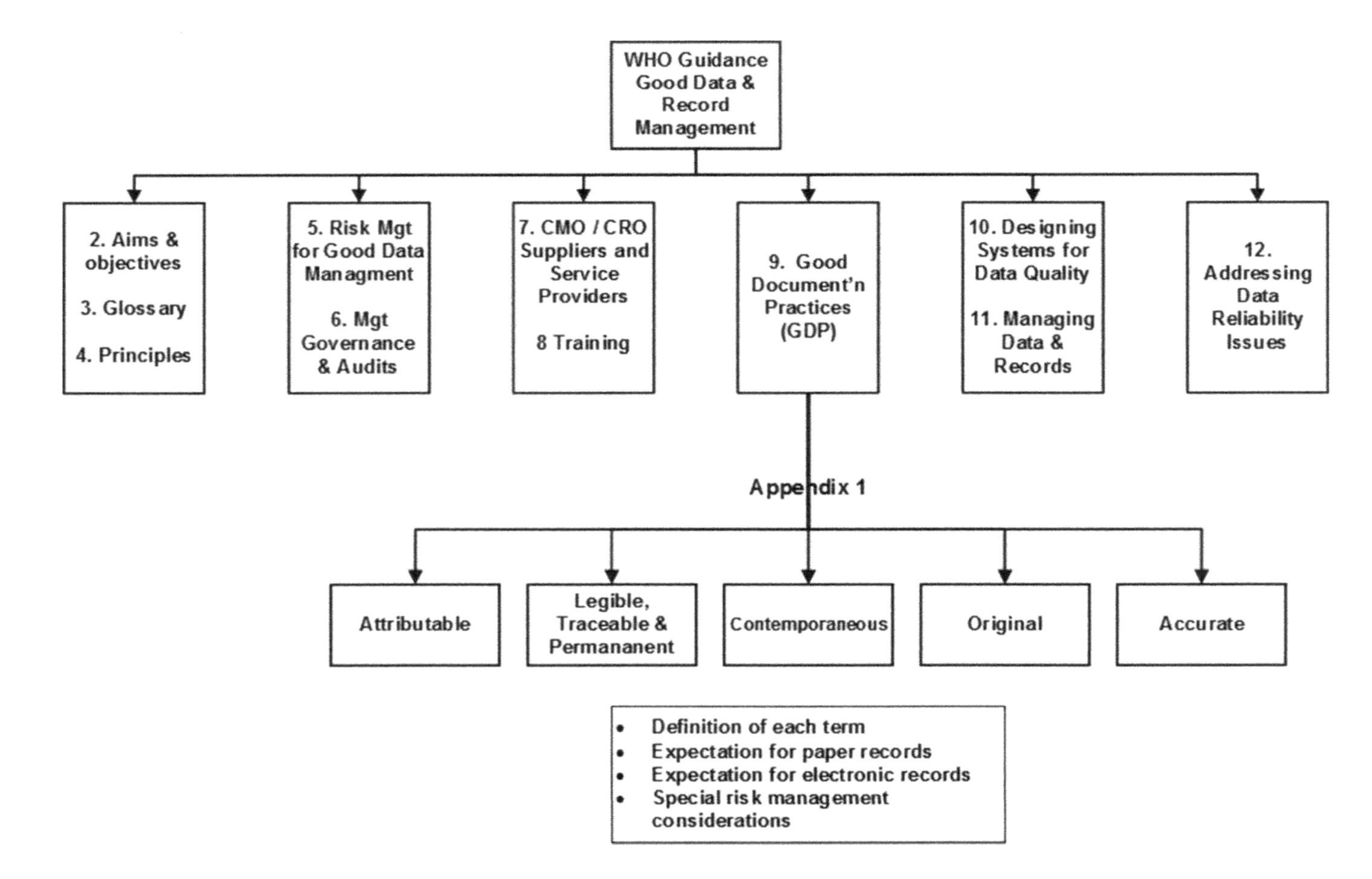

Figure 3.2 Structure of the WHO guidance on good data and record management practices.

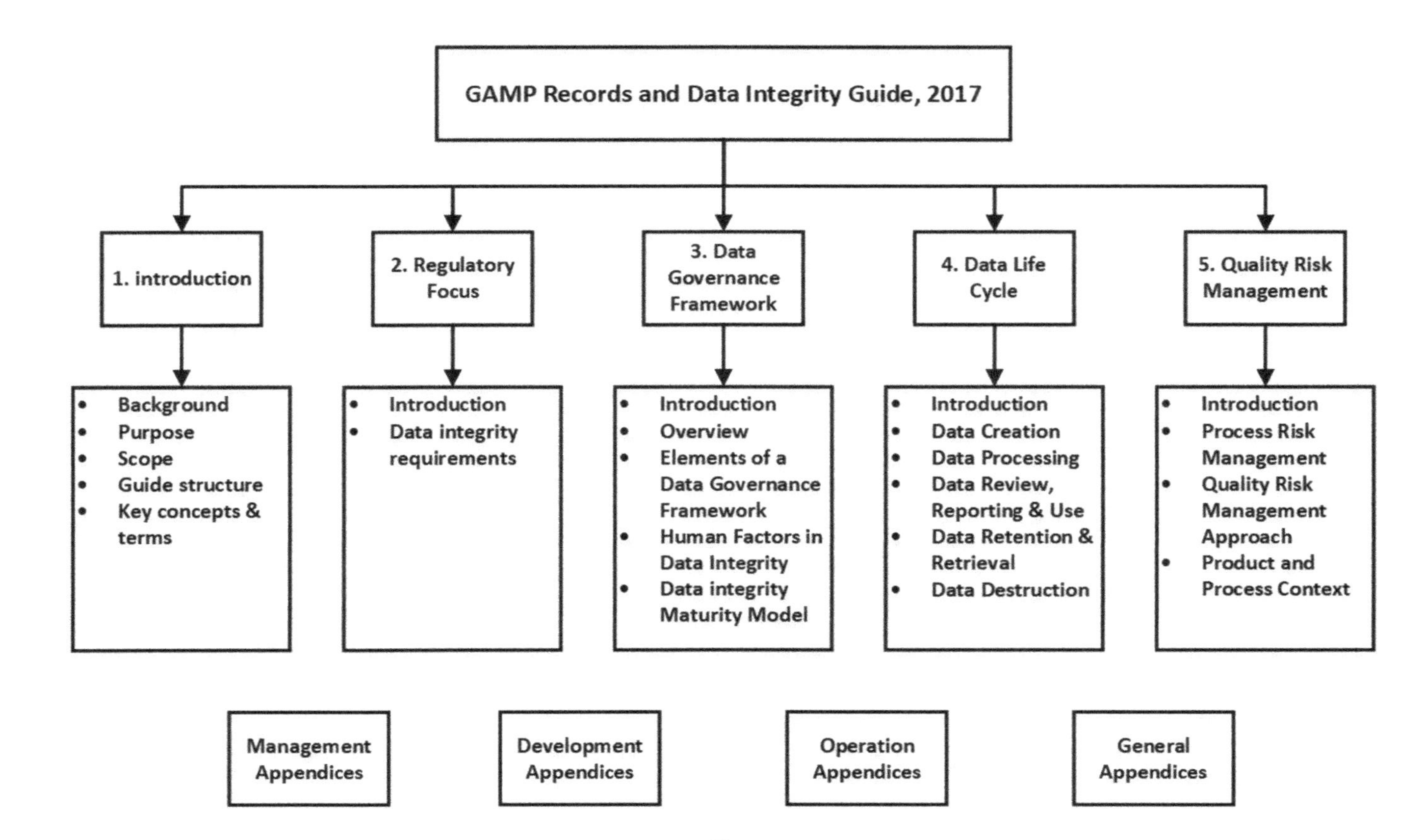

Figure 3.3 Overview of the GAMP records and data integrity guide.[20]

The seven management appendices are shown in Figure 3.4. The main appendices of interest are:

- M2 Data Integrity Maturity Model consisting of five levels with characteristics for each level.
- M3 Human Factors involved in data integrity.
- M4 Audit Trail and Review (this will be discussed in Chapter 17 and not in this chapter).

These appendices M2 and M3 provide more information about the range of human issues when dealing with poor data management practices to falsification and a five-level data integrity maturity model.

The data integrity maturity model in Appendix M2 appears to be based upon the Capability Maturity Model (CMM) and CMM Implementation (CMMI) developed by the Software Engineering Institute (SEI) a department of Carnegie Mellon University of Pittsburgh, Pennsylvania, USA. Another publication of the CMMI Institute is the Data Management Maturity (DMM) model[35] which is another five-level model for describing how well an organisation manages data, information and knowledge. We will discuss these two models as well as a data integrity model for explaining an overall approach to data integrity in Chapter 5.

3.9 PDA Technical Report 80

The Parental Drug Association (PDA) published Technical Report 80 entitled Data Integrity Management System for Pharmaceutical Laboratories in 2018 that was written by a team of industry experts, consultants and FDA inspectors.[36] This data integrity guidance document has the following main sections:

- Glossary that gathers applicable definitions from the main data integrity guidance documents. It is similar to this book's glossary where the same term has the different definitions presented for comparison.
- Regulatory trends for data integrity issues in pharmaceutical laboratories, see Section 3.9.1.
- General considerations for the control of data integrity in the laboratory covers establishing controls covering the whole data life cycle in parallel with ensuring that human factors that could result in falsification are mitigated culture and ethics permeate the laboratory.
- Data integrity in the pharmaceutical microbiology laboratory, see Section 3.9.2.
- Data integrity in the analytical quality control laboratory, see Section 3.9.3.
- Risk management of data governance systems.
- How to remediate breaches in data integrity, see Section 3.9.4.

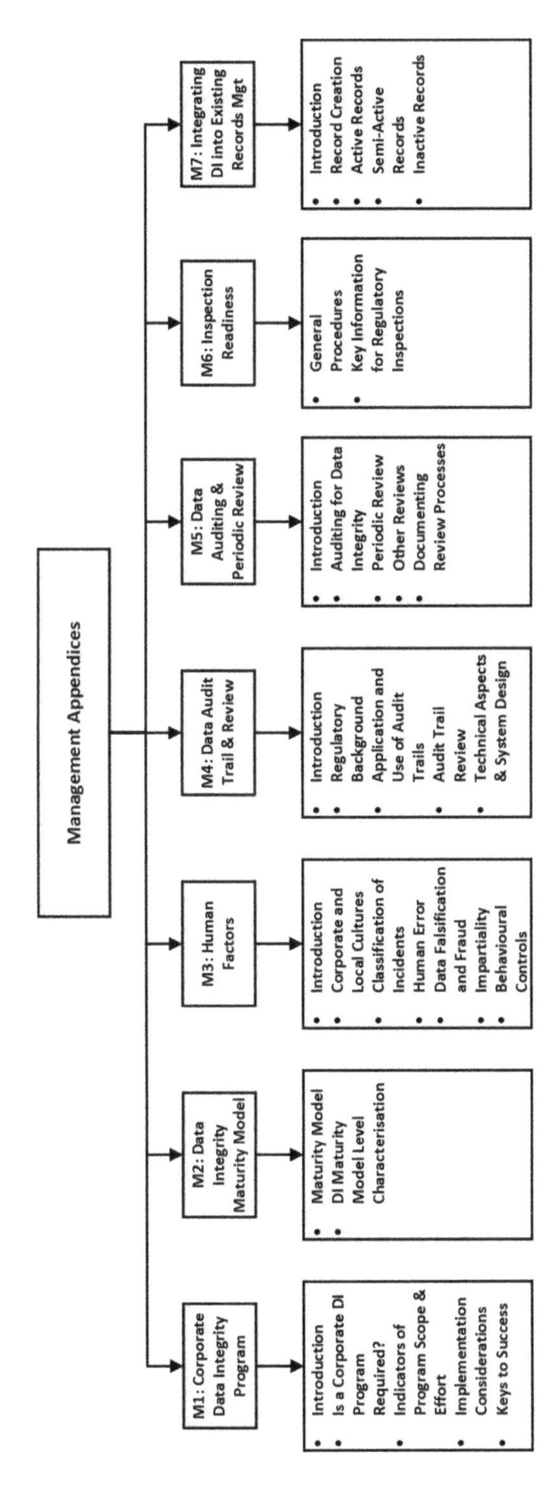

Figure 3.4 Overview of the GAMP records and data integrity guide management appendices.[20]

Overall, the document varies in tone between guidance for interpretation and a set of instructions that runs through the whole document. Data governance is hidden in the section on QC laboratories but should really be an upfront discussion as the topic has a greater scope than just laboratories.

3.9.1 Regulatory Trends for Data Integrity Issues

A summary of regulatory citations for lapses in data integrity and outright falsification between 2010 and 2017 is given in this section and these are:

- Failure to perform required testing;
- Falsification of critical data and GMP records;
- Data deletion;
- Deficiencies in reports and investigations of data integrity issues;
- Failure to ensure complete data;
- Failure to configure computerised system to meet requirements for security and control of data;
- Failure to document laboratory records contemporaneously and/or deliberate falsification of manual records;
- Performing unreported sample test injections;
- Failure to validate analytical methods.

In all, this makes grisly reading and would put the thought in the mind of the reader – don't become sick. However, these findings and observations reflect a minority of pharmaceutical companies but badly tarnish the industry as a whole.

3.9.2 Data Integrity in Microbiology Laboratories

The best part of this guidance document is that this is the first data integrity guidance for microbiology, as previously all guidance regulatory and industry documents have focused on the analytical laboratory. The number one guidance in a field of one.

After an introductory discussion about general considerations and risks for microbiological testing based on regulatory observations, the guidance launches straight into a discussion on interviewing analysts for data integrity issues. This is a rather strange way of presenting a discussion on data integrity as one would typically begin a description of good practices to avoid inspection observations. This downward spiral then focuses on sampling, worksheet review, contract laboratories, equipment and instrument review and reads like a guide for inspectors. The guidance continues in the same vein through environmental monitoring, sterility and endotoxin testing. Section 17.4.1 in this book discusses reading microbiological plates briefly.

3.9.3 Data Integrity in Analytical QC Laboratories

In contrast, this section begins from the establishment of policies, procedures and governance to ensure data integrity. One may ask, why this should apply for analytical but not for microbiology laboratories? In this way there is not a logical approach to data integrity. As we shall see in Chapter 5 with a presentation and discussion of a data integrity model, there is a way to approach data integrity and data governance in a simple and logical manner.

There is a discussion on what is a hybrid system, the risks associated with them and finally an approach to qualification of these systems. The section on qualification reads more like validation but without reference to the new version of USP <1058> or the need for a user requirements specification. One of the regulatory observations quoted at the end of the qualification section is not even relevant to a computerised system as it concerns sample weights created after the analysis.

Then there is a section on Analytical Laboratory Computerised Systems (ALCS) covering the common analytical instruments with attached data systems *e.g.* USP <1058> Group C instruments. As most of these systems operate in hybrid and standalone mode in most laboratories, an option could have been to merge this section with that on hybrid systems. A section on sever qualification is rather simplistic as it does not consider the layers involved: physical server, database (if applicable) and then the application. The qualification of the physical server, operating system and any software utilities should be a relatively simple task especially if this is performed from a prequalified image. Most of the qualification and validation effort should be at the application layer. There is a good discussion on Software as a Service (SaaS) but this should must be accompanied by rigorous risk assessments carried out by both the laboratory user and the service provider.

The poorest part of this section is the use of the abysmal acronym COTS – defined as Commercial off the Shelf software. The problem is that this term is misleading as much of the software referenced in the guide is configurable COTS. Owing to an apparent attack of not invented here syndrome, GAMP software categories would have been a far better approach and more specific but this was developed by a different society.

Section 6.3.4 presents a discussion on data governance, followed by Section 6.6 on electronic data governance. The questions that come to mind are:

- Why is this topic split into two?
- Why is it placed in the middle of an analytical chemistry section and not at the beginning of the document?
- Does data governance only apply to chemical and not microbiological analysis?

There is a large section on chromatographic data processing and peak integration with associated audit trail that extends over 10 pages of the guidance document. There are acceptable and unacceptable examples of integration of chromatograms given and is a good discussion on the subject. Spreadsheets are also covered in the guidance again with good and bad practices.

3.9.4 How to Remediate Breaches in Data Integrity

The guidance finishes with how to investigate and remediate data integrity violations: discovering the issue, considerations for the investigation, a comprehensive assessment of systems and the CAPA plans. This is covered in Chapter 22 of this book.

3.10 Understanding the Meaning of Raw Data and Complete Data

As mentioned in Section 3.5.4, we will now discuss what is meant by raw data as it is germane to the data integrity debate and later we will see how it compares with complete data for laboratory records under FDA GMP.[3] Raw data is a term that is easily said in both GMP and GLP laboratories but can create confusion and misunderstanding. What exactly does raw data mean and what electronic and paper records are within the scope of the term?

3.10.1 Are Raw Data First-capture or Original Observations?

As we have already mentioned in Section 3.5.4, MHRA defines raw data as:

> ... the original record (data) which can be described as the first-capture of information, whether recorded on paper or electronically.[14]

There is also the rider in the explanation section that *raw data should permit full reconstruction of the activities.*[14] In my view this definition is misleading as first-capture of information is only the start of the definition of raw data as many people will not read the explanation. The problem is that if the term raw data is not fully understood it can lead to poor decision making and regulatory non-compliance. This problem is compounded by the failure of the current version of EU GMP Chapter 4[31] to define the term.

3.10.2 In the Beginning ...

Raw data is not a GMP term as there is no definition of the term in any regulation or regulatory guidance document until the MHRA's GXP guidance. This phrase first saw the light of day with the FDA's Good Laboratory Practice regulations in 1978[6] where the term is defined in Section 21 CFR 58.3(k) as:

> Raw data means any laboratory worksheets, records, memoranda, notes, or exact copies thereof, that are the result of original observations and activities of a nonclinical laboratory study and are necessary for the reconstruction and evaluation of the report of that study.
> In the event that exact transcripts of raw data have been prepared (*e.g.*, tapes which have been transcribed verbatim, dated, and verified accurate by signature), the exact copy or exact transcript may be substituted for the original source as raw data.
> Raw data may include photographs, microfilm or microfiche copies, computer printouts, magnetic media, including dictated observations, and recorded data from automated instruments.

In the regulation there is just a single paragraph but I have added the breaks above to aid reading and understanding. Note that the OECD GLP regulations define raw data a little differently but the meaning is similar[32] but as the FDA definition is more encompassing, the 21 CFR 58 definition has been used here.

You can see how up to date the US regulations are when there are references to microfilm and microfiche. However, the first bullet point contains the beginning of the term original observations. When framed in the context of laboratory data involving a computerised system, *e.g.* LC, GC, UV, NIR, FTIR, MS, NMR, the first thought is to focus on the data files generated from the samples during an analysis: these are my raw data. If writing observations on an approved paper record these are also raw data.

From this misconception, some of the current problems that we have with data integrity begin.

The issue from many analytical scientists working in GLP regulated laboratories is that they never read the regulations as they pertain to their work. The regulations have been interpreted for them by the great and the good and handed down like tablets of stone to the laboratory staff to implement and follow. For the raw data debate, much GLP training emphasises that raw data are original observations that must be captured, secured and protected.

Look back to the definition of raw data. Yes, it talks about original observations. But there is more, much more.

> Raw data means ... records ... that are the result of original observations and activities of a nonclinical laboratory study and are necessary for the reconstruction and evaluation of the report of that study

Do you see the problems? It is not just original observations but also "activities" that comprise raw data. Therefore, there are more records that must be included under the heading of raw data. There is an additional requirement for raw data they are necessary for the reconstruction and evaluation of the report for that study. This is the same as the explanation in the MHRA guidance document as noted above.

3.10.3 Later, Much Later in Europe ...

Let us move the clock forward from 1978 to 2011 when the European Union issued an updated version of EU GMP Chapter 4 on documentation.[2] Unlike US GMP where documentation requirements are spread throughout the regulation like honey on a slice of bread, Chapter 4 has the main requirements for documentation in a single location and can be quite explicit in expectations for specifications, instructions and records. In the Principle of the chapter, under the subject of records we have the following regulatory requirements for records[2]:

- Provide evidence of various actions taken to demonstrate compliance with instructions, *e.g.* ... manufactured batches a history of each batch of product,....
- Records include the raw data which is used to generate other records.
- For electronic records regulated users should define which data are to be used as raw data.
- At least, all data on which quality decisions are based should be defined as raw data.

The problem is when a new term is introduced into a regulated environment there should be a definition of that term so that organisations can interpret it and apply it to their processes and systems. However, EU regulators have failed to provide the definition of raw data to enable the industry to begin any interpretation.

Also, you will note, the fourth bullet point above contains the ever popular "at least" phrase. What do inspectors interpret this as? This is the minimum but we would expect more. How does the industry interpret this phrase? This is all we will do.

3.10.4 The GLP Quality System – The Proposed 21 CFR 58 Update

Section 3.2 discussed the proposed update of 21 CFR 58 published for public comment in 2016.[5] As part of the revision of the regulation there is an update of the raw data definition to address copying requirements, computerised systems, and to include the pathology report. The proposed raw data definition reads:

> Raw data means all original nonclinical laboratory study records and documentation or exact copies that maintain the original intent and meaning and are made according to the person's certified copy procedures.
> Raw data includes any laboratory worksheets, correspondence, notes, and other documentation (regardless of capture medium) that are the result of original observations and activities of a nonclinical laboratory study and are necessary for the reconstruction and evaluation of the report of that study.
> Raw data also includes the signed and dated pathology report.[5]

What the Agency has done, by adding other documentation regardless of capture medium and copying, is to eliminate the examples in the original definition. By eliminating the examples, it takes away the media that make the current definition appear so out of date now. The specific inclusion of *the signed and dated pathology report* to what is considered as raw data changes the definition of raw data from mere *original observations* to emphasise the whole process from analysis to reporting is included under the term raw data.

3.10.5 Extracting Principles for Laboratory GXP Raw Data

How can we interpret raw data for both GLP and GMP laboratories? Let us look at either the proposed or current GLP definition of raw data. Let us begin with original observations. How do we make the original observations? We need:

- a sampling plan (GMP) or study protocol (GLP) that document how samples will be taken, stored and transported;
- an analytical procedure describing the work to be performed;
- a sample with relevant information: identity, study, batch or lot number, analysis request, *etc.*, *etc.*;
- analytical instruments that are qualified and calibrated and software that is validated;
- an appropriate and validated analytical procedure including the preparation of the sample for presentation to the instrument;
- reference standards;
- spectral library (if we are using a spectrometer for identification);
- qualified and trained staff to perform the work.

From these pre-requisites, the analysis is undertaken and one or more files will be generated and saved by the instrument data system. These are the first part of raw data. Not just the data files themselves but all the other associated contextual metadata that must be linked together to support the generated files containing the identity of the sample, instrument and method used, the analyst performing the work, date and time stamps on the files, audit trail entries, *etc.*

As we have only acquired the original data files – we now need to interpret the data in accordance with the analytical procedure that we are using, for example:

- comparison with a reference material or standard for confirmation of identity;
- identification using a spectral library and chemometrics;
- identification of an unknown by interpretation of the spectra generated;
- quantification of the analyte using a calibrated curve.

These will generate more records of the work that would include laboratory notebook entries, completed blank or template form (see Chapter 11), together with further contextual metadata, *e.g.* identity of the library or reference standards, the person who carried out the work, date and time of work and audit trail entries, *etc.*

At the completion of the analytical work, a draft report can be generated at this point for review by a second person reviewer. This may result in changes required made due to say typographical errors or misinterpretation of a spectrum that will result in more metadata and possibly more files being created. At the end of the process, a final approved report is available.

Now we have a better understanding of what constitutes raw data: all records, including any contextual metadata, generated from the sample to the report. Simple! This is similar to the reportable result and the associated contextual metadata described in Chapter 1 and shown in Table 1.3 and Figure 1.4.

3.10.6 Visualising What Raw Data Mean

Too see what this means look at Figure 3.5. This shows what constitutes raw data for a GLP/GMP spectroscopic analysis. Please understand that this is a generic representation and aims to present both qualitative and quantitative analysis, to represent one or the other a little interpretation is required by the reader. You should also read Chapter 9 on the Analytical Data Life Cycle to gain a complete understanding of raw and complete data.

You will see from Figure 3.5 that the raw data trail starts from the sampling through the sample preparation with work documented in laboratory notebooks, worksheets for which there is accountability (see Chapter 11) before the sample is presented to the instrument. If we are dealing with identity testing and the analyst puts a probe into a drum in a warehouse, then this stage is minimal or omitted altogether. The next stage in the figure is the actual analysis where the right method to control the instrument and acquire the data is used along with sample specific information that is entered to uniquely identify the analysis. Of course, the analyst is logged on to the instrument data system and all work including correction of typographical errors is recorded in the audit trail is attributed to this individual.

Next, the sample spectrum is interpreted by a variety of ways such as:

- Fit against a composite spectrum in a spectral library for identification (the name of the library and version is part of the metadata supporting the raw data).
- Interpretation of spectra for structure elucidation, here there may be notes in a laboratory notebook associated with the thinking associated with the interpretation that forms part of the raw data for the analysis.
- Quantification of analytes *via* calibrated curves or comparison with reference standards depending on the spectroscopic technique being used. Here, we can find that these calculations could be carried out in the instrument data system or a spreadsheet. All part of life's rich raw data tapestry.

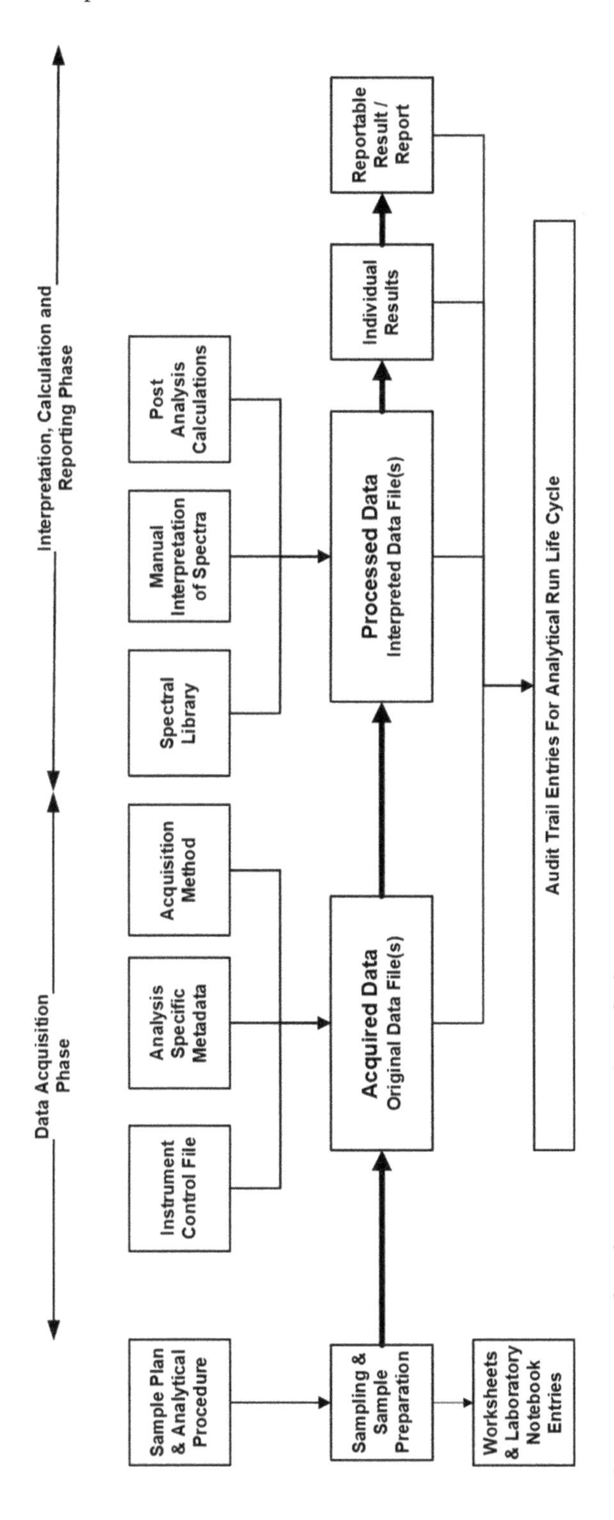

Figure 3.5 Raw data for a spectroscopic analysis.

Finally, the reportable result will be generated either in the data system or outside it and the report written. As noted above in the GLP definitions of raw data,[6,37] the report itself is part of the raw data for the work.

3.10.7 Summary: Raw Data Is the Same as Complete Data

Therefore, raw data is more than just original observations. The term includes all records created from sampling to reporting and that all stages of the process should be transparent. It also means that an auditor or reviewer can track back from a result in the report to the original observations or forward from the sample to a result in the report.

What should also be apparent, as we have done throughout this discussion, is the similarity between raw data in a GLP context and complete data for GMP as per 21 CFR 211.194(a).[3] In my view, the two terms mean the same thing regardless of the GXP discipline that a laboratory is working to. This also means that raw data is equivalent to the requirement in the FDA data integrity guidance that when created to support a GMP activity all data become a GMP record.[10]

This is a very important subject for data integrity that staff need to be trained in the concept and practical interpretation that will be presented in Chapter 7.

3.11 Regulations and Data Integrity Guidance Summary

From a regulatory perspective, EU GMP Chapter 1 regulations require that analytical work is carried out and records are created for work that actually took place. Under EU GMP Chapter 4 that work is documented in such a way that the accuracy and integrity of the record are preserved.

The regulatory guidance and industry guidance that has been issued merely goes into more depth on the detail that is required to comply with these regulations, *e.g.*

- Integration of data integrity and data governance into the pharmaceutical quality system.
- Management leadership, active involvement and review of the data integrity programme.
- Communication to all staff about the importance of ensuring data integrity in normal work.
- Development of an open culture that allows mistakes to be made and corrected without intimidation or blame.
- Automatic generation of metrics to monitor how various aspects of data integrity are performing for management review.
- Policies and procedures for data integrity and good records management for paper, hybrid and electronic processes and effective training in them.

- Assessment of processes and systems to identify risks to records and mitigate them, which may include automation and improvement to ensure the integrity of records and data. Implement short term remediation of existing systems to mitigate risks to records as well as develop long term solutions. The latter should focus on replacement of hybrid systems and automation of processes where technical controls are in place to ensure the integrity of data.
- Recording of work correctly, accurately and completely. All data generated in the course of an analysis must be retained and assessed. If any data is invalidated, there must be a scientific rationale for this. Raw data and complete data are equivalent terms and mean that all data generated in the course of a regulated analysis must be retained and cannot be deleted.
- Handling of out of trend (OOE), out of expectation (OOE) and out of specification (OOS) results to avoid testing into compliance.
- Quality oversight that monitors normal work, performs data integrity audits of processes and systems, operates a system of raising data integrity concerns and conducts investigations into data integrity violations.

All this comes as the pharmaceutical industry has not interpreted the existing regulations correctly as data integrity is implicit in GMP and GLP regulations. Nor has the industry kept current with technical advances to automate processes. In many laboratories, the ways of working are little changed in 25 years since the Barr Laboratories case.

References

1. *EudraLex – Volume 4 Good Manufacturing Practice (GMP) Guidelines, Chapter 1 Pharmaceutical Quality System*, European Commission, Brussels, 2013.
2. *EudraLex – Volume 4 Good Manufacturing Practice (GMP) Guidelines, Chapter 4 Documentation*, E. Commission, Brussels, 2011.
3. *21 CFR 211 Current Good Manufacturing Practice for Finished Pharmaceutical Products*, Food and Drug Administration, Sliver Spring, MD, 2008.
4. *EudraLex – Volume 4 Good Manufacturing Practice (GMP) Guidelines, Annex 11 Computerised Systems*, European Commission, Brussels, 2011.
5. 21 CFR Parts 16 and 58 Good Laboratory Practice for Nonclinical laboratory Studies; Proposed Rule, *Fed. Regist.*, 2016, **81**(164), 58342–58380.
6. *21 CFR 58 Good Laboratory Practice for Non-Clinical Laboratory Studies*, Food and Drug Administration, Washington, DC, 1978.
7. *Inspection of Pharmaceutical Quality Control Laboratories*, Food and Drug Administration, Rockville, MD, 1993.
8. *FDA Questions and Answers on Current Good Manufacturing Practices, Good Guidance Practices, Level 2 Guidance – Records and Reports*, 2010 [27 May 2016], Available from: http://www.fda.gov/Drugs/GuidanceCompliance-RegulatoryInformation/Guidances/ucm124787.htm.

9. *FDA Compliance Program Guide CPG 7346.832 Pre-Approval Inspections*, Food and Drug Administration, Silver Springs MD, 2010.
10. *FDA Draft Guidance for Industry Data Integrity and Compliance with cGMP*, Silver Spring, MD, USA, 2016.
11. *MHRA GMP Data Integrity Definitions and Guidance for Industry 2nd Edition*, Medicines and Healthcare Products Regulatory Agency, London, 2015.
12. *MHRA GMP Data Integrity Definitions and Guidance for Industry 1st Edition*, Medicines and Healthcare Products Regulatory Agency, London, 2015.
13. *MHRA GXP Data Integrity Definitions and Guidance for Industry, Draft Version for Consultation July 2016*, Medicines and Healthcare Products and Regulatory Agency, London, 2016.
14. *MHRA GXP Data Integrity Guidance and Definitions*, Medicines and Healthcare Products Regulatory Agency, London, 2018.
15. *WHO Draft Guidance on Good Data and Record Management Practices*, World Health Organisation, Geneva, 2015.
16. *WHO Technical Report Series No.996 Annex 5 Guidance on Good Data and Records Management Practices*, World Health Organisation, Geneva, 2016.
17. *EMA Questions and Answers: Good Manufacturing Practice: Data Integrity*, 2016, Available from: http://www.ema.europa.eu/ema/index.jsp?curl=pages/regulation/general/gmp_q_a.jsp&mid=WC0b01ac058006e06c#section9.
18. *PIC/S PI-041 Draft Good Practices for Data Management and Integrity in Regulated GMP/GDP Environments*, Pharmaceutical Inspection Convention/Pharmaceutical Inspection Co-Operation Scheme, Geneva, 2016.
19. *GAMP Good Practice Guide a Risk-based Approach to Compliant Electronic Records and Signatures*, International Society for Pharmaceutical Engineering, Tampa, FL, 2005.
20. *GAMP Guide Records and Data Integrity*, International Society for Pharmaceutical Engineering, Tampa, FL, 2017.
21. C. Burgess, *et al.*, *Guidance Document GMP Data Governance and Data Integrity Edition 2*, European Compliance Academy, Heidelberg, 2018.
22. *Barr Laboratories: Court Decision Strengthens FDA's Regulatory Power*, 1993, Available from: https://www.fda.gov/Drugs/DevelopmentApprovalProcess/Manufacturing/ucm212214.htm.
23. C. L. Burgess, Issues related to United States versus Barr Laboratories Inc., in *Development and Validation of Analytical Methods*, ed. C. L. Riley and T. W. Rosanske, Pergamon Press, Oxford, 1996, p. 352.
24. R. J. Davis, Judge Wolin's interpretation of current good manufacturing practice issues contained in the Court's ruling United States versus Barr Laboratories, in *Development and Validation of Analytical Methods*, ed. C. L. Riley and T. W. Rosanske, Pergamon Press, Oxford, 1996, p. 352.
25. *FDA Guidance for Industry Out of Specification Results*, Food and Drug Administration, Rockville, MD, 2006.
26. *Able Laboratories Form 483 Observations*, 2005 [1 Jan 2016], Available from: http://www.fda.gov/downloads/aboutfda/centersoffices/officeofglobalregulatoryoperationsandpolicy/ora/oraelectronicreadingroom/ucm061818.pdf.

27. *FDA Guidance for Industry Circumstances that Constitute Delaying, Denying, Limiting, or Refusing a Drug Inspection*, Food and Drug Administration, Rockville, MD, 2014.
28. *FDA Warning Letter Wockhardt Limited (WL 320-13-21)*, Food and Drug Administration, Silver Springs, MD, 2013.
29. *FDA Warning Letter Fresenius Kabi Oncology (WL: 320-13-20)*, Food and Drug Administration, Silver Springs, MD, 2013.
30. *MHRA Expectation Regarding Self Inspection and Data Integrity*, 2013 [cited 2013 01 Jan 2016], Available from: http://webarchive.nationalarchives.gov.uk/20141205150130/http://www.mhra.gov.uk/Howweregulate/Medicines/Inspectionandstandards/GoodManufacturingPractice/News/CON355490.
31. *EudraLex – Volume 4 Good Manufacturing Practice (GMP) Guidelines, Chapter 9 Self Inspection*, European Commission, Brussels, 2001.
32. *OECD Series on Principles of Good Laboratory Practice and Compliance Monitoring Number 1, OECD Principles on Good Laboratory Practice*, Organisation for Economic Co-Operation and Development, Paris, 1998.
33. *PIC/S Aide-Memoire Inspection of Pharmaceutical Quality Control Laboratories (PI-023-2)*, Pharmaceutical Inspection Convention/Pharmaceutical Inspection Co-Operation Scheme, Geneva, 2007.
34. *PIC/S Computerised Systems in GXP Environments (PI-011-3)*, Pharmaceutical Inspection Convention/Pharmaceutical Inspection Co-Operation Scheme (PIC/S), Geneva, 2007.
35. M. Mecca, R. Young and J. Halcomb, *Data Management Maturity (DMM) Model*, CMMI Institute, Pittsburgh, PA, 2014.
36. *Technical Report 80: Data Integrity Management System for Pharmaceutical Laboratories*, 2018, Parenteral Drug Association (PDA), Bethesda, MD.
37. *OECD Series on Principles of Good Laboratory Practice and Compliance Monitoring Number 17 on Good Laboratory Practice Application of GLP Principles to Computerised Systems*, Organisation for Economics Co-Operation and Development, Paris, 2016.

CHAPTER 4

What Is Data Governance?

The definition of data governance is *the sum total/the arrangements to ensure that data, irrespective of the format in which it is generated, is recorded, processed, retained and used to ensure a complete, consistent and accurate record throughout the data lifecycle.*[1–4] However, this definition tells a reader very little, as how can the "*arrangements or sum total of arrangements*" be interpreted adequately? Data governance provides the overall framework to ensure data integrity and encompasses all who work in an organisation from the boardroom to the laboratory bench. This chapter explores the various aspects of data governance to provide a full picture of what is required for an organisation and a regulated laboratory.

This chapter is an overview before detailed discussions of individual topics in the following chapters, *e.g.* a model for data integrity in Chapter 5, the roles and responsibilities for data governance in Chapter 6, policies procedures and training for data integrity in Chapter 7, establishing and maintaining an open culture in Chapter 8, an analytical data life cycle in Chapter 9 and quality metrics in Chapter 19.

4.1 What Do the Regulators Want?

Interestingly, the FDA[5] shows no interest in data governance, as evidenced in their draft data integrity guidance document, in contrast to the MHRA, PIC/S and WHO guidances[1–4] that discuss the topic in some depth as we shall now see.

4.1.1 EU GMP Chapter 1 Pharmaceutical Quality System

Updated in 2013 following publication of ICH Q10,[6] EU GMP Chapter 1 sets out requirements for a pharmaceutical quality system (PQS)[7]:

Data Integrity and Data Governance: Practical Implementation in Regulated Laboratories
By R. D. McDowall

Published by the Royal Society of Chemistry, www.rsc.org

In the Principle, it requires a PQS that incorporates GMP and is documented and effectiveness monitored:

> To achieve this quality objective reliably there must be a comprehensively designed and correctly implemented Pharmaceutical Quality System incorporating Good Manufacturing Practice and Quality Risk Management. It should be fully documented and its effectiveness monitored.

The responsibility for the PQS is placed firmly on the shoulders of senior management of the organisation:

> 1.5 Senior management has the ultimate responsibility to ensure an effective Pharmaceutical Quality System is in place, adequately resourced and that roles, responsibilities, and authorities are defined, communicated and implemented throughout the organisation. Senior management's leadership and active participation in the Pharmaceutical Quality System is essential. This leadership should ensure the support and commitment of staff at all levels and sites within the organisation to the Pharmaceutical Quality System.
>
> 1.6 There should be periodic management review, with the involvement of senior management, of the operation of the Pharmaceutical Quality System to identify opportunities for continual improvement of products, processes and the system itself.

Focusing on the Quality Control laboratory:

> 1.9 Quality Control is that part of Good Manufacturing Practice which is concerned with sampling, specifications and testing, and with the organisation, documentation and release procedures which ensure that the necessary and relevant tests are actually carried out and that materials are not released for use, nor products released for sale or supply, until their quality has been judged to be satisfactory. The basic requirements of Quality Control are that:
>
> (iv) Records are made, manually and/or by recording instruments, which demonstrate that all the required sampling, inspecting and testing procedures were actually carried out. Any deviations are fully recorded and investigated;

4.1.2 FDA Proposed GLP Quality System Update

In 2016, the FDA issued a proposed update[8] to the Good Laboratory Practice regulations that had been unchanged since promulgated in 1978.[9] The update, the GLP Quality System, is to introduce a complete quality system approach with increased responsibilities for senior management.

> § 58.31 Testing facility management with executive responsibility.
>
> Management with executive responsibility is ultimately responsible for the GLP Quality System and must establish policy and objectives for a GLP Quality System and a commitment to quality, as defined in § 58.3. Management with executive responsibility must ensure that the quality policy, as defined in § 58.3, is implemented and maintained at all levels of the organization.

This aligns this proposed GLP regulation with ICH Q10 and EU GMP Chapter 1.

4.1.3 MHRA GXP Data Integrity Guidance

Under item 6.5 of the MHRA GXP Guidance for Industry on Data Integrity[2] there are the following paragraphs following the definition of data governance:

> The arrangements to ensure that data, irrespective of the format in which they are generated, are recorded, processed, retained and used to ensure the record throughout the data lifecycle.
> Data governance should address data ownership throughout the lifecycle, and consider the design, operation and monitoring of processes/systems in order to comply with the principles of data integrity including control over intentional and unintentional changes to information.
> Data Governance systems should include staff training in the importance of data integrity principles and the creation of a working environment that enables visibility of errors, omissions and aberrant results.
> Senior management is responsible for the implementation of systems and procedures to minimise the potential risk to data integrity, and for identifying the residual risk, using risk management techniques such as the principles of ICH Q9. Contract Givers should ensure that data ownership, governance and accessibility are included in a contract/technical agreement. The Contract Giver should also perform a data governance review as part of their vendor assurance programme.
> Routine data review should evaluate the integrity of an individual data set, compliance with established organisational and technical measures and any data risk indicators (*e.g.* data amendment). Periodic review of data governance measures (for example audit) should assess effectiveness of established organisational and technical measures, and also consider the possibility of unauthorised activity.
> Data governance systems should also ensure that data are readily available and directly accessible on request from national competent authorities.

4.1.4 WHO Guidance on Good Records and Data Management Practices

There are multiple references to data governance within the WHO guidance document,[3] the key ones are shown below:

4.2 The data governance programme should include policies and governance procedures that address the general principles listed below for a good data management programme.
4.6 Management governance. To establish a robust and sustainable good data management system it is important that senior management ensure that appropriate data management governance programmes are in place (for details see Section 6).

Elements of effective management governance should include:

- Application of modern QRM principles and good data management principles that assure the validity, completeness and reliability of data;
- Application of appropriate quality metrics;
- Assurance that personnel are not subject to commercial, political, financial and other organizational pressures or incentives that may adversely affect the quality and integrity of their work;
- Allocation of adequate human and technical resources such that the workload, work hours and pressures on those responsible for data generation and record keeping do not increase errors;
- Ensure staff are aware of the importance of their role in ensuring data integrity and the relationship of these activities to assuring product quality and protecting patient safety.

6.1 Assuring robust data integrity begins with management, which has the overall responsibility for the technical operations and provision of resources to ensure the required quality of GXP operations. Senior management has the ultimate responsibility for ensuring that an effective quality system is in place to achieve the quality objectives, and that staff roles, responsibilities and authorities, including those required for effective data governance programmes, are defined, communicated and implemented throughout the organization. Leadership is essential to establish and maintain a company-wide commitment to data reliability as an essential element of the quality system.
6.2 The building blocks of behaviours, procedural/policy considerations and basic technical controls together form the foundation of good data governance, upon which future revisions can be built. For example, a good data governance programme requires the necessary management arrangements to ensure personnel are not subject to commercial, political, financial and other pressures or conflicts of interest that

may adversely affect the quality of their work and integrity of their data.

Management should also make staff aware of the relevance of data integrity and the importance of their role in protecting the safety of patients and the reputation of their organization for quality products and services.

6.3 Management should create a work environment in which staff are encouraged to communicate failures and mistakes, including data reliability issues, so that corrective and preventive actions can be taken and the quality of an organization's products and services enhanced. This includes ensuring adequate information flow between staff at all levels. Senior management should actively discourage any management practices that might reasonably be expected to inhibit the active and complete reporting of such issues, for example, hierarchical constraints and blame cultures.

6.4 Management reviews and regular reporting of quality metrics facilitate meeting these objectives. This requires designation of a quality manager who has direct access to the highest level of management and can directly communicate risks, so that senior management is made aware of any issues and can allocate resources to address them. To fulfil this role the quality unit should conduct and report to management formal, documented risk reviews of the key performance indicators of the quality management system. These should include metrics related to data integrity that will help identify opportunities for improvement.

6.5 Quality audits of suppliers, self-inspections and risk reviews should identify and inform management of opportunities to improve foundational systems and processes that have an impact on data reliability. Allocation of resources by management to these improvements of systems and processes may efficiently reduce data integrity risks. For example, identifying and addressing technical difficulties with the equipment used to perform multiple GXP operations may greatly improve the reliability of data for all of these operations. Another example relates to identifying conflicts of interests affecting security. Allocating independent technical support personnel to perform system administration for computerized systems, including managing security, backup and archival, reduces potential conflicts of interest and may greatly streamline and improve data management efficiency.

6.6 All GXP records held by the GXP organization are subject to inspection by the responsible health authorities. This includes original electronic data and metadata, such as audit trails maintained in computerized systems. Management of both contract givers and contract acceptors should ensure that adequate resources are available and that procedures for computerized systems are available for inspection. System administrator personnel should be available to readily retrieve requested records and facilitate inspections.

Although this is long quotation it provides a good overview of the various strands required for a data governance that must be within an overall pharmaceutical quality system. It is also a key point that data integrity is not just about computerised systems.

4.1.5 PIC/S PI-041 – Good Practices for Data Management and Integrity in Regulated GMP/GDP Environments

Section 5.2 of the PIC/S data integrity guidance[4] describes data governance systems:

> 5.2.1 Data governance systems should be integral to the pharmaceutical quality system described in PIC/S GMP/GDP. It should address data ownership throughout the lifecycle, and consider the design, operation and monitoring of processes/systems in order to comply with the principles of data integrity, including control over intentional and unintentional changes to, and deletion of information.
>
> 5.2.2 The data governance system should ensure controls over data lifecycle which are commensurate with the principles of quality risk management. These controls may be:
>
> Organisational
>
> - procedures, *e.g.* instructions for completion of records and retention of completed paper records;
> - training of staff and documented authorisation for data generation and approval;
> - data governance system design, considering how data is generated recorded, processed retained and used, and risks or vulnerabilities are controlled effectively;
> - routine data verification;
> - periodic surveillance, *e.g.* self-inspection processes seek to verify the effectiveness of the data governance policy.
>
> Technical
>
> - computerised system control,
> - automation
>
> 5.2.3 An effective data governance system will demonstrate Management's understanding and commitment to effective data governance practices including the necessity for a combination of appropriate organisational culture and behaviours (Section 6) and an understanding of data criticality, data risk and data lifecycle. There should also be evidence of communication of expectations to personnel at all levels within the organisation in a manner which ensures empowerment to

report failures and opportunities for improvement. This reduces the incentive to falsify, alter or delete data.

5.2.4 The organisation's arrangements for data governance should be documented within their Quality Management System and regularly reviewed.

The PIC/S guidance emphasises the inclusion of data governance within the PQS and the role of senior management in leading, establishing, maintaining and communicating the requirements for data integrity and data governance within an organisation.

4.1.6 EMA Questions and Answers on Good Manufacturing Practice – Data Integrity

In the EMA Q&A guidance,[10] Question 12 asks: Is it required by the EU GMP to implement a specific procedure for data integrity?

> There is no requirement for a specific procedure, however it may be beneficial to provide a summary document which outlines the organisation's total approach to data governance.

However, this is contrasted with the PIC/S guidance Section 5.2.4 above that requires that data governance arrangement is documented in the PQS and regularly reviewed.

4.1.7 Summary of Regulatory Guidance

Although the quotations from the regulatory guidances take up a significant proportion of this chapter, this is deliberate. It is intended to illustrate that data governance is wide-ranging in scope and encompasses all in an organisation from the boardroom to the laboratory bench. Section 4.4 will present and discuss all elements of a data governance programme abstracted from these guidance documents.

4.2 The Rationale for Data Governance: Regulatory Boot or Business Imperative?

As we can see above, data governance is a topic of the data integrity guidance documents from MHRA, PIC/S and WHO[2–4]: the regulatory boot provides the impetus for implementation. These guidance documents provide the "what" has to be done but organisations need to interpret the requirements outlined in Section 4.1 to understand "how" to implement them.

Pause and take some time to consider the following questions:

- Should data governance be a business imperative?
- Should business needs pull data governance and hence data integrity rather than regulations push?

A pharmaceutical company, and a regulated laboratory both require representative samples followed by the acquisition and processing of reliable data to generate reliable information and knowledge to take decisions in a timely manner.[11] This, of course, will involve documented evidence of regulated activities using a variety of media. When using laboratory computerised systems, remember that the cost of compliance is always, always, always cheaper than the cost of non-compliance[12] and as outlined in Chapter 2, Section 2.5 in this book. Doing it right first time may take a little more time to think things through before implementing and validating adequate technical controls for records but this is much preferable and far cheaper than the cost of corrective actions. Companies either find the money the second time around when there is a regulatory boot driving their corrective actions or they cease to exist.

4.3 Perspectives of Data Governance Outside the Pharmaceutical Industry

Viewed from the insular world of the pharmaceutical industry data governance may appear to be a new subject. In the world outside our industry, we find that the concept of data governance is not new. The subject has been around for over 10–15 years but not in a GXP context. The problem comes when you compare data governance definitions inside and outside the pharmaceutical industry as shown in Table 4.1. In the left-hand column are the data governance definitions from WHO and MHRA, which are very similar and have essentially the same meaning. The problem is how to interpret the "*sum total of arrangements*", "*the arrangements*" or "*totality of arrangements*" directly from the definitions? The answer, when you read the three guidance

Table 4.1 Regulatory and non-regulatory definitions of data governance.

GXP data governance definitions	Non-GXP data governance definitions
The sum total of arrangements to ensure that data, irrespective of the format in which it is generated, is recorded, processed, retained and used to ensure a complete, consistent and accurate record throughout the data lifecycle.[3]	Data governance is a system of decision rights and accountabilities for information-related processes, executed according to agreed-upon models that describe who can take what actions with what information, and when, under what circumstances, using what methods.[13]
The arrangements to ensure that data, irrespective of the format in which they are generated, are recorded, processed, retained and used to ensure the record throughout the data lifecycle.[2]	
The totality of arrangements to ensure that data, irrespective of the format in which they are generated, are recorded, processed, retained and used to ensure a complete, consistent and accurate record throughout the data life cycle.[4]	The execution and enforcement of authority over the management of data assets and the performance of data functions.[14]

documents in detail[2–4] and the abstracts presented in Section 4.1, you discover what the regulators want is the engagement of the whole regulated organisation from senior management down to the laboratory bench. Data governance in the pharmaceutical context is all encompassing.

Looking outside of the regulated GXP world, the term data governance is given a different and more focused approach, as shown in the right-hand column of Table 4.1. Here, the emphasis is on data, perhaps this is not surprising as the subject is data governance? Thomas identifies the need to define who can generate, manipulate and take actions based on the derived information[13] and the more recent definition by Seiner is a simpler definition for using authority levels to the same end.[14] These two definitions focus on the data – and less on the wider context implied by the MHRA and WHO definitions.[2,3]

Does this mean that we should not look outside of the pharmaceutical industry and gain from other industries experience? No, because when you explore the wider data governance subject outside of the pharmaceutical industry you will find the need to engage senior management and control access to data and information in much the same way as required by the current crop of pharmaceutical data integrity guidance documents. Indeed, both Thomas and Seiner acknowledge the need for executive management involvement to establish and maintain effective data governance when discussing the subject in more detail.[13,14]

For the avid reader who wants to find out more about data governance the following books are suggestions:

- Data Stewardship, An Actionable Guide to Effective Data Management and Data Governance by D. Plotkin.[15]
- Data Governance Simplified, Creating and measuring Trusted Data for Business, H. Starling.[16]
- The Chief Data Officer Handbook for Data Governance, S. Soares.[17]
- Measuring data Quality for Ongoing Improvement, L. Sebastian-Coleman.[18]
- Data Governance: How to Design, Deploy and Sustain an Effective Data Governance Program, J. Ladley.[19]

Following a look outside of regulated environment, we need to turn inward and consider the subject of this chapter – data governance within the pharmaceutical industry.

4.4 Key Data Governance Elements

4.4.1 Summary of Regulatory Guidance for Data Governance

From the regulatory requirements quoted in Section 4.1 we can now extract the items comprise data governance and is shown in Figure 4.1. This presents a visual framework for data governance within a pharmaceutical quality or quality management system.

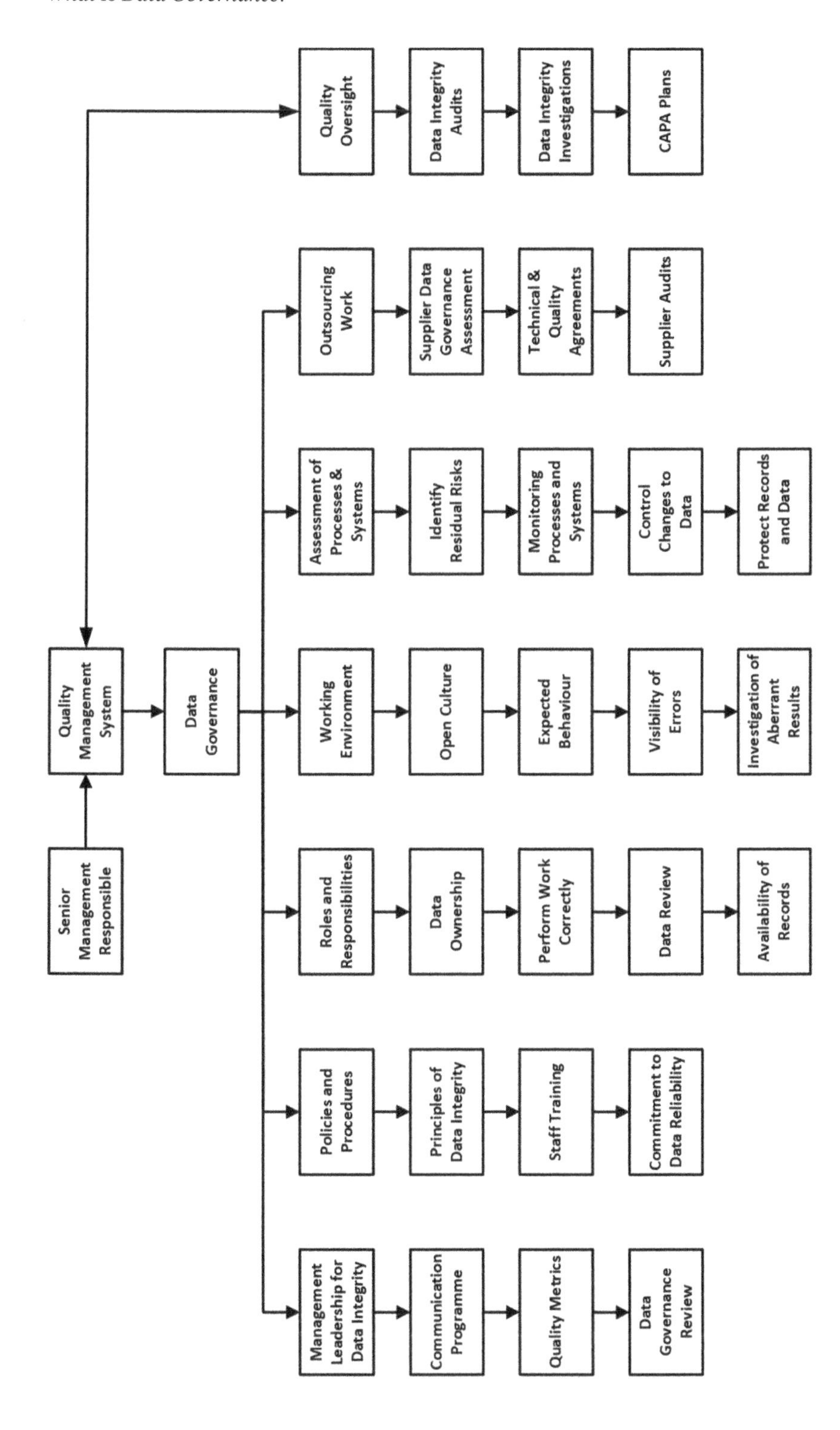

Figure 4.1 Overview of data governance elements abstracted from regulatory guidance documents.

4.4.2 Main Data Governance Areas

From Figure 4.1, the following data governance areas can be described:

- Management Leadership for Data Integrity. Senior management has overall responsibility for quality and compliance with GMP within the PQS as defined in EU GMP Chapter 1.[7] The leadership of the data governance and data integrity programme is required that includes a communication programme that includes setting performance objectives for data integrity in every staff member's objectives. Generation of quality metrics to monitor how the data integrity programme is performing are inputs into regular management reviews of the programme.
- Policies, Procedures and Effective Training. A data integrity policy and procedures for ensuring data integrity for all activities must be followed by training in these procedures for all staff is essential. Data integrity must be included in the regulatory requirement for on-going GMP training to reinforce the message. Staff must understand the importance of data integrity and sign a commitment to generating reliable data. One question that arises is should Research staff be subject to data integrity principles and training? The simple answer is yes. The rationale is that data generated in Research can be submitted as part of a product application and hence under be under 21 CFR 11 regulations[20] and subject to PAI under CPG 7346.832.[21]
- Roles and Responsibilities for Data integrity. One of the main roles in data integrity is a data owner as the MHRA, PIC/S and WHO guidances make clear.[2–4] Rather that create another role, possibly duplicating some responsibilities, the existing process owner in the laboratory for each system or process should also be responsible for the integrity of data generated and managed within their process/systems and should be designated as the data owner as well. Trained users must perform work correctly, followed by an effective second person review. Finally, all records (both paper and electronic) should be readily available for audit and inspection.
- The Working Environment. This is the most important area that senior management can establish and foster. What is required is the creation and maintenance of an open culture that includes the establishment of a no-blame culture to enable staff to raise data integrity issues. Part of this culture is the ability of staff to raise data integrity issues without fear of retribution *via* reporting mechanism to quality assurance. Errors should be seen as an opportunity to improve rather than hide under the carpet, as should the reasons for aberrant results. In addition, management should not place undue pressure on staff to perform work as this may compromise both data quality and data integrity. The creation and maintenance of the correct working

environment and culture is the most difficult and challenging aspect of data governance.

- Assessment of Processes and Systems: Each process and system needs to be formally assessed for data integrity issues. The records generated, their criticality as well as their vulnerability is documented along with plans for short term remediation and long term solutions. These plans should determine, what if any, additional controls are required to ensure data integrity. Once remediated, monitoring should be established to ensure that the plans work.
- Outsourcing work: If considering outsourcing laboratory work to a CRO or CMO, an assessment of the data governance and data integrity controls in place with the supplier is essential. Are they adequate? Roles and responsibilities and access to paper and electronic records generated on behalf of the contract-giver/sponsor, the right to audit and communications must be included in the technical or quality agreements between the two organisations.
- Quality oversight: This includes interpretation of regulations and guidance documents involving data integrity, monitoring routine analytical work, planning and carrying out data integrity audits and carrying out data integrity investigations. From the last two items, there may be CAPA plans written to ensure that the processes and systems be compliant.

4.4.3 Further Data Governance Chapters in this Book

In this and other chapters of this book there will be more detailed discussions of these elements as follows:

- Management leadership and the communication programme for data integrity (this chapter and Chapter 6).
- Quality metrics (Chapter 19).
- Data governance review (this chapter).
- Policies and procedures (Chapter 7).
- Principles of data integrity (Chapters 1, 7 and 9)
- Staff training (Chapter 7).
- Commitment to data reliability (Chapters 7, 11, 12, 13 and 14).
- Roles and responsibilities, data ownership (Chapter 6).
- Performing work correctly (Chapters 7 and 16).
- Data review (Chapter 17).
- Availability of Records (Chapter 18).
- Working environment, open culture, expected behaviours and visibility of errors (Chapter 8).
- Investigation of aberrant results (Chapters 7, 16 and 17).
- Assessments of processes and systems and identifying residual risks (Chapter 10).
- Monitoring processes and systems (Chapter 19).

- Controlling changes to data and protecting records and data (Chapters 7, 13, 14, 16, 17 and 18).
- Quality oversight and data integrity audits (Chapters 21 and 24).
- Data integrity investigations and CAPA plans (Chapter 23).
- Outsourcing works, supplier data governance assessment, technical and quality agreements and supplier audits (Chapters 23 and 24).

We will start on the data governance and data integrity journey by understanding the full scope of the data integrity programme with a data integrity model in Chapter 5.

References

1. *MHRA GMP Data Integrity Definitions and Guidance for Industry 2nd Edition*, Medicines and Healthcare Products Regulatory Agency, London, 2015.
2. *MHRA GXP Data Integrity Guidance and Definitions*, Medicines and Healthcare Products Regulatory Agency, London, 2018.
3. *WHO Technical Report Series No. 996 Annex 5 Guidance on Good Data and Records Management Practices*, World Health Organisation, Geneva, 2016.
4. *PIC/S PI-041 Draft Good Practices for Data Management and Integrity in Regulated GMP/GDP Environments*, Pharmaceutical Inspection Convention/Pharmaceutical Inspection Co-Operation Scheme, Geneva, 2016.
5. *FDA Draft Guidance for Industry Data Integrity and Compliance with cGMP*, Silver Spring, MD, USA, 2016.
6. *ICH Q10 Pharmaceutical Quality Systems*, International Conference on Harmonisation, Geneva, 2008.
7. *EudraLex – Volume 4 Good Manufacturing Practice (GMP) Guidelines, Chapter 1 Pharmaceutical Quality System*, European Commission, Brussels, 2013.
8. 21 CFR Parts 16 and 58 Good Laboratory Practice for Nonclinical laboratory Studies; Proposed Rule, *Fed. Regist.*, 2016, **81**(164), 58342–58380.
9. *21 CFR 58 Good Laboratory Practice for Non-Clinical Laboratory Studies*, Food and Drug Administration, Washington, DC, 1978.
10. *EMA Questions and Answers: Good Manufacturing Practice: Data Integrity*, 2016, Available from: http://www.ema.europa.eu/ema/index.jsp?curl=pages/regulation/general/gmp_q_a.jsp&mid=WC0b01ac058006e06c#section9.
11. R. D. McDowall and C. Burgess, The Ideal Chromatography Data System for a Regulated Laboratory, Part 1: The Compliant Analytical Process, *LCGC North Am.*, 2015, **33**(8), 554–557.
12. R. D. McDowall, *Validation of Chromatography Data Systems: Ensuring Data Integrity, Meeting Business and Regulatory Requirements Second Edition*, Royal Society of Chemistry, Cambridge, 2017.
13. G. Thomas, *Definition of Data Governance*, 2006, Available from: http://www.datagovernance.com/adg_data_governance_definition/.

14. R. S. Steiner, *Non-Invasive Data Governance. The Path of Least Resistence and Greatest Success*, Technics Publications, Basking Ridge, NJ, 2014.
15. D. Plotkin, *Data Stewardship. An Actionable Guide to Effective Data Management and Data Governance*, Morgan Kaufman, Waltham, MA, 2014.
16. H. Starling, *Data Governance Simplified: Creating and Measuring Trusted Data for Business*, CreateSpace Independent Publishing Platform, 2015.
17. S. Soares, *The Chief Data Officer's Handbook for Data Governance*, MC Press Online, Boise, ID, 2014.
18. L. Sebastian-Coleman, *Measuring Data Quality for Ongoing Improvement*, Morgan Kaufman, Waltham, MA, 2013.
19. J. Ladley, *Data Governance: How to Design, Deploy and Sustain an Effective Data Governance Program*, Morgan Kaufmann, Waltham, MA, 2012.
20. *21 CFR 11 Electronic records; electronic signatures, final rule*, in *Title 21*, Food and Drug Administration, Washington, DC, 1997.
21. *FDA Compliance Program Guide CPG 7346.832 Pre-Approval Inspections*, Food and Drug Administration, Silver Springs, MD, 2010.

CHAPTER 5

A Data Integrity Model

Understanding the scope of a data integrity and data governance programme is essential to ensure that all work is undertaken and that there are no regulatory gaps left that could result in a citation during an inspection. The full scope of a data integrity and data governance programme is very extensive, as we have seen in Chapter 4. One of the tasks of senior management is to communicate the DI programme and expectations throughout a regulated organisation. As the data integrity guidances from regulatory agencies have little in the way of a structured approach to the topic there is no easy way to understand the full scope of the problem. This is where a Data Integrity Model is useful to visualise the subject and should be the vehicle to understand the topic that will be presented and discussed in this chapter.

As we shall see again, data integrity is not just about numbers. It is much more than that, as can be seen with the scope of the Data Integrity Model. You will recall that the laboratory portion of the Data Integrity Model was introduced in Chapter 1 where the chapters of the book were plotted against the various levels, however, this chapter provides the full understanding of the Model.

5.1 A Data Integrity Model

From the scope of data governance and data integrity presented in Chapter 4 and shown in Figure 4.1 we have many elements that it is important to understand the full scope of a data integrity and data governance programme for an organisation. One way to achieve this is to develop a model that allows everybody to understand what is involved. It is better if the model is visual as it aides understanding and is easy to communicate: a picture is worth a thousand words.

Data Integrity and Data Governance: Practical Implementation in Regulated Laboratories
By R. D. McDowall

Published by the Royal Society of Chemistry, www.rsc.org

The topics of Figure 4.1 can be taken and classified into a four-layer model that will be described in this chapter.

5.1.1 A Logical Organisation of Data Integrity Elements

Although the guidance documents about data integrity have been issued by the, PIC/S, MHRA, WHO, EMA and FDA,[1–6] they lack a rigorous holistic structure for a regulated laboratory to fully understand and implement. Typically, these guidance documents do not have any figures to explain in simple terms some of the concepts that regulators want to see. Instead, the poor reader must hack through a jungle of words to figure out what is needed – this being the analytical equivalent of an Indiana Jones archaeological expedition. Chapter 4 shows that the subject is not just numbers but a subject that involves:

- management leadership;
- involvement of all the staff in an organisation;
- open culture;
- data integrity procedures and training amongst others;
- assessment and remediation of processes and systems;
- technical controls for computerised systems.

However, these guides do not, in my view, go far enough nor are sufficiently well organised to present the whole subject in a logical manner. Thus, it is important to understand that laboratory data integrity must be thought of in the context of analysis of samples within an analytical process that is operating under the auspices of a pharmaceutical quality system. Data integrity does not exist in a vacuum. In formalising a holistic approach to data integrity both within a pharmaceutical organisation and focusing on regulated laboratories I have chosen to look at four layers consisting of a foundation and three layers above it in a Data Integrity Model shown in Figure 5.1.[7]

Although it is GMP focused, the model is relatively easy to modify to a GLP or GCP model. There are three organisation elements:

- Analytical laboratory (for GMP this is this is Quality Control or Analytical Development);
- Process Development or Production;
- Quality Assurance.

5.1.2 Descriptions of the Four Levels in the Model

The Model, shown in Figure 5.1, consists of four levels that must be present to ensure data governance and data integrity within an organisation and within regulated laboratories in particular.[7] The levels are:

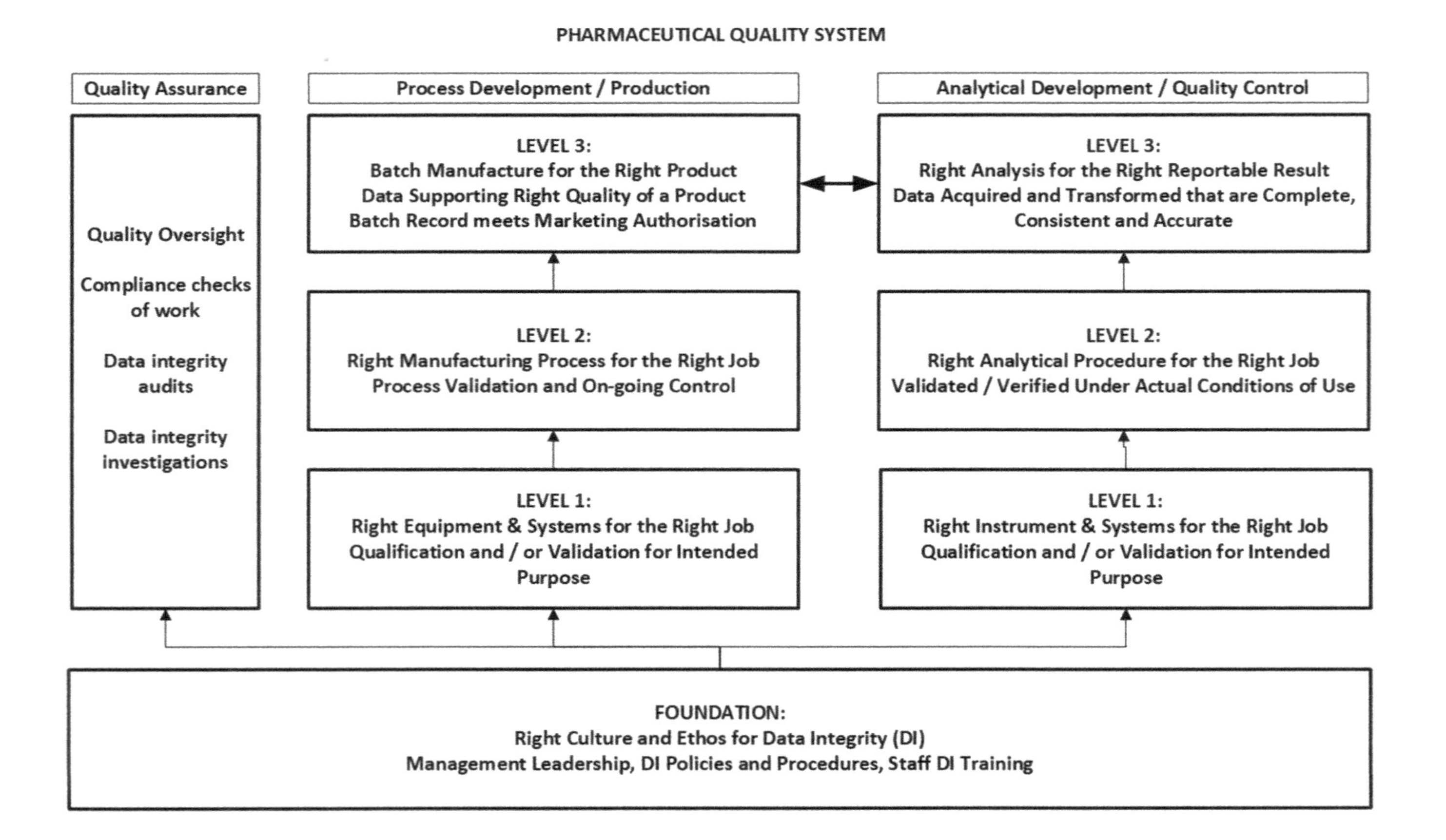

Figure 5.1 A corporate Data Integrity Model. Reproduced from ref. 7 with Permission from the Royal Society of Chemistry.

- Foundation: Right Corporate Culture for Data Integrity. The foundation goes across all elements in an organisation and is the Data Governance layer and must contain management leadership, data integrity policies including data ownership, staff training in these procedures, management review including quality metrics and an open culture.
- Level 1: Right Instrument or System for the laboratory or the Right Equipment for the Right Job for production. Qualification of analytical instruments and validation of application software including spreadsheets. Included here are calibration, point of use checks or system suitability test samples to confirm that the analytical instrument of laboratory computerised system is within user specifications before use.
- Level 2: Right Analytical Procedure or the Right Process for the Job for production and the laboratory, respectively. For a laboratory this is validation of analytical procedures or verifying the performance under actual conditions of use.
- Level 3: Right Analysis for the Right Reportable Result or Right Manufacturing Process for the Right Product. Here, production provides the laboratory samples for analysis taken to demonstrate adequate product quality and conformance with the product specification in the Marketing Authorisation (MA).
- Quality Assurance: although shown on the left in Figure 5.1 to illustrate the link between production and quality control, the QA function is pervasive throughout the Data Integrity Model to provide quality oversight, *e.g.* ensure compliance with regulations, policies and procedures as well as performing data integrity audits and data integrity investigations.

Each level feeds into and interacts with the layer above it. Like building a house, if the foundation is not right, the levels above it will be suspect and liable to collapse, often despite the best efforts of the staff who want to do a good job. We will explore each layer of the data integrity model with the foundation and then focus on the requirements for Levels 1–3 for the laboratory only.

As shown in Figure 5.1, the Data Integrity Model does not exist in a vacuum but within an existing pharmaceutical quality system (PQS). This is important as the MHRA note in their guidance that the data governance system should be integral to the pharmaceutical quality system as described in EU GMP Chapter 1.[8] For US companies the equivalent situation is described in the FDA guidance for industry on Q10 pharmaceutical quality systems.[9]

5.1.3 An Analogy of Building a House

As can be seen from Figure 5.1, the bottom of the model is the Foundation level. This is deliberately named and the image that I want to convey with the model is building a house. The Foundation level contains essential items such as management leadership, policies and procedures, training, culture, *etc.*

The way the Data Integrity Model works is that each level above is totally reliant on the levels underneath it to operate correctly. Therefore, without the Foundation level working correctly, the best qualified analytical instruments, validated analytical procedures and sample analysis is in vain if staff can manipulate data and falsify data or are put under too much pressure by management to produce results.

Houses without adequate foundations collapse.

5.1.4 Focus on the Laboratory Levels of the Data Integrity Model

Figure 5.1 presents the Data Integrity Model for a regulated manufacturing organisation but as this book is focused on the data integrity and data governance in regulated laboratories, we need to look at the laboratory specific elements of the Model as shown in Figure 5.2. Here, the Foundation level and the three laboratory levels are shown along with the Quality Oversight of laboratory operations. The QA functions are consistent across the production and laboratory areas in this Data Integrity Model.

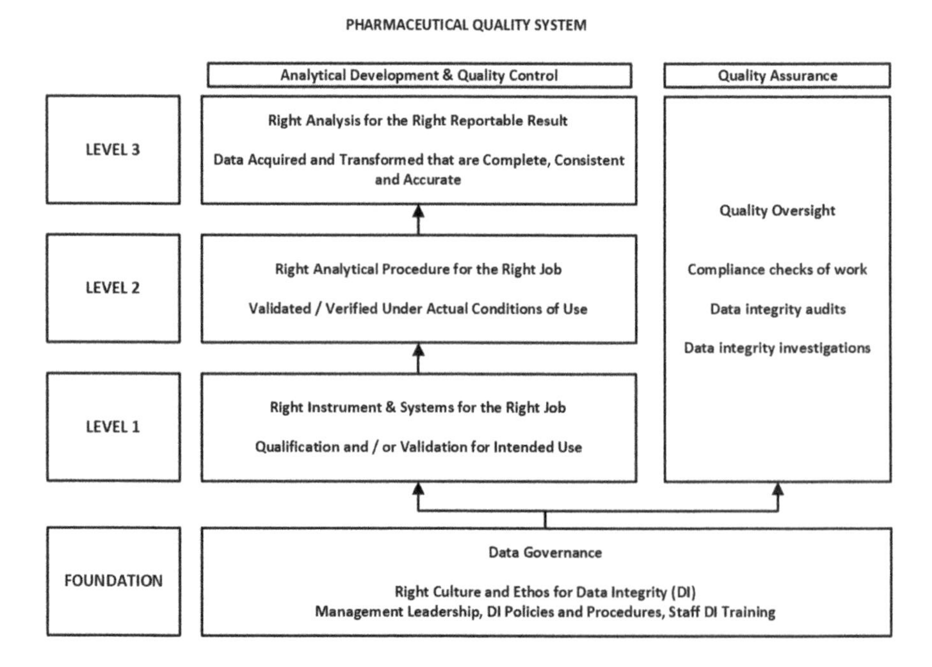

Figure 5.2 Laboratory focus of the Data Integrity Model.

5.2 Foundation Level: The Right Corporate Culture for Data Integrity

5.2.1 Role of Senior Management

The Foundation level is the core of data governance. It can only begin with the engagement and involvement of executive and senior management within any regulated organisation. Data integrity is not a project as there are many strands that need to be co-ordinated together. It is a programme of work. Therefore, to oversee and manage the data integrity programme senior management must be actively involved as this is essential to ensure that data integrity/data governance is set firmly in place within the context of a pharmaceutical quality system. Therefore, there must be management leadership, corporate data integrity policies that cascade down to laboratory data integrity procedures and finally to staff who have initial and on-going data integrity training. We will discuss data integrity training in Chapter 7.

Engagement of executive and senior management in ensuring that data integrity is in place is essential. FDA in the PQS guidance and EU GMP Chapter 1[8] make it crystal clear that executive management is responsible for quality within an organisation and the scope of quality includes data integrity. Just to ensure regulatory completeness, guess to whom the FDA address their warning letters? Yes, the Chief Executive Officer who has executive responsibility for the QMS.

However, both the MHRA[2,3] and WHO[4] guidance documents talk blithely about the need for data governance but fail to mention any substantial guidance about what is required other than the need for data owners and *the arrangements for ensuring integrity of data.*

5.2.2 Data Governance Functions in the Foundation Level

Figure 4.1 in Chapter 4 presented all the areas that were pertinent to data governance. Figure 5.3 is a modification of this diagram and represents the elements of data governance that must be present to enable a robust Foundation layer for the Data Integrity Model. We will discuss the key areas now.

Senior Management involvement:

- Senior Management leadership for data integrity that includes a communication programme for all staff.
- Generation and use of quality metrics for monitoring data integrity. This is the subject of Chapter 19 later in this book.
- As part of the Pharmaceutical Quality System, a review of the effectiveness of the data governance and data integrity projects within the overall programme. EU GMP Chapter 1[8] mandates that management must review the QMS and this is part of that review process.

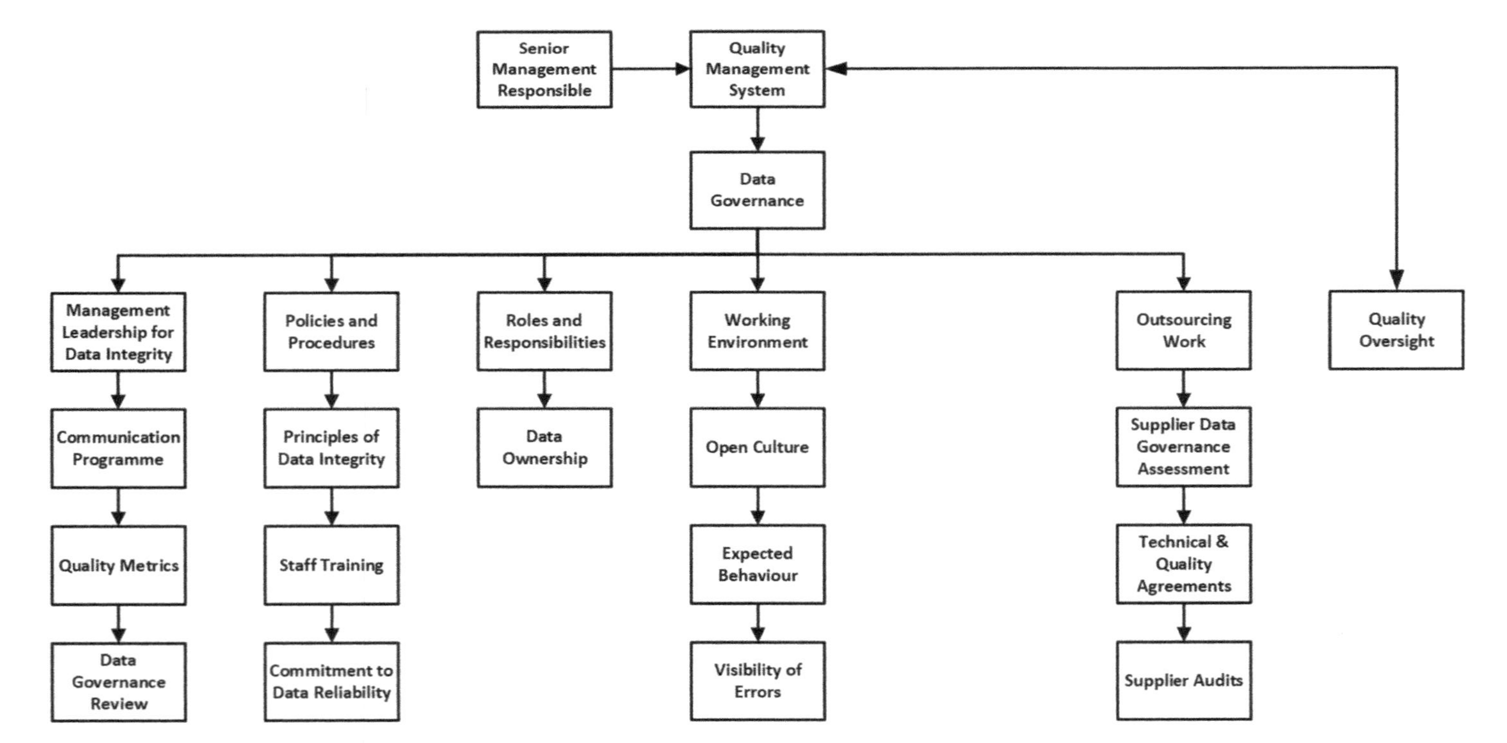

Figure 5.3 Data governance functions of foundation level of the Data Integrity Model.

Policies and Procedures need to be in place as follows:

- Writing a data integrity policy with initial and on-going training in its contents. After being trained each member of staff should sign a commitment to data reliability for their work. Although the EMA Q&A[5] notes that there is no regulation for a data integrity policy, MHRA, WHO and PIC/S guidances all state that one is needed.
- Good Documentation Practices covering paper, hybrid and electronic processes coupled with training in the procedure.
- Interpretation of analytical data, *e.g.* chromatographic integration, comparison of spectra using libraries specifically what are analysts allowed to do and when and what activities are not allowed.
- Training in these procedures with demonstrable understanding of the contents evidenced by using questionnaires or practical execution of the procedure.
- A suggested list of procedures to support data integrity will be given in Chapter 7.

Who does what:

- The roles and responsibilities of all staff involved in a data governance and data integrity programme will be discussed in more detail in Chapter 6, including data ownership.
- Roles and responsibilities must be reinforced by appropriate sections in job or position descriptions.

Quality Culture and the Working Environment:

- Management needs to create an open quality culture where there are standards for expected behaviour.
- Incorporation of data integrity goals in personnel objectives.
- Accordingly, there is an expected behaviour and an open culture where mistakes can be admitted without blame.

Outsourcing work:

- Any outsourced work requires an assessment of the outsourcing organisation's or laboratory's data governance and data integrity status before technical and quality agreements are written and signed. This will be discussed in more detail in Chapter 23.

Quality Oversight by ensuring:

- compliance with data integrity procedures *via* data integrity audits;
- investigation of data integrity violations;

- findings from audits and investigation have CAPA plans generated and their effectiveness monitored by QA.

Having completed the overview of the Foundation level, we can move on to Level 1.

5.3 Level 1: The Right Analytical Instrument and Computer System for the Job

The next level of the Data Integrity Model involves using the right analytical instruments and computerised systems used for carrying out an analysis.

5.3.1 Analytical Instrument Qualification and Computerised System Validation (AIQ and CSV)

There is little point in carrying out an analysis if an analytical instrument is not adequately qualified or the software that controls it or processes the data generated is not validated. Therefore, at Level 1, the analytical instruments and computerised systems used in the laboratory must be qualified for the specified operating range and validated for their intended purpose, respectively. There are the following reference sources:

- Updated version of USP <1058> for Analytical Instrument Qualification.[10]
- GAMP Good Practice Guide for Validation of Laboratory Computerised Systems.[11]
- Validation of Chromatography Data Systems.[7]

These provide guidance and advice on these two inter-related subjects. Indeed, the new version of USP <1058> integrates instrument qualification and computer validation for analytical equipment.[10] In Chapter 14 of this book, we will discuss an integrated approach for conducting analytical instrument qualification and computerised system validation. A user requirements specification must be written for both instruments and software to define the intended use and against which the instrument will be qualified and the software validated. Where the software application must be configured to protect electronic records generated by the system, this must be reflected in the validation documents for the application software. By implementing suitable controls to transfer, mitigate or eliminate the risk posed to these records they can be adequately protected.

Failure to ensure that an analytical instrument is adequately qualified or software is adequately validated means that all work in the top two levels of the Data Integrity Model above is wasted as the quality and integrity of the reportable results is compromised by unqualified instrumentation and unvalidated and uncontrolled software.

5.3.2 Data Governance Functions in Level 1

The following data governance functions are from Figure 4.1 in Chapter 4, the applicable ones at Level 1 of the Data Integrity Model are:

- assessment of analytical processes and computerised systems for data integrity;
- short-term remediation of processes and computerised systems;
- long-term solutions for processes and computerised systems.

These aspects will be discussed in more detail in Chapter 10 that in turn will lead into Chapters 11–13 that consider the data integrity issues with paper, the hybrid system problem and getting rid of paper – why electronic processes are better, respectively.

5.4 Level 2: The Right Analytical Procedure for the Job

5.4.1 Validation of Analytical Procedures

Using qualified analytical instruments with validated software, an analytical procedure is developed or established and then validated or verified. The GMP requirement is that analytical methods must be verified under actual conditions of use as per 21 CFR 211.194(a)(2).[12]

There are several published references for this from ICH Q2(R1)[13] and the respective chapters in the European Pharmacopoeia (EP) and United States Pharmacopoeia (USP). However, the focus of these publications is on validation of an analytical procedure that has been already developed. Method development is far more important as it determines the overall robustness or ruggedness of the analytical procedure receives scant attention in these publications. However, this analytical world is changing, following the publication in 2012 by Martin *et al.*[14] there is a draft USP <1220> on The Lifecycle Procedure Lifecycle[15] issued for comment. This will mean a move from chapters focused only on validation, verification or transfer of a method to a life cycle approach to analytical chapters that encompass development, validation, transfer and continual improvement of analytical methods.

A life cycle approach to analytical procedures validation means that following definition of an Analytical Target Profile (ATP) leads to good scientifically sound method development that ends with the definition of the procedure's design space. This now becomes important as changes to a validated method within the validated design space would be deemed to be validated *per se*. There will be a transition period where the old approach is phased out while the new one is phased in. There is also a revision of ICH Q2(R1) planned to ensure global harmonisation in this area in the future.

5.4.2 Verification of Pharmacopoeial Methods

Given the vague descriptions of most analytical methods in pharmacopoeias it is amazing that any laboratory can get a method working at all. One of the reasons is that if an HPLC method is developed using a specific supplier's C18 column, the only information about the column that appears in the monograph is a description of the type of packing and the column dimensions. For gradient methods there is no information if the gradient is formed using either a low pressure or high pressure mixing pump. For these reasons, pharmacopoeial methods need to be developed from the "methods" in the pharmacopoeia and verified under actual conditions of use. The pharmacopoeia simply provides an indication of where to start but the details are left to the individual laboratory to work out.

5.4.3 Bioanalytical Method Validation Guidance

For bioanalytical work there are guidance documents for bioanalytical method validation from the FDA[16–18] and EMA.[19] These guidance documents outline what needs to be done to validate a bioanalytical method but there is little mention of a life cycle approach or method development. Only in the 2018 FDA guidance is method development mentioned to any extent.[18] This means that the most crucial part of the method life cycle is omitted, which may result in poorly performing methods.

Therefore, a properly developed and validated or transferred analytical method is required at Level 2 of the Data Integrity Model and is a pre-requisite for ensuring data integrity of the reportable results generated in Level 3. In addition, there is also a need for the other lower levels of the Data Integrity Model to be in place and fully functioning at the top level to work correctly to obtain the right analytical results.

You will notice that there is no mention in Figure 4.1 in Chapter 4 of validated and verified analytical methods in the scope of data governance from the various regulatory authorities. This is an omission that needs to be corrected as analytical procedures are at the heart of the operations of regulated laboratories.

5.4.4 Manual Analytical Procedures Must Be Designed for Data Integrity

Like the integrated AIQ-CSV approach described at Level 1, manual or paper analytical procedures also need to be designed to ensure data integrity. Such procedures can be sampling, sample preparation or simple observational tests such as colour, odour or appearance. When conducting these procedures or methods the following need to be considered to ensure data integrity:

- Blank forms and their master templates need to be controlled and reconciled.
- Recording observations on paper can be subject to error by the tester.

- Objective evidence, where possible, is required for a second person review.

These topics will be discussed in Chapter 11.

5.5 Level 3: Right Analysis for the Right Reportable Result

Finally, at Level 3 of the Data Integrity Model, the analysis of sample will be undertaken using the right method and right process, instrument or data system. This requires analytical staff who are working in an open environment that enables data will be generated, interpreted and the reportable result will be calculated. Staff should be encouraged to admit any mistakes and there must be a no-blame culture in place based on the leadership of Senior Management from the Foundation Level of the model. It is also important not to forget the importance of the overall pharmaceutical quality system in providing the umbrella for quality such as the investigation of out of specification results, managing deviations and developing corrective and preventative actions.

From Figure 4.1 we have the following elements of data governance:

- Performing the work correctly including any deviations contemporaneously.
- Effective and comprehensive second person reviews of the work.
- Visibility of errors.
- Investigation of aberrant results.
- Availability of both paper and electronic records for audit or inspection.
- Monitoring the development and maintenance of operational data integrity procedures and training.

These complete the laboratory levels of the Data Integrity Model shown in Figure 5.2, now we move onto the Quality Oversight portion of the model.

5.6 Quality Oversight for Data Integrity

Shown on the side of Figure 5.2, quality oversight covers the Levels 1–3 of the Data Integrity Model, responsible for ensuring adherence to the regulations and company procedures for all work done in these levels. In addition, there are further data integrity responsibilities for data integrity audits and investigations that will be discussed in more detail in Chapters 21 and 22, respectively.

5.6.1 Quality Oversight of Laboratory Procedures and Work

Normally there will be quality oversight of qualification and validation work, validation of methods and routine work, but with an increased emphasis on data integrity there needs to be increased vigilance in this area. This can be achieved by QA highlighting the data integrity risks as well as recommending that higher risk processes and systems undergo remediation, upgrade or replacement.

5.6.2 Data Integrity Audits

As part of the QA audit programme, data integrity audits of processes and systems in the laboratory need to be planned, performed, reported and, where appropriate, CAPAs raised from the audit findings. When performing a data integrity audit of a computerised system, there is potential overlap with computerised system validation periodic reviews carried out under EU GMP Annex 11.[20] Therefore, in the planning the scope of data integrity audits, it is essential to separate what will be covered in a data integrity audit (typically carried out more frequently than a periodic review) and what will be checked in a periodic review. This will be covered in Chapter 21 and is linked to Chapter 24 that contains data integrity audit aide memoires.

5.6.3 Data Integrity Investigations

The data governance system should have a process for raising data integrity concerns by members of staff. This should be confidential and with a non-retaliation policy for the person raising the concern. Typically, the issue would be raised through line management. However, if the potential violation is caused by an individual's supervisor or Head of Department, then an alternative route would be *via* Quality Assurance. If valid, then a data integrity investigation is started and this is discussed in more detail in Chapter 22, along with two case studies to illustrate the process.

5.7 Linking the Data Integrity Model to the Analytical Process

At this point you may be thinking, great in theory but how does this fit into a regulated analytical laboratory? Enter stage left – the big picture.

5.7.1 The Data Integrity Model in Practice

It is important to understand that the four layers of the Data Integrity Model do not exist in isolation. They interact with each other as shown in Figure 5.4, this shows the four layers of the Data Integrity Model in a column down the left-hand side against the various tasks in an analytical process from sample management to generation of the reportable result.

- The foundation layer shows in outline what is required at the corporate layer with management leadership, culture, ethos and data integrity

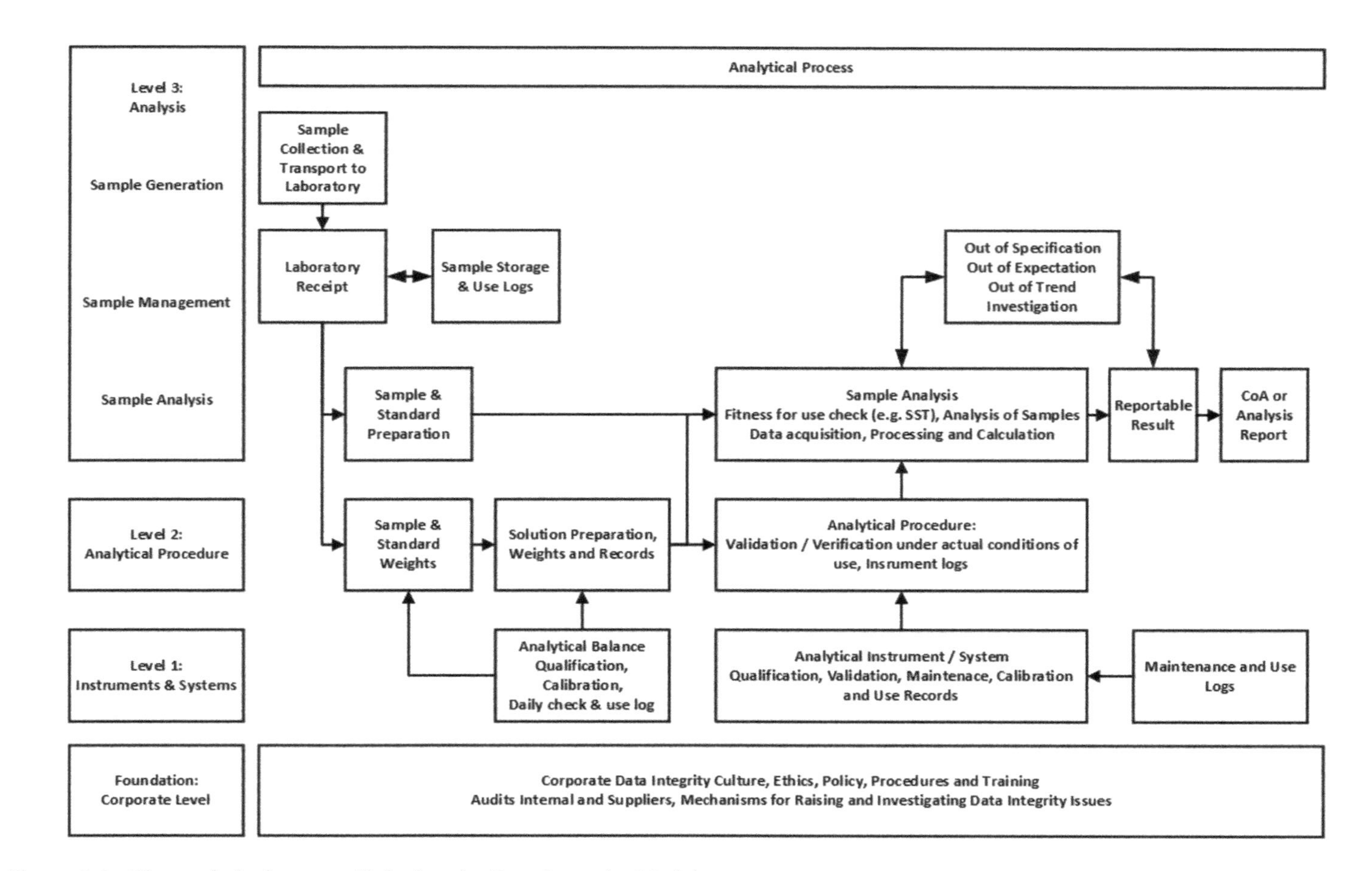

Figure 5.4 The analytical process linked to the Data Integrity Model.

policies, procedures and planning. Above the Foundation is an analytical process with the various requirements at the three Levels of the Data Integrity Model.

- Layer 1 shows qualification and use of an analytical balance as well as analytical instrument such as a chromatograph coupled with the validation of the controlling chromatography data system software. In addition, we have the regulatory requirements for calibration as well as instrument maintenance and use logs.
- Layer 2 is represented by the preparation of reference standard solutions, sample preparations, and the development and validation of the analytical procedure.
- Layer 3 The qualified instruments and validated CDS and analytical procedure are applied to the analysis of a sample. The flow starts with the sampling, transporting the sample to the laboratory, sample management, application of the analytical procedure using the validated CDS, calculation of the reportable result as well as any out of specification investigation.

5.7.2 Quality Does Not Own Quality Anymore

Figure 5.4 shows far better how the layers of the laboratory Data Integrity Model interact together. However, without the Foundation layer, how can the three other layers hope to succeed? Without qualified analytical instruments and validated software how can you be assured of the quality and integrity of the data used to calculate the reportable result? And so on up the upper layers. It is less important where an individual activity is placed in the various layers, the primary aim of this Model is to visualise for analytical scientists what data integrity actually involves.

If the data integrity model works from the foundation through the three layers that exist on top, it means that the responsibilities for data integrity and data quality are now dispersed throughout the laboratory and organisation. Whilst the overall accountability for quality oversight remains with a quality assurance function, it is not the role of Quality Assurance to fix other people's mistakes. The responsibility for data integrity and data quality in the laboratory is with the analytical staff performing the work showing that quality does not own quality anymore. Everyone in the laboratory and the whole organisation does.

5.8 Mapping the WHO Guidance to the Data Integrity Model

A useful check of the laboratory portion of Data Integrity Model is to map it *versus* the main contents of the WHO guidance document.[21] The rationale is to see how the comprehensive the Model is *versus* the WHO guidance and *vice versa*, in short this is a mapping and verification exercise for the Data Integrity Model. The results of the mapping are shown in Figure 5.5. A note of caution, this mapping of the WHO document only focuses on Chapters 5–12 inclusive of the guidance,[21] this is a crude level of mapping and does not include any detail within each of the 8 sections shown. Similarly, any data integrity requirements contained in the introductory chapters of the guidance such as the section entitled Principles have been omitted from this figure.

Looking at Figure 5.5 let us assess the distribution of the main chapters:

- The main focus of the WHO guidance is on the Foundation layer that is aimed squarely at management and corporate level efforts at data integrity. Four out of eight chapters are positioned in this layer: management governance, quality audits, an extension to external suppliers, training and addressing data reliability issues.
- Level 1 focuses on designing computerised systems for data quality and reliability as well as risk management for the protecting and managing the records generated by these systems.
- Level 2 is not represented in the WHO guidance.
- Level 3 has good documentation practices coupled with managing data across the data life cycle but no mention of the actual analysis of samples.

Again, reiterating the message earlier, the data integrity model needs to be integrated with the overall pharmaceutical quality system of a

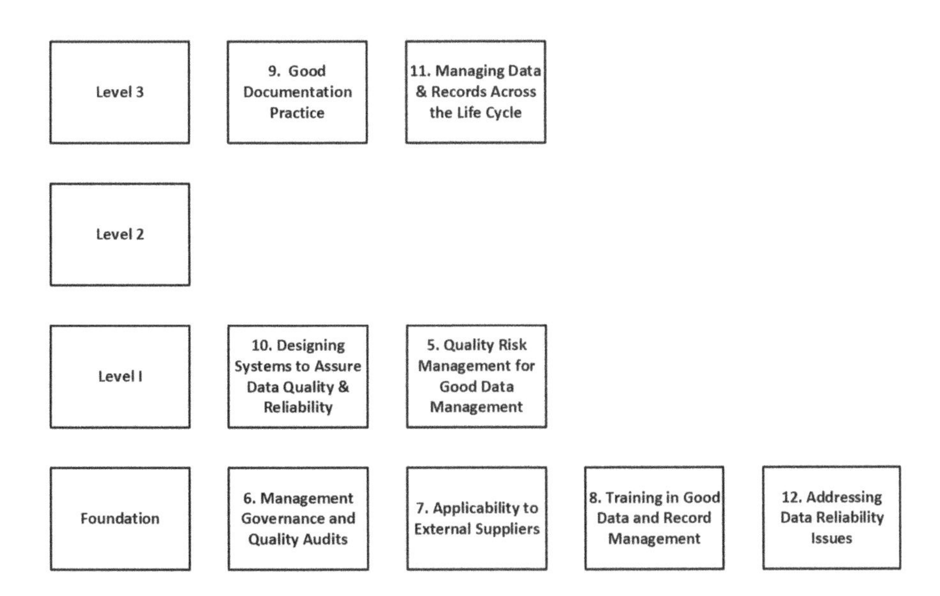

Figure 5.5 Mapping the contents of the WHO guidance *versus* the levels of the Data Integrity Model.

regulated organisation. The rationale for mapping the WHO guidance against the Data Integrity Model is to illustrate that if regulatory authorities are to provide an encompassing perspective and effective guidance on data integrity, then they need to ensure that all aspects of the subject are covered.

My perspective, *via* the Data Integrity Model, is a personal one but based on a holistic approach to analytical science in a regulated context in the first instance. However, if regulators are to provide effective guidance then they need to consider the big picture and present their requirements in a logical way.

5.9 Assessment of Data Integrity Maturity

The Data Integrity Model discussed up until now has no mechanism to determine the level of understanding of an organisation to data integrity. Audit aide memoires can be prepared, as you can see in Chapter 24, but there is no yardstick to assess the competence of an organisation. There are two other data management/data integrity models that would help with such an assessment. These are:

- Data Management Maturity Model (DMM) from the CMMI institute.[22]
- Data Integrity Maturity Model contained in the GAMP Guide on Records and Data Integrity, Appendix M2.[23]

Each will be discussed here in outline but the reader is recommended to read further the references for further detail and information.

5.9.1 Data Management Maturity Model

The Software Engineering Institute (SEI) of Carnegie Mellon University of Pittsburgh developed a Capability Maturity Model (CMM) for an organisation's approach to developing bespoke software. As application software became more common and configurable/customisable the norm rather than bespoke software, the model was extended to Capability Maturity Model Implementation (CMMI). The CMMI Institute has now been spun off as a not for profit organisation.

Having developed CMM and CMMI processes for bespoke and commercial software, the Institute turned to the data that the applications managed as the concern that an organisation's intellectual property and collective knowledge about processes and products was not being managed effectively. In 2014 the CMMI Institute published the Data Management Maturity Model (DMM) available as an e-book for purchase.[22]

Similar to CMM and CMMI models there are five levels of competence in the DMM, see Table 5.1:

Table 5.1 Description of the Data Management Maturity Model levels.[22]

DMM level	Characterisation of level	Overview
1. Performed	• Processes performed *ad hoc* (typically within projects) • Processes are typically not applied across business areas • Process discipline is reactive, *e.g.* repair *versus* prevention	Data are managed as a requirement for projects
2. Managed	• Processes are planned and executed in accordance with procedures • Have skilled people with adequate resources producing controlled outputs • Involve relevant stakeholders • Processes are monitored and controlled and evaluated for adherence to defined procedures	Awareness of managing data as a critical asset
3. Defined	• Standard processes are used and consistently followed • Specific needs are tailored to the set of standard processes	Data treated at the organisation level as critical for successful performance
4. Measured	• Process metrics have been defined and used for data management, *e.g.* management of variance, prediction • Analysis of data by statistical and other techniques • Process performance managed across the life of the process	Data treated as a source of competitive advantage
5. Optimised	• Process performance is optimised through applying Level 4 analysis or identification of improvement opportunities • Best practices are shared with peers and industry	Data seen as critical for survival in a dynamic and competitive market

- Performed.
- Managed.
- Defined.
- Measured.
- Optimised.

Looking at these levels and seeing some of the citations in FDA warning letters and 483 inspectional observations, you may be forgiven for wondering where some regulated entities would be placed on this scale. There are requirements in EU GMP Chapter 6 – specifically Clauses 6.9 and 6.16 for trending data – which is a Level 4 activity. However, how many in the industry do this on a regular and consistent basis?

The DMM model also has six functional areas within an organisation, as shown in Figure 5.6. Note that there are some items in each functional area that are similar to the pharmaceutical laboratory such as:

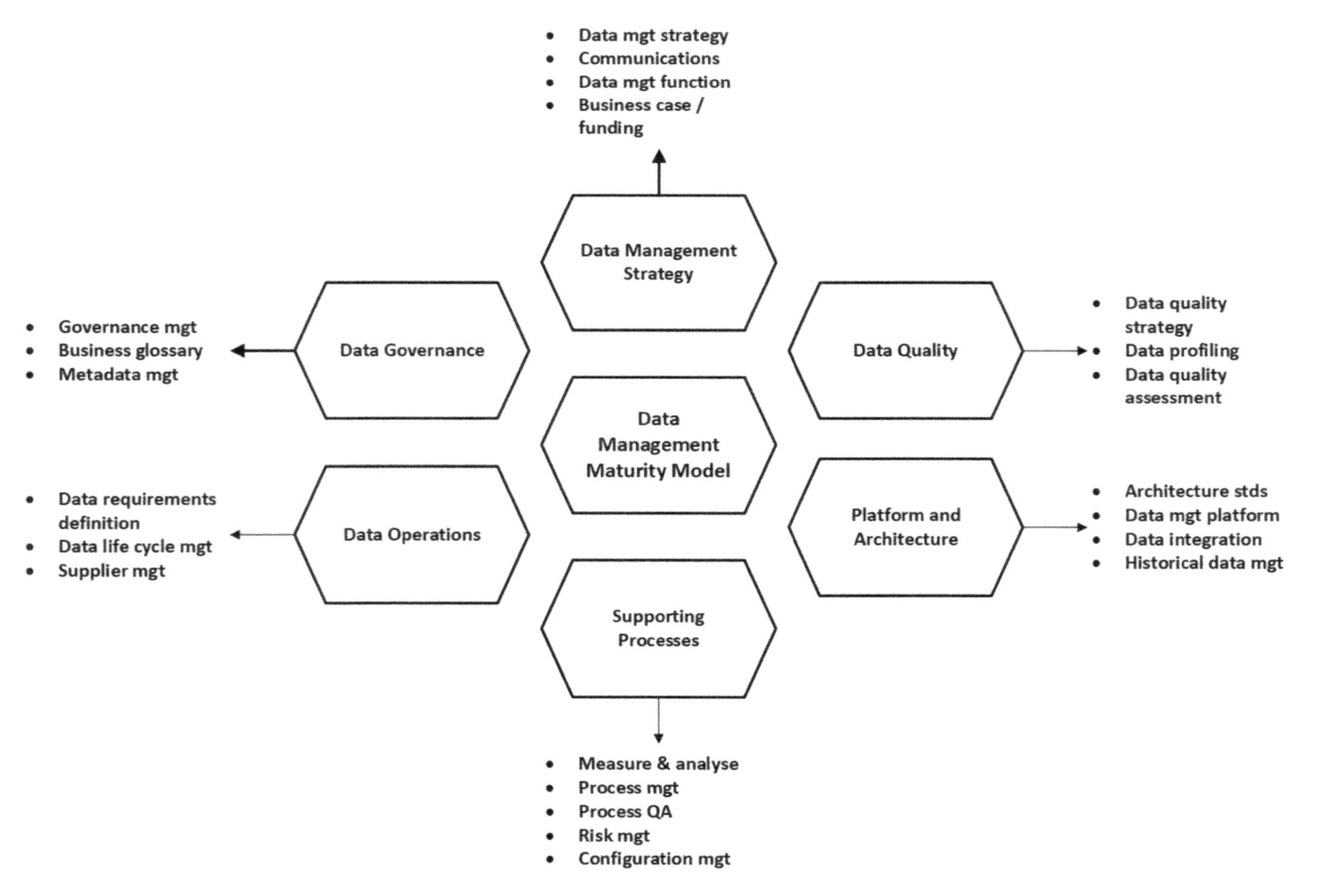

Figure 5.6 The Data Management Maturity Model with functional areas.[22]

- Governance management (Data governance);
- Data lifecycle management;
- Historical data management;
- Risk management;
- Configuration management.

However, there is no mention of data integrity at this level of the DMM, only data quality. As we discussed in Chapter 1, data quality is can I use the data. The implicit assumption is that the data have integrity as well.

5.9.2 Data Integrity Maturity Model

The publication of the GAMP Guide for Records and Data Integrity[23] in 2017 contains a Data Integrity Maturity Model in Appendix M2. It also defines five levels of data integrity maturity, like the DMM, as shown in Figure 5.7.

Within Appendix M2 [23] there are a number of maturity factors defined such as:

- Culture;
- Governance and Organisation;
- Strategic Planning and Data Integrity Programme;
- Regulatory Knowledge;
- Data Life Cycle and Supporting Processes.

Within the tables of the Appendix, characterisation of the five levels are given, for example consider culture. One of the factors is:

- Data integrity understanding and awareness
  For which a maturity factor should be:
  Awareness of the importance of data integrity and understanding of data integrity principles

There then follows a characterisation of the five levels for this factor, *e.g.*

- Level 1: Low awareness limited to subject matter experts and specialists.
- Level 2: General awareness of the topic but not fully reflected in working practices.
- Level 3: Principles reflected in working practices, but not consistently applied.
- Level 4: Data integrity principles fully incorporated and applied in established processes and practices.
- Level 5: Formal on-going awareness programme proactively keeping abreast of industry developments.

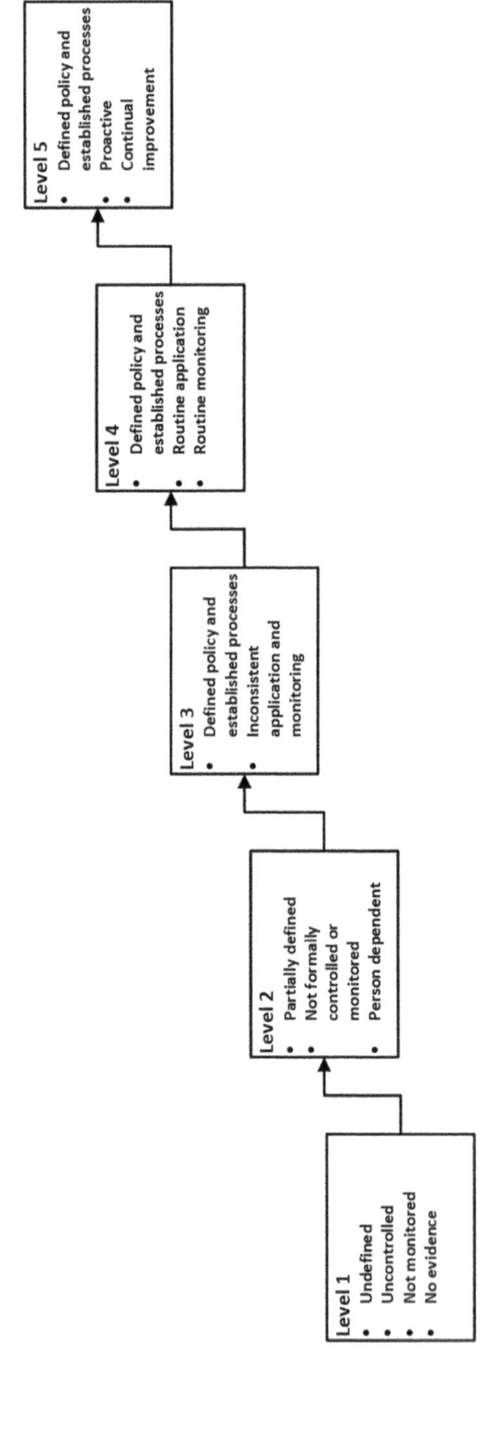

Figure 5.7 GAMP Data Integrity Maturity levels.[23]

The five levels shown in Figure 5.7 are colour coded using a traffic light approach: levels 1 and 2 are red, level 3 is yellow and levels 4 and 5 are green. The obvious implication is that laboratories and their parent organisations need to be at level 4 as a minimum.

Having looked at various models for data integrity and data management, the next chapter discusses data governance roles and responsibilities.

References

1. *PIC/S PI-041 Draft Good Practices for Data Management and Integrity in Regulated GMP/GDP Environments*, Pharmaceutical Inspection Convention/Pharmaceutical Inspection Co-Operation Scheme, Geneva, 2016.
2. *MHRA GMP Data Integrity Definitions and Guidance for Industry 2nd Edition*, Medicines and Healthcare Products Regulatory Agency, London, 2015.
3. *MHRA GXP Data Integrity Guidance and Definitions*, Medicines and Healthcare Products Regulatory Agency, London, 2018.
4. *WHO Technical Report Series No. 996 Annex 5 Guidance on Good Data and Records Management Practices*, World Health Organisation, Geneva, 2016.
5. *EMA Questions and Answers: Good Manufacturing Practice: Data Integrity*, 2016, Available from: http://www.ema.europa.eu/ema/index.jsp?curl=pages/regulation/general/gmp_q_a.jsp&mid=WC0b01ac058006e06c#section9.
6. *FDA Draft Guidance for Industry Data Integrity and Compliance with cGMP*, Silver Spring, MD, USA, 2016.
7. R. D. McDowall, *Validation of Chromatography Data Systems: Ensuring Data Integrity, Meeting Business and Regulatory Requirements Second Edition*, Royal Society of Chemistry, Cambridge, 2017.
8. *EudraLex – Volume 4 Good Manufacturing Practice (GMP) Guidelines, Chapter 1 Pharmaceutical Quality System*, European Commission, Brussels, 2013.
9. *ICH Q10 Pharmaceutical Quality Systems*, International Conference on Harmonisation, Geneva, 2008.
10. *USP 41 General Chapter <1058> Analytical Instrument Qualification*, United States Pharmacopoeia Convention, Rockville, MD, 2018.
11. *GAMP Good Practice Guide a Risk Based Approach to GXP Compliant Laboratory Computerised Systems, Second Edition*, International Society for Pharmaceutical Engineering, Tampa, FL, 2012.
12. *21 CFR 211 Current Good Manufacturing Practice for Finished Pharmaceutical Products*, Food and Drug Administration, Sliver Spring, MD, 2008.
13. *ICH Q2(R1) Validation of Analytical Procedures: Text and Methodology*, International Conference on Harmonisation, Geneva, 2005.
14. G. P. Martin, *et al.*, Lifecycle management of analytical procedures: method development, procedure performance qualification and procedure performance verification, *Pharmacopoeial Forum*, 2012, **38**(1).

15. G. P. Martin, *et al.*, Stimuli to the Revision Process: Proposed New USP General Chapter: The Analytical Procedure Lifecycle ⟨1220⟩, *Pharmacopoeial Forum*, 2017, **43**(1).
16. *FDA Guidance for Industry, Bioanalytical Methods Validation*, Food and Drug Administration, Rockville, MD, 2001.
17. *FDA Draft Guidance for Industry, Bioanalytical Methods Validation*, Food and Drug Administration, Rockville, MD, 2013.
18. *FDA Guidance for Industry: Bioanalytical Methods Validation*, Food and Drug Administration, Silver Spring, MD, 2018.
19. *EMA Guideline on Bioanalytical Method Validation*, European Medicines Agency, London, 2011.
20. *EudraLex – Volume 4 Good Manufacturing Practice (GMP) Guidelines, Annex 11 Computerised Systems*, European Commission, Brussels, 2011.
21. *WHO Technical Report Series No. 996 Annex 5 Guidance on Good Data and Records Management Practices*, World Health Organisation, Geneva, 2016.
22. M. Mecca, R. Young and J. Halcomb, *Data Management Maturity (DMM) Model*, CMMI Institute, Pittsburgh, PA, 2014.
23. *GAMP Guide Records and Data Integrity*, International Society for Pharmaceutical Engineering, Tampa, FL, 2017.

CHAPTER 6

Roles and Responsibilities in a Data Governance Programme

Data governance to ensure GXP data integrity in laboratories requires a multi-layered approach that runs throughout a regulated organisation, as can be seen in Chapter 4. Although data governance features in many data integrity guidance documents from both regulators and industry, any approach taken should be business rather than regulatory driven. It is important to ensure the integrity and protection of records for reasons of intellectual property protection as well as supporting products already on the market. In this chapter we discuss the roles and responsibilities of all people involved in a data integrity and data governance programme within a regulated pharmaceutical organisation at both the corporate and laboratory levels.

6.1 What Do the Regulators Want?

6.1.1 ICH Q10 Pharmaceutical Quality Systems

To begin this discussion, we start with Section 2.1 of ICH Q10 on Pharmaceutical Quality Systems that is entitled Management Commitment[1] that states:

(a) Senior management has the ultimate responsibility to ensure an effective pharmaceutical quality system is in place to achieve the quality objectives, and that roles, responsibilities, and authorities are defined, communicated and implemented throughout the company.

In a joint publication of the regulators and industry of the EU, USA and Japan we have a definitive statement that senior management is responsible

Data Integrity and Data Governance: Practical Implementation in Regulated Laboratories
By R. D. McDowall

Published by the Royal Society of Chemistry, www.rsc.org

for the quality objectives throughout a pharmaceutical organisation and for communicating them to the staff. One of those quality objectives is data integrity. Therefore, Senior Management must ensure that any data supporting a marketing authorisation or used for batch release are complete, consistent and accurate, thus meeting regulatory requirements of the authorities.

6.1.2 EU GMP Chapter 1

This approach is incorporated in the Principle of Chapter 1[2] where there is the following statement:

> The holder of a Manufacturing Authorisation must manufacture medicinal products so as to ensure that they are fit for their intended use, comply with the requirements of the Marketing Authorisation or Clinical Trial Authorisation, as appropriate and do not place patients at risk due to inadequate safety, quality or efficacy.
>
> The attainment of this quality objective is the responsibility of senior management and requires the participation and commitment by staff in many different departments and at all levels within the company, by the company's suppliers and by its distributors.
>
> To achieve this quality objective reliably there must be a comprehensively designed and correctly implemented Pharmaceutical Quality System1 incorporating Good Manufacturing Practice and Quality Risk Management.
>
> It should be fully documented and its effectiveness monitored.

Furthermore, Section 1.5 reinforces the responsibilities of senior management for the Pharmaceutical Quality System[2]:

> 1.5 Senior management has the ultimate responsibility to ensure an effective Pharmaceutical Quality System is in place, adequately resourced and that roles, responsibilities, and authorities are defined, communicated and implemented throughout the organisation.
>
> Senior management's leadership and active participation in the Pharmaceutical Quality System is essential. This leadership should ensure the support and commitment of staff at all levels and sites within the organisation to the Pharmaceutical Quality System.

The key elements that can be abstracted from the Principle and clause 1.5 are[2]:

- The manufacture of quality, safe and effective pharmaceutical products is the responsibility of senior management.

- The vehicle for ensuring this is a Pharmaceutical Quality System.
- A PQS requires senior management leadership and review.
- A PQS applies to all staff in an organisation and the supply chain but requires management leadership to achieve effectiveness.
- Senior management is responsible for communication of the aims and objectives of the PQS to all staff.
- A PQS incorporates GMP and thus implicitly includes data integrity to ensure product quality, efficacy and patient safety.

There are further requirements in Chapter 1 for data integrity in a regulated laboratory in Clause 1.9 on Quality Control[2]:

> Quality Control is that part of Good Manufacturing Practice which is concerned with sampling, specifications and testing, and with the organisation, documentation and release procedures which ensure that the necessary and relevant tests are actually carried out and that materials are not released for use, nor products released for sale or supply, until their quality has been judged to be satisfactory. The basic requirements of Quality Control are that: ....
>
> (iv) Records are made, manually and/or by recording instruments, which demonstrate that all the required sampling, inspecting and testing procedures were actually carried out. Any deviations are fully recorded and investigated;

Clause 1.9 requires work to be "actually carried out" and that the quality of any product must be established before release. Both are implicit requirements of data integrity.

As Chapter 1 is derived from ICH Q10,[1] there are equivalent requirements for senior management in the United States and Japanese regulations.

6.1.3 PIC/S-041 Good Practices for Data Management and Integrity in Regulated GMP/GDP Environments

The PIC/S guidance has a specific section on data governance[3]:

> 5.2.2 The data governance system should ensure controls over the data lifecycle which are commensurate with the principles of quality risk management. These controls may be:
>
> - Organisational
>   - procedures, *e.g.* instructions for completion of records and retention of completed records;
>   - training of staff and documented authorisation for data generation and approval;

  - data governance system design, considering how data is generated, recorded, processed, retained and used, and risks or vulnerabilities are controlled effectively;
  - routine data verification;
  - periodic surveillance, *e.g.* self-inspection processes seek to verify the effectiveness of the data governance policy.

- Technical

  - computerised system validation, qualification and control,
  - automation

5.2.3 An effective data governance system will demonstrate Senior Management's understanding and commitment to effective data governance practices including the necessity for a combination of appropriate organisational culture and behaviours and an understanding of data criticality, data risk and data lifecycle. There should also be evidence of communication of expectations to personnel at all levels within the organisation in a manner which ensures empowerment to report failures and opportunities for improvement. This reduces the incentive to falsify, alter or delete data.

5.2.4 The organisation's arrangements for data governance should be documented within their Pharmaceutical Quality Management System and regularly reviewed.

6.1.4 WHO Guidance on Good Data and Record Management Practices

There are several relevant clauses in this data integrity guidance that impacts roles and responsibilities for data integrity[4]:

4.6 Management governance. To establish a robust and sustainable good data management system it is important that senior management ensure that appropriate data management governance programmes are in place (for details see Section 6).

Elements of effective management governance should include:

- application of modern QRM principles and good data management principles that assure the validity, completeness and reliability of data;
- application of appropriate quality metrics;
- assurance that personnel are not subject to commercial, political, financial and other organizational pressures or incentives that may adversely affect the quality and integrity of their work;

- allocation of adequate human and technical resources such that the workload, work hours and pressures on those responsible for data generation and record keeping do not increase errors;
- ensure staff are aware of the importance of their role in ensuring data integrity and the relationship of these activities to assuring product quality and protecting patient safety.

> 4.7 Quality culture. Management, with the support of the quality unit, should establish and maintain a working environment that minimizes the risk of non-compliant records and erroneous records and data. An essential element of the quality culture is the transparent and open reporting of deviations, errors, omissions and aberrant results at all levels of the organization, irrespective of hierarchy. Steps should be taken to prevent, and to detect and correct weaknesses in systems and procedures that may lead to data errors so as to continually improve the robustness of scientific decision-making within the organization. Senior management should actively discourage any management practices that might reasonably be expected to inhibit the active and complete reporting of such issues, for example, hierarchical constraints and blame cultures.
>
> 6.1 Assuring robust data integrity begins with management, which has the overall responsibility for the technical operations and provision of resources to ensure the required quality of GXP operations. Senior management has the ultimate responsibility for ensuring that an effective quality system is in place to achieve the quality objectives, and that staff roles, responsibilities and authorities, including those required for effective data governance programmes, are defined, communicated and implemented throughout the organization. Leadership is essential to establish and maintain a company-wide commitment to data reliability as an essential element of the quality system.

6.1.5 Update of the US GLP Regulations

The proposed update of the US GLP regulations (21 CFR 58) was published in 2016 and the title is the GLP Quality System.[5] In this proposed update there are several changes:

> § 58.3: Management with executive responsibility means those senior employees of a testing facility or test site who have the authority to establish or make changes to the quality policy and GLP Quality System at the testing facility and test site, respectively.
>
> § 58.31 Testing facility management with executive responsibility. Management with executive responsibility is ultimately responsible for the GLP Quality System and must establish policy and objectives for a GLP Quality System and a commitment to quality, as defined in § 58.3.

In essence, there is no hiding place for senior management in either GLP or GMP, they are ultimately responsible for all work performed and the quality outputs of any regulated laboratory.

6.1.6 GAMP Guide Records and Data Integrity

The GAMP guide on Records and Data Integrity is a companion volume for GAMP 5 for data integrity[6] and has the following sections for data governance:

> Section 3 contains a data governance framework covering the elements of data governance, the description of a data integrity maturity model and human factors in data integrity. The last two items have appendices M2 and M3 respectively for further detail on these two subjects.

An earlier ISPE publication, entitled Considerations for a Corporate Data Integrity Program,[7] is also worth reading for help in implementing data integrity from the top down in an organisation.

6.1.7 A Summary of Regulatory and Industry Guidance Documents

Taking the key elements of ISPE document,[7] the WHO[4] and the PIC/S PI-041[3] guidances, we can come up with the following elements of a data governance structure within an organisation, as follows:

- Executive sponsorship providing corporate leadership, resources and oversight of the whole data governance and data integrity programme.
- Changing the organisation's culture to ensure that mistakes can be discussed openly and blame is not attached to individuals who own up to errors (this will be discussed in Chapter 8).
- A cross functional steering committee to co-ordinate the data integrity work across the whole organisation including system assessments, allocating budgets for remediation and communication about the overall programme of data integrity work (this will be dependent on the size of the organisation).
- Data integrity policies and procedures with associated training including the policy itself, good documentation practices together with process and system specific procedures for generating processing, and reviewing data throughout the life cycle locally.
- Identification of data owners for processes and systems who are responsible for defining the quality, integrity and security of the data generated under their remit.
- Quality assurance oversight including data integrity audits and conducting investigations of potential data integrity violations.
- Providing a mechanism for individuals to raise concerns about data integrity without retaliation (discussed in Chapter 20).

That is a snapshot of data governance elements at a high level in an organisation but we still need to know:

- Who are involved?
- What are the organisational structures required?
- What are everybody's responsibilities?

Let us start at the top of a regulated organisation and work down to the analytical laboratory.

6.2 Data Governance Roles and Responsibilities – Corporate Level

As ICH Q10[1] makes clear, senior management are responsible for quality and hence data integrity. An extreme example of a data governance organisation can be found in 2012 Ranbaxy consent decree, where a post of Chief Data Integrity Officer with associated staff was created[8] to investigate the systematic falsification within the organisation. This approach may not be appropriate for many regulated companies, as we should avoid creating parallel organisational structures. That said, there will be some additional roles, committees and responsibilities but as much as possible should be an extension or reinforcement of each individual's normal working practices. The guiding principles here are:

- Data integrity is more than just numbers.
- Quality does not own quality anymore.
- Take responsibility for your own work: right first time, every time.
- Tell somebody if you make a mistake and document it.

Figure 6.1 shows the main organisational elements involved in data governance at a corporate level within a regulated organisation that involves the following individuals and organisational elements:

- Senior Management;
- Executive Sponsor;
- Corporate Data Governance/Data Integrity Steering Committee;
- Site/Division Data Governance/Data Integrity Committee (optional);
- Line Management;
- Quality Assurance;
- Information Technology.

The overall approach needs to be tailored to the size of the regulated organisation. A smaller organisation with a single site will probably not require a corporate steering committee as it can be co-ordinated with a small number of individuals with the executive sponsor. The responsibilities

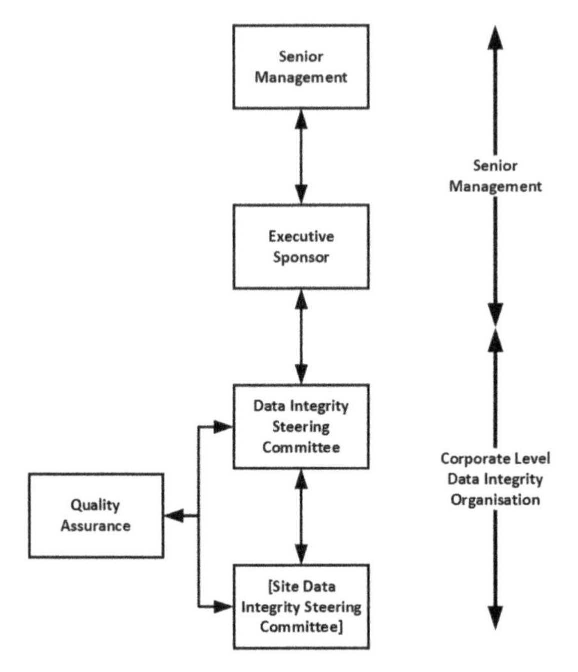

Figure 6.1 Data governance organisation at the corporate level.

of each area or individual are presented in Table 6.1 and discussed below.

- Senior Management must provide the leadership, resources and direction for the overall data integrity and data governance programme. Note the use of the word programme, there will be multiple work streams or projects involved to ensure the integrity of data within an organisation.
- Executive Sponsor: One member of the senior management team should be the Executive Sponsor who monitors overall progress of the programme. This individual sits on the corporate data integrity steering committee and liaises between it and the senior management team.
- A Data Integrity Steering Committee oversees devising, implementing, managing and monitoring the corporate data integrity and data governance programme. It is important at this stage to set and manage expectations. You will recall that in the overview of this chapter, I believe that an overall approach should bring business benefit and if designed well should provide regulatory compliance as well. Some of the other main responsibilities of this committee are outlined in Table 6.1 but please note that this list is not exhaustive. Depending on the size of the organisation, there should be an option for site or division data governance committees to manage the data integrity programme

Table 6.1 Corporate level data governance roles and responsibilities.

Role	Responsibilities
Senior Management	• Sets overall objectives for the data integrity and data governance programme • Leadership for the data integrity programme • Defines open culture • Allocates budget and resources for the data governance programme • Management review of the whole data integrity and governance programme
Executive Sponsor	• Member of senior management who participates in the Data Integrity Steering Committee meetings • Liaison between Senior Management and Data Integrity Steering Committee • Promotes best practices data integrity and an open culture • Approves the corporate data integrity policy • Empowers and authorises the Data Integrity Steering Committee's work
Data Integrity Steering Committee	• Co-ordinates and monitors the overall data governance and data integrity programme across sites/divisions: defines the road map, priorities, *etc.* • Defines criteria for assessment and remediation of systems: improved data integrity and business efficiency • Establishing procedures for mapping and monitoring of data processes • Prepares progress reports for management review • Approves and monitors data integrity projects within the programme • Reviews business and compliance benefits of data integrity projects and reviews after completion to see benefits have been obtained • Allocates resources and budgets to projects • Implements and monitors data integrity training at all levels, *e.g.* corporate data integrity policy, good documentation practice for paper, hybrid and electronic systems • Defines an open culture where mistakes are admitted so that management has an understanding of risk in each process
[Divisional/ Site Data Integrity Committee]	• Depending on the size of an organisation, a divisional or site data integrity committee takes on the local co-ordination of the data integrity programme and projects from the corporate steering committee with the same responsibilities • Reports work to the Corporate Data Integrity Steering Committee
Line Management	• Interactions with data owners and data stewards on specific process and system data integrity issues • Ensure that there is not undue pressure on staff to complete work, which can lead to data integrity problems • Ensure staff are trained in all aspects of data integrity including second person review • Ensure data integrity principles are incorporated into staff performance objectives • Ensure that an open culture is established and maintained to avoid cover ups, *e.g.* reward staff for admitting mistakes, spotting poor practices, *etc.* • Ensure data integrity requirements are in all staff objectives and reviewed in performance appraisals

(*continued*)

Table 6.1 (*continued*)

Role	Responsibilities
Quality Assurance	• Interpretation of current regulations and regulatory guidance to feed into the data integrity programme • Compliance overview of the data integrity and data governance programme • Compliance review of projects and their deliverables • Defines the data integrity requirements for quality and technical agreements with suppliers • Data integrity audits: internal processes and systems as well as suppliers • Conduct data integrity investigations
Corporate IT	• Understanding of the corporate data integrity programme • Technical assessment of IT infrastructure and computerised systems for data integrity • Technical Involvement in remediation and improvement projects involving computerised systems and data storage • Support operational systems to ensure data integrity, *e.g.* record management, application administration and user account management

and project at a local level. The responsibilities of a site or divisional committee are broadly similar to the corporate steering committee to whom they report a site or division's progress in the overall programme of work.

- Line Management takes the policies and procedures from the Corporate Data Governance Steering Committee and ensures that they are implemented in the areas for which they are responsible. At the corporate level, there will be policies and procedures for data integrity and good documentation practices but in the analytical laboratory there will be requirements for interpretation of data including integration of chromatograms, interpretation of spectra and second person review that are typically a local issue unless there is global harmonisation within an organisation. However, the most important role of line management is to take the corporate requirement for an open culture and implement this in their own laboratories. Allowing analytical scientists to admit their mistakes in an open way and without finger pointing in some laboratories is a big ask – but it must be implemented and maintained. In addition, line management must ensure that there is not undue pressure placed on staff to perform work that could lead to working practices that compromise the integrity of data.
- Quality Assurance provides the interpretation of regulations and quality oversight in the form of interpretation of regulations and advice, approval of updated computer validation documents, data integrity audits and data integrity investigations is quality assurance. As much of the supply chain is globalised, QA must provide the data integrity input to quality and technical agreements for suppliers and audit them against these agreements.

- Corporate IT also need to be involved at the Steering Committee level as the function will be responsible for storing, managing and protecting the electronic records generated by the user departments. This means that the IT infrastructure must be resilient with redundant data storage and be secure from internal and external threats. In addition, for networked systems there will be the management of user account and configuration settings.

6.3 Data Integrity Policy

One of the functions of the Corporate Data Integrity Steering Committee and the Executive Sponsor is to ensure that there is a corporate data integrity policy in association with training material available for communication to all employees including part-time and temporary staff. Interestingly, the EMA Data Integrity Question and Answers notes that there is no regulatory requirement for such a policy,[9] but without one, how do staff know what is expected of them? An outline structure of such a data integrity policy is shown in Figure 6.2, the document should be approved by the Executive Sponsor on behalf of the Senior Management team.

There needs to be effective training for all members of staff in what the policy means for them and at the end there should be a test to check each person's understanding of the policy, as discussed in Chapter 7.

In addition, the data integrity policy and data governance structures described above must be integrated within the pharmaceutical quality system of an organisation.

6.4 Management, Monitoring and Metrics

To help manage and monitor the various data integrity programme streams, there will be the need for metrics, such as:

- percentage of staff trained in the corporate integrity policy against the target timescale;
- percentage of processes and systems assessed;
- number of remediation projects on schedule.

However, do not forget that, as with all compliance projects, data integrity is a journey and not an event. As processes and systems are assessed and remediated, the data integrity programme moves into the operational phase and the metrics change from remediation to monitoring the effectiveness of the new processes and systems, *e.g.*:

- percentage of out of specification results as a percentage of samples analysed;
- number of repeat analyses;
- number of reported mistakes.

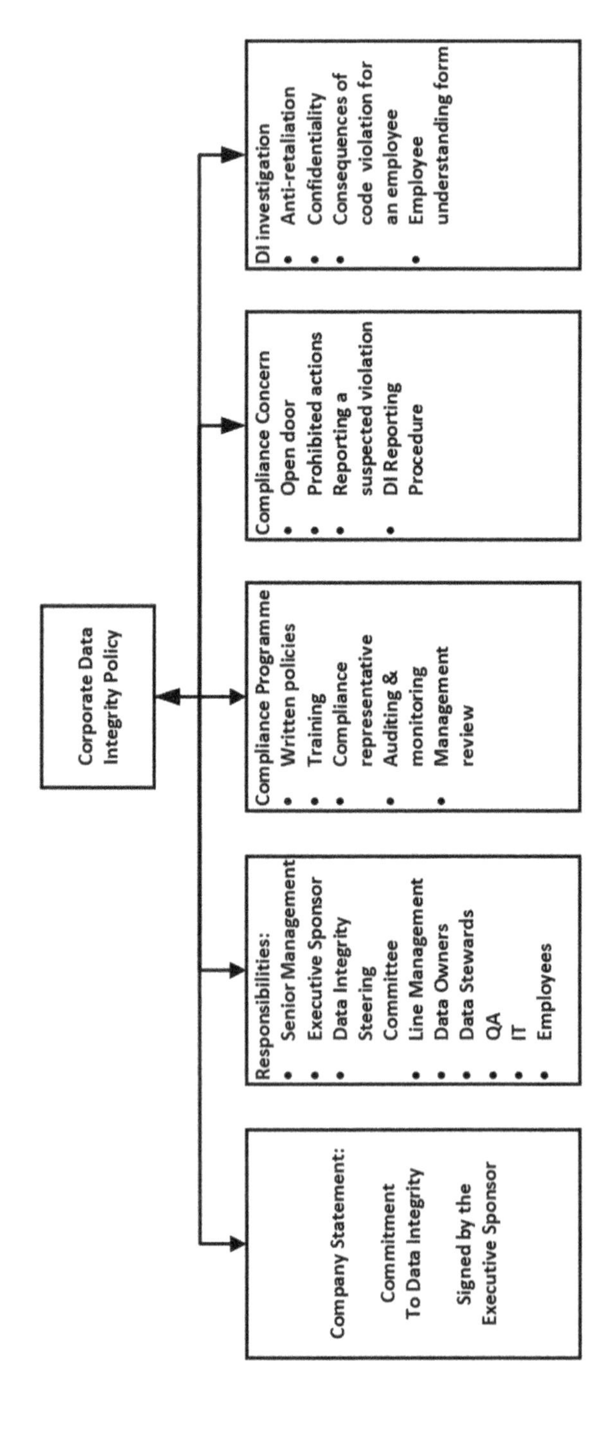

Figure 6.2 Overview of a corporate data integrity policy.

The aims of these metrics are to keep senior management aware of the residual risk associated with a process or computerised system. The topics of data integrity and quality metrics will be discussed further in Chapter 19 of this book.

6.5 Data Integrity and Data Governance Roles and Responsibilities – Process and System Level

Not wishing to bore you to death discussing data integrity and data governance at the corporate level, we now need to focus on the laboratory and discuss the impact of the data integrity programme as it cascades down to the processes and systems at the laboratory bench. Line management will be responsible for ensuring that the corporate data integrity policies and procedures as well as the requirements for an open culture are communicated down the line to the operational staff as outlined in Table 6.1. The focus here will be on data ownership and data stewards as the ISPE paper only has a single passing reference to data owners[7] but data ownership is mentioned in the MHRA and WHO data integrity guidance documents.[4,10,11]

6.5.1 From Data Governance to Data Ownership

How do we get from data governance to data ownership? The WHO data integrity guidance[4] states in Section 4:

> 4.10*To ensure that the organization, assimilation and analysis of data into a format or structure that facilitates evidence-based and reliable decision making,* ***data governance should address data ownership and accountability for data process(es)*** *and risk management of the data life cycle.*

As we can see from Chapter 4, Table 4.1 and the two references quoted therein,[12,13] a part of data governance is data ownership and accountability for data processes and associated risk management in any data life cycle. Here, we need to consider both manual, hybrid and electronic processes to ensure the integrity of all records generated in a regulated laboratory.

Questions now arise, how do we know who is a data owner and what are their responsibilities? The guidance documents are not much help here, so to help, let us return to the definitions of data governance in Table 4.1. You will see from the non-pharmaceutical definitions in the right-hand column that data governance focuses on defining the rules, roles and responsibilities for acquisition and management of data. In my view, the data owner should be responsible for these functions as they should know the process that is automated by the spectrometer. However, there are already two roles for a computerised system defined in GAMP 5 and EU GMP Annex 11: the process owner and the system owner.[14,15]

6.5.2 Process Owner and System Owner

These two roles were originally defined in GAMP 5 [14] and were simplified in EU GMP Annex 11 [15] and are:

- Process Owner:
  The person ultimately responsible for the business process or processes being managed.[14]
  The person responsible for the business process.[15]
- System Owner
  The person ultimately responsible for the availability, support and maintenance of a system, and for the security of the data residing on the system.[14]
  The person responsible for the availability, and maintenance of a computerised system and for the security of the data residing on that system.[15]

These are the two main roles that are defined in the EU GMP regulations. The split in responsibilities above, gives rise to the concept of process owner in the business, in our case the laboratory, and system owner responsible for the computing platform and the data that resides upon it in IT. Note that for standalone laboratory computerised systems, the two sets of responsibilities are merged and the process owner is also the system owner.

6.5.3 Can a Process Owner Be a Data Owner?

The question is how should these regulatory roles fit within a data integrity and data governance framework? We will start the discussion with the data owner. The obvious fit is for the process owner to be the data owner as well because data are generated and converted to information in the business. Therefore, we could modify the EU GMP Annex 11 definition of process owner and encompass data ownership as follows:

> The person responsible for the business process including the data generated, transformed and reported by a manual process or a computerised system.

Who should be a process/data owner? GAMP 5 gives some guidance, which is self-explanatory[14]:

> *This person is usually the head of the functional unit or department that uses the system, although the role should be based on specific knowledge of the process rather than position in the organisation.*

One potential area of confusion concerns the name "data owner", which implies that it is an individual rather than the organisation that owns the system and the data generated by it. This is not so, the organisation owns the data, the data owner is merely the custodian of the data in the system who acts on behalf of the organisation. Perhaps data custodian may be a better title for this individual?

6.5.4 Other Data Governance Roles at the System Level

We will look in some more detail within our regulated analytical laboratory at the other roles involved in data ownership. In addition to the data owner, there are two other roles as follows:

- Data stewards or Laboratory Administrators: enabling the requirements of the data owner for the system. (These people would typically be the power users or system administrators in the laboratory.)
- Technology stewards or IT Staff: enabling the IT requirements of the data owner and is a person or persons who, for a networked system is or reports to the system owner. This role is essential for segregation of duties and to avoid conflicts of interest when administering the system. Note that this role will not be found in a paper based process as it is only where a computerised system is involved.

The responsibilities of each role are listed in Table 6.2 and the interactions between these two roles and the data owner are shown in Figure 6.3. Subject matter experts, some of whom may be either data or technology stewards are also mentioned in Table 6.2 and quality assurance input is only shown in Figure 6.3 but the responsibilities are many of those presented in Table 6.1. To map from data governance roles to those used currently in laboratories, I have put titles in brackets in Table 6.2 *e.g.* a data steward is a laboratory administrator.

Looking at the three main roles above, you will find two of them are not mentioned in any data integrity guidance document: data steward and technology steward. The role of these stewards, and others, can be seen in the book by Plotkin.[16] Why is there the need for data and technology stewards? The answer can be found in the discussion of who should be a process owner in GAMP 5[14] and quoted in Section 6.5.3 above. Typically, the data owner is the head or a section head of a laboratory but often the work to be done in administering the system will be devolved to others such as super users or lab administrators and IT support staff if the system is connected to the network. This is the normal business process and these are the data and technology stewards. Furthermore, as the data owner will specify the security, data quality and data integrity requirements for the system who better to ensure that they have been implemented than the second person reviewers, another group of data stewards. You will remember my comment

Table 6.2 Process and system level roles and responsibilities.

Role	Responsibilities
Data Owner/ Process Owner	• Implements directives and procedures from the Corporate Data Integrity Steering Committee • Responsible for one or more workflow processes that may be manual, hybrid or electronic or a combination of all three • Risk assessment of the processes and systems to determine record vulnerability • Remediation efforts in the short term to fix record vulnerability and a longer-term fix for business benefit • Approves, modifies or rejects all data integrity requests with the process or system for which they are responsible • Defines and authorises the data quality, integrity and security of the process • Defines and authorises the record security, user roles and access privileges of computerised systems • Defines the criticality of the data and hence the data backup and recovery strategy for computerised systems • Reporting to the data governance steering committee on the progress of work and process and system metrics
Data Steward (Laboratory Administrator)	• Implements Data Owner's requirements for the process and systems • Ensures and helps users to follow the security, data quality and data integrity requirements implemented in the system or process • Review data generated by users using the system and process • Co-ordinate IT activities for supporting networked systems • Identify improvement ideas for discussion with the Data Owner • Generate automatic metrics for the data owner to report
System Owner (Head of IT)	• A written agreement with the Data Owner for the scope of services carried out • Monitoring of the system performance and availability of the system and IT platform • Hardware and software maintenance of the system, *e.g.* patching the application, database and operating system • Physical and logical security of the data, *e.g.* antivirus, backup/recovery, disaster recovery plan is current and works • Delegation of the work to Technology Stewards
Technology Steward (IT Administrator/IT Support)	• Ensure that the computer platform is adequate for data integrity needs defined by the Data Owner • Support computerised systems that create, process, report and store data • Configure the application to meet the Data Owner's specifications • Granting, modifying and revoking user access following receipt of a request from the Data Owner • Backup and recovery of data on computerised systems according to the Data Owner's requirements • Propose hardware and software changes to the system • Use change control process for implementing changes to ensure that data are protected during the change • Ensure the security of the system • Generate metrics for the data owner to report
Subject Matter Experts	• Process understanding • Operate the process or system • Technical understanding of the computerised systems used in the process

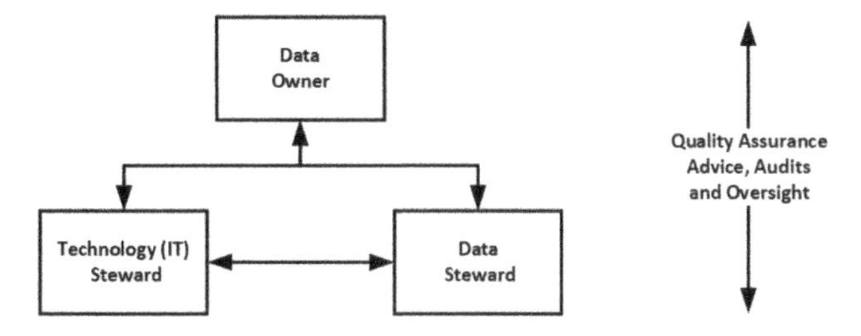

Figure 6.3 Data governance at the process and system level.

earlier in this chapter about the need to integrate data integrity roles with normal business operations. Here is a very good example of that. Take the normal situation and simply reinforce or overlay the data integrity responsibilities on top.

6.5.5 Data Owner

It is important to realise that data integrity and data quality begins at the point of data acquisition by the process (typically sampling) and not in the computer or data centre. If data acquisition is compromised by poor working practices, sampling or using an uncalibrated instrument, data integrity is lost from this point forward. Therefore, the data owner's responsibilities for a regulated computerised system from the business side include:

- Define what is required of the process and system in terms of data quality, data integrity and data security. This will result in inputs to the sampling plan, configuration specification for the setting of application policies, writing of SOPs for using the system or the agreement with IT to support the system (*e.g.* backup, user account management, *etc.*). This begins from the start of the analytical procedure and a flexible analytical data life cycle is discussed in more detail in Chapter 9.
- Assessment of the computerised system to determine if there are vulnerabilities of the records contained therein, presented in Chapter 10. Although a system may be validated, record vulnerabilities may exist that must be managed, as we shall see in Chapter 14.
- Development of a remediation plan with the data and technology stewards to secure the records and reduce or eliminate data vulnerabilities following the assessment. Although a remediation plan may solve immediate data integrity vulnerabilities, long term solutions are also required to improve the business process and eliminate paper, as we shall see in Chapter 13.
- Approve access to the system for new users and changes in access privileges for existing ones for IT administrators to implement.
- Approval or rejection of change control requests.

- Approval for archiving data and removing them from the system.
- Receive feedback from the data stewards of the system of issues involving quality, integrity and security of the application data and implement any modifications of procedures, *etc.* for the data stewards to implement.

That is the good news for data owners.

6.5.6 Data Steward

The concept of data stewardship is defined in the literature as the enabling capability of data governance. Defining different types of stewardship addressing different aspects of the data governance process are also described in the literature[16] but the focus in this chapter is only on data and technology stewards.

As the data owner probably will not have the time or the training to implement the requirements for data integrity and quality that they have mandated, this is the role of the data stewards for the system.

- The data stewards, in the form of power users or super users in the laboratory, are the first point of contact for user questions for help with the system.
- The stewards will also be instrumental in ensuring the smooth running of the system such as developing custom reports or custom calculations.
- As expert users of the system, they will be responsible for ensuring that the requirements for data integrity and data quality set by the data owner have been implemented and are working.
- They are also responsible for data queries and monitoring data integrity from a system perspective, *e.g.* regular review of system level audit trails for system related issues rather than data integrity problems or aiding QA data integrity audits.

In monitoring the system from the business perspective, they can raise issues for discussion with the data owner to resolve, as noted in Section 6.5.5.

6.5.7 Is a Lab Administrator a Data Steward?

Although data stewards are involved with data governance outside of the pharmaceutical industry, we need to see how this role can be overlaid onto the situation in a regulated laboratory. For most systems, either standalone or networked, the data owner may not be involved in some of the technical administration of the application such as generation of spectral libraries, the creation of custom reports, custom calculations or macros. This is the role of the laboratory administrators. You can see that adding the responsibilities involved for data stewardship in Table 6.2 to a laboratory administrator or power user would be a logical step within an overall data integrity and data governance framework and to integrate governance within the current roles typically used in a laboratory.

6.5.8 Is a Technology Steward a System Owner?

Recall the definition of system owner from Annex 11 and Section 6.5.2 as the person who is responsible for the availability, and maintenance of a computerised system and for the security of the data residing on that system. This, in my view, is a role that is performed by the IT Department. In particular, the system owner could be the head of the functional IT group responsible for system support. An alternative approach is that the system owner could be in the laboratory or business, however, that would only work with larger organisations and have the capacity for the headcount. In smaller organisations the role of system owner would need to be delegated directly to the IT department under an agreement with the Process/Data Owner.

For a networked application, technology stewards are members of the IT department who would be responsible for the administration of a networked application, logical security, backup/recovery and other support functions and carries out the responsibilities outlined in Table 6.2. However, if the laboratory computerised system is standalone and has few users, then the data owner is also the data steward and the technology steward. Three hats and no pay rise.

6.5.9 Segregation of Roles and Duties

In this data governance framework, it is important that there is a segregation of duties between the different roles in a computerised system, *e.g.* administration of the application, access to the data and to the system clock from the users and laboratory administrators from the configuration of the system. For networked systems this can be achieved relatively easily but there is a problem with standalone systems where IT will probably not be involved. Here, the individuals who will be laboratory administrators need to have two roles: the first would be as an admin with no user privileges and the second as a user but with no admin privileges.

6.6 The Short Straw ......

6.6.1 Where Are We Now?

Look around any regulated laboratory and you will see a data integrity disaster in operation. There are your lovely toys, sorry analytical instruments, that are operated by a software operating on a standalone workstation. Do you want the good news or the bad news? As there is no good news, the bad news comes in a variety of forms:

- You, the data owner, are now also the system owner as the workstation is not connected to the network – you are now doing user account management, documenting the configuration of the software and backing up the data and this is a conflict of interest.
- It is highly likely that you are using the system as a hybrid by generating electronic records and signing paper printouts.
- Most of your electronic records are stored in directories within the operating system that any user can access by going outside of the operating system and deleting files. Moreover, such users can also access the system clock that may not be checked as well as the recycle bin.

How to resolve this situation?

6.6.2 The Hybrid System Nightmare

A hybrid system is the worst possible situation as the data owner must ensure that the electronic records created by the software are linked to the signed paper printouts. Managing and coordinating two disparate media forms over the life time of the data is difficult. The WHO data integrity guidance[4] says:

- Data integrity risks are likely to occur and to be highest when data processes or specific data process steps are ...... hybrid,......
- The use of hybrid systems is discouraged....
- The hybrid approach is likely to be more burdensome than a fully electronic approach; ....
- Replacement of hybrid systems should be a priority.
- In the hybrid approach, which is not the preferred approach, paper printouts of original electronic records ....

Are you getting the message?

Whilst the aim of short term remediation is to ensure that the record vulnerabilities are reduced to acceptable levels. Longer term, the replacement of hybrid systems is essential. Again, the aim here is to provide substantial business benefit to the laboratory with the remediation and not just to replace one hybrid system with another, as was the case with Part 11 remediation. Automation of a laboratory process must have the aim of improving working practices, speeding up work as well as ensuring data integrity. However, if the new laboratory process fails to eliminate most paper records or the way the software works is not managed correctly then all that happens is that the process becomes an automated mess and data integrity cannot be ensured. Back to square one.

The problem is that most computerised analytical systems are very poorly designed, often before the problems of data integrity surfaced, such as:

- Designed to work as a hybrid with no electronic signatures.
- Electronic signatures are not implemented properly in that the signature is not applied to the record but recorded only as an audit trail entry.

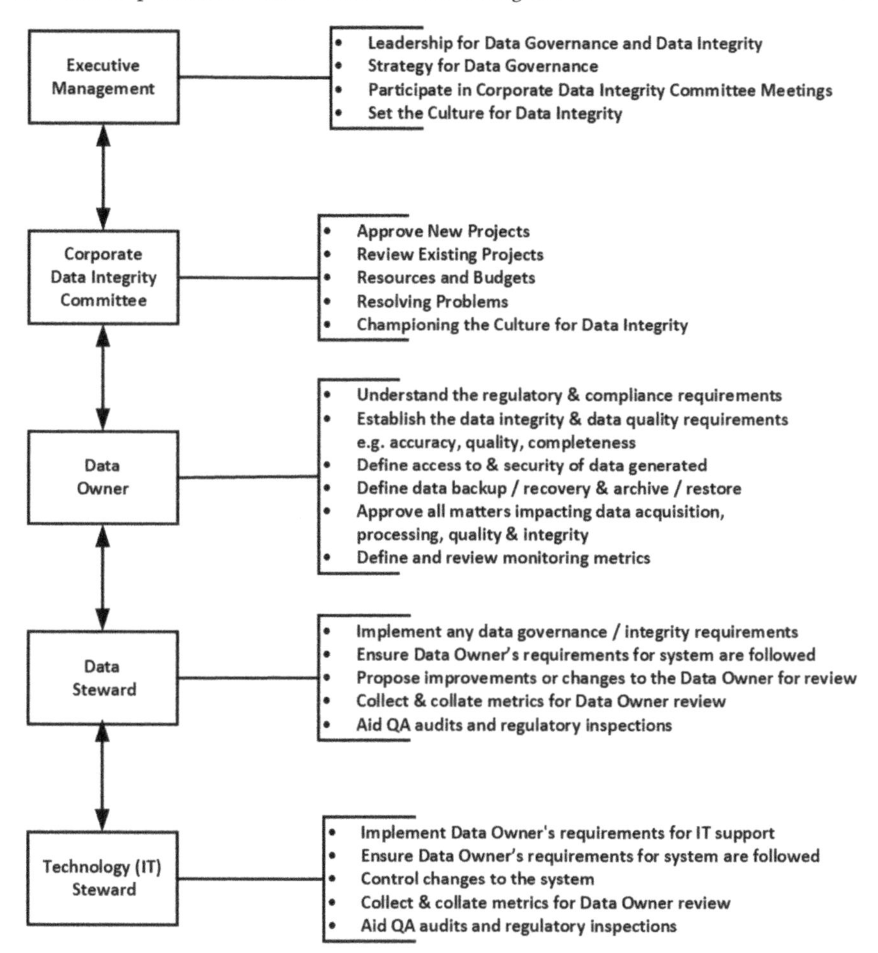

Figure 6.4 Cascade of roles and responsibilities from the Corporate Boardroom to the Laboratory Bench.

- Operates as a standalone system with no option to work as a networked system, *e.g.* being able to review data at a second workstation.
- Data are acquired to a single hard drive and not to a secure network drive.
- Data are managed in directories in the operating system rather than in a database.
- Poor audit trail review functions that highlight if GMP critical data have been modified or deleted.

Unfortunately, users have not pressured suppliers to ensure that the software is adequate to meet changing regulatory requirements and suppliers only react to market forces. We will address the assessment of computerised systems in more detail in Chapter 11 and the hybrid system problem in more detail in Chapter 12.

6.7 Cascade of Roles and Responsibilities: from Boardroom to Bench

Now that we have finished our discussion on the roles and responsibilities of a data governance and data integrity programme, we can see that everybody in a regulated pharmaceutical organisation has a responsibility for data integrity as can be seen in Figure 6.4 from the Executive Sponsor of the Corporate Data Integrity programme to every analyst working on GXP work as well as the IT personnel who support the infrastructure and regulated applications with the associated data.

Data integrity must be pervasive throughout an organisation and can never be considered as being an IT or laboratory issue. One of the means to achieve this is by training that is discussed in Chapter 7.

References

1. *ICH Q10 Pharmaceutical Quality Systems*, International Conference on Harmonisation, Geneva, 2008.
2. *EudraLex – Volume 4 Good Manufacturing Practice (GMP) Guidelines, Chapter 1 Pharmaceutical Quality System*, European Commission, Brussels, 2013.
3. *PIC/S PI-041 Draft Good Practices for Data Management and Integrity in Regulated GMP/GDP Environments*, Pharmaceutical Inspection Convention/Pharmaceutical Inspection Co-Operation Scheme, Geneva, 2016.
4. *WHO Technical Report Series No. 996 Annex 5 Guidance on Good Data and Records Management Practices*, World Health Organisation, Geneva, 2016.
5. 21 CFR Parts 16 and 58 Good Laboratory Practice for Nonclinical Laboratory Studies; Proposed Rule, *Fed. Regist.*, 2016, **81**(164), 58342–58380.
6. *GAMP Guide Records and Data Integrity*, International Society for Pharmaceutical Engineering, Tampa, FL, 2017.
7. J. Avellanet and E. Hitchings, *Considerations for a Corporate Data Integrity Program (A Concept Paper by the ISPE GAMP Community of Practice)*, International Society of Pharmaceutical Engineering, Tampa, FL, 2016.
8. *Ranbaxy Laboratories Ltd & Ranbaxy Inc: Consent Decree of Permanent Injunction*, 2012.
9. *EMA Questions and Answers: Good Manufacturing Practice: Data Integrity*, 2016,Availablefrom:http://www.ema.europa.eu/ema/index.jsp?curl=pages/regulation/general/gmp_q_a.jsp&mid=WC0b01ac058006e06c#section9.
10. *MHRA GMP Data Integrity Definitions and Guidance for Industry 2nd Edition*, Medicines and Healthcare Products Regulatory Agency, London, 2015.
11. *MHRA GXP Data Integrity Guidance and Definitions*, Medicines and Healthcare Products Regulatory Agency, London, 2018.
12. G. Thomas, *Definition of Data Governance*, 2006, Available from: http://www.datagovernance.com/adg_data_governance_definition/.

13. R. S. Steiner, *Non-invasive Data Governance. The Path of Least Resistence and Greatest Success*, Technics Publications, Basking Ridge, NJ, 2014.
14. *Good Automated Manufacturing Practice (GAMP) Guide Version 5*, International Society for Pharmaceutical Engineering, Tampa, FL, 2008.
15. *EudraLex – Volume 4 Good Manufacturing Practice (GMP) Guidelines, Annex 11 Computerised Systems*, European Commission, Brussels, 2011.
16. D. Plotkin, *Data Stewardship. An Actionable Guide to Effective Data Management and Data Governance*, Morgan Kaufman, Waltham, MA, 2014.

CHAPTER 7

Data Integrity Policies, Procedures and Training

Owing to either poor data management practices or data falsification in the industry, regulatory agency inspectors are now trained to detect the data integrity violations. The pharmaceutical industry needs to respond and not simply by paying lip service to the subject but by having effective data integrity policies and procedures in place throughout an organisation from an overall data integrity policy, good documentation practices for paper, hybrid and computerised systems but also how to interpret laboratory data. To do this, the key issue for laboratories working GMP or GLP is to understand the terms complete data and raw data, which is explained in detail here.

Coupled with the policies and procedures there must be effective staff training and, where necessary, evidence of the effectiveness of that training. Typical poor documentation practices and record keeping failures are highlighted here as a means of ensuring that they are avoided. This chapter will only focus on training for data integrity and will not consider the wider scope of training in GXP regulations as applicable for staff to perform their jobs.

7.1 What Do the Regulators Want?

7.1.1 EU GMP Chapter 4 on Documentation

In the Principle of the chapter[1] it states:

> Good documentation constitutes an essential part of the quality assurance system and is key to operating in compliance with GMP requirements.

Data Integrity and Data Governance: Practical Implementation in Regulated Laboratories
By R. D. McDowall

Published by the Royal Society of Chemistry, www.rsc.org

> There are two primary types of documentation used to manage and record GMP compliance: instructions (directions, requirements) and records/reports.
>
> The term "written" means recorded, or documented on media from which data may be rendered in a human readable form.
>
> Records: Provide evidence of various actions taken to demonstrate compliance with instructions

Good Documentation Practices are listed in Chapter 4[1] as:

> 4.7 Handwritten entries should be made in a clear, legible, indelible way.
>
> 4.8 Records should be made or completed at the time each action is taken and in such a way that all significant activities concerning the manufacture of medicinal products are traceable.
>
> 4.9 Any alteration made to the entry on a document should be signed and dated; the alteration should permit the reading of the original information. Where appropriate, the reason for the alteration should be recorded.

As noted in the Principle of the Chapter, documentation can be on any medium and therefore Clauses 4.7–4.9 apply to any records supporting GXP activities.

7.1.2 WHO Guidance on Good Data and Record Management Practices

In Section 8 of the WHO guidance there are the following requirements for training in good data and record management[2]:

> 8.1 Personnel should be trained in data integrity policies and agree to abide by them. Management should ensure that personnel are trained to understand and distinguish between proper and improper conduct, including deliberate falsification, and should be made aware of the potential consequences.
>
> 8.2 In addition, key personnel, including managers, supervisors and quality unit personnel, should be trained in measures to prevent and detect data issues. ......
>
> 8.3 Management should also ensure that, at the time of hire and periodically afterwards, as needed, all personnel are trained in procedures to ensure GDocP for both paper and electronic records.

Appendix 1 of the WHO guidance[2] provides a detailed discussion of the ALCOA principles for both paper and electronic records plus some special risk factors that need to be considered by organisations. This is recommended reading to understand fully the ALCOA principles for both paper and electronic systems.

7.1.3 PIC/S PI-041 Good Practices for Data Management and Integrity in Regulated GMP/GDP Environments

Section 6.2 of PIC/S PI-041 guidance[3] refers to a code of ethics and policies:

> 6.2.1 A Code of Values and Ethics should reflect Management's philosophy on quality, achieved through policies (*i.e.* a Code of Conduct) that are aligned to the quality culture and develop an environment of trust, where all individuals are responsible and accountable for ensuring patient safety and product quality.
>
> 6.2.2 Management should make personnel aware of the importance of their role in ensuring data integrity and the implication of their activities to assuring product quality and protecting patient safety.
>
> 6.2.3 Code of Conduct policies should clearly define the expectation of ethical behaviour, such as honesty. This should be communicated to and be well understood by all personnel. The communication should not be limited only to knowing the requirements, but also why they were established and the consequences of failing to fulfil the requirements.
>
> 6.2.4 Unwanted behaviours, such as deliberate data falsification, unauthorised changes, destruction of data, or other conduct that compromises data integrity should be addressed promptly. Examples of unwanted behaviours and attitudes should be documented in the company Code of Conduct policies. Actions to be taken in response to unwanted behaviours should be documented. However, care should be taken to ensure that actions taken, (such as disciplinary actions) do not impede any subsequent investigation. Conforming behaviours should be recognised appropriately.

In addition, Section 7.2 recognises that Good Documentation Practice is essential for ensuring data integrity as follows[3]:

> 7.2 To ensure that decision making is well informed and to verify that the information is reliable, the events or actions that informed those decisions should be well documented. As such, Good Documentation Practices (GDocPs) are key to ensuring data integrity, and a fundamental part of a well-designed Pharmaceutical Quality Management System.

7.1.4 Regulatory Requirements Summary

Although there is no regulatory requirement for a data integrity policy,[4] both PIC/S and the WHO guidances[2,3] advocate writing one as an essential document for staff training and assure inspectors that a company is taking data integrity seriously. As well as a policy, there needs to be an effective training and an assessment of an individual's understanding of the policy before they sign a form stating that they will comply with it and that they know the consequences if they do not follow the policy principles.

This is just the beginning of the data integrity journey, there also needs to be training in:

- Good Documentation Practices for paper, hybrid and electronic processes. The key to this is understanding what complete and raw data really mean so that the procedure is effective and all who are trained in it are apply the principles in every day work.
- In the laboratory there is a need for procedures for interpretation of scientific data, *e.g.* spectra or chromatograms, which will be discussed in the overview in this chapter but presented in more detail in Chapter 16. The practical meaning of complete and raw data, as discussed in Chapter 3, should have a training associated with it to ensure understanding by laboratory staff.

7.2 Environmental Analysis and an Approach to Data Integrity

Rather than derive data integrity policies and procedures from first principles and the definitions in the various guidance documents, it is far easier to see what other industries have done first and then copy or modify their approaches. To this end, we can turn to the environmental analysis carried out in the United States where problems with data integrity surfaced long before those in the pharmaceutical industry.

7.2.1 Background to EPA and Data Integrity

The US Environmental Protection Agency (EPA) regulates environmental analysis through their version of Good Laboratory Practice regulations.[5] In 1990, with increasing computerisation it was apparent that there was little in the EPA GLP regulations to control these systems. Therefore, the EPA initiated a program to ensure the integrity of computer-resident data in laboratories by developing standards for automated laboratory processes. At the start of this programme, the EPA commissioned a survey of automated clinical laboratories under the assumption of a high degree of data integrity as they were regulated and inspected by a variety of State authorities and professional organisations. The survey, carried out by consulting firm Booz Allen & Hamilton, highlighted a wide range of controls from very lax to very stringent within the six laboratories surveyed.[6]

This survey was an input to the development of the Good Automated Laboratory Practice (GALP) guidelines issued by the EPA in draft for comment in 1992 with the final version being released in 1995.[7] Interestingly the draft version was far more stringent than the released version. GALP was archived;

but the document is still available on the EPA web site[8] and reading it provides an interesting compare and contrast with the current GXP approach to data integrity.

7.2.2 NELAC and Laboratory Accreditation

Let us move the narrative forward into the early years of the 21st century, a US national accreditation programme for environmental laboratories was established in 2006 with the formation of The NELAC Institute (TNI) (http://www.nelac-institute.org). Formed from two organisations, one of which was NELAC or National Environmental Laboratory Accreditation Conference. In 2003 NELAC developed a standard for laboratory accreditation[9] that includes specific requirements for training in laboratory data integrity. This standard is being superseded by a consensus one currently under development by TNI. For this discussion on data integrity training, the 2003 standard is fit for purpose as it contains requirements that pertain to data governance, data integrity and specifically the requirements for data integrity training. For this reason, it will help regulated GXP organisations and laboratories understand the requirements for data integrity training and they can adapt the approach for their own uses.

7.2.3 NELAC Quality System

The NELAC standard document contains much about the conference that established it, the organisational structure of the organisation but our immediate interest is in Section 5 dealing with the Quality System Requirements. Figure 7.1 shows the four main data integrity elements within the quality system of the NELAC standard[9]:

- Data integrity procedures;
- Data integrity training;
- Signed data integrity forms for all staff;
- Periodic in-depth monitoring.

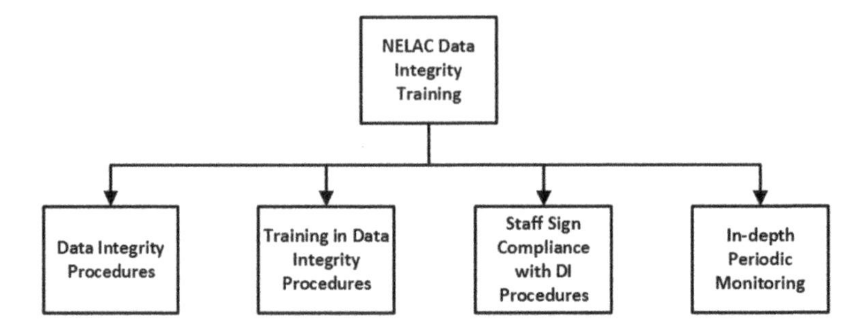

Figure 7.1 Four main elements for data integrity of the NELAC standard.[9]

Table 7.1 presents a précis of the NELAC requirements for data integrity. These sections of the standard take the four elements above and break them down into more detail. The two main elements to consider for an approach to data integrity are:

- Data integrity starts at the top:
  Senior management is responsible for the establishment and maintenance of the data integrity procedures within the scope of the quality management system.

  Management commitment and support for data integrity is required throughout an organisation.

  There must be the ability of staff to raise data integrity issues confidentially for discussion with management.

  This is reflected in the roles of senior management in data governance outlined in Chapter 6 of this book, but also in ICH Q10,[10] EU GMP Chapter 1,[11] and Good Laboratory Practice regulations from the FDA and OECD, respectively.[12–14]
- Data integrity reviews are essential:
  Routine assessment (second person review) of data generated in the laboratory must include checks for data integrity of the data generated.

  In-depth data integrity checks as part of the internal audits conducted by an organisation.

  Documented data integrity investigations in cases of suspected breaches of the data integrity procedures with follow-up disciplinary and corrective actions as appropriate.

These areas are similar to the requirements outlined in the various data integrity guidance documents from the pharmaceutical industry regulators.

7.2.4 NELAC Data Integrity Training

Missing from the list above are the detailed requirements for data integrity training as these are presented in Table 7.2 and diagrammatically in Figure 7.2. As you can see there are very specific requirements for data integrity training for both new and existing employees. The NELAC standard also mentions ethics as well as data integrity training, my interpretation is that the two terms are interchangeable in the context of the right analytical result is supported by the right data records and documentation with the correct attitude of the trained staff. Right in the second use of the word in the last sentence encompasses all data, all calculations and all processed data that have been generated following the correct analytical procedure and the applicable data integrity procedures.

Table 7.1 A summary of NELAC main requirements for data integrity.[9]

Section	Précised NELAC requirements
5.1.7	• Data integrity is an integral part of a Quality System. • Data integrity SOP provides assurance that a highly ethical approach to testing is a key component of all laboratory planning, training and implementation of methods. • Further data integrity requirements are: Management responsibilities: 5.4.2.6, 5.4.2.6.1 and 5.4.2.6.2 Training 5.5.2.7 Control and Documentation 5.4.15
5.4.15	• Data integrity reviews are part of internal auditing to assess evidence of inappropriate actions. • Issues will be handled confidentially until investigation or other appropriate actions are completed and the issues clarified. • Investigations finding inappropriate activity are documented including any disciplinary action, corrective actions and all appropriate notifications of clients. • All documentation will be maintained for at least five years.
5.4.2.6	• Establish and maintain data integrity procedures. • There are four required elements within a data integrity system: 1) data integrity training, 2) signed data integrity documentation for all laboratory employees, 3) in-depth, periodic monitoring of data integrity and 4) data integrity procedure documentation. • Senior management will approve data integrity procedures • Procedures and the associated implementation records will be maintained and be available for assessor review. • Data integrity procedures will be annually reviewed and updated by management.
5.4.2.6.1	• Management will provide a confidential data integrity reporting mechanism. • This will assure confidentiality and a receptive environment for raising ethical issues.
5.4.2.6.2	• The mechanism will inform laboratory management of the need to investigate data integrity further.
5.5.4.1.1	• Establish and maintain procedures that accurately reflect all phases of current laboratory activities including methods for assessing data integrity.
5.4.15	• Part of the overall internal auditing programme, reviews should be conducted if inappropriate actions or vulnerabilities related to data integrity are found. • Potential issues shall be handled in a confidential manner until such time as an investigation or other appropriate actions have been completed and the issues clarified. • All investigations that result in finding of inappropriate activity shall be documented and shall include any disciplinary actions involved, corrective actions taken and all appropriate notifications of clients. • All documentation of these investigations and actions taken shall be maintained for at least five years
5.5.2.7	• Training requirements see Table 7.2

Table 7.2 NELAC requirements for data integrity training.[9]

Section	Précised NELAC requirements
5.5.2.7	• Data integrity training will be given to new starts and on an annual basis for all current employees. • The training must include organisational mission, the critical need for honesty and full disclosure in all analytical reporting, plus how and when to report data integrity issues and record keeping. • Training will include discussion of all data integrity SOPs and training documentation including how to document analytical records. • Employees must understand that failure to follow the laboratory data procedures will result in a detailed investigation that could lead to very serious consequences. • A copy of all training materials will be given to each trainee. • All data integrity training will have signature attendance sheets or equivalent. • Attendees must demonstrate that they understand their obligations related to data integrity. • Senior managers actively support and implement the data integrity procedures. • Specific examples of breaches of unethical behaviour should be discussed including improper data manipulations, adjustment of instrument time clocks and inappropriate changes in concentrations of standards. • Data integrity training requires emphasis on the importance of proper recording of data with examples of both poor and acceptable records. • Information about external ethics training and other external resources should be available to employees.

As shown in Figure 7.2, the training must be introduced by a senior manager as this is the only way to demonstrate that the management team is serious about data integrity, see also Chapter 6 on the data governance roles and responsibilities. As senior management are responsible for establishing and maintaining data integrity procedures as well as supporting data integrity in the organisation, then there should not be a problem for one of them to introduce, at least, any data integrity training session.

Note: the contents of Tables 7.1 and 7.2 are my summary of the NELAC standard requirements and are not quoted verbatim, therefore if you wish to see the original wording that I have précised please read the standard itself.

7.3 Corporate Data Integrity Policy Coupled with Effective Training

As noted in Chapter 4, a major component of the data governance framework is a data integrity policy. In the EMA data integrity question and answer guidance available on-line,[4] Question 12 asks: Is it required by the EU GMP to implement a specific procedure for data integrity? To which the answer is:

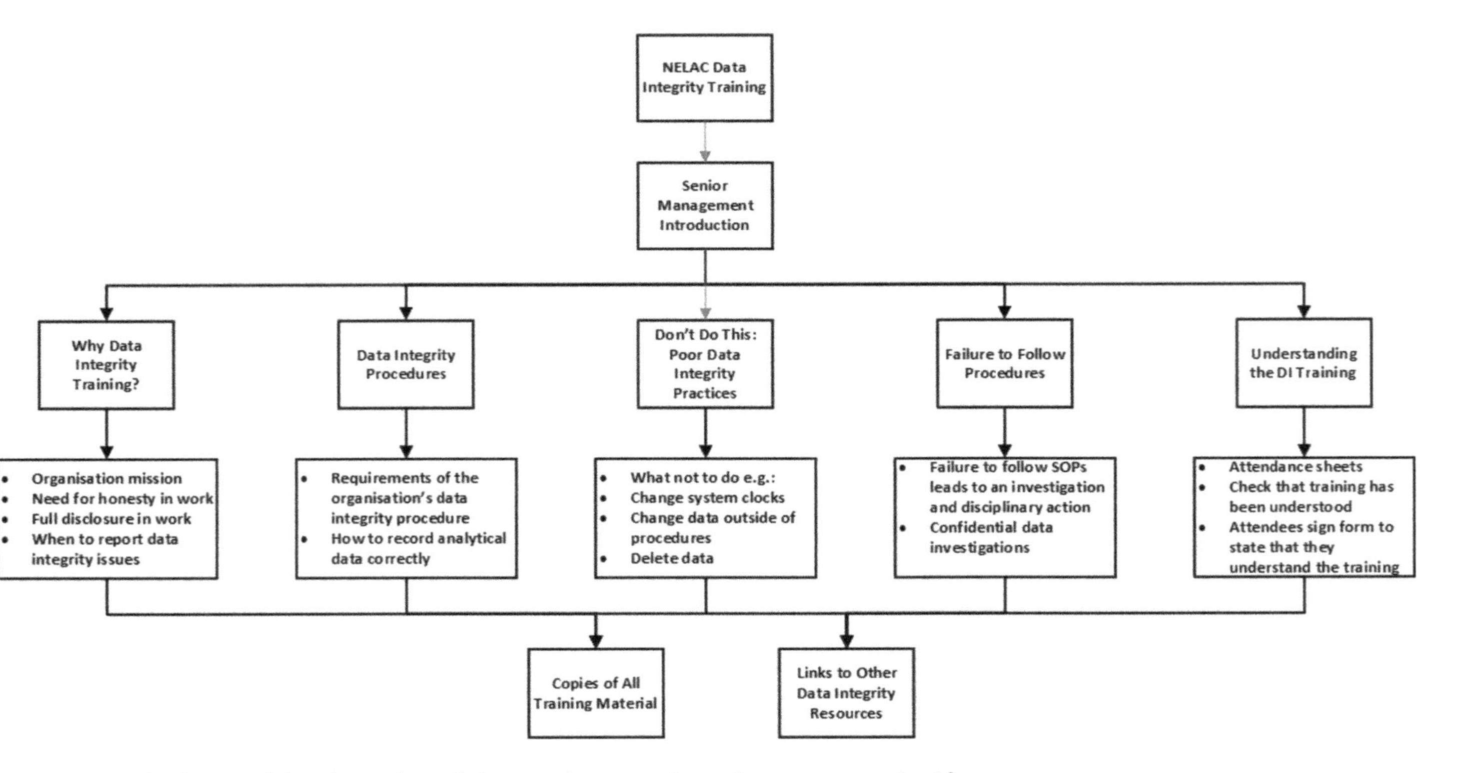

Figure 7.2 Derived general data integrity training requirements from the NELAC standard.[9]

There is no requirement for a specific procedure, however it may be beneficial to provide a summary document which outlines the organisation's total approach to data governance.

This is a no but yes but answer and not very helpful. If an organisation does not put their approach to data integrity into a written policy, how can staff be expected to ensure the integrity of the data they generate? How can an organisation judge if analytical staff are following the rules if they are not defined? Therefore, a corporate data integrity policy is a mandatory requirement of data governance in the author's opinion. One of the first work streams within the data integrity and data governance programme should be the writing of this document, which will impact the whole organisation including the analytical laboratory, with the subsequent training for all staff. This policy must be approved by the Executive Sponsor within the senior management team, although the writing of the document may be undertaken by more qualified staff experienced in data integrity and good documentation practices.

7.3.1 Contents of a Corporate Data Integrity Policy

With the case for writing a corporate data integrity policy established, the next question is what should the policy contain? The outline contents were shown in Figure 6.2 and are described below, this is not a long document as it will have principles and summary statements that will point to procedures containing more detail that will exist underneath it. For example, within the data integrity programme the data integrity auditing conducted by the QA department described in Chapter 21 and supported by the audit aide memoire in Chapter 24 will require an SOP or linked to an existing auditing SOP.

The main areas of a data integrity policy are:

- Commitment to data integrity and data reliability:

  A clear policy statement from the company covering the intent, the expectations of all employees to work in an ethical way that ensures and maintains data integrity.

- Signed by the Executive Sponsor

  The policy must be signed by the Executive Sponsor who heads the Corporate Data Integrity Committee (see Chapter 6) on behalf of the senior management team to demonstrate the importance and management involvement in the data integrity programme to all staff throughout the company.

- Responsibilities of All Staff

  The responsibilities of all personnel at all levels within the company with respect to data ownership, data integrity and ethical operation must be outlined in the document. As noted by the NELAC standards[9] there should be examples of both what to do and what not to do, *e.g.* contemporaneously recording results, not being selective by using only data that is within specification. Responsibilities will include the allocation of data owners and data stewards for all computerised systems and processes within an analytical laboratory. Data owners will be responsible for writing procedures for ensuring data integrity in the systems for which they are accountable and for effective training of users to use it.

- Compliance Programme

  The compliance programme is laid out with sections on procedures for good record keeping for ensuring data integrity, training in such procedures, auditing and monitoring with feedback into the Data Integrity Steering Committee for regular review of progress.

- Quality Oversight

  Quality oversight includes the establishment of effective internal data integrity audits and, where necessary, data integrity investigations.

- Open Culture

  Within the responsibilities section, there is the requirement for management to establish and maintain an open culture as discussed in Chapter 8. This will allow an individual to admit mistakes rather than hide them.

- Raising Compliance Concerns

  The compliance concern section covers open culture and no-blame culture with an open door for raising compliance concerns as well as how to investigate data integrity issues and problems. Breaches by any member of the staff of the policy will result in disciplinary action up to and including dismissal.

It is fine having a company data integrity policy as described above, but how should company employees be trained and are there any regulatory requirements for this?

7.3.2 Training in the Data Integrity Policy

Training in the Data Integrity Policy is too important a subject to leave an employee in the corner of a room to read and understand the document. This will result in misunderstanding and multiple interpretations of the document's intent. Instead, to ensure a consistent understanding and application of the policy, face to face or instructor led (ILT) training is the preferred option for data integrity training. However, in larger organisations this is impracticable and computer based training will be used. Regardless of the means, the session must be introduced by a senior management to reiterate company commitment to data integrity. One scope of data integrity policy training is shown in Figure 7.3, which can be compared with the NELAC approach in Figure 7.2 and must include what actions are permitted and those that are unethical and therefore prohibited.

Most importantly, after the training there *MUST* be a means of demonstrating competence and understanding of the policy. The rationale for this is based on the NELAC standard[9] that requires demonstrable evidence of understanding as an employee will then sign a form stating their understanding and the consequences of their actions if they commit a data integrity violation. If an individual commits a violation they could claim that they do not understand the data integrity policy. However, if there is demonstrable evidence that after the training they successfully completed an assessment of understanding, then there is no opportunity to use this as an excuse.

Training can be done either by an open book questionnaire or a computer based test. As this is a critical component of the overall data integrity programme the pass mark for the test must be 100% if an open book test is employed. If an attendee does not achieve this standard, then they must attend another training session and assessment. In contrast, if a closed book test is not used, then the pass mark can be lowered to 80–85% as the test would have to be completed by memory. As a personal opinion, an open book test is better as it encourages the trainees to read the policy, the training materials and their notes to check the answer before answering a question, which also reinforces the training and understanding.

The competence assessment is typically a multiple choice open book questionnaire, *i.e.* the attendee can refer to the training material to answer the questions. The attendee should answer between 15 and 20 questions. Three example questions are shown in Table 7.3 and there is a single answer of A, B or C for each attendee to write in the answer column. If the assessment is undertaken using a Learning Management System (LMS), these applications can hold a repository of say 30 questions of which 20 are selected in random order for each person to answer. Alternatively, a paper questionnaire needs to be completed by everyone undergoing this training to demonstrate understanding. If many people need to be trained and assessment uses a paper questionnaire, I would advise having the same questions but in three different orders just to avoid cheating. In multinational companies it may be preferable to deliver the corporate data integrity policy training in the local language to ensure maximum understanding of the content.

Every existing employee working in a GXP environment including Research should be given this training now. New employees joining the company or transferring from a non-GXP area should also undergo this training as part of their induction training.

For the foreseeable future, data integrity will become an integral part of the annual GXP refresher training material.

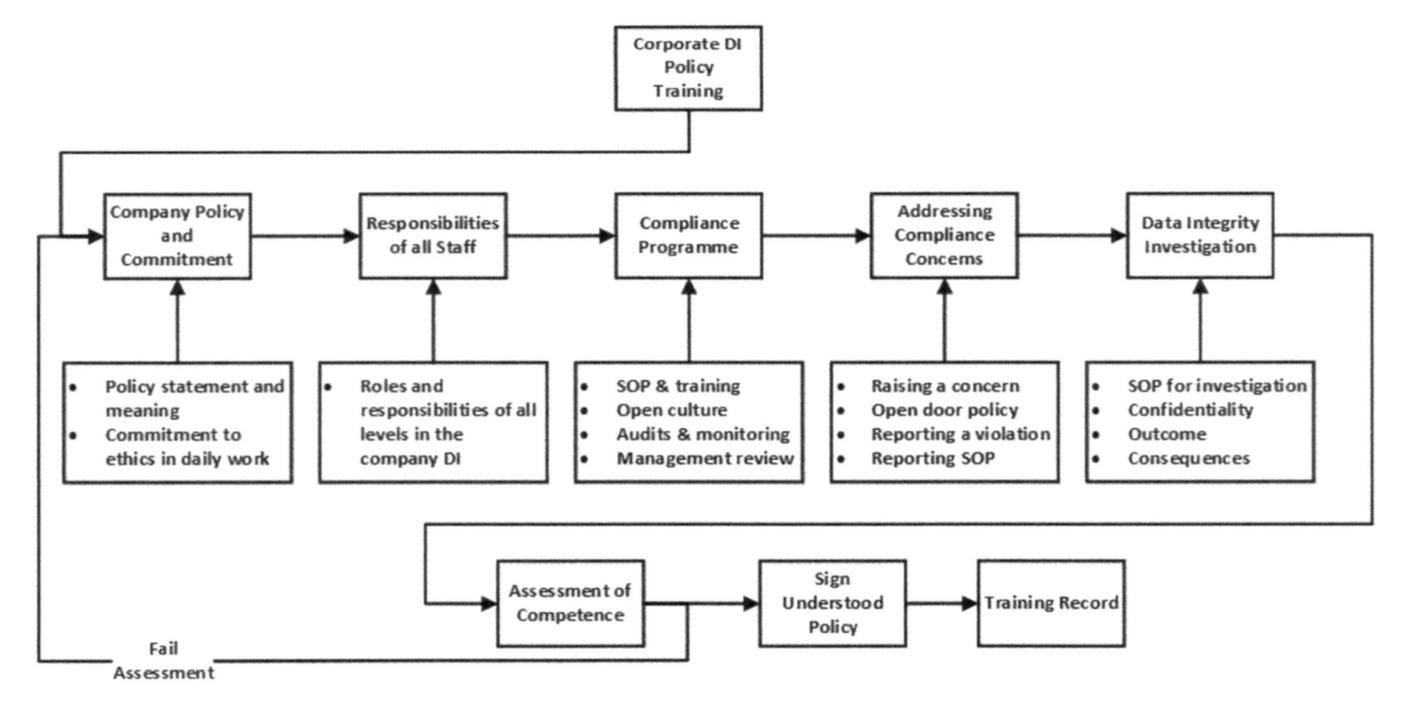

Figure 7.3 Training in the corporate data integrity policy.

Table 7.3 An example of a questionnaire for checking the understanding of data integrity policy training.

No.	Question	Answer
1.	What does the first A in ALCOA mean? A = Attributable B = Accurate C = Alarming	
2.	Why did the Company write the data integrity and ethics policy? A = There is an FDA inspection due and we want to impress the inspector B = To prevent or detect and remediate violations of rules, regulations and policies C = There is no regulatory requirement to write such a policy	
3.	Who is responsible for data integrity in the Company? A = Senior Management B = Quality Assurance C = Everyone	

7.3.3 Agreeing to Comply with the Policy

Following the determination of understanding, each employee signs a form that states that they have read and understood the corporate data integrity policy and consequences of failing to comply with it. This is a specific requirement of the NELAC standard[9] and should also be part of the data integrity training in any regulated laboratory and organisation.

7.4 Suggested Data Integrity Procedures

Underneath the corporate data integrity policy will be many procedures that will be the detailed basis for ensuring data integrity in the laboratory as well as the overall organisation. These will fall into two main areas:

1. Laboratory data integrity procedures;
2. Quality Assurance data integrity procedures.

Table 7.4 contains a list of the suggested procedures for both laboratory and quality assurance areas. Key ones will be discussed in this chapter and others in Chapters 14–17.

7.5 Principles of Good Documentation Practice

One of these key procedures, applicable not only in the laboratory but throughout a regulated organisation, is on Good Documentation Practice (GDocP). The principles of Good Documentation Practice are very simple and are shown in Figure 7.4. These principles hold for paper based records as well as for hybrid and electronic computerised systems. EU GMP Chapter 4[1]

Table 7.4 Main laboratory and quality assurance procedures for ensuring data integrity.

Laboratory data integrity procedures	Quality assurance data integrity procedures
• Qualification and validation of analytical instruments and software • Analytical procedure development and validation and performance monitoring • Preparation of reference standards, buffers, mobile phases, media, *etc.* • Instrument set up, *e.g.* chromatographs, spectrometers, *etc.* • Instrument maintenance and use logs: completion and review • Handling complete data: paper, hybrid and electronic systems • Chromatographic integration • Interpretation of scientific data • Second person review of analytical work • Review of audit trails • Trending of analytical data and results • Investigating out of specification results (OOS)	• Good Documentation Practice for paper, hybrid and electronic records • Management of staff training • Change management • Deviation management and CAPA actions • Record retention • Raising compliance concerns • QA data integrity audits • QA data integrity investigations • Backup and restore (networked systems) • Disaster recovery • Business continuity management (BCM) • Audit and quality agreements with contract laboratories

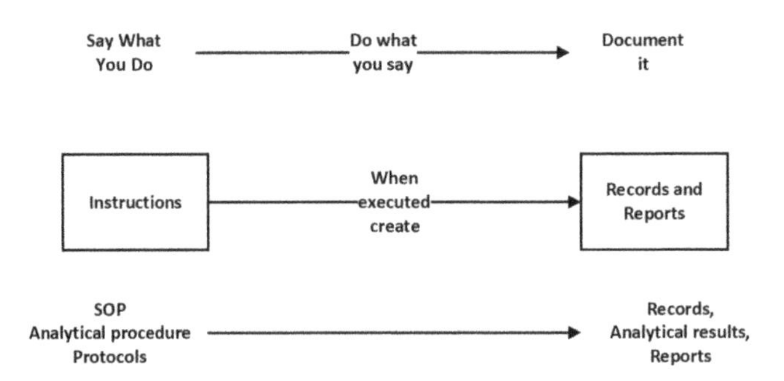

Figure 7.4 Principles of Good Documentation Practice.

talks about instructions that when executed create records. Instructions as shown in Figure 7.4 can be procedures (standard operating procedures or work instructions), analytical procedures or protocols (non-clinical, clinical or stability for example).

Apart from demonstrating compliance with the principles of GXP regulations, good, well written documentation enables troubleshooting of problems, root cause analysis and laboratory investigations.

7.5.1 Say What You Do

Using the statements in the Principle of EU GMP Chapter 4[1] there will be instructions that may take the form of a standard operating procedure (SOP) or work instruction (WI), an analytical procedure or a protocol. Whatever the type, this will be a controlled and approved document within the quality management system and will tell the person executing the instruction what to do.

The FDA Guide to Inspection of Pharmaceutical Quality Control Laboratories notes for procedures and analytical procedures that[15]:

> SOPs should be complete and adequate and the operations of the laboratories should conform to the written procedures. Specifications and analytical procedures should be suitable and, as applicable, in conformance with application commitments and compendial requirements.

Only the current versions of instructions should be available for staff to read. All other versions must be retired and a copy of each one archived.

7.5.2 Do What You Say

When the instruction is executed, the person follows the contents of the document. Typically, a trained person may execute such instructions from memory but there must always be copies of the instruction available for reference in case the individual is uncertain of what to do. Alternatively, it may be the first occasion the person has executed the procedure for some time and needs to refresh their memory. Copies can be available either on paper or electronically but regardless of the format of the instruction, they need to be easily available, especially where the procedure is executed.

If there is a deviation from the procedure, then this must be documented during execution and needs to be discussed with the person's supervisor. In the light of data integrity, this may be the start of a deviation process that could result in an improvement or change to the procedure. Alternatively, if a mistake has been made it may result in stopping the procedure, documenting the reason for this and starting again. A deviation may be raised depending on the mistake.

7.5.3 Document It

During the execution of the instructions, there will be requirements to record information on authorised media or creation of records during an analysis. This is the most important part of executing the instruction as it creates the documented evidence to demonstrate that the procedure was followed. There must be procedural requirements for recording and managing data when executing a paper, hybrid or electronic process that will be covered in later sections in this chapter.

7.5.4 Automating Procedure Execution

To ensure that a procedure is followed correctly, Laboratory Execution Systems (LES) may take an analytical procedure or written procedure and convert it into an electronic checklist that states the steps to follow. The steps are presented on a tablet or wall mounted screen and when a step is completed or data have been acquired the step is noted as completed by the analytical scientist performing the work. There is the ability to note a deviation from the procedure that can be reviewed by a supervisor. Given the current data integrity focus by regulatory authorities, implementation of an LES or equivalent system may avoid questions such as was the work performed at all and this will be discussed in Chapter 13.

7.6 Training to Collect and Manage Raw Data and Complete Data

Chapter 3 had a discussion of what constituted complete data and how it was the same as raw data, *i.e.* all data and records from the sampling to the reportable results. In this section, we need to consider the training necessary to ensure all analytical staff understand complete data and raw data. Please note that this section focuses on raw/complete data, later sections focus on Good Documentation Practice for paper, hybrid and electronic systems and therefore this section and the GDocP sections are complimentary and should be read together.

7.6.1 Principles for GXP Laboratory Raw Data and Complete Data

From the discussion in Chapter 3 we can derive what constitutes raw data/complete data in a GXP laboratory, in general, we would need the following:

- Sampling plan (GMP) or study protocol (GLP) that document what work will be undertaken, how samples will be taken, stored and transported. Depending on the scope of the work the plan or protocol may also detail how data will be treated, analysed and reported, *e.g.* study protocol and method validation or method transfer plans.
- Samples with relevant information: identity, study, batch or lot number, analysis request, *etc.*, *etc.*
- Where appropriate, qualified and validated analytical instruments and computerised systems.
- A validated analytical procedure including the preparation of the sample for presentation to the instrument.
- Reference standards and/or a library (if we are using a spectra library for identification)
- Qualified staff to perform the work.
- Records of work performed and data acquired (paper, hybrid and electronic).
- Instrument log books updated with records of any calibration undertaken and the work performed.

- Raw data and any subsequent transformation and calculations including spreadsheet files and printouts.
- Reportable results in a suitable format, *e.g.* study report or Certificate of Analysis (COA).

From these pre-requisites, the analysis is undertaken and results observed and recorded and/or one or more files will be generated and saved by the instrument. These are the first part of raw data. Not just the data files themselves but all the other associated contextual metadata that must be linked together to support the generated files containing the identity of the instrument and method used, the analyst performing the work, date and time stamps on the files, audit trail entries, *etc.*

7.6.2 Approach to Training for Complete and Raw Data in the Laboratory

As we shall see in Chapter 9, an analytical data life cycle needs to be very flexible as you cannot fit all analytical procedures into a single generic model. Therefore, there needs to be a flexible approach both to documenting an analytical procedure life cycle and also to the training.

For example, consider Figure 1.3 in Chapter 1 that shows all the raw data/complete data for CDS analysis but this figure would not be appropriate for a test involving observation such as colour or appearance. Thus, we will look at three examples of analytical measurement to understand what raw and complete data mean in practice: paper records associated with weighing and preparing an analytical reference solution, a hybrid spectroscopic method and an electronic process with a LIMS interfaced with a CDS.

7.6.3 Example 1 – Paper Records from a Manual Test

An example of a manual test can be seen in Figure 7.5 where an analytical reference standard is weighed and a reference solution is prepared for use in multiple analyses. The process features the following:

- A uniquely numbered blank form or a controlled laboratory notebook is used for recording the work as outlined in the applicable analytical procedure (control of blank forms is discussed in more detail in Chapter 11).
- The user retrieves the required reference standard from the appropriate store and checks the expiry date and documents the material used on the form.
- Then, the analytical balance to be used is identified and the current qualification status is recorded and a balance calibration is performed to see if the instrument is within acceptance limits. The mass set and the individual masses used are recorded on the form and then the user logs into the balance and performs the calibration. The measured and

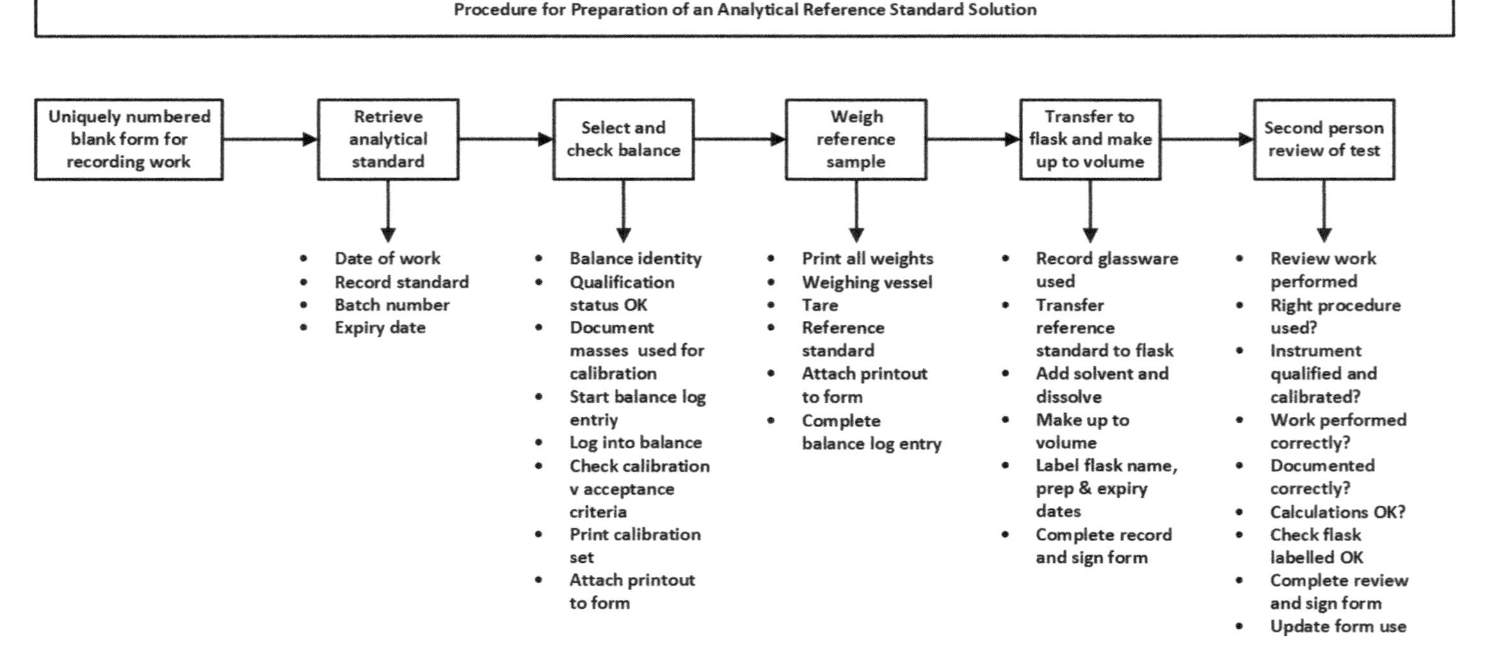

Figure 7.5 Raw and complete data associated with preparation of an analytical reference solution.

calibrated masses are calculated to determine if acceptance criteria have been met.

- A weighing vessel is placed on the balance pan, weighed and the value tared for the required weight of reference material to be taken. All weights should be recorded sequentially on the printout together with the date, time and user identity of the individual performing the work. The printout is stuck onto the partially completed form. At the completion of the weighing sequence, the log book entry that was started with the balance calibration is completed by the tester.
- The appropriate volumetric flask and glassware are made ready and the weighing vessel is transferred to the flask. This description assumes an aluminium boat is used, if a glass boat is used the material will be transferred to the flask and the boat reweighed to determine the amount of material taken. If this is the case, then the process in Figure 7.5 needs to be modified accordingly with a further calculation.
- The flask is labelled with the name, solution ID, reference standard, who prepared the solution, storage temperature and the dates of preparation and expiry. Then, the material in the flask has a proportion of solvent added and the material dissolved, then the flask is made up to volume.
- All the paperwork using the uniquely numbered form is completed by the individual including a check of any calculations and to see that all data are complete. If there are any blank spaces, these are struck through then dated and initialled.
- The last stage of the process is the second person review and, as this is the subject of Chapter 17, this topic will not be discussed here.

The training for this should highlight all elements required for ensuring data integrity as well as reinforcing the need for complete data that would include any additional records if a mistake was made during the process.

7.6.4 Example 2 – Spectroscopic Analysis Using a Hybrid System

For the components of raw data and complete data for a spectroscopic analysis used as a hybrid system see Figure 7.6. Please note that this is a generic representation and aims to present both qualitative and quantitative analysis, to represent one or the other a little interpretation is required by the reader.

You will see from Figure 7.6 that the raw/complete data trail starts from the sampling plan through the sample preparation with work documented in laboratory notebooks or blank forms for which there is accountability before the sample is presented to the instrument. If we are dealing with identity testing and all that occurs is that an operator is putting a probe into a drum in a warehouse, then this stage is minimal or omitted altogether as discussed in Chapter 9. The next stage in the figure is the actual analysis where the right method to control the instrument and acquire the data is used along

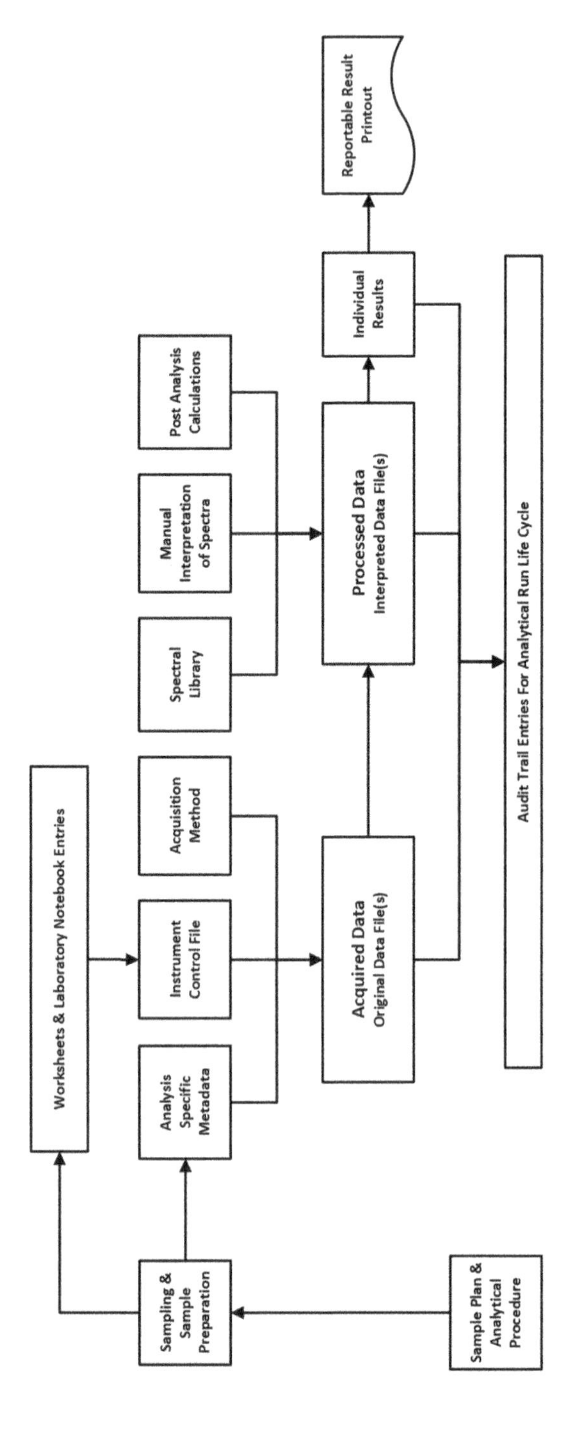

Figure 7.6 Raw data and complete data for an analysis involving a hybrid spectroscopic analysis.

with sample specific information that is entered to uniquely identify the analysis. Of course, the spectroscopist has logged on and all work including correction of typographical errors is recorded in the audit trail.

Next, the sample spectrum is interpreted by a variety of ways such as:

- Fit against a composite spectrum in a spectral library for identification (the name of the library and version are both parts of the metadata supporting the raw data).
- Interpretation of spectra for structure elucidation, here there may be notes in a laboratory notebook associated with the thinking associated with the interpretation that forms part of the raw data for the analysis.
- Quantification of analytes *via* calibrated curves or comparison with reference standards depending on the spectroscopic technique being used. Here, we can find that these calculations could be carried out in the instrument data system, LIMS or a spreadsheet.

Finally, the reportable result will be generated in the data system and the printout signed. As stated in the GLP definitions of raw data, the printout is part of the raw/complete data.

7.6.5 Example 3 – Chromatographic Analysis with a CDS Interfaced with a LIMS

For the third example to enable effective training to understand raw and complete data, let us consider a CDS and LIMS interfaced with each other as shown in Figure 7.7. Again, compare this with Figure 1.3 to see the whole analytical process starting from sampling. The CDS and the LIMS have a two-way communication with each other with sample information passed to the CDS from the LIMS and the reportable result from the CDS to the LIMS. In addition, the analytical balance for weighing reference materials and samples, discussed in Section 7.6.3, is now interfaced with the LIMS and weights can be linked to the correct sample identities in the CDS sequence file or LIMS database:

- Electronic transfer of data from the LIMS to the CDS to help set up the analytical run can include a download of sample identities and sample weights used for the analysis and possibly reference standard purities and water content depending on the type of analysis performed.
- Dilutions made during the sample preparation phase of the analysis will probably need to be recorded on paper and entered manually into the CDS sequence file as done now. As these values should be classified as critical data, they will need to be checked by a second person for correctness and accuracy.

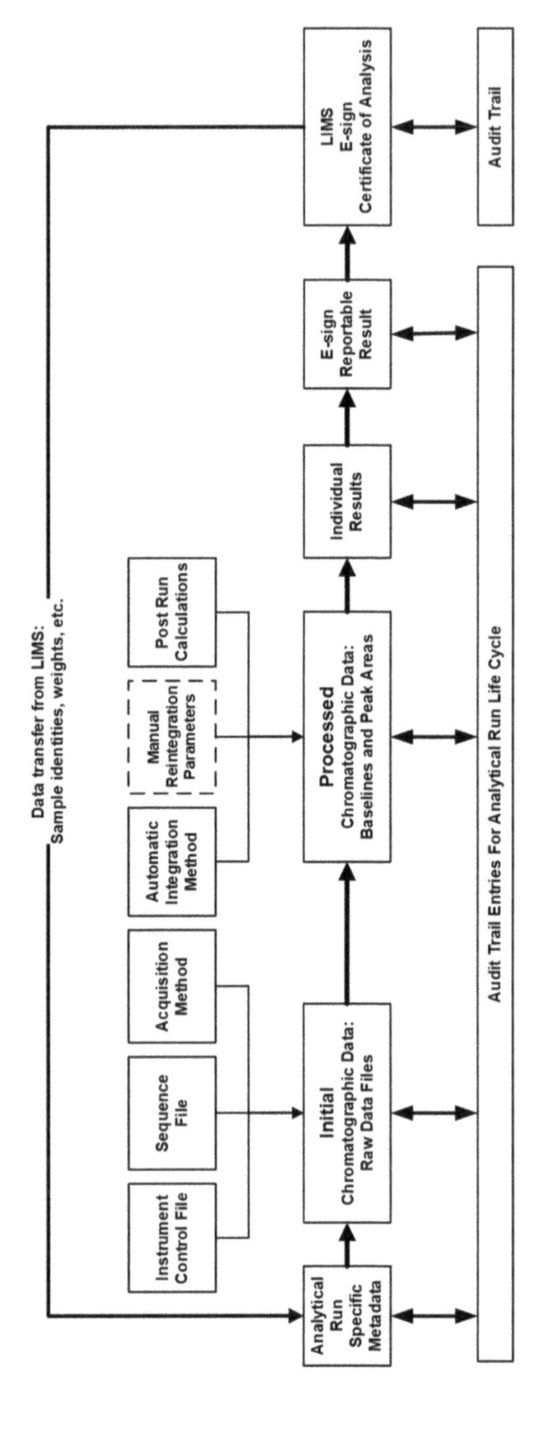

Figure 7.7 Raw and complete data for a LIMS and CDS working electronically.

- The CDS data files are integrated and interpreted by the analyst, *e.g.* the integration parameters are modified and, if permitted, the baselines manually positioned as described in Chapter 16 and ref. 16.
- Following the completion of integration, a transfer from the CDS to the LIMS can vary from laboratory to laboratory. This transfer can vary from all integrated peak areas per sample identity of all injections so that all subsequent calculations are performed in the LIMS (typically this is for bioanalytical LIMS) to a single reportable result that is compared with the product specification in the LIMS.
- If a file is transferred between the CDS and a LIMS that is parsed to extract the requisite data, the file must be protected from tampering between transfer and parsing as it is part of the complete data or raw data of the analysis. Ideally, such a mechanism would not be used – direct transfer between the two systems is much better.

7.6.6 Additional Raw Data?

A CDS can also collect additional data such as column temperature or flow pressure. It is also possible to collect more data than you need "just in case" of an investigation. In each case, the laboratory needs to determine if these data are GMP-relevant or not. For example:

- If the column pressure is simply a by-product of the way the CDS works and no GMP decisions are taken with these data, then these readings are not part of the raw data or complete data. However, this must be justified and documented.
- In contrast, if an analytical procedure requires a single wavelength to be measured by a UV detector and the laboratory chooses to collect a full spectrum, just in case of a problem with a resultant investigation. The spectrum could be used to make a GMP decision and therefore the spectral data should be classified as part of complete/raw data. This decision should be documented and the analytical procedure updated to reflect the actual working practice.

Although the data and metadata shown in Figure 7.7 would usually be considered the core of raw and complete data in a CDS, some working practices could require further records to be included. A data owner of a CDS, or indeed any laboratory computerised system, needs to be vigilant to ensure that all situations are covered and documented.

7.7 Good Documentation Practice for Paper Records

It is important that for recording work in a compliant way that all staff know and understand the need for Good Documentation Practice when generating or completing paper records. In this section, we will discuss GDocP expectations as well as good and bad practice, see Table 7.5.

Table 7.5 Good Documentation Practice for paper records.

GDocP factor	Good Documentation Practice expectation
Company requirements	• The company must maintain a signature and initial log so that entries can be traced to a specific individual • Each new employee must sign and initial the log so that it is kept up to date • The signature log must be reviewed regularly to ensure that it is current
Record format	• Use only printed authorised media to record data, *e.g.* controlled blank forms • Ensure that the pagination is in the format, *e.g.* Page *X* of *Y* • Multiple pages must be bound securely to prevent removal, photocopying and replacement • Regularly reviewed and revised as appropriate
Creation of the record	• All handwritten entries must be legible and understandable • Entries must be written in indelible ink • Entries must be initialled or signed by the person performing the work • Entries must be dated and where appropriate timed in an unambiguous format according to company standards • Entries must be in chronological order • Entries must be made contemporaneously • Blank spaces must be struck through with a *Z* or single line, dated and initialled as N/A or not applicable.
Prohibited actions	• Pencil and erasable pens must never be used • Typewriter correction fluid or whiteout must never be used • Never use ditto, do or same as above – the observation must be written in each time • Blank spaces should not be left blank. • Date stamps should not be used (date stamps may or may not be acceptable depending on an inspector) • Other employees cannot sign on behalf of another person or "pp" a signature • Signatures and initials must never be forged • Never overwrite entries • Never copy pages to mask repeated work • Never use unauthorised media to record GXP data, *e.g.* scrap paper • Avoid using asterisks linked to an annotation for making changes to entries • Never back date or forward date a record
Correcting entries	• Correction of entries must be made by striking through the value so that it can still be read • Entering the new value with initials of the person making the change • Write the date and reason for the change

7.7.1 Recording Observations and Results

Good Documentation Practice for paper records is shown in Table 7.6 and should be self-explanatory. These principles can be included in a procedure that should highlight what is allowable and what is prohibited. When

Table 7.6 Examples of good and bad documentation practice for handwritten records.

Practice	Poor or non-compliant example	Good example
Recording an observation	1.35 mg	1.35 mg RDM 3 Apr 2019
Duplicate entries	1. White powder 2. Ditto 3. Ditto	1. White powder 2. White powder 3. White powder
Date	3/4/19	3 APR 2019
Time	1.36	1.36 pm or 13.36
Correcting an error	~~3.124~~ 3.142	~~3.124~~ 3.142 RDM 3 Apr 2019 Transcription error

considering the prohibited actions, the procedure needs to focus on both covering up mistakes such as back dating an omitted signature to wilful falsification.

Further information about good documentation practices for paper records *versus* the ALCOA principles along with special risk management can be found in Appendix 1 of the WHO data integrity guidance.[2]

7.7.2 Examples of Good and Poor Documentation Practice for Handwritten Records

Training in Good Documentation Practice should be illustrated with both good and unacceptable practices so that staff are fully aware of what is expected of them when working in a regulated laboratory. Some of these examples can be seen in Table 7.6.

The following are non-compliant documentation practices that will not ensure data integrity and staff must be trained not to perform these acts:

- Never use notebooks or paper pads with easily removed pages as you cannot guarantee that pages have been removed.
- Never write GXP records or observations on post it notes, scrap paper, envelopes or even the sleeve of your laboratory coat (do not even think of using your hand).
- Altering or destroying GMP records to hide or falsify data is fraud and is a data integrity violation that will lead to serious disciplinary action.
- Never throw a GMP record away if you have made a mistake. The record forms part of the complete data/raw data and must be retained.

  Within an open culture, it could be the start of a corrective action to improve a procedure because it is too complex.

- Never destroy GMP records even if the record retention period has elapsed without documented management approval.

Some of these examples are data integrity violations that would trigger a data integrity investigation with possible disciplinary action for the individual(s) involved.

7.7.3 Fat Finger, Falsification and Fraud – Take 1

When does human error slide down the slippery slope of falsification and fraud and how can training prevent this? A central component of data integrity in any laboratory is the human element, the analytical chemist who will be involved with developing and validating methods or performing analysis on samples. Mistakes or fat finger moments are part of human nature, but where is the dividing line between this and falsification and fraud?

In the sub-heading I have suggested that there are three types of data integrity deviation: fat finger, falsification, and fraud. Here are my definitions of the terms:

- Fat Finger: An inadvertent mistake made by an analyst during their work that can be made either on paper or electronically.
- Falsification: An individual who deliberately writes or enters data or results with the intention to deceive.
- Fraud: Collusion between two or more individuals who deliberately write or enter data or results with the intention to deceive.

I have drawn the distinction between the definitions of falsification and fraud: Falsification is perpetrated by an individual and fraud by two or more people. However, the impact of both is the same: the intent to deceive.

This brings us to a common issue that we all have experience with signing a record. Typically, this is a bound book with prenumbered pages, just there to prevent you tearing out a page to write down the shopping list. This is the first stage of ensuring data integrity in the laboratory. At the bottom of each page is space for you to sign as a reviewer after you have checked the work and accepted it as complete and accurate.

Here is the situation: You are a supervisor and you are checking a laboratory notebook for some current work, and in turning the page you notice that your signature is missing from when you reviewed some earlier work. Three out of four pages of the old work are signed and dated but you have neglected to sign one of the pages — so what do you do? Temptation time! You have the following options:

1. Ignore the problem and wait for somebody else to discover it.
2. Sign the page and date it the same as the other pages.
3. Sign the page but date it with the current date and add a note that you have just noticed the problem.

Most analysts should reject the first option as this is unprofessional, remember quality is everybody's responsibility. So, we are down to options

2 and 3. Option 2 is accompanied by a little voice whispering in your ear "nobody will know if you put the same date that the other pages were signed on." You are now on the brink of the abyss — on the plateau is ethics and integrity and down the slippery slope is falsification and fraud. May I suggest that option 3 is the only option worth considering that will establish credibility for you and the laboratory? As an auditor, I expect to see mistakes on paper and if none are found this raises suspicions.

7.7.4 Original Records and True Copies

In discussing GXP data there is the concept of original records and true copies[12,17] and this area is a key requirement for good documentation practice training. The original record is the original observation but a true copy is defined by WHO as:

> A true copy is a copy of an original recording of data that has been verified and certified to confirm it is an exact and complete copy that preserves the entire content and meaning of the original record, including, in the case of electronic data, all essential metadata and the original record format as appropriate.[2]

The classic example of a true copy is a printout from an analytical balance that uses thermal paper that will darken over time and will not be readable throughout the record retention period. In this instance, the balance printout would be photocopied and the copy annotated by the analyst as a true copy then initialled and dated. There needs to be sufficient information on the balance printout to enable the printout to be traceable in the work process, *e.g.* sample identify or reference standard, batch or study number, *etc.* The original and the verified copy are then stuck into the laboratory notebook or controlled analytical worksheet, see Figure 7.8.

The paper example is relatively easy but it becomes more complex when considering electronic data. From the WHO definition above. For not only must the data files plus all associated metadata including the applicable audit trail entries for the work package be capable of easy copying from the data system they must be abstracted in a form that enables them to be read later if required. This comes back to the fundamental design of many laboratory computerised systems that we use. This process must be defined and validated when it works.

7.8 Good Documentation Practice for Hybrid Records

As we shall discover in Chapter 12, hybrid systems are the worst possible situation to be in. There are electronic records in the computerised system with signed paper printouts that must be synchronised. Therefore, Good Documentation Practice must follow two sets of requirements:

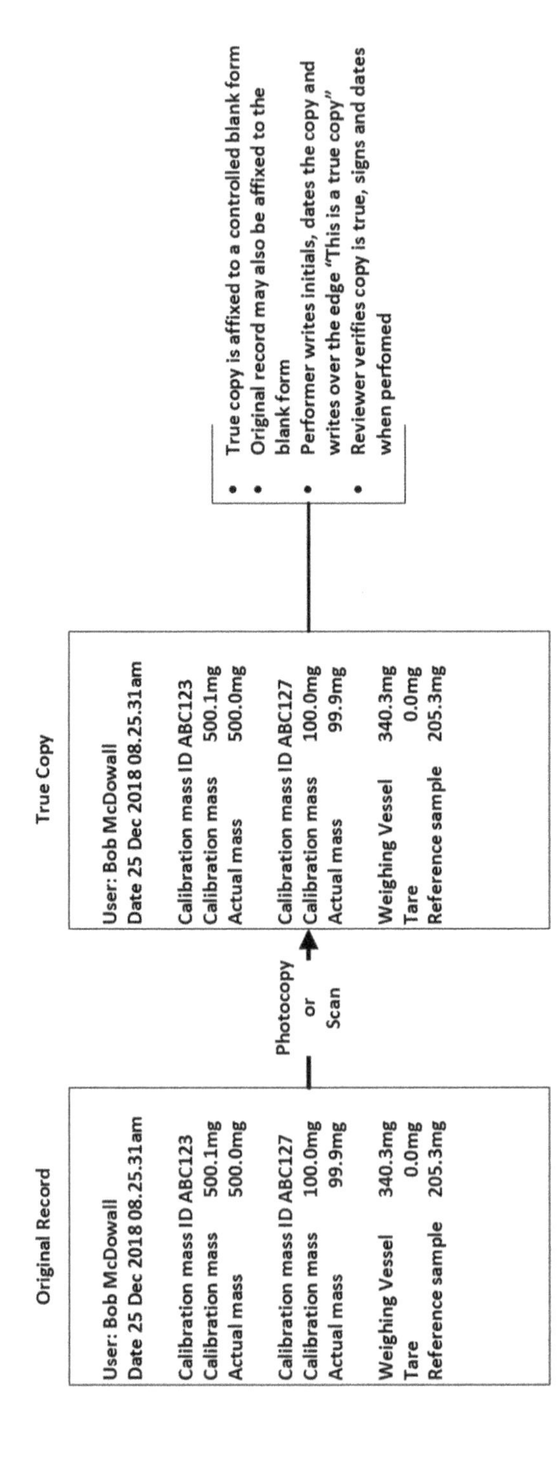

Figure 7.8 Creating a true or verified copy of an original paper record.

- Paper records must be managed as described in Section 7.7.
- Electronic records must be managed as described in Section 7.9.

Paper and electronic records are in two incompatible record formats, *i.e.* physical and virtual, and the major problem comes with the record signature linking required by 21 CFR 11.[18]

7.8.1 Record Signature Linking for Hybrid Systems – Spreadsheet Example

21 CFR 11 has a requirement for linking electronic records with the hand signed paper printouts in §11.70 of the regulation[18]:

> Electronic signatures and handwritten signatures executed to electronic records shall be linked to their respective electronic records to ensure that the signatures cannot be excised, copied, or otherwise transferred to falsify an electronic record by ordinary means.

This is a key area for Good Documentation Practice training to ensure that records and signatures are linked to comply with regulations. Given the wide use of spreadsheets in regulated laboratories for calculations, it is important that the signed paper printout is linked to a saved spreadsheet file for the work.

Figure 7.9 shows the principle for linking the electronic spreadsheet file with the signed printout. During the validation process of the spreadsheet template, the six fields in the top and bottom margins of the spreadsheet can be filled in. These are labelled A–F in Figure 7.9 and are completed as follows:

- Fields A–C inclusive will be completed during the development of the spreadsheet with the name and version of the spreadsheet template, the

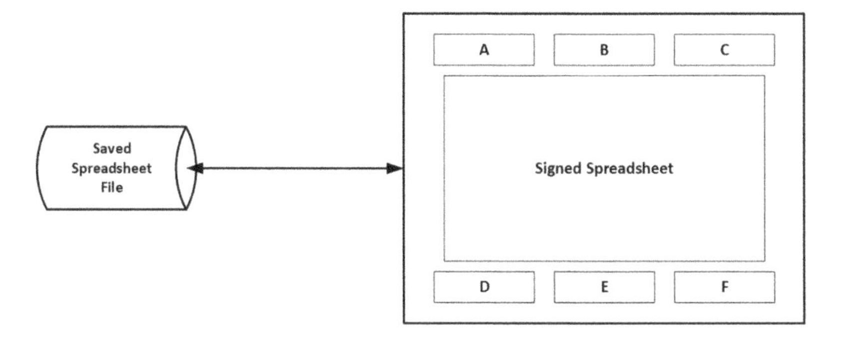

Figure 7.9 Record linking for a spreadsheet used for GXP regulated work.

SOP to which it relates and the effective date. These will be fixed and will not change until the spreadsheet is updated.
- Field D will contain a variable field for the filename. When the validated working template is completed in an analysis and saved, this field will be populated automatically by the spreadsheet application. There is an option for a field name with the directory path but as IT infrastructure can be changed during the retention period this option is not advised.
- Field E will have the create date and time that will be auto completed upon saving the completed spreadsheet.
- Field F will contain the print date and time that is also auto completed by the application.

This approach gives better control but the reader is referred to Chapter 12 to understand more fully the problems with hybrid systems and the records that they generate.

Ideally spreadsheets should be eliminated from an analytical process and the calculations incorporated into an instrument data system or other informatics application such as an ELN or LIMS as discussed in Chapters 12 and 13. However, due to the ease of availability of the spreadsheet application and the laziness of analytical staff to incorporate calculations in an informatics solution, uncontrolled spreadsheets for regulated calculations are rife in many laboratories.

7.9 Good Documentation Practice for Electronic Records

When considering Good Documentation Practice for electronic records, there are several assumptions we must make:

- The laboratory computerised system is validated and is fit for intended use.
- The application is configured to protect electronic records, this is documented and the configured application is the instance validated.
- There is a correctly functioning audit trail that complies with the technical requirements of EU GMP Annex 11 and 21 CFR 11.[18,19]
- Electronic signatures are linked to the records through a report not hidden in the audit trail. Once a record is signed the underlying records should be locked so that changes cannot be made unless the signature is revoked by an authorised user. There will be records of such actions both on the report and in the audit trail.
- Access to the operating system, data directories, system clock and recycle bin is restricted to authorised people and this does not include analysts and their supervisors.
- If possible, workflows within the application are configured to enforce compliance and data integrity so that missing entries or signatures should not be an issue.

7.9.1 Good Documentation Practice for Electronic Records

The expectations for Good Documentation Practice for electronic records are shown in Table 7.7. Further information about good documentation practices for electronic records *versus* the ALCOA principles along with special risk management considerations can be found in Appendix 1 of the WHO data integrity guidance.[2]

7.10 Good Documentation Practice Training

In the Principle of EU GMP Chapter 4 there are the following statements[1]:

> Good documentation constitutes an essential part of the quality assurance system and is key to operating in compliance with GMP requirements.
>
> There are two primary types of documentation used to manage and record GMP compliance: instructions and records/reports.
>
> Records provide evidence of various actions taken to demonstrate compliance with instructions.
>
> Documentation may exist in a variety of forms, including paper-based, electronic or photographic media.

From the above elements of Chapter 4, the scope of the good documentation practice training can be developed to include the following topics, which are also shown in Figure 7.10:

- Principles of the training are applicable throughout all GXP environments.
- Responsibilities of staff, *e.g.* tester and reviewer. This is important as quality is everybody's job now.
- Application of Good Documentation Practice principles for each of the three types of records as outlined in Sections 7.7, 7.8 and 7.9: Illustrate the training with acceptable and unacceptable practices and good and poor examples of documenting data.
- Understand and apply the ALCOA+ principles for all applicable record keeping activities.

Owing to the importance of this training as an essential part of data integrity, there needs to be an assessment of understanding or competence and a similar approach should be taken as with the data integrity policy.

7.11 Role of the Instrument Log Book

One of the essential documents for data integrity is the instrument log book, which is typically a bound and paginated paper book. There must be a procedure and training for using an instrument log book as this is an important

Table 7.7 Good Documentation Practice for electronic systems.

GDocP factor	Good Documentation Practice expectation
Company requirements	• The company must maintain a list of users and their identities for each system so that entries can be traced to a specific individual • Each new employee must be allocated a unique user identity that should never be reused • A policy covering complexity, strength, expiry and replacement of passwords must be devised and where possible enforced technically. The list must be reviewed regularly to ensure that it is current • Disable unused user accounts or before a user leaves the company
Access control	• Each system will have the user types and the access privileges for each one defined and documented • Larger systems will have user groups enabled to compartmentalise work performed, *e.g.* Analytical Development and QC or method development and routine analysis
Application administration	• IT will independently administer networked systems and control user accounts, access privileges and application configuration to prevent turning audit trails on or off or overwriting or deleting data • There may also be laboratory administrators responsible for implementing application workflows, calculations, custom reports and first line error resolution
Record format	• Defined by the computerised system • Records can be protected *via* checksums or other means to deter changes to a record • Libraries, *e.g.* spectra must be controlled and maintained under a procedure and/or change control
Signing records	• Use of unique user identities (see company requirements above) • Electronic signature components, *e.g.* user identities, password, tokens or PINs must be managed and checked regularly • A scanned handwritten signature is not an electronic signature • Electronic signatures must be securely liked to the electronic records that they pertain to
Creation and storage of a record	• When records are created they must be stored on media – manipulation of records in memory without saving is not allowed[20] but the system must support this • Records should be auto-numbered by the system using a user defined scheme to prevent test and prep informal testing • Records should be stored in a well-designed database rather than in directories where users could access them • Data storage locations should be on secure and resilient network storage devices • Technical controls, if available, should restrict where data can be stored
Prohibited actions	• Never share passwords with another user • Never overwrite entries • Never copy data to be used to falsify records • Never share accounts • Never time travel (typically on standalone systems) • Never delete data and repeat work • Never use samples to check to see if they pass a test

Table 7.7 (*continued*)

Correcting entries and audit trail	• Data corrections must only be made through the application • Changes are captured in a computer-generated audit trail that should identify the user who made the change, old and new values with date and time of the change • The audit trail must be secure and searchable • The reason for change should be given; this can be *via* context sensitive pre-defined reasons and/or free text entries
Verification of data entry	• There should be technical controls to verify, where possible, data entries, *e.g.* Context sensitive dropdown lists/pick lists of allowable entries Verification of the correct data format of manually entered data (note this does not ensure that the data are correct, only that they are in the correct format)
Backup and recovery	• There must be a regular backup performed on the records generated by a system • The extent of the backup (full, differential or incremental) depends on the criticality of the records generated by each system • Backup and recovery must be validated • Recovery must be tested regularly to ensure that the records can be recovered • Backup is the first part of a disaster recovery and/or business continuity plan(s)
Archive and restore	• If an archive of records is used, it must be possible to read the data throughout the record retention period • Archived records must be secure • Records must be indexed to allow ready retrieval • The archive must include all associated contextual metadata including all pertinent audit trail entries
System architecture	• System availability for users where they work to ensure activities are recorded contemporaneously • Criticality of the system will determine the hardware resilience required for operation and record storage • Secure time and date stamp linked to a trusted time source • Time zones implemented on applications where appropriate • Storage and protection of the records produced must be ensured • Standalone workstations lack resilience and will suffer keyboard contention • Hardware and software protection against malware and malicious attack must be in place

document from a regulatory perspective. This is an introductory discussion about laboratory log books, see Chapter 11 for a more detailed discourse on the subject.

7.11.1 EU GMP Chapter 4 on Documentation

In EU GMP Chapter 4 on documentation Clause 4.31 states[1]:

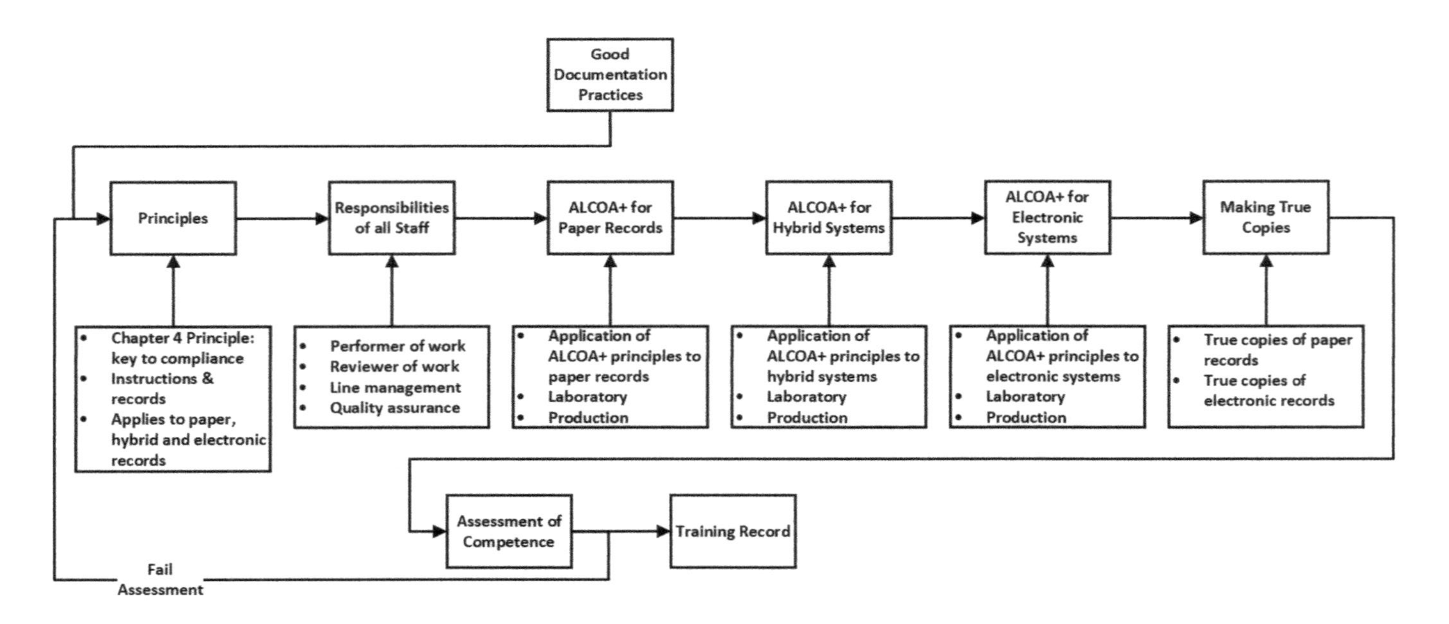

Figure 7.10 Outline of training in Good Documentation Practices.

Logbooks should be kept for major or critical analytical testing, production equipment, and areas where the product has been processed. They should be used to record in chronological order, as appropriate, any use of the area, equipment/method, calibrations, maintenance, cleaning or repair operations, including the dates and identity of people who carried these operations out.

7.11.2 FDA Good Laboratory Practice 21 CFR 58

In section 58.63 on the maintenance and calibration of equipment[12] there is the following requirement:

> (c) Written records shall be maintained of all inspection, maintenance, testing, calibrating and/or standardizing operations. These records, containing the date of the operation, shall describe whether the maintenance operations were routine and followed the written standard operating procedures. Written records shall be kept of nonroutine repairs performed on equipment as a result of failure and malfunction. Such records shall document the nature of the defect, how and when the defect was discovered, and any remedial action taken in response to the defect.

7.11.3 FDA 21 CFR 211 cGMP for Finished Pharmaceutical Products

The US GMP regulations have a requirement in §211.182 for an equipment cleaning and use log[17]:

> A written record of major equipment cleaning, maintenance (except routine maintenance such as lubrication and adjustments), and use shall be included in individual equipment logs that show the date, time, product, and lot number of each batch processed. .... The persons performing and double-checking the cleaning and maintenance (or, if the cleaning and maintenance is performed using automated equipment under 211.68, just the person verifying the cleaning and maintenance done by the automated equipment) shall date and sign or initial the log indicating that the work was performed. Entries in the log shall be in chronological order.

7.11.4 FDA Inspection of Pharmaceutical QC Laboratories

The FDA guidance for inspectors for the Inspection of Pharmaceutical QC Laboratories has an interesting approach that is very useful for data integrity approaches and audits[15]:

> Laboratory records and logs represent a vital source of information that allows a complete overview of the technical ability of the staff and of overall quality control procedures.

7.11.5 Instrument Lag Books in Practice

Typically, the majority of instrument log books are paper based using bound books with sequentially numbered pages to prevent pages being removed and replaced, requiring users to complete the entries manually, consistently and contemporaneously. The way the use of an instrument is recorded will depend on the interpretation of the regulations above by each individual organisation. As the log book is paper it is always difficult to know if an entry has been made contemporaneously.

As an auditor, one item that is usually overlooked is that each instrument log book in any analysis should be subject to second person review to ensure that the work is correctly documented.

When we move attention to laboratory computerised systems we return to the mediaeval age. We have the ridiculous situation where an amazing amount of money has spent on an analytical instrument and associated software application to control it. What about the instrument log book? Paper! The software has in the sequence file a record of the samples, the analysis method, *etc.* Why can we not have an electronic instrument log book that is generated automatically by the software? It would be protected and secure like an audit trail and would be able to be viewed as part of a second person review with a function to document specific entries have been reviewed.

7.12 Training for Generating, Interpreting and Reviewing Laboratory Data

The last level of data integrity training are those procedures for the actual work being carried out in a regulated laboratory, regardless if the work is research, development or quality control analysis. Please note that this discussion excludes the analytical procedures or methods.

7.12.1 Data Integrity Training for a Chromatography Data System: Operational SOPs

Whilst training in the corporate data integrity policy and good documentation practices provides a background to all individuals working in a GXP environment there also needs to be specific training for the systems and processes used within the regulated laboratory for use of a chromatography data system. The minimum operational SOPs for a CDS operating in a regulated environment to ensure data integrity are shown in Figure 7.11. These five SOPs will operate under the umbrella of the corporate data integrity policy, the general good documentation practice SOP and the data governance defined by the data owner and implemented and monitored by the data stewards of the system.

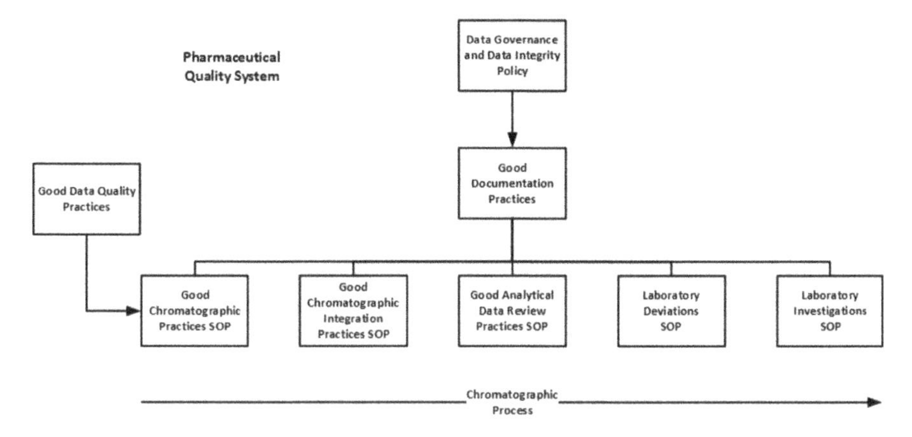

Figure 7.11 Data integrity SOPs for a chromatography data system.

- Good Chromatographic Practices SOP;
- Chromatographic Integration SOP (For other analytical techniques this would be replaced by say a procedure for interpretation of spectra, *etc.*);
- Analytical Data Review SOP;
- Laboratory Deviations SOP;
- Laboratory Investigations SOP.

These will be discussed in detail in Chapters 16 and 17 later in this book.

7.12.2 Training Is of Little Value without an Open Culture

Although we have spent much time presenting and discussing training for various aspects of GXP regulations and data integrity, it is of little value if organisational or managerial pressures subvert it. Therefore, we move to the next part within the Foundation layer of the Data integrity Model and discuss the establishment and maintenance of an open culture.

References

1. *EudraLex – Volume 4 Good Manufacturing Practice (GMP) Guidelines, Chapter 4 Documentation*, E. Commission, Brussels, 2011.
2. *WHO Technical Report Series No. 996 Annex 5 Guidance on Good Data and Records Management Practices*, World Health Organisation, Geneva, 2016.
3. *PIC/S PI-041 Draft Good Practices for Data Management and Integrity in Regulated GMP/GDP Environments*, Pharmaceutical Inspection Convention/Pharmaceutical Inspection Co-Operation Scheme, Geneva, 2016.
4. *EMA Questions and Answers: Good Manufacturing Practice: Data Integrity*, 2016, Available from: http://www.ema.europa.eu/ema/index.jsp?curl=pages/regulation/general/gmp_q_a.jsp&mid=WC0b01ac058006e06c#-section9.

5. *40 CFR 160 Good Laboratory Practice Standards*, Environmental Protection Agency, Washington, DC, 1997.
6. *Automated Laboratory Standards: Evaluation of the Standards and Procedures Used in Automated Clinical Laboratories, (Contract 68-W9-0037)*, Booz Allen Hamilton, Washington, DC, 1990.
7. *Good Automated Laboratory Practice Guidelines (GALP)*, Environmental Protection Agency, Washington, DC, 1995.
8. *FDA Warning Letter Apotex Research Pvt Ltd.*, 2015, Available from: http://www.fda.gov/ICECI/EnforcementActions/WarningLetters/ucm432709.htm.
9. *NELAC Quality Standard*, National Environmental Laboratory Accreditation Conference (NELAC), Weatherford, TX, 2003.
10. *ICH Q10 Pharmaceutical Quality Systems*, International Conference on Harmonisation, Geneva, 2008.
11. *EudraLex – Volume 4 Good Manufacturing Practice (GMP) Guidelines, Chapter 1 Pharmaceutical Quality System*, European Commission, Brussels, 2013.
12. *21 CFR 58 Good Laboratory Practice for Non-clinical Laboratory Studies*, Food and Drug Administration, Washington, DC, 1978.
13. 21 CFR Parts 16 and 58 Good Laboratory Practice for Nonclinical laboratory Studies; Proposed Rule, *Fed. Regist.*, 2016, **81**(164), 58342–58380.
14. *OECD Series on Principles of Good Laboratory Practice and Compliance Monitoring Number 1, OECD Principles on Good Laboratory Practice*, Organisation for Economic Co-Operation and Development, Paris, 1998.
15. *Inspection of Pharmaceutical Quality Control Laboratories*, Food and Drug Administration, Rockville, MD, 1993.
16. R. D. McDowall, *Validation of Chromatography Data Systems: Ensuring Data Integrity, Meeting Business and Regulatory Requirements Second Edition*, Royal Society of Chemistry, Cambridge, 2017.
17. *21 CFR 211 Current Good Manufacturing Practice for Finished Pharmaceutical Products*, Food and Drug Administration, Sliver Spring, MD, 2008.
18. *21 CFR 11 Electronic records; electronic signatures, final rule*, in *Title 21*, Food and Drug Administration, Washington, DC, 1997.
19. *EudraLex – Volume 4 Good Manufacturing Practice (GMP) Guidelines, Annex 11 Computerised Systems*, European Commission, Brussels, 2011.
20. *FDA Draft Guidance for Industry Data Integrity and Compliance with cGMP*, Silver Spring, MD, USA, 2016.

CHAPTER 8

Establishing and Maintaining an Open Culture for Data Integrity

Of all the components of a data integrity and data governance programme within any organisation, establishing and maintaining an open culture is the most difficult. First and foremost, this requires leadership from senior management to disseminate the open culture message throughout the organisation along with setting up the on-going communication and management effort to maintain it. The essential elements of an open culture are honesty, a willingness to admit mistakes and a no-blame approach by the organisation. However, in some regions of the world, an open culture may be difficult to achieve due to local customs and social behaviour, resulting in situations where it is difficult to admit mistakes or question management.

8.1 What Do the Regulators Want?

8.1.1 WHO Guidance on Good Data and Record Management Practices

There are several references to an open culture in the WHO guidance document:[1]

- Section 1.4: adoption of a quality culture within the company that encourages personnel to be transparent about failures so that management has an accurate understanding of risks and can then provide the necessary resources to achieve expectations and meet data quality

Data Integrity and Data Governance: Practical Implementation in Regulated Laboratories
By R. D. McDowall

Published by the Royal Society of Chemistry, www.rsc.org

standards: a reporting mechanism independent of management hierarchy should be provided for;

- Section 4.7 Quality culture. Management, with the support of the quality unit, should establish and maintain a working environment that minimizes the risk of non-compliant records and erroneous records and data.

  An essential element of the quality culture is the transparent and open reporting of deviations, errors, omissions and aberrant results at all levels of the organization, irrespective of hierarchy. Steps should be taken to prevent, and to detect and correct weaknesses in systems and procedures that may lead to data errors so as to continually improve the robustness of scientific decision-making within the organization.

  Senior management should actively discourage any management practices that might reasonably be expected to inhibit the active and complete reporting of such issues, for example, hierarchical constraints and blame cultures.
- Section 6.3 Management should create a work environment in which staff are encouraged to communicate failures and mistakes, including data reliability issues so that corrective and preventive actions can be taken and the quality of an organization's products and services enhanced. This includes ensuring adequate information flow between staff at all levels.

  Senior management should actively discourage any management practices that might reasonably be expected to inhibit the active and complete reporting of such issues, for example, hierarchical constraints and blame cultures.

8.1.2 PIC/S PI-041 Good Practices for Data Management and Integrity in Regulated GMP/GDP Environments

Section 6.3 of the PIC/S guidance is focused on quality culture in a regulated organisation:[2]

> 6.3.1 Management should aim to create a work environment (*i.e.* quality culture) that is transparent and open, one in which personnel are encouraged to freely communicate failures and mistakes, including potential data reliability issues so that corrective and preventative actions can be taken.
>
> Organisational reporting structure should permit the information flow between personnel at all levels.
>
> 6.3.2 It is the collection of values, beliefs, thinking, and behaviours demonstrated consistently by management, team leaders, quality personnel and all personnel that contribute to creating a quality culture to assure data integrity.

6.3.3 Management can foster quality culture:

- Ensure awareness and understanding of expectations (*e.g.* Code of Ethics and Code of Conduct);
- Lead by example, management should demonstrate the behaviours they expect to see;
- Ensure accountability for actions and decisions;
- Stay continuously and actively involved;
- Set realistic expectations, consider the limitations that place pressures on employees;
- Allocate resources to meet expectations;
- Implement fair and just consequences and rewards that promote good cultural attitudes towards ensuring data integrity; and
- Be aware of regulatory trends to apply lessons learned to the organisation.

8.1.3 MHRA "GXP" Data Integrity Guidance and Definitions

The GXP Data Integrity Guidance and Definitions published by the MHRA contains the following in Section 3.3:[3]

The impact of organisational culture, the behaviour is driven by performance indicators, objectives and senior management behaviour on the success of data governance measures should not be underestimated.

8.1.4 Regulatory Guidance Summary

It is clear from reading the selected excerpts from these three guidance documents that senior management bears the responsibility for ensuring an open culture (as well as the overall data integrity and data governance programme). It is important to realise that establishing and maintaining an open culture is a journey and not an event. The PIC/S guidance lists several elements that must be present to ensure a quality culture within an organisation. Management must approach the issue of an open culture in a positive and not sceptical frame of mind, back the initiative in private as well as in public and, most importantly of all, lead by example.

A quality culture consists of several elements:

- Management responsibility for the Pharmaceutical Quality System and the allocation of resources and budgets for data integrity. These responsibilities were discussed in Chapter 6 earlier in this book.
- Leadership and communication of the data integrity initiatives to all in the organisation including the changes required for an open culture.

- Management review of the PQS and the data integrity projects specifically.
- Open and honest approach to analysis.
- Willingness to admit mistakes.
- Most importantly, it is a set of beliefs and approaches to identifying problems, discussing them and resolving them openly that should be at the forefront of thinking when considering data quality and data integrity.

8.2 Bad Culture: Cressey's Fraud Triangle and Organisational Pressure

How can the culture of an organisation impact both data quality and data integrity? In this section we discuss are two of the areas that contribute to an organisation's culture for the worse: Cressey's fraud triangle and organisational pressure.

8.2.1 Cressey's Fraud Triangle

The fraud triangle was developed by Donald Cressey in the 1950's when investigating financial fraud and embezzlement in a publication entitled "Why Do Trusted Persons Commit Fraud?". The principles of Cressey's fraud triangle are directly relevant to the present-day data integrity issues in the pharmaceutical industry and are shown in Figure 8.1.

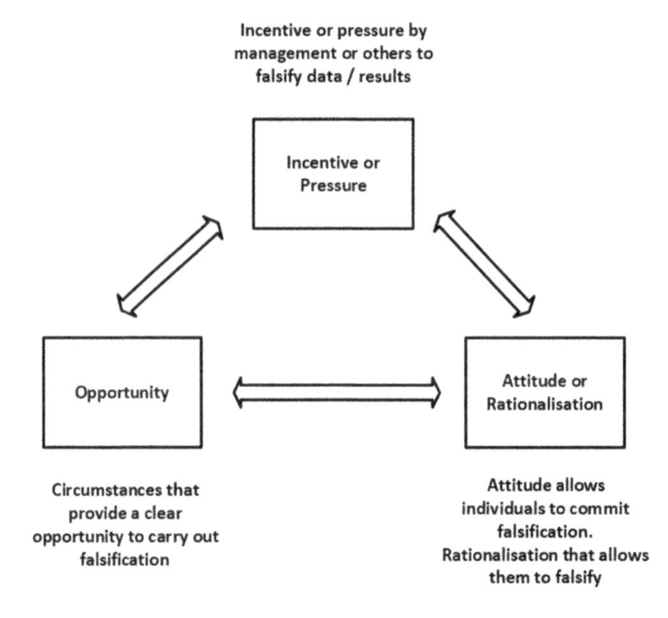

Figure 8.1 Cressey's fraud triangle.

The three elements of Cressey's fraud triangle are:

- Incentive or Pressure: why commit falsification of data? Managerial pressure or financial gain are the two main drivers here to push people to commit fraud. Setting unrealistic objectives such as stretch goals, turnaround time or key performance indicators (KPIs) that are totally divorced from reality especially when these are linked to pay or advancement will only encourage staff to falsify data to receive rewards. These goals coupled with poor analytical instruments and methods will only ensure that corners will be cut to meet deadlines or targets.
- Rationalisation or Incentive: to commit fraud people must either have an incentive or can rationalise that this is an acceptable practice within an organisation or department.
- Opportunity: the opportunity to falsify data can be due to encouragement by management as a means of keeping cost down or a combination of lax controls or poor oversight of activities that contribute to staff being able to commit fraud.

All three elements must be present for fraud or falsification to occur.

8.2.2 Breaking the Fraud Triangle

To ensure that the triangle is broken, some actions are essential within an organisation:

- Remove the Incentive or Pressure: Here management must lead by example. Not just through communication or establishing data governance structures but by ensuring the pressure to falsify data is removed, *e.g.* by setting realistic expectations that are compatible with the organisation's capacity and process capability. According to the MHRA, perhaps a better KPI than "Turn Around Time" would be "Released When Correct" as this removes organisational pressure.[4]
- Rationalisation: Staff working in a regulated laboratory need to understand how their actions can impact the health of the patient by releasing poor quality product can remove the element of rationalisation from the triangle. As well as having an understanding of the importance of reliable and accurate data to the wellbeing of the patient as well as the business health of the company.
- Opportunity: Remove the opportunity by implementing a process that is technically controlled so that there is little, if any, opportunity to commit falsification of data in a regulated laboratory.

When creating a culture that reduces the rationalisation of data manipulation, there should be a way for personnel to challenge well established processes or colleagues seen as "subject matter experts", and see positive outcomes from reporting problems, as we see in the next section.

8.2.3 Managerial and Peer Pressures Can Influence Analytical Results

Within a regulated laboratory there can be organisational pressure from staff where if a junior analyst carries out an analysis that is out of specification, more senior staff can view this as a case of inexperience in carrying out the analysis rather than a proper result that requires investigation. This peer pressure can force an analyst to ensure that analytical data meet expectations rather than real results. Such a scenario was published in the MHRA Inspectorate Blog[5] about a new analyst and this is reproduced below to illustrate the issues:

> A problematic result.......
> Jane has always tried to do the right thing. She has been alarmed recently by some of her senior colleagues making questionable decisions when unexpected laboratory results are encountered. One Friday afternoon, Jane obtains a low assay result (83%) during a routine product release test. She asks her senior colleagues if they have ever encountered something similar, but no instances had been reported. She is told that this product has been manufactured for years with no issues. It is suggested that she retest the samples and report the results before the start of the weekend so that the batch can be released by QA.
> Jane starts to second guess herself thinking "maybe I contaminated the sample" or "maybe the instrument experienced a disturbance like an air bubble". Jane realises that some of the instruments and analytical methods are problematic. As a new analyst, she is eager to impress her managers by making them look good in company performance metrics – the most products released in the shortest period of time. Jane has been relatively slow as a new employee, already far behind her colleagues, and is fearful of losing her position in the upcoming corporate restructure. She feels valued by her manager, but everyone has been under extreme pressure to perform since the restructure was announced two months ago. What should she do now?[5]

Here, we have an example of both organisational and managerial pressure through metrics that compromise both quality and integrity of data, coupled with peer pressure that contributes to the issue. Management at all levels in an organisation and a laboratory has a duty to train staff to do their jobs but also to ensure that concerns of front-line analytical staff are addressed and where appropriate action can and is seen to be taken. Any regulated organisation must ensure that data integrity and data quality are valued over efficiency and getting the job done at any cost. What management should have been done is to ensure that there was the culture in place to take the analyst's result and start a laboratory investigation rather than rubbish the result and test into compliance.

8.3 ISPE Cultural Excellence Report

ISPE has published an extensive Cultural Excellence Report in 2017[6] in which it proposes that there are six elements of cultural excellence within any regulated organisation:

- Leadership and Vision: As noted in Chapter 6 senior management must develop the strategies and approaches for data governance and ensuring data integrity within the PQS of the organisation. Communication of the vision for data integrity to all staff through formal and informal means is the beginning of a quality culture and the means of reinforcing it.
- Mindset and Attitudes: These are the key to a quality culture that must be developed and fostered but this requires the leadership of senior management, engagement of line management to inform all staff of what is required. This is winning hearts and minds but must also be followed up by more formal means, *e.g.* incorporation of data integrity objectives in position descriptions and personal objectives.
- Gemba Walk: Managers must get out of their offices and walk around laboratories on a regular basis, not as a one-off exercise. The purpose of the Gemba walk is two-fold. The first is for management to listen to laboratory staff and understand issues and suggestions for improvement first hand without filtering. The second is for management to communicate aims of the data integrity programme.
- Quality Metrics to Measure and Monitor. Develop metrics that monitor the way the organisation works. Care needs to be taken that the metrics do not influence behaviour to the detriment of data integrity.[7] This topic is the subject of Chapter 19 and will not be discussed further here.
- Oversight and Reporting: Management oversight is a part of management review of the Quality System. These practices should involve management and all staff in a quality culture where all issues can be raised openly and transparently to ensure improvement in working practices that ensure quality and integrity.
- Cultural Enablers for Learning and Improvement: To reinforce the process and improve working and data integrity change to a learning organisation, solve problems proactively before they impact integrity, quality, compliance and identify true root causes.[6]

In addition to a discussion of these topics, there are questionnaires and assessment tools in the report that can be used to gauge where an organisation is currently with respect to an open quality culture.[6] In the author's opinion, this publication represents the goal to aim for and needs to be adapted to an individual organisation, however, it is a worthwhile read.

8.4 Management Leadership

This is the most important part of generating and maintaining the open culture of an organisation because if senior management is not engaged in developing and leading the debate, then any further data integrity work is a waste of resources. Similarly, communication of the open culture and the objectives can never be accomplished in a single fire and forget e-mail to all staff.

8.4.1 Generate and Communicate the Data Integrity Vision

The senior management team and the Executive Sponsor need to understand both data integrity and data governance and then set up the organisational structures, policies, procedures, projects and training for these. Once developed, there need to be a means of communicating the vision for quality and integrity to the whole organisation. This can be *via*:

- formal presentations including video recordings for staff unable to attend training sessions in person;
- written communications, although care needs to be taken to ensure that these are read and not deleted as e-mail spam;
- face to face discussions with staff, *e.g.* as part of a Gemba walk.

Reiterating, it is vitally important for management to realise that data integrity is a journey and not an event. Communication needs to be an on-going process, as regulations and guidance change or the data integrity projects come to fruition, then the message will probably change. However, once started you cannot stop.

8.4.2 Talk the Talk and Walk the Walk

There is little point in senior management developing the vision and communication for data integrity for an organisation but then acting differently. For example, in one organisation where I have trained the staff in data integrity, senior management and QA staff would regularly backdate document approvals. With this cavalier approach, how can management ever convince staff to act ethically? Therefore, senior management must talk the talk AND walk the walk, because if they do not act in a way that is consistent with the things they say, they will defeat the whole aim of any data integrity programme.

However, this goes further than just the senior management team, the attitude and belief in the data integrity programme must go right down the line – directors, managers, supervisors must also talk the talk as well as walk the walk. This should be carefully monitored *via* personal objectives and performance appraisal of line managers. For example, if a manager does not agree with the programme, then their attitude can impact all staff below them. The higher up in the organisation the individual is, the great the number of staff that will be influenced adversely.

8.4.3 Reinforcing an Open Culture for Data Integrity

An element of a data governance programme is that senior management must set an expectation of a no-blame culture and communicate it to all staff. However, the approach also needs to be reinforced and this can be achieved in several ways such as:

- Ensure that compliance with the data integrity policy is written into all job descriptions.
- All staff need to have data integrity written into their personal objectives.
- Performance assessment must include an element of how has this individual contributed towards the data integrity programme?
- Formal communication by senior managers for the need to ensure data integrity.
- Informal communication by senior managers as well as line managers to remind staff of their responsibilities.

The objective of these strands is to ensure that staff can admit mistakes freely and without any comeback. Rather than go to the end of the analysis, the analyst should stop work and inform their supervisor then document the issue and any resolution.

8.4.4 FDA Expectations for Analysts

One of the outcomes of the Barr Laboratories count case was the publication in 2006 of the FDA's guidance for industry on Investigating Out of Specification Results.[8] In the section on the responsibilities of the analyst, there is the following paragraph:

> *If errors are obvious*, such as the spilling of a sample solution or the incomplete transfer of a sample composite, *the analyst should immediately document what happened.*
> Analysts should not knowingly continue an analysis they expect to invalidate at a later time for an assignable cause (*i.e.*, analyses should not be completed for the sole purpose of seeing what results can be obtained when obvious errors are known).

The italic text above is my emphasis to highlight the fact that owning up to a mistake is a regulatory expectation and failure not to do this will not be treated leniently.

8.5 Mind Set and Attitudes

Once senior management has got their act together, the next major hurdle is to win hearts and minds of the staff. Under the umbrella of positive senior management engagement, the changes in mind set and attitudes of the staff can begin. Depending on the current organisational culture this may be relatively simple or the start of a long journey (in many organisations it will probably be the latter).

8.5.1 Quality Does Not Own Quality Anymore

The first item is to let staff know that they are both responsible and accountable for performing or reviewing work correctly throughout an organisation and the laboratories. It is not the job of the Quality Assurance department to identify and fix your mistakes. In any regulated laboratory, everyone is responsible and accountable for:

- the quality of their work;
- following written procedures;
- documenting work accurately, completely and honestly;
- ensuring that any errors or deviations are documented, investigated where appropriate and corrected as soon as possible.

For example, if you are preparing an analytical reference standard solution and when making up the meniscus in the volumetric flask you overfill it. What do you do? Nothing as it is only a small error? Remember, a reference solution could be stable for some time and could have a detrimental impact on many analyses. You should document what you have done and state the reason why you cannot use the solution. Then reweigh the standard and do make up to volume correctly. Keep all the records as part of the complete records of the solution. This is also a requirement in the FDA OOS guidance discussed in Section 8.4.4.[8]

8.5.2 The Iceberg of Ignorance

There is the apocryphal statistic that senior management only knows 4% of the problems in an organisation. Middle management does little better at 9% and supervisors know 74% of the problems. This means that management is at the top of an iceberg with the majority of problems hidden from their view. Thus, there is a need for managers to become more proactive and aware of issues and resolve them before they impact the work of the laboratory or the regulatory status of the organisation.

8.5.3 How Do I Raise Problems to Management?

Staff should also be empowered that when they find a problem they can discuss it with management without any resentment or bad temper. Ideally, if there is a small issue a single person could generate a potential resolution. However, given the complexity of laboratory operations and informatics solutions, it may be that finding a problem is the work of one person but the solution to it may require a team. The open culture within an organisation needs to encourage and empower staff to identify issues but this is only the start:

- Speak Up. Analytical scientists should be empowered to identify problems for resolution or make suggestions for improved quality with management without the possibility of management rejecting or labelling a person as a troublemaker or complainer. They must not hide problems but be honest if they make a mistake and own up without fear of retaliation.
- Management Approachability. In allowing staff to speak up, management must reciprocate to ensure that if a problem is raised it can be discussed openly and in some detail. Managers must ensure that they are approachable and must not react badly if given bad news, instead they must empathise with the person raising the problem to elicit more information. Reacting badly will result in this being the last problem being raised and missing current and future opportunities for improvement. Widening the discussion by talking with others can often result in discussions that look at problems differently and obtain a better outcome. An attitude of don't bring me problems or bad news will inhibit the discussion of them and keep them hidden and unresolved. Who would you like to discover these issues? The laboratory staff or an inspector? Approachability in combination with Gemba walks and fostering an environment that encourages raising problems and suggestions will help develop and maintain a quality culture.
- Understand the Problem. When presented with a problem, managers and staff must attempt to understand it further by asking questions. One way of understanding an issue is to generate problem statements based on facts to separate a valid concern from a complaint. From the problem statement, it should be possible to identify causes and factors that cause the issue. For example, if a turnaround time for a specific analysis has been missed consistently, is this due to an unreliable instrument or method, are there enough resources or is there poor communication from the group generating the samples? Root cause analysis using the 5 whys could be used here to identify root causes of a data integrity problem.
- Who can Resolve the Issue? Once the problem is understood, you can now consider who is best able to resolve it; this might be one individual or a team, depending on the complexity. As the pharmaceutical industry is regulated, it may require a change request or deviation to be raised or will the resolution require informing Regulatory Authorities *via* the Regulatory Affairs Department, *e.g.* would a modification to a registered analytical procedure trigger marketing authorisation change?

A different approach is found with the NELAC accreditation standard[9] that requires management to put in place a separate reporting mechanism for raising data integrity issues, usually *via* the Quality Assurance department. This is one mechanism for raising a data integrity concern that is discussed in Chapter 20. However, if the culture is right many data integrity issues can be raised directly with management on a Gemba walk.

8.6 Gemba Walks

A Gemba walk is the term used to describe the personal observation of work by management. It is not just management by walking about but engaging and discussing with analysts actually doing the work in the laboratory and not a meeting room. The origin of the term comes from the Japanese word gembutsu, meaning *real thing or place*.

Gemba walks, enable management to get out of their offices and see life in the laboratory with the aim of understanding what actually happens, meeting, talking and hearing about problems first hand rather than being filtered through the organisational structure. A function of a Gemba walk is to reduce the iceberg of ignorance, as discussed in Section 8.5.2. Problems become visible as they are discussed face to face with management. Management sees things for themselves and there is no second-hand information or selective filtering, allowing management to make decisions based on first-hand information.

8.6.1 Where Does a Gemba Walk Fit in a QMS?

Part of the continual improvement is a process called plan, do, check and act, as shown in Figure 8.2, the Gemba walk is part of the third stage under check. Checks are there to ensure that procedures are followed, deviations are recorded then investigated and CAPAs generated to resolve the issue.

The aim is a health check of laboratory processes by management engaging analysts by:

- visiting a laboratory;
- look at the analytical process;
- talk with the analysts.

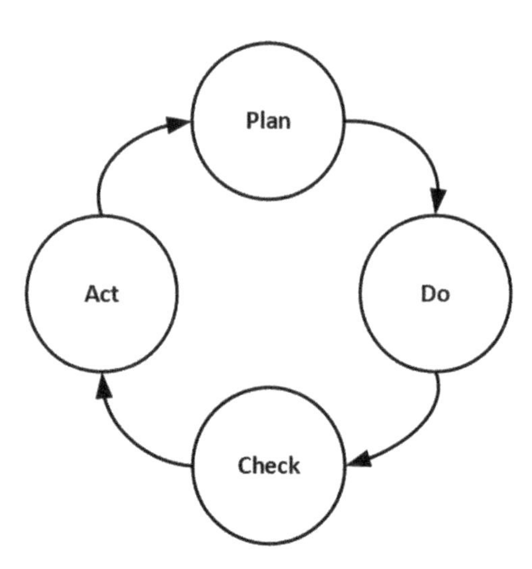

Figure 8.2 Plan, do, check and act for continual improvement.

Understand the processes first-hand and talk with the analytical scientists performing the work. It is an opportunity for staff to show how their process works, the successes and the issues.

8.6.2 What Gemba Walks Are and Are Not

According to the ISPE Cultural Excellence report,[6] a Gemba walk is:

- an enabler for cultural change in management style and philosophy;
- a role-modelling opportunity for leaders;
- a way to empower laboratory staff;
- an enabler for continual improvement through problem solving in the laboratory with the analysts who experience them;
- an opportunity to find the root cause of issues, spot quality risks, and request managers to remove obstacles to improvement;
- a coaching/mentoring opportunity to build and/or enhance capabilities and behaviours and recognize and reinforce desired behaviours;
- an enabler for communication of site priorities/challenges and how the unit's performance contributes to the overall success of the site;
- an opportunity for the laboratory staff to show their pride and excellence in their jobs.

A Gemba walk also provides a means for management to praise staff for the good work that they do. One of the biggest complaints employees have is they feel that their work is not appreciated or recognised by management. This is an opportunity to correct this complaint.

In contrast, Gemba walks are not:[6]

- audits;
- complaint sessions;
- a debating society on different views;
- trouble shooting exercises.

It can be conducted on a frequent basis by laboratory management or on a longer scheduled basis by senior management. Some of the advantages of a Gemba walk are:

- Management can be sure the work that needs to be done is getting done.
- If not, goals and objectives can be clearly communicated to staff face-to-face.
- Increase the engagement of the analysts in the laboratory in the data integrity programme.

Gemba walks should replace, or at least substantially reduce, traditional conference-style meetings and hence minimise the production of the many charts and reports created, just for such meetings, to communicate progress related to laboratory activities. Because Gemba walks facilitate stand-up style meetings in the laboratory, they tend to be much shorter and more efficient than the typical meeting room presentations. Furthermore, decisions are often made more quickly because all participants have all the necessary information right in front of them.[6]

8.6.3 Why Bother with a Gemba Walk?

There are several reasons why Gemba walks are important for senior management. If senior management is present it is a non-verbal message that this is important for the organisation and the laboratory. It should not be seen by staff as a threat. Some of the advantages of a Gemba walk are:

- Building relationships between the analysts who do the work and senior managers.
- Coaching analysts and helping them to do their job better is essential for effective teamwork.
- Regularly talking with laboratory staff gives management the opportunity to find out about any problems faced, and to ensure that the process and systems problems are addressed appropriately.
- Preventing problems from developing into critical issues later as management attention is given earlier.
- Fostering the key role of management as leaders by coaching analytical staff through questioning.

Done well, it not only results in leaders teaching the system, but it establishes norms of how leaders are expected to behave and spend their time, and it helps employees understand what is important at both a technical and management process level. More importantly, it provides managers with direct experience of what is happening in the organisation.

8.6.4 Activation Energy for a Gemba Walk

Any chemical reaction needs an activation energy for the reaction to commence, this is always more than the energy required to maintain the reaction. Like a chemical reaction, there needs to be a greater effort at the start of a Gemba walk programme to overcome resistance, suspicion and scepticism of the laboratory analysts (and there are some analytical chemists that take Sir Robert Boyle's approach to the limit). When a member of senior management walks into a laboratory area there is immediate suspicion: what do they want? what is wrong? It can be seen as a threat rather than its true purpose of finding out how things work and how to encourage and coach people in good data integrity practices.

From personal experience, I invited a senior manager to visit our laboratories on a regular basis. My objective was less about data integrity and more about ensuring that if I was explaining an issue, he had first-hand information about it and could understand the problem. He was sceptical at first and so were the analytical staff when he first appeared in a laboratory coat. They thought he had taken the wrong turn on the way to a fancy-dress party. However, after the third and fourth times he had visited the laboratories and spoken to the staff, both had a good understanding of what was required and the analysts appreciated him being in the laboratories.

The data integrity communication programme should give advance notification of Gemba walks such as:

- what they are;
- objectives and purpose;
- who will be involved;
- when they will occur.

Regardless of all communication and advance notification, the first two or three walks will be treated carefully by the laboratory staff. We do not do this round here will be a common theme in talks amongst analysts. It is important that the communication around and more importantly the outcomes of the first one or two walks show that management is taking the process seriously, listening to analyst's concerns and acting upon them, this is why changing the culture is a journey and not an event.

8.6.5 Performing the Gemba Walk

Laboratory staff is typically very concerned and very suspicious when a senior manager rolls up and starts asking questions. One of the key issues will be how questions are asked by a senior manager and how the question is perceived by the laboratory analysts. Could the answer they give be seen as a threat to them or the laboratory? This is in marked contrast to the intention of the manager of being supportive and trying to elicit information about the process and working practices.

- Senior managers are not there to tell analysts how to do their job. In contrast, a manager should be asking questions to understand what instruments and systems they need to do their job.
- Managers should challenge analysts to think about a problem. In a Gemba walk, the role of a manager should be as a leader and an enabler for problem solving the problems in an analytical process and to engage analysts to start thinking in the same way.
- The manager in these twin roles leader and enabler asks guiding and probing questions that help process owners and teams to identify their own problems and solutions by asking what, why, what if, and why not, in that order.

When organising a Gemba walk, especially if by a senior manager, let the facility and department know in advance that a Gemba walk will be carried out so that they can be prepared. The manager must turn up on time and allocate sufficient time to have a good discussion. Some of the approaches that can be used are:

- An introduction to all staff and setting the scene for the walk.
- Set the expectation that staff input is required as well as managerial feedback after the activity.
- Be practical: have a reasonable group to discuss and ensure that the laboratory manager (*e.g.* the process owner) accompanies the walk.
- The process should be to ask open questions (*e.g.* what, why, when, how, *etc.*) and importantly listen to the answers and follow up with appropriate questions where necessary.
- Honesty is essential in this process and notes should be taken to ensure that nothing is missed or misinterpreted.
- The feedback mechanism to the attendees of the Gemba walk is a simple way to check that management has understood the issues but also reiterated to them that management cares about their concerns.

8.6.6 Keep the Focus on the Process

When conducting a Gemba walk it is important that everybody involved follow an established protocol, and are aware of the process and the purpose of the walk, as presented in Section 8.6.2. The original aim of Gemba from Toyota was to eliminate non-value added work and to focus on creating value for the end customer. However, in the context of the analytical laboratory, the focus must be on data integrity, data quality, business efficiency to align with the data integrity programme:

- Understand the processes carried out the laboratory.
- Managers are not there to provide solutions to problems they find but to coach analytical staff to take ownership and develop solutions to eliminate data integrity problems.

8.6.7 Generic Questions for a Gemba Walk

The questions presented in this section are, of necessity, generic and need to be adapted to an individual organisation and its culture.

- What is the aim of the analytical process with respect to data integrity?
- What is the current situation?
- What are the obstacles preventing you from reaching the aim?
- Which of these obstacles is being addressed now?
- What are you doing now?
- What procedures describe this process?
- How complex is it?
- How automated is the process?

- What are the problems with the process?
- What can you fix?
- What can you not fix?
- What happens when something goes wrong?

8.6.8 Let Management See Analytical Instruments First Hand

In many laboratories, especially those in generic, CMO and API manufacturers or in remote laboratories of large ethical pharmaceutical companies, there are many older analytical instruments that are operated with or without software that is unsupported. In fact, many of these instruments should be replaced or be exhibited in a museum due to a lack of investment over time. One of the main issues in data integrity is convincing management to spend the money to update or replace existing instrumentation as they suffer serious data integrity issues and procedural controls are only an error-prone temporary solution. As input into a Gemba walk, analytical staff should discuss some of the most critical laboratory systems with data integrity issues. The input to management should be for long-term solutions that will deliver not only data integrity compliance but also business benefits for the investment, as will be presented in more detail in Chapter 13.

8.7 Fat Finger, Falsification and Fraud – Take 2

Whilst various regulatory guidance documents state the requirement for an individual to admit mistakes and for the organisation to be open, there is little written about prevention of these mistakes in the first instance. Correction of errors made on paper records is relatively easy to identify and corrected. However, data entered into a computerised system is often difficult to spot and sometimes difficult to correct. Here, we consider some ways to prevent errors being introduced into computer records in the first place. This has two benefits: it saves some time in second person review and saves time correcting the error.

8.7.1 To Err Is Human

You can have analysts who have an appropriate level of education, training and experience and who follow laboratory SOPs and analytical procedures. In addition, the individual's organisation has also trained them in the ethical and professional standards it expects of its staff at their induction and *via* regular training sessions thereafter. Yet, despite all of this, people will still make mistakes.

Mistakes or fat finger moments? If we are honest, we all make them. That is why any GXP quality system has the four eyes principle: one individual to perform the work and a second one to review the data produced to see that the procedure was carried out correctly and that there are no typographical errors or mistakes with calculations. Many errors and mistakes we make are self-corrected. For example, as you enter a number into a spreadsheet cell or an application field, often you will know that while the brain says enter a pH value of "12.3" your fingers magically enter "13.2." This is a fat finger moment, but before committing the number to the cell or database you can correct this as you can see and have realized your error. The equivalent moment on paper is when you write the wrong numbers down in your laboratory notebook and then correct them by striking through the original entry so as not to obscure it and then entering the correct value along with your initials, the date, and possibly the reason for the change. This is the paper version of an audit trail.

8.7.2 Verification of Data Entry

Some other mistakes that are not noticed by an analyst can be detected by the software application you are using such as a spell checker or by verification of the field. Some examples of verification are:

- data entered fail to meet the pre-defined format;
- outside of a predefined range;
- data are within a predefined range;
- data meet a predefined format, e.g. XX.YY.

With our example above, if the data verification range was 11.0–13.0 the software would have picked up the problem even if you had not. However, that still leaves the mistakes you do not realise you have made. For example, if the entry in the case above was 11.3, data verification would be useless and the error would have been entered without you or the software realising that there was a problem.

Do not assume that you will spot all of your mistakes—they are human mistakes, which is why we need the second pair of eyes to check the analytical data and calculated results. From my experience as a laboratory manager and an auditor, supervisors know which members of their staff are diligent about their work and how well they check it and which members are slapdash, and the supervisor will adjust the review accordingly. So, if you do not want a dubious reputation to precede you, be diligent and try your best to find and correct your own errors before passing your work to be checked.

8.7.3 What Is the Fat Finger Rate in a Laboratory?

To a certain extent, this section is about airing dirty laundry, which may not be a particularly interesting problem to all but is the heart of any good quality management system: empowering staff and self-audits coupled with effective corrective and preventative action planning. Quality is everybody's problem, and it is not the sole responsibility of the quality assurance group to pick up the errors that the analytical laboratory has made.

However, finding published papers on how often we make mistakes in an analytical laboratory is difficult—probably because we do not really want to go there. This, however, is the wrong approach to take and we should encourage studies to investigate this.

8.7.4 Learning from Health Service Studies

Luckily, help is at hand from clinical chemists working in hospitals who have published many studies on error rates in laboratories. For those that do not know, clinical chemistry is involved in the analysis of blood, urine, and other bodily outputs to help the diagnosis and management of diseases. Mistakes in this area can have a critical impact on the health of a patient and therefore the reduction in errors is essential.

One paper, entitled "The Blunder Rate in Clinical Chemistry," measured the rate of detected analytical errors before and after the introduction of a laboratory information management system (LIMS) and found that they were reduced from about 5% to less than 0.3% following the introduction of the computer system.[10]

Manual transcription errors in patient records were assessed for blood results recorded in a critical care setting by comparing the handwritten and printed laboratory results in 100 consecutive patients in the intensive care unit of a UK hospital. Out of 4664 individual values, 67.6% were complete and accurate, 23.6% were not transcribed at all, and 8.8% were inaccurate transcriptions of the results. Interestingly transcription was significantly more accurate in the morning ($p = 0.02$). This study highlights that the current system of recording blood results was unreliable. These results strengthen the case for computerisation of the patient record in terms of data retrieval and transcription accuracy.[11] The same is true of a manual based process in an analytical laboratory.

An Australian study of transcribing and inputting hand-written pathology request forms into a computer system and chemical analysis of the samples found that error rates were both in the 1–3% range in the best laboratories. The worst laboratories, however, had error rates of up to 39% in transcription and 26% in analytical results.[12]

Let us extrapolate from the clinical chemistry laboratory and suggest that a range of error rates in an analytical laboratory is in the range of 0.3–3% depending on the degree of automation you have. The more manual input there is the more transcription error checking is required and the greater the number of errors that need to be detected and captured. Therefore, laboratory errors are expected by external quality audits and regulatory inspections. Not finding these detectable errors raises suspicion of problems with the analytical process and this would result in delving further into laboratory records.

8.8 Maintaining the Open Culture

Similar to a chemical reaction, the activation energy to reach an open culture is large but maintaining it, if managed correctly, is much smaller. Maintenance of an open culture requires active management input as this is not a fix and forget situation. Active management leadership and involvement is essential to maintain the open culture using the same approaches as outlined in this chapter.

It is also important to understand that a minor slip or comment on the perception of management's commitment to an open culture can damage it, sometimes irreparably. Verbal comments, often off-the-cuff can alter staff perception in an instant. It is important for senior management to think before they talk. This is the reason that establishing and maintaining an open culture is one of the most difficult parts of a data integrity and data governance programme.

References

1. *WHO Technical Report Series No. 996 Annex 5 Guidance on Good Data and Records Management Practices*, World Health Organisation, Geneva, 2016.
2. *PIC/S PI-041 Draft Good Practices for Data Management and Integrity in Regulated GMP/GDP Environments*, Pharmaceutical Inspection Convention/Pharmaceutical Inspection Co-Operation Scheme, Geneva, 2016.
3. *MHRA GXP Data Integrity Guidance and Definitions*, Medicines and Healthcare products Regulatory Agency, London, 2018.
4. D. Churchward, *MHRA Inspectorate Blog: Too Much Pressure: A Behavioural Approach to Data Integrity (Part 1)*, 2017, Available from: https://mhrainspectorate.blog.gov.uk/2017/03/10/too-much-pressure-a-behavioural-approach-to-data-integrity-part-1/.
5. D. Churchward, *MHRA Inspectorate Blog: Too Much Pressure: A Behavioural Approach to Data Integrity (Part 2)*, 2017, Available from: https://mhrainspectorate.blog.gov.uk/2017/03/30/too-much-pressure-a-behavioural-approach-to-data-integrity-part-2/.
6. *ISPE Cultural Excellence Report*, International Society of Pharmaceutical Engineering, Tampa, FL, 2017.
7. *PIC/S Recommendations on Validation Master Plan, Installation and Operational Qualification, Non-Sterlie Process Validation and Cleaning Validation (PI-006–3)*, Phamaceutical Inspection Convention/Pharmaceutical Inspection Co-operation Scheme (PIC/S), Geneva, 2007.
8. *FDA Guidance for Industry Out of Specification Results*, Food and Drug Administration, Rockville, MD, 2006.
9. NELAC Quality Standard, *National Environmental Laboratory Accreditation Conference (NELAC)*, Weatherford, TX, 2003.

10. A. M. Chambers, J. Elder and D. St. J. O'Reilly, The blunder rate in clinical chemistry, *Ann. Clin. Biochem.*, 1986, **23**, 470–473.
11. R. Black, P. Woolman and J. Kinsella, Variation in the transcription of laboratory data in an intensive care unit, *Anaesthesia*, 2004, **59**(8), 767–769.
12. M. Khoury, L. Burnett and M. A. Mackay, Error rates in Australian chemical pathology laboratories, *Med. J. Aust.*, 1996, **165**, 128–130.

CHAPTER 9

An Analytical Data Life Cycle

Many of the regulatory guidance documents mention a data life cycle from initial generation through processing to reporting, retention and finishing with the destruction of the records. These discussions involve outline definitions or generic life cycles that are not specific to a regulated laboratory and can be rather vague. In this chapter, a flexible analytical data life cycle that is sample centric is presented and discussed. When considering an analytical data life cycle, the concept also includes raw data and complete data that were discussed in Chapters 3 and 7 and it is on the basis of understanding of the meaning of these two terms that the discussion of an analytical data life cycle will be presented.

The aim of this chapter is to ensure that a complete data life cycle from an analytical perspective is presented to ensure that all data are captured and managed correctly. It also allows a laboratory to identify the areas that have a greater risk where controls and data audits should place more emphasis to ensure that data integrity is ensured. Understanding the phases of an analytical data life cycle allows an effective assessment and remediation of processes and systems that will be discussed in Chapter 10.

9.1 What Do the Regulators Want?

9.1.1 MHRA GXP Data Integrity Guidance

The definition of data life cycle from the MHRA GXP Guidance for Industry[1] is:

> All phases in the life of the data from generation and recording through processing (including analysis, transformation or migration), use, data retention, archive/retrieval and destruction.

Data Integrity and Data Governance: Practical Implementation in Regulated Laboratories
By R. D. McDowall

Published by the Royal Society of Chemistry, www.rsc.org

The guidance then explains that data governance, presented earlier in Chapters 4–8 inclusive, must be applied across the whole data lifecycle to provide assurance of data integrity including data ownership.

9.1.2 WHO Guidance on Good Data and Record Management Practices

Section 11 of the WHO Guidance[2]:

> 11.1 Data processes should be designed to adequately mitigate and control and continuously review the data integrity risks associated with the steps of acquiring, processing, reviewing and reporting data, as well as the physical flow of the data and associated metadata during this process through storage and retrieval.
>
> 11.2 QRM of the data life cycle requires understanding the science and technology of the data process and their inherent limitations. Good data process design, based upon process understanding and the application of sound scientific principles, including QRM, would be expected to increase the assurance of data integrity and to result in an effective and efficient business process.
>
> 11.3 Data integrity risks are likely to occur and to be highest when data processes or specific data process steps are inconsistent, subjective, open to bias, unsecured, unnecessarily complex or redundant, duplicated, undefined, not well understood, hybrid, based upon unproven assumptions and/or do not adhere to Good Document and Records Practice.

The simple message is that you need to know your analytical processes, the risks associated with them and mitigation of them short-term and ideally plans for long-term solutions. This means that the data life cycle must be known in some detail, as we shall discuss in Section 9.2 and apply this knowledge in Chapter 10.

9.1.3 PIC/S PI-041 Good Practices for Data Management and Integrity in Regulated GMP/GDP Environments

The PIC/S guidance has several clauses on the data life cycle[3]:

> 5.3.2 The effort and resource assigned to data governance should be commensurate with the risk to product quality, and should also be balanced with other quality resource demands. Manufacturers and analytical laboratories should design and operate a system which provides an acceptable state of control based on the data integrity risk, and which is fully documented with supporting rationale.

5.3.3 Where long term measures are identified in order to achieve the desired state of control, interim measures should be implemented to mitigate risk and should be monitored for effectiveness. Where interim measures or risk prioritisation are required, residual data integrity risk should be communicated to senior management, and kept under review. Reverting from automated/computerised to paper-based systems will not remove the need for data governance. Such retrograde approaches are likely to increase administrative burden and data risk and prevent the continuous improvement initiatives referred to in paragraph 3.5.

5.3.4 Not all data or processing steps have the same importance to product quality and patient safety. Risk management should be utilised to determine the importance of each data/processing step. An effective risk management approach to data governance will consider:

- Data criticality (impact to decision making and product quality) and
- Data risk (opportunity for data alteration and deletion, and the likelihood of detection/visibility of changes by the manufacturer's routine review processes).

From this information, risk proportionate control measures can be implemented.

5.4.1 Data may vary in importance and impact depending on its intended use. For example, the decision that data influences may differ in importance, and the impact of the data on a decision may also vary. Points to consider regarding data criticality include:

- Which decision does the data influence?

For example: when making a batch release decision, data which determines compliance with critical quality attributes is normally of greater importance than warehouse cleaning records.

- What is the impact of the data on product quality or safety?

For example: for an oral tablet, active substance assay data is of generally greater impact on product quality and safety than tablet friability data.

9.1.4 Regulatory Requirements Summary

From selections quoted above out of the MHRA, WHO and PIC/S guidance documents[1–3] we can summarise:

- There needs to be a data life cycle defined and the records generated in a process identified, however, the guidance documents vaguely mention generation to destruction with no interpretation for a laboratory.

- The risk associated with each process and the vulnerability of the records generated and processed must be known. It is my contention that there needs to be an analytical data life cycle that is flexible to accommodate several different analytical procedures and techniques.
- Records have different importance, impact and criticality to product quality or data integrity, *e.g.* appearance or colour is not as critical as API purity or product stability.
- Risk mitigation measures need to be in place to protect the records and to ensure the integrity of them.
- Risk mitigation must be addressed in both the short term and long term. For example, if a process is to be fully automated to eliminate risk and the solution will take 12 months to specify, implement and validate. It is not acceptable to do nothing with the existing process as short-term remediation needs to be put in place now to close any data integrity and compliance gaps.

9.2 Published Data Life Cycles

The regulatory guidance documents above expect a data life cycle but they do not give any clue about what one should be. There are two published data life cycles one in the GAMP Guide Records and Data Integrity[4] and one published by myself in the second edition of Validation of Chromatography Data Systems.[5]

9.2.1 GAMP Guide on Records and Data Integrity

To expand the regulatory guidance definitions, the GAMP Guide on Records and Data Integrity[4] has in Section 4 a generic life cycle consisting of five phases as follows:

- Creation
  Acquisition of data.
- Processing
  Transforming and interpreting the data and calculation of the reportable result.
- Review, Reporting and Use
  Consisting of data review, audit trail review, data reporting, data distribution. The recipients of the report then use the information to make decisions either in drug development or manufacturing.
- Retention and Retrieval
  Covering the availability of the records, the security to ensure that unauthorised changes are not made, backup and recovery, physical storage and security and archiving as needed.
- Destruction

The formal process of destroying the records at the end of the record retention period.

A generic model is good in that it provides a simple basis for understanding a data life cycle but it must be adapted to a specific situation, *e.g.* analysis of samples from toxicological/clinical studies or analytical development/quality control product analysis to understand and control fully all laboratory processes. However, this model is too simplistic to apply to all situations. We need to think differently as we need to modify and adapt this model to an analytical centric data life cycle that accommodates both sample and protocol based assays.

9.2.2 Validation of Chromatography Data Systems

An alternative data life cycle is shown in Figure 9.1 and was published in the second edition of the author's book on Validation of Chromatography Data Systems.[5] One of the important concepts of this alternative data life cycle is that of phases. The life cycle has two phases – active and inactive. The active phase is where most of the action occurs from acquisition to data use and short-term retention but this is the shortest part of the life cycle. The inactive portion is where the data and records are stored for the remainder of the record retention period.

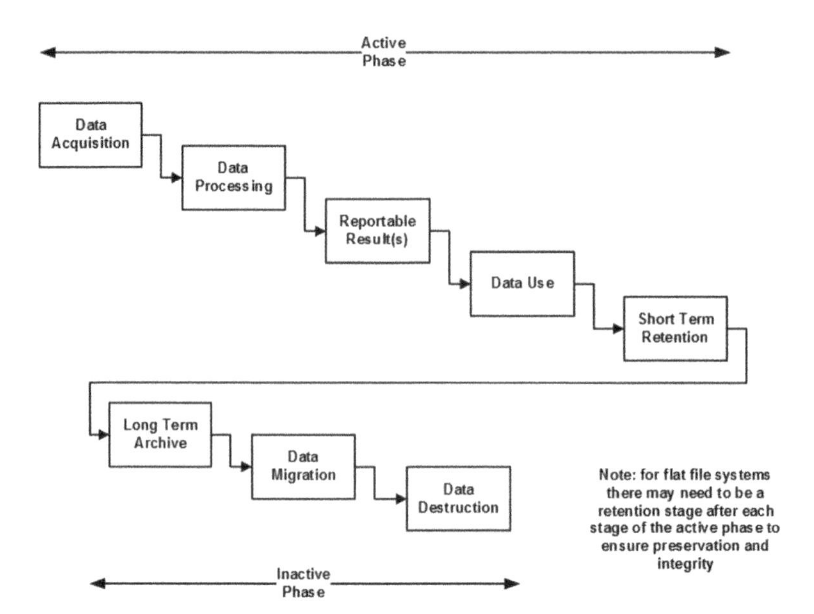

Figure 9.1 A generic data life cycle. Reproduced from ref. 5 with permission of the Royal Society of Chemistry.

The active phase of the data life cycle consists of the following activities:

- Data acquisition
  The process of controlling and recording the observation or generating the data from the analytical procedure.
- Data processing
  Interpretation or processing of the original data.
- Generate reportable result
  Calculation of the reportable result for comparison *versus* specification.
- Information and Knowledge Use
  Use of the result for the immediate purpose but also over a longer time for trending and preparation of Annual Quality Reviews and Annual Product Reviews.
- Short Term Retention
  Storage of the data and information in a secure but accessible environment for any further use, *e.g.* annual quality reviews, complaints, investigations as well as audits and/or inspections.

Note that for many laboratory systems where electronic records are stored in directories within the operating system there may need to be a retention process performed after each stage of the active phase to ensure the preservation of the record and the integrity.

The inactive phase of the data life cycle consists of the following stages:

- Long Term Archive – the movement of the records into a secure archive for long term retention.
- Data Migration – if necessary or required there may be one or more migrations of data from one system/repository to another over the retention period. Migration will only be necessary for electronic records and not paper.
- Data/Record Destruction – when the retention period has elapsed then a formal process to destroy the data/records should be executed providing that there is no litigation pending.

This life cycle does not account for any other use of the data, *e.g.* trending over time or product quality reviews where the information generated during an analysis is used as the input data for generation of additional information or knowledge abstraction.

9.2.3 Critique of the Two Life Cycle Models

Whilst these two data life cycles are a start, they are not perfect for a laboratory situation. To ensure that all data are captured and managed both life cycles are not sufficiently granular to enable them to be fully transposed into a laboratory environment. For example, there is no consideration of the sampling or sample preparation phases of the analytical life cycle in either of these models.

We also need to consider that a one size fits all approach may need to be changed into a flexible approach that can be expanded and contracted as required to fit an individual analytical process. For example,

- If a sample does not need any preparation stage before testing, will a data life cycle accommodate this?
- If the data acquisition is the final result there is no need for interpretation or calculation phases.
- Can the life cycle accommodate paper, hybrid and electronic processes?
- How will protocol based work such as method validation, stability and bioanalytical non-clinical and clinical studies be accommodated by an analytical focused data life cycle model?

You may think that this is being very critical of the two generic data life cycle models, however, if you are going to manage the data integrity in a laboratory you need to have a good understanding of each process down to a data and record level. These two models are not sufficiently detailed and therefore we need a flexible analytical data life cycle. In this way, all process laboratory computerised systems and the associated records can be managed more effectively.

9.3 An Analytical Data Life Cycle

From the critique of data life cycles in the previous section, a new data life cycle is required that is designed specifically around the analytical process and this analytical data life cycle will be described in this section.

9.3.1 Overview of an Analytical Data Life Cycle

An analytical data life cycle is essentially an expansion of the life cycle shown in Figure 9.1 and adapted to chemical analysis. As with the previous life cycle, there are two phases:

- Active Phase: this consists of three parts:
  - Control: A study plan, an analytical procedure or an analytical request to define the work to be performed or a combination of these elements.
  - Analyse: The analytical process to generate the data from sampling to reporting.
  - Review: Second person review to ensure data quality and data integrity.
- Inactive Phase: this consists of long term archiving; data migration and the destruction of the records as is as described in Section 9.2.2

and Figure 9.1. The difference in this data life cycle is the separation of paper and electronic records in the inactive phase. Typically, paper records are stored in an archive or record storage and are not migrated or moved during the retention period. Electronic records may have to undergo at least one data migration during the retention period. Throughout the inactive phase there is a need to check that the records are available and can be accessed and this applied to both paper and electronic records. The synchronisation of paper and electronic records from hybrid systems is particularly important here.

Although these two phases are not shown explicitly in the visualisation of the analytical data life cycle in Figure 9.2 they are implied but to include them would make the diagram too complex to explain easily. Each portion of the life cycle will be described in more detail in the following sections of this chapter.

9.3.2 Controlling the Analytical Data Life Cycle

In all instances, there needs to be control of the active phase of the analytical life cycle and this is shown in Figure 9.2 as an analytical request, study/sample plan and/or an analytical procedure. The plan or procedure will determine how samples will be collected, how samples will be analysed and how data will be generated and reported during the analytical data life cycle. As

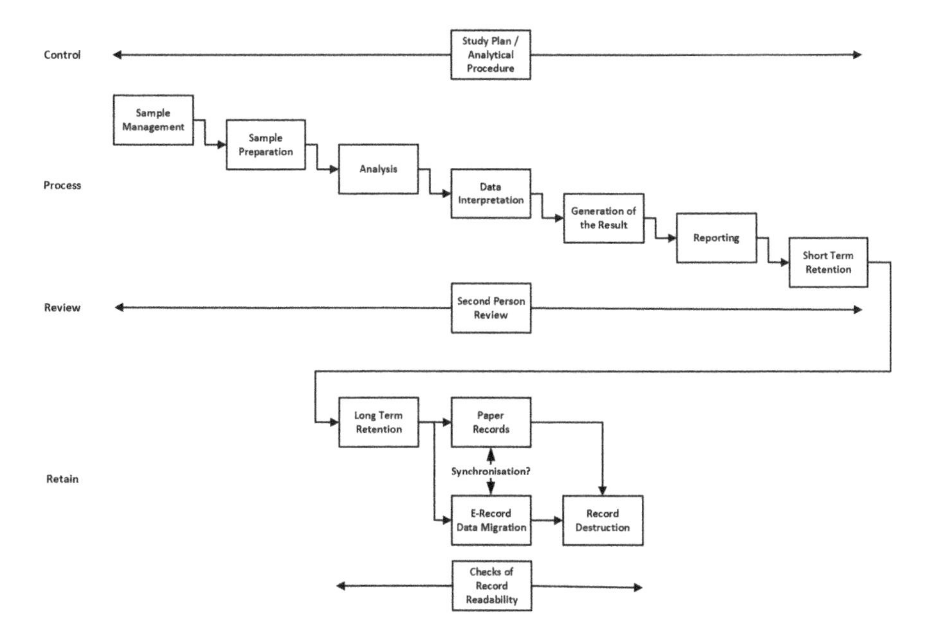

Figure 9.2 An analytical data life cycle.

such, the plan or procedure must be considered as an essential part of the analytical data life cycle.

Examples of these study plans include:

- method validation plan;
- method transfer protocol;
- non-clinical or clinical protocol;
- stability study;
- analytical procedure;
- sampling plan.

These are all formal GXP documents that need to be reviewed and approved. In some instances, *e.g.* study protocols, can have formal amendments issued after the work has started to modify the approach. These amendments, if issued, also form part of the analytical data life cycle as they control specific tasks in the analytical process in a specific study.

In addition to these controlling documents for the analytical life cycle, there will be standard operating procedures or work instructions for performing component tasks within the overall life cycle. These are not considered in this discussion but are an essential part of performing and reviewing the work carried out in any regulated analytical laboratory.

9.3.3 Phases of the Analytical Data Life Cycle

Underneath the study plan or the analytical procedure is the process of the main analytical data life cycle. It is based upon the analytical process as throughout the process data are collected that are used to support both data integrity and data quality of the final outcome of the analysis, *e.g.* a QC certificate of analysis containing the reportable result or a validation report, *etc.*

The stages of the analytical process, as shown in Figure 9.2, are:

- Sample Management
  This covers the sampling plan (if not included in the study plan), sampling, sample containers to be used, any sample preservation requirements, transport and storage in the laboratory. Throughout this activity of the analytical process, data will be acquired to ensure that the most important part of the analysis is correctly documented. The integrity of the data gathered here is essential to support the final results of the analysis and this will include any environmental monitoring records of storage conditions during transport and/or storage. The major problem that much of the sampling process is usually manual and can contain errors as well as be falsified easily. This phase is never discussed in data integrity guidance documents.
- Sample Preparation
  Preparing the samples for analysis can be as simple as transferring a liquid sample to a vessel and presenting it to an instrument, through

dissolving with dilution, to complex liquid–liquid or solid phase extraction. Although some sample preparation techniques can be automated, many of the steps are manual and are typically recorded on paper. The scope of work may include preparation of reference solutions, buffers, mobile phases using apparatus and instrumentation such as sonic baths, analytical balances, pipettes, volumetric glassware, homogenisers and pH meters. Data demonstrating that this work has been performed is essential for demonstrating both the integrity and quality of the work including appropriate instrument calibration checks and associated instrument log book entries. As with sampling, the preparation phase is not covered in regulatory or industry guidance documents on data integrity.

- Analysis
  This activity involves the set-up of the analytical instrument with appropriate calibration and point of use checks. This is followed by the acquisition of data from the sample by following the applicable analytical procedure. The spectrum of analytical techniques applied can vary from observation (colour, appearance or odour) through wet chemistry such as loss on drying and water content to instrumental techniques such as spectroscopy or chromatography. The data collected here will include, as appropriate, the result, setting up the instrument, running any system suitability tests or point of use checks before committing the samples for analysis and followed by data values or data files for interpretation in the next stage of the analytical data life cycle.
- Data Interpretation
  Here, the data acquired during the analysis are interpreted to obtain processed data. Where an instrument such as a chromatograph or spectrometer is used, the data need to be interpreted by an analyst to obtain an identity, absorbance at a specific wavelength or peak area counts. This is a key area of the analytical data life cycle and needs to be controlled carefully to ensure the integrity of the data. This area is also the subject of rigorous regulatory scrutiny and is the source of many data integrity citations and warning letters.
- Generation of the Result
  Following the interpretation of the data, the result is calculated. This can be by a variety of means such as manually using a calculator, using a spreadsheet or incorporated into an instrument data system or other informatics application. Where possible calculations should be performed by a validated software application, avoiding manual data entry whenever possible.
- Reporting
  Following the calculation of the result(s), the report will be drafted. There are many forms that the report can take such as a method validation or transfer report, certificate of analysis or study report. At this stage, the outliers can be identified, *e.g.* out of specification results or a value on a time *versus* drag concentration curve. These will invoke a procedure to investigate the anomalous result in both GMP or GLP analyses.[6]

- Second Person Review
  Before a report or COA is formally issued, the complete data package needs to be subject to a second person review. The laboratory reviewer needs to be suitably trained and the review will include any instruments and computerised systems involved in the analysis. The aim of the second person review is to ensure that the work has been carried out correctly, procedures have been followed, data have been interpreted correctly, results generated accurately and the report is complete. In addition, the second person reviewer needs to check that there have not been any data falsification or poor data management practices. Second person review is discussed in Chapter 16.
- Short Term Retention
  Data are retained in a secure manner regardless if the records are paper or electronic. If the computerised system generated the records is a hybrid there will be paper and electronic records that need to be synchronised if any changes are made subsequently after the analysis is completed and reported, *e.g.* complaints or regulatory questions.

When the short-term data retention period has elapsed, the data are retained for the applicable retention period mandated either by the regulations or company policy as described in Section 9.2.2.

Implicit throughout the whole of the analytical process is the applicable standard operating procedures, work instructions, that describe how work should be conducted by the analytical staff. These should include how results are trended in compliance with EU GMP Chapter 6 requirements[7] and the identification of out of specification (OOS) or out of trend or expectation (OOE and OOE) results.[6]

9.3.4 Generic Data Life Cycles Do Not Work in the Laboratory

Comparing the data life cycles discussed in Sections 9.2.1 and 9.2.2 with the analytical data life cycle shown in Figure 9.2 and described in Section 9.3.1 you will see that the former are inappropriate for most analytical procedures as records that impact data integrity could be omitted or not even considered. As stated earlier, a generic or one-size fits all approach to an analytical data life cycle is not appropriate as there are many ways to analyse a sample.

9.3.5 The Requirement for Flexibility to Adapt to Different Analytical Procedures

Generic data life cycle models are fine as they introduce analysts to the topic of a data life cycle for the laboratory and get people thinking but, as discussed in Sections 9.2.3 and 9.3.4 they do not fit into a laboratory scenario. To cope with analytical procedures that vary from simple to complex an analytical data life cycle must be flexible and scalable.

The analytical data life cycle in Figure 9.2 is a better approach but it is still not perfect and needs to be refined further. In this respect, different analyses should have a variant of the analytical data life cycle to reflect what happens in a regulated laboratory. To illustrate this, we will consider four different analytical procedures that are typical in the pharmaceutical industry as follows:

1. analysis by observation;
2. sample preparation and instrumental analysis;
3. instrumental analysis followed by data interpretation;
4. sample preparation, instrumental analysis followed by data interpretation.

The next section will look at how the analytical data life cycle can be adapted to fit these different analyses. We will revisit these four different analyses in Chapters 15 and 16 when we consider the analysis of samples and second person review, respectively.

Each type of analytical procedure has its own version of the analytical data life cycle as shown in Figure 9.3. The one consistent factor across all four analyses is the second person review of the data generated by each procedure.

1. Analysis by observation
   Typically, a sample will be taken and an aliquot transported to the laboratory for a test based on observation such as colour, appearance or odour. As such, sample preparation, data interpretation and generation of the result are not needed as the analysis is the reportable result. This means that for these types of analysis a simpler and more appropriate data life cycle is indicated.
2. Sample Preparation and Instrumental Analysis
   Typified by a loss on drying (LOD) analysis there is a slightly more complex data life cycle. Sample management is applicable but the sample preparation is relatively minimal and is restricted to weighing the sample aliquots for analysis within the limits specified by the pharmacopoeial method. The analysis consists of weighing the sample before and after heating and then calculating the percentage loss. As such, there is no requirement for interpretation of data.
3. Instrumental Analysis and Data Interpretation
   Here the sample management and sample preparation tasks of the analytical data life cycle are not required and this is typified by NIR identity analysis. Instead, the sampling and analysis phases are combined into one as the analytical instrument is connected directly into an analytical process when a shipment of material is received in the warehouse. The spectrum of the sample is compared to a composite spectrum in a spectral library to verify the identity of the material under test.
4. Instrumental analysis with Data Interpretation

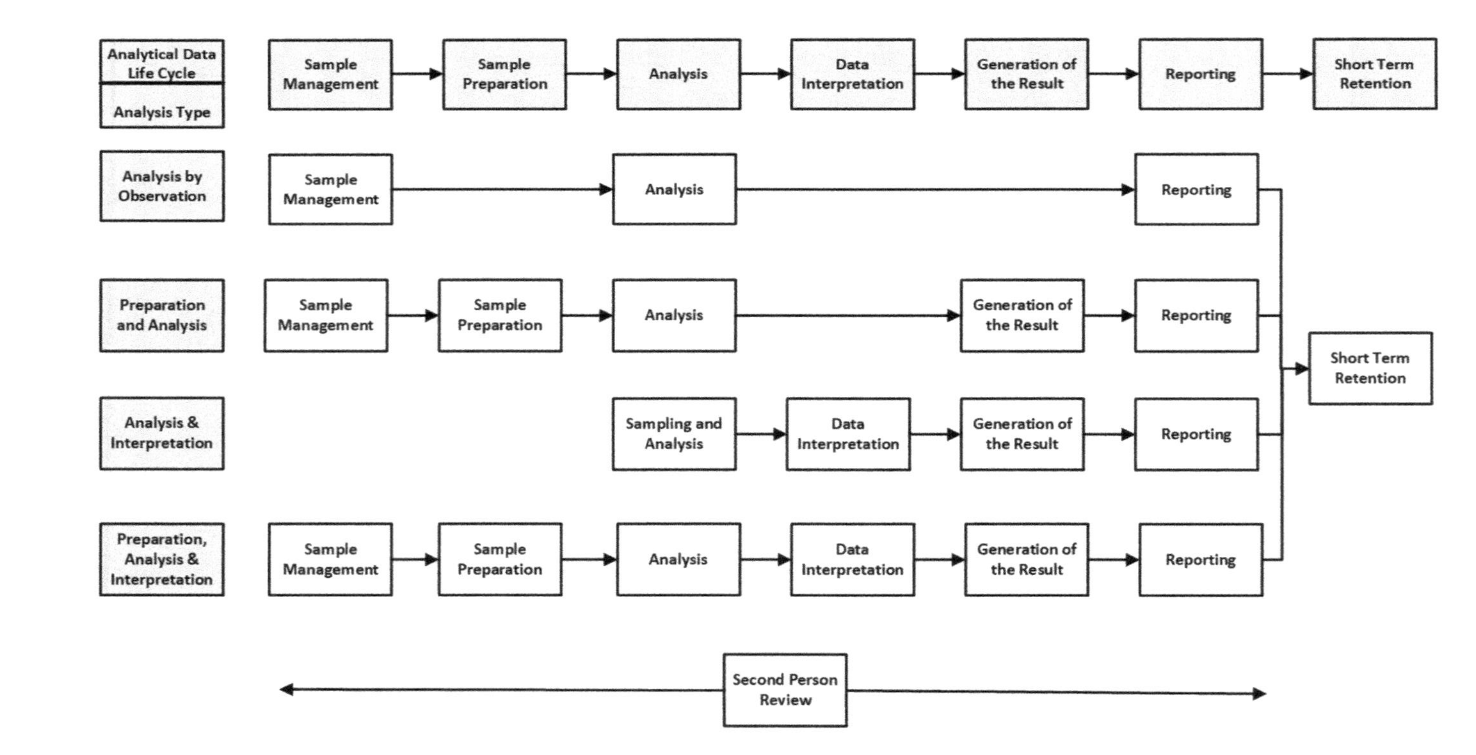

Figure 9.3 The type of analysis impacts the phases of the analytical data life cycle required.

The whole analytical data life cycle is applicable to instrumental analysis. This is typified by a chromatographic analysis that requires sample management and preparation, setting of the instrument, acquisition of the data and calculation of the reportable result.

As can be seen, the analytical data life cycle in Figure 9.3 is flexible and can be adapted to meet any analytical procedure and the data generated by it. Therefore, map and adapt the analytical data life cycle to any assay procedure to identify which tasks in the life cycle are applicable. The four options described above are only a selection of the combinations possible so fit the life cycle to each analytical procedure. This approach is far preferable to forcing all processes to fit a single model.

A one size data life cycle does not fit all.

9.4 Establishing Data Criticality and Inherent Integrity Risk

Apart from the analytical data life cycle, we also need to consider data integrity from another perspective. In the MHRA data integrity guidance documents[8,9] there is a section entitled establishing data criticality and inherent integrity risk that is essential to consider before we consider risks associated with individual instrument types later in this chapter and data process mapping in Chapter 10.

9.4.1 Spectrum of Analytical Instruments and Laboratory Computerised Systems

In the 2015 and 2016 versions of the MHRA guidances, there was a figure of instruments and computerised systems.[8,9] This figure has been updated for all analytical processes in the laboratory and is presented in Figure 9.4.

- The horizontal axis at the top of the figure are the different processes that can be used in a laboratory environment to generate data; these vary from observation, through simple instruments such as balances and pH meters, chromatography data systems to LIMS (Laboratory Information Management Systems).
- The vertical axis consists of the attributes of each process such as if software is used and if so the GAMP classification, the mode of data recording, the raw data produced and the main data integrity issues of each process. Note that Figure 9.4 refers to firmware as GAMP Category 2 software, although this has been discontinued in GAMP version 5 (ref. 10) it equates to Group B instruments in USP <1058> on Analytical Instrument Qualification (AIQ).[11,12] When mapping USP <1058> groups *versus* GAMP software categories[13] if Category 2 software were reinstated there would be equivalence between Category 2 software and Group B instruments.

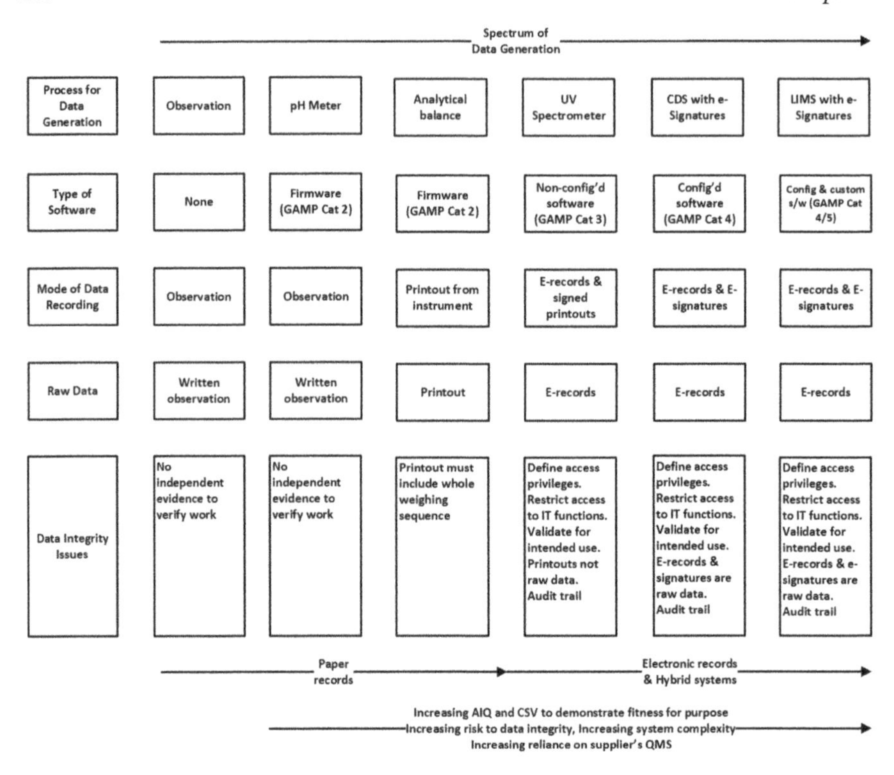

Figure 9.4 Laboratory data integrity issues in a spectrum of manual processes, instruments and systems.

The first three processes from observation to analytical balance have paper records and the remaining four items have electronic records. Dependent on how the latter four computerised systems are used they can either be hybrid or electronic by using electronic signatures. Furthermore, the pH meter and analytical balance are discussed here from the perspective of being standalone instruments rather than being interfaced to a LIMS, instrument data system or ELN (Electronic Laboratory Notebook).

Figure 9.4 also shows that for analytical instruments and laboratory computerised systems the following items hold true:

- Going from left to right there is increasing system complexity and more records generated for interpretation and review.
- Increasing AIQ and/or CSV work is required to demonstrate fitness for intended use as one goes from a simple instrument with integrated software to a complex computerised system.
- There is increasing risk to data integrity from either inadvertent error by users or deliberate falsification going from left to right.
- There is increasing reliance on a laboratory on a supplier's quality management system the further to the right one goes.

Let us look at four examples of data gathering from Figure 9.4:

- Observation
  Manual observations may be found in many laboratories for tests such as colour or odour of samples as well as recording data from some instruments, as shown in the first column on the left of Figure 9.4. As noted here, the data integrity issue is that there is no independent evidence to verify that the value or result recorded is correct, has suffered from a transcription error (value only) or has been falsified. Therefore, each process using observation only needs to be risk assessed to determine the criticality of the data being generated: for example, is an odour determination the same criticality as the pH determination of HPLC mobile phase?
- Instrument
  The example used in Figure 9.4 is an analytical balance with a printer. Given the importance of accurately measuring reference materials and samples and the impact that an analytical balance can have on a regulated laboratory, it is important that the integrity of measurement is maintained. At a minimum, as a balance does not contain any electronic records a printer is essential for an analytical balance as the MHRA guidance makes clear.[8] However, we need to consider more detail, what data need to be recorded when making a weighing measurement? In my view, the printer needs to record all the weights captured during any weighing operation, *e.g.* any calibration checks, weight of weighing vessel, tared weight and the weight of the material along with the date and time of activities and the identity of the user who performed the work.
- Hybrid System
  The hybrid system, typified by a UV spectrometer using GAMP Category 3 software, is the worst of both worlds as the laboratory must manage and synchronise two different and incompatible media types: paper records with handwritten signatures and electronic records. The issues are that paper cannot be defined as raw data, as noted by the WHO, MHRA and FDA.1,2,8,14 Note that the FDA level 2 guidance[15] is a much better discussion of why paper cannot be raw data. Other data integrity issues are that configuration of the software must be recorded, including definitions of user types and the access privileges for each type and validation of this configured software for the intended use. Many hybrid systems consist of the instrument connected to a standalone workstation where there are potential issues of access to the operating system, clock, the data files themselves *via* the operating system and effective and validated backup and recovery. Systems using the operating system to store the data files in open access directories can suffer from the stupidity of operators performing unintended deletions and well as attempts at falsification from individuals. However, the use of an adequately designed database should protect data from many falsification attacks but in reality data need to be acquired and stored securely in the network when using flat file systems.

- Electronic System
  Using a chromatography data system with GAMP Category 4 software with electronic signatures enabled as an example. In this instance, the raw data are electronic records with electronic signatures. To ensure data integrity the application must be configured for security and access control (definition of user types and access privileges) and also for the use of electronic signatures. Data are acquired to the network and are secured with a database. Validation for intended use will demonstrate that the configured system works as intended and according to the specification documents.[5] The audit trail documents changes made by authorised individuals. The issue now is the separation of system administration roles from that of the use of the system by chromatographers.

This approach can only be a generalisation: know your instrument or system, how it operates and the records and data that are generated when integrated into the respective analytical data life cycle is the maxim here. For example, modern balances can have clocks and their screens can access software such as electronic laboratory notebooks or LIMS to act as a terminal as well as an analytical instrument. Simply having a balance connected to such an application may not be enough – where is the time and date stamp applied in such cases: at the balance or in the software application? Can anybody change the clock in the balance and impact the time stamp in the application? Indeed, in response to the data integrity issues faced by the pharmaceutical industry some of the newest analytical balances from one supplier cannot operate as a standalone instrument but must be operated through an instrument data system to ensure that all data are captured and audit trailed. This is an essential approach for ensuring the integrity of electronic records stored on a standalone workstation.

9.5 Risks to Data Over the Data Life Cycle

Having looked at the risk associated with the analytical instruments and systems, let us look to see what risks there are when the instrument is combined with the four analytical procedures that were first presented and discussed in Section 9.3.3 and Figure 9.3. Before we start let us see what the PIC/S guide says about risks to data over the data life cycle.

9.5.1 PIC/S PI-041 Good Practices for Data Management and Integrity in Regulated GMP/GDP Environments

In this PIC/S guidance, as presented at the beginning of this chapter, there is the following section[3]:

> 5.3.4 Not all data or processing steps have the same importance to product quality and patient safety. Risk management should be utilised to

determine the importance of each data/processing step. An effective risk management approach to data governance will consider:

- Data criticality (impact to decision making and product quality) and
- Data risk (opportunity for data alteration and deletion, and the likelihood of detection/visibility of changes by the manufacturer's routine review processes).

From this information, risk proportionate control measures can be implemented.[3]

9.5.2 Initial Assessment of Risk of the Analytical Data Life Cycle Phases

The PIC/S guidance notes not all phases of a data life cycle have the same importance,[3] but what does this mean in practice? You will recall the four analytical procedures presented in Figure 9.3, each one of these has different phases of the analytical data life cycle applicable.

Figure 9.5 presents the four procedures and their life cycle phases arranged vertically.

In presenting the phases of the analytical data life cycle in this way we can apply the same principles of the Data Integrity Model apply. If the lowest applicable phase of the analytical life cycle of a procedure is compromised

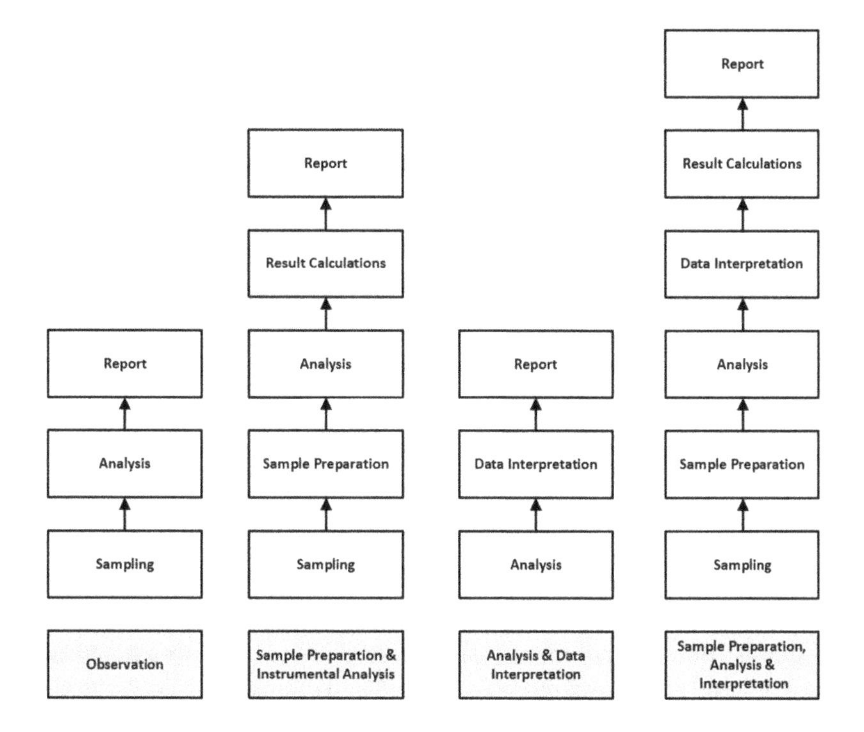

Figure 9.5 The analytical data life cycle phases of four analytical procedures.

either through poor data management practices, a data integrity violation or the use of a non-calibrated instrument, it is immediately apparent that the remainder of the life cycle phases is also compromised.

9.5.3 Phases of the Data Life Cycle are Equal but Some are More Equal than Others

Using the life cycle phases from the four analytical procedures in Figure 9.5 we will review the risks associated with each one.

Observation: This type of analysis is typified by colour, odour or appearance and has three phases: sampling, analysis and reporting. It is a manual process, with a manual sampling phase and is non-destructive testing with no analytical instrument involved. The results are usually subjective with no objective evidence available unless there is a colour palette used. However, as the test is non-destructive it is possible for the second person reviewer to repeat the test. Therefore, although it is fully manual, such tests should be considered low risk and the test can be repeated at any time using the retained sample.

Loss on Drying: A pharmacopoeial analysis applicable to many API samples consisting of sampling, sample preparation, instrumental analysis, results calculation and reporting. Usually, this analysis is performed manually unless it has been automated using an analytical balance coupled to a halogen lamp for heating the sample. As the whole process is manual and may involve transcription of values into a validated spreadsheet or manually with a calculator it should be considered high risk. An alternative to lower risk, is to connect the balance to an instrument data system, ELN or LIMS where the calculations are verified and automated and eliminate transcription error checks.

The risk could be lowered further by implementing an alternate analytical approach of a halogen lamp integrated with an analytical balance. After cross validating against the pharmacopoeial method, the halogen drying is faster and can calculate the LOD automatically with no need to print and transcribe data.

NIR Identity: A test performed in the goods inwards warehouse using a probe place in the container or through the plastic bag surrounding the sample, as such there is just analysis and data interpretation using an in-house developed spectral library to confirm the identity of the substance. Assuming the library has been developed and validated correctly and there is no access to the data files, system clock and recycle bin, the identification should be automatic with little operator interpretation or interference. This analytical procedure should be considered medium risk.

Chromatographic Analysis: A chromatographic analysis consisting of sampling, sample preparation, analysis, data interpretation, calculation and reporting is the most complex analysis of the four procedures we consider. Sampling and sample preparation are typically manual and therefore high risk, furthermore, values such as weights and dilutions are usually entered manually into the CDS sequence file and must be checked to comply with

EU GMP Annex 11 clause 6.[16] Integration is performed by the CDS software and we will revisit this subject later in Chapter 15. If the system is used electronically then all subsequent calculations of SST parameters and reportable values will use validated functions in the CDS and will be low risk, providing functions cannot be altered.

Alternatively, if the CDS is used as an electronic ruler and paper is printed out after integration, followed by manual input into spreadsheets for the calculation of SST parameters, calibration curves and reportable results, then the risk moves from low to high.

9.5.4 Summary Risks in the Analytical Data Life Cycle

The summary risks for an analytical data life cycle that we can derive from the above analysis are:

- When a data integrity issue occurs in a life cycle phase, that phase and all subsequent ones are compromised and the reportable result is questionable.
- Manual life cycle phases without any independent evidence or ability to repeat analysis, *e.g.* sampling and sample preparation are high risk.
- When allowed, human interpretation of analytical data, *e.g.* spectra, chromatograms can be subjective and could be subject to manipulation and therefore need to be carefully controlled and reviewed.
- When the same process is enforced by technical controls the risk is reduced to medium or low depending on the type of technical control and the sophistication of the application's audit trail.
- Procedural controls are operated by analysts and are inconsistent and error-prone, hence they are high risk.
- Technical controls enforce compliance and should be used where possible following validation to demonstrate that they work, these processes are low risk.

To understand the risks associated with a manual, hybrid or electronic process a methodology is required to assess both processes and electronic workflows to identify risks to data and records and then implement controls to protect them. We will find what approaches are available in Chapter 10.

References

1. *MHRA Gxp Data Integrity Guidance and Definitions*, Medicines and Healthcare products Regulatory Agency, London, 2018.
2. *WHO Technical Report Series No.996 Annex 5 Guidance on Good Data and Records Management Practices*, World Health Organisation, Geneva, 2016.
3. *PIC/S PI-041 Draft Good Practices for Data Management and Integrity in Regulated GMP/GDP Environments*, Pharnaceutical Inspection Convention/Pharmaceutical Inspection Co-Operation Scheme, Geneva, 2016.

4. *Gamp Guide Records and Data Integrity*, International Society for Pharmaceutical Engineering, Tampa, FL, 2017.
5. R. D. McDowall, *Validation of Chromatography Data Systems: Ensuring Data Integrity, Meeting Business and Regulatory Requirements*, Royal Society of Chemistry, Cambridge, 2nd edn, 2017.
6. *FDA Guidance for Industry Out of Specification Results*, Food and Drug Administration, Rockville, MD, 2006.
7. *EudraLex - Volume 4 Good Manufacturing Practice (GMP) Guidelines, Chapter 6 Quality Control*, European Commission, Brussels, 2014.
8. *MHRA Gmp Data Integrity Definitions and Guidance for Industry*, Medicines and Healthcare products Regulatory Agency, London, 2nd edn, 2015.
9. *MHRA Gmp Data Integrity Definitions and Guidance for Industry*, Medicines and Healthcare products Regulatory Agency, London, 1st edn, 2015.
10. *Good Automated Manufacturing Practice (GAMP) Guide Version 5*, International Society for Pharmaceutical Engineering, Tampa, FL, 2008.
11. *USP 31 General Chapter <1058> Analytical Instrument Qualification*, United States Pharmacopoeial Convention, Rockville, MD, 2008.
12. *USP 41 General Chapter <1058> Analytical Instrument Qualification*, United States Pharmacopoeia Convention, Rockville, MD, 2018.
13. L. Vuolo-Schuessler, *et al.*, Harmonizing USP <1058> and GAMP for Analytical Instrument Qualification, *Pharm. Eng.*, 2014, **34**(1), 46–56.
14. *FDA Draft Gudance for Industry Data Integrity and Compliance with cGMP*, Silver Spring, MD, USA, 2016.
15. *FDA Questions and Answers on Current Good Manufacturing Practices, Good Guidance Practices, Level 2 Guidance - Records and Reports. 2010*, 27 May 2016, Available from: http://www.fda.gov/Drugs/GuidanceComplianceRegulatoryInformation/Guidances/ucm124787.htm.
16. *EudraLex - Volume 4 Good Manufacturing Practice (GMP) Guidelines, Annex 11 Computerised Systems*, European Commission, Brussels, 2011.

CHAPTER 10

Assessment and Remediation of Laboratory Processes and Systems

A key part of any data integrity programme is the examination and understanding of system and process risk for data and records created and maintained during GXP activities. This discussion was started in Chapter 9 with an exploration of the flexible analytical data integrity life cycle and the risk that some activities have compared with others. In this chapter, we will discuss ways of formally assessing both manual processes as well as those involving computerised systems to identify record vulnerability and propose remediation in the short term and well as develop long-term solutions to eliminate paper and generate business benefit. However, some companies focus only on computerised systems for this and ignore manual processes that leave potential compliance gaps.

Process and system assessments based on questionnaires can have limitations as this can be a proscriptive approach and where there is a gap in the questionnaire there will be a gap that could be overlooked. An alternative approach is to use data flow mapping (a combination of process mapping combined with the identification of data and records generated and modified during the execution of a process). This is a methodology that allows the visualisation of the current process and following any remediation. Data process mapping is applicable to both manual processes and computerised systems.

Data Integrity and Data Governance: Practical Implementation in Regulated Laboratories
By R. D. McDowall

Published by the Royal Society of Chemistry, www.rsc.org

10.1 What Do the Regulators Want?

10.1.1 WHO Guidance on Good Data and Record Management Practices

There are various sections of this guidance[1] that cover understanding and mitigating risk of processes and systems:

> 1.4: mapping of data processes and application of modern Quality Risk Management (QRM) and sound scientific principles throughout the data life cycle;
>
> 5.2 Within the quality management system, the organization should establish the appropriate infrastructure, organizational structure, written policies and procedures, processes and systems to both prevent and detect situations that may impact on data integrity and, in turn, the risk-based and scientific robustness of decisions based upon those data.
>
> 5.6 A data management programme developed and implemented on the basis of sound QRM principles is expected to leverage existing technologies to their full potential. This, in turn, will streamline data processes in a manner that not only improves data management but also the business process efficiency and effectiveness, thereby reducing costs and facilitating continual improvement.

10.1.2 PIC/S PI-041 Good Practices for Data Management and Integrity in Regulated GMP/GDP Environments

The PIC/S guidance has the following section covering processes and risk management:[2]

> 5.3.4 Not all data or processing steps have the same importance to product quality and patient safety. Risk management should be utilised to determine the importance of each data/processing step. An effective risk management approach to data governance will consider:
> - Data criticality (impact to decision making and product quality) and
> - Data risk (opportunity for data alteration and deletion, and the likelihood of detection/visibility of changes by the manufacturer's routine review processes).

From this information, risk proportionate control measures can be implemented.

10.1.3 MHRA GXP Data Integrity Guidance and Definitions

In the section of principles of data integrity, Clause 3.4 (ref. 3) notes:

> Organisations are expected to implement, design and operate a documented system that provides an acceptable state of control based on the data integrity risk with supporting rationale.

An example of a suitable approach is to perform a data integrity risk assessment (DIRA) where the processes that produce data or where data is obtained are mapped out and each of the formats and their controls is identified and the data criticality and inherent risks documented.

10.1.4 Regulatory Guidance Summary

Understanding the regulatory risk of processes and systems is an important part of an overall data governance framework within a laboratory. As noted by the PIC/S PI-041 guidance not all steps in a process have the same risk and importance when it comes to the end point of product quality and patient safety, but knowing where data integrity can be compromised is essential within a data governance and regulatory compliance framework.[2] Both the MHRA and WHO guidance documents recommend mapping the process to understand the data integrity risks.[1,3]

10.2 Business Rationale for Assessment and Remediation

There are some major business reasons for conducting assessment and subsequent remediation or improvement of processes and systems.

10.2.1 Improve Business Processes

Despite many publications and conferences on paperless and smart laboratories, many regulated laboratories have changed little in the last 25 years. Even with the 21 CFR 11 assessment and remediation, the software was updated to technically compliant applications but usually the business process was unchanged with a focus on paper records. Therefore, much money was spent but for the little business benefit. A data integrity and data governance programme will be more expensive than the Part 11 project as the scope of the programme is far wider than the project that was just focused on computerised systems.

The problem is that many laboratories are starved of investment and after process and system remediation are condemned to a perpetual state of drowning in paper. Jump to Chapter 13 to understand why eliminating paper is critical for laboratories to survive. If you want to be depressed first, read Chapters 11 and 12 to understand the reasons why you need to eliminate paper.

10.2.2 Ensure Regulatory Compliance

As shown in Figure 2.1 the cost of compliance (the cost of doing it right the first time) is always less than the cost of non-compliance (the cost of getting caught). Use the assessment and remediation work to fix immediate problems now but plan to improve the process and implement long-term solutions. Improvements will not happen overnight and must be planned. A well thought out long term plan with demonstrable evidence of progress will ensure that an inspector recognises that the laboratory has identified the problems, implemented as many short-term remediation measures as necessary and is planning for the future.

The alternative, if data integrity violations are found, can generate grave long-term business and financial consequences. For example, it may be necessary to hire a data integrity consultant as shown in the majority of FDA warning letters, undergo rapid remediation and change, have a drug application refused or suffer an import ban.

10.2.3 Release Product Earlier

If business processes can be improved one of the business benefits is the potential for faster release of the product to the marketplace. This has substantial business benefits to an organisation in terms of cash flow. Faster laboratory working alone could self-fund many data integrity improvement projects provided all analytical procedures for a specification could be executed faster. This target cannot be achieved with either paper processes or hybrid systems, processes must be electronic with technical controls to ensure data integrity and compliance.

10.2.4 The Problem is Management

The problem and a rate limiting factor in any business-driven data integrity assessment and remediation is management. In part, some in management think that data integrity is only a computer issue and fails to see the whole scope of work. Other managers only want to paper over the cracks in processes and systems by remediating as cheaply as possible by using procedural rather than technical controls. These are naïve approaches that merely delay the inevitable because regulators will start to force change through further rule making and regulatory citations to push the industry to ensure compliance with regulations including data integrity.

10.3 Current Approaches to System Assessment and Remediation

10.3.1 The Rationale for Current Approaches?

As stated in the WHO data integrity guidance[1] in Section 1.3 discussing the rationale for writing the guidance:

> Contributing factors include *failures by organizations to apply robust systems that inhibit data risks*, to improve the detection of situations where data reliability may be compromised, and/or to investigate and address root causes when failures do arise.

> For example, organizations subject to medical product good practice requirements have been *using validated computerized systems for many decades but many fail to adequately review and manage original electronic records and instead often only review and manage incomplete and/or inappropriate printouts.*

From a GMP perspective, the pharmaceutical industry has not moved and is still considered a 2 or 3 sigma industry.[4] This is in stark contrast to the US and EU GMP regulations that mandate that the industry adopt new technologies that support data quality and patient safety.[5,6]

10.3.2 Assessment of Validated Computerised Systems

Owing to the data integrity issues and poor data management practices discussed earlier in Chapters 1 and 2 there is a need to assess current systems even although they may be validated and operational to assess if the electronic records and signatures have any vulnerabilities. The assessment will be formal, documented and any issues found will need remediation planning up to and including upgrade or replacement.

The assessment process will probably be initiated by the data integrity committee and involve a generic assessment checklist written to cover all systems. Using the inventory of computerised systems for the laboratory, there should be a prioritisation of assessments based on system criticality. The analytical laboratory will be the focus as the FDA CPG 7346.732 mandates a data integrity audit of data in the application.[7] Typical systems for inspection include:

- Chromatography data systems as a CDS and chromatographic analysis is involved throughout pharmaceutical R&D and manufacturing for both GLP and GMP.
- Karl Fischer titration.
- Spectroscopic techniques, especially UV, NIR and FTIR systems.

A small multi-disciplinary team, consisting of the process owner, a data integrity expert, an expert laboratory user and, the system is networked, an IT administrator, will usually be involved in making the assessment using a typical outline shown in Table 10.1. This provides a framework for consistent assessment of all systems in an organisation, although there needs to be flexibility for the assessor(s) to deviate if a specific system has some issues that need further investigation.

Underneath most of the areas on the right-hand column of Table 10.1 will be further questions to understand the system operation and to identity any record vulnerabilities. For example, some of the questions to assess the access control of the system are shown in Table 10.2:

The aim of the assessment is to identify record vulnerabilities and to determine remediation options and potential costs. Some issues can be resolved relatively quickly under change control, *e.g.* providing all users

Table 10.1 An outline of a typical data integrity assessment checklist.

Checklist section	Contents
System description	• Identification of the system • Location • Description of the system function • Interfaces to other systems/standalone system • Data transfer to other systems
Assessment summary	• The overall result of the data integrity assessment: Record vulnerabilities summary Short term remediation plans Long term solution plans
How is the system used	• What are the raw data: Paper Hybrid (signed paper printouts and electronic records) Electronic (electronic records and signatures)
User account management and system access	• User profiles defined? • Independence of system administration? • What is the number of system administrators? • Physical security? • Login – unique user identities? • No shared accounts (including read-only)?
System validation	• Workflows defined? • Workflows validated? • Database or operating system files? • Standalone or networked system? • Application settings enabled to protect electronic records? • Application settings enabled for use of electronic signatures?
Data integrity assessment	• Logical and physical security of data? • Access to system clock, data in directories and recycle bin? • Using defined workflows acquire, process and report data? • Changes to data recorded in the audit trail?
System support	• Who backs up the data? • Data backup frequency and records? • Data recovery and restore? • Change control and level of revalidation?
Data retention	• Where are the data generated stored? Paper? E-records?

with unique user identities, establishing access privileges for each user type, segregation of duties, turning on e-record protection such as audit trails, *etc.* For more involved remediation or long-term solutions, it will be for the Data Integrity Committee to allocate resources to allow a team to perform any process redesign, software updates and revalidation as necessary.

Table 10.2 Some additional questions for access control of a computerised system.

Subject	Questions	Answers
Security and access control	Who requests access to the system? How is user access to the system granted? Are a user's name or company initials used as their user identity? Are any accounts shared? What are the user roles defined in the system? Where are the user roles documented? Are there any conflicts of interest in the roles, *e.g.* administration and user rights shared?	

10.4 Data Process Mapping

10.4.1 The Problem with Checklists

Checklists are fine but the majority of them that I have seen are aimed at computerised systems and omit paper based processes. In addition, the checklist does not provide a visual description of a process (manual or automated) that allows analysts, quality assurance, auditors or inspectors a common understanding of the process and the associated record vulnerabilities. There needs to be a better way to present all processes in a laboratory in a common way.

One such approach that can be used is data process mapping.

10.4.2 What is Data Process Mapping?

Put simply, this is a variation of process mapping that can be used for implementing computerised systems to ensure that the automated process is efficient and streamlined. It has been described previously as a pre-requisite for the validation of a chromatography data system.[8] In this section, we will consider several scenarios to show its usefulness in a data integrity assessment and remediation programme.

One basic process mapping methodology that can be used is IDEF or integrated definition that is a hierarchical process decomposition from a single process description is broken down into an adequate level of description to identity data integrity issues. This is shown simplistically in Figure 10.1 and shows the breakdown of a process from a description into successive levels of detail.

As shown in Figure 10.1 there are a number of levels in an IDEF process map:

- Level 0: A single box describing the overall process description, *e.g.* method validation, routine analysis, *etc.*
- Level 1: The process description is broken down into 6–12 main activities that comprise the process description as shown in Figure 10.1.

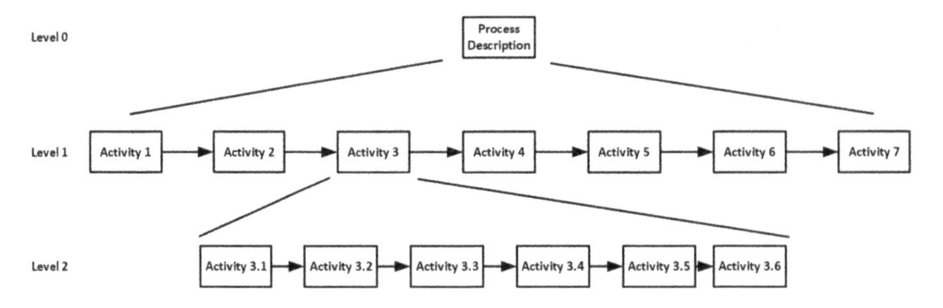

Figure 10.1 An outline of IDEF process mapping.

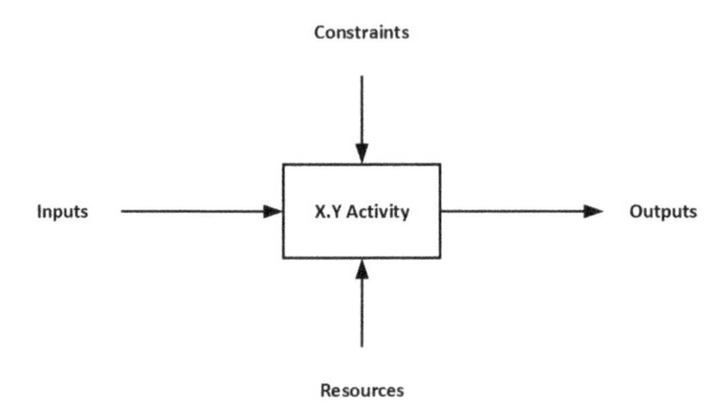

Figure 10.2 Input, outputs, constraints and resources of an IDEF activity.

- Level 2: Each activity at Level 1 is broken down into between 4–8 sub-activities.
- The process can be decomposed further into more detail if required. Normally, if dealing with a single process with computerised systems linked by manual data input then decomposition to Level 2 is adequate. Where systems are interfaced, it may be necessary to decompose the process to a further level focused on the interfacing to understand the data and records transferred and also the detail to discover if there are any record vulnerabilities.
- Correction between activities at each level is ensured by the numbering scheme. As shown in Figure 10.1, activity 3 at Level 1 is decomposed into activities 3.1–3.6 in Level 2. Each other Level 1 activity will also have a similar decomposition to Level 2. The same principles apply to numbering Level 3 activities.

You may be asking what is the relevance of this type of process mapping with data integrity?

This is a fair point. The answer is in the detail of the activity boxes as shown in Figure 10.2. For each activity at any level of the process map, there must be inputs and outputs as well as the resources necessary to perform the activity and

Table 10.3 Use of IDEF methodology for process improvement and data process mapping.

IDEF activity	Process mapping and improvement	Data integrity vulnerability
Inputs	• Vials for injection • List of vial numbers *versus* identification (*e.g.* samples, system suitability samples, standards, blanks and quality controls) • Columns and column log • Mobile phase components, *e.g.* aqueous buffers and organic solvents • Analytical conditions to be used (typically in a procedure) • Instrument log book • Data acquisition and instrument control methods to be used	• Unique user identities? • User roles defined? • No conflicts of interest? • Who set up the instrument? • Who made changes? • Where are the data stored? • Has the data storage location been changed by the user? • Who modified data acquisition and instrument control methods? • Who modified sequence file entries?
Outputs	• Data files (1 per injection or 1 sequence file, depending on the CDS used) • Modifications to the method as the run progresses if permitted	• Are files identified correctly? • How are the data stored? • Who can access the data? • Are data files protected? • Can the files be substituted, deleted or changed outside of the application? • Audit trail entries for all changes?
Resources	• Qualified chromatograph • Trained staff	• Is the instrument qualified? • Is the instrument calibrated or the day or SSTs run? • Is the software validated? • Are staff trained to use the instrument?
Constraints	• GXP regulations • Pharmacopoeial requirements (if GMP) • Internal procedures and work instructions	• Identification of the technical controls to protect e-records? • Restricted access to the system clock, recycle bin and e-records?

to identify any constraints. If we take an example for preparing a chromatograph for analysis and running the samples as the activity in Figure 10.2 we can list the main components of the four IDEF requirements when validating a data system shown in the Process mapping and improvement column in Table 10.3.

When IDEF is used for process mapping prior to the implementation and validation of a computerised system, the main aim is to identify constraints in the process such as printing from an instrument and manually entering some of the data into a spreadsheet for calculating results. When these constraints are identified, the aim is to eliminate the bottleneck, in this case, the spreadsheet, and incorporate the calculations in the instrument data system.

To extend IDEF for data integrity additional questions need to be asked as shown in the data integrity vulnerability column in Table 10.3

10.4.3 Instrument Data System with Spreadsheet Calculations

To understand what this means in practice, let us consider a common situation in many regulated laboratories today. An instrument data, say a standalone chromatography data system, is used to generate peak areas, then the analytical scientist prints out the results as shown in Figure 10.3. This generates printout 1 and covers activities 1 to 3 inclusive. The CDS is running on a workstation with a single disk, where data files are stored in directories in the operating system, the storage location is chosen by the analyst performing the work and backup is performed by the laboratory staff when they have time available. All users share the same user identity and password that is written down on a sticky note underneath the keyboard in case a user forgets the admin/admin combination.

Then the peak areas are manually typed into a spreadsheet to calculate the system suitability test injections and the results of the individual aliquots and the reportable result, see Figure 10.3. I will be charitable and assume that the spreadsheet template is validated. However, the working practice of the laboratory is that the completed file is only printed out but the electronic spreadsheet file is not saved. Finally, as shown in Figure 10.3, the reportable result is manually entered into a LIMS as part of the process to generate the certificate of analysis for the sample.

From a process perspective, this way of working is very inefficient. The calculations performed by the spreadsheet can easily be performed by the CDS, eliminating the need for one printout and if the CDS was interfaced to the LIMS so that the result was transferred electronically, both printouts and the spreadsheet become redundant.

However, look at the current data integrity issues in the various parts of the process:

- Activities 1–3: CDS
  Data stored on a single hard drive in the laboratory.
  Local data backup, there is no guarantee that backup will be performed regularly by laboratory staff.

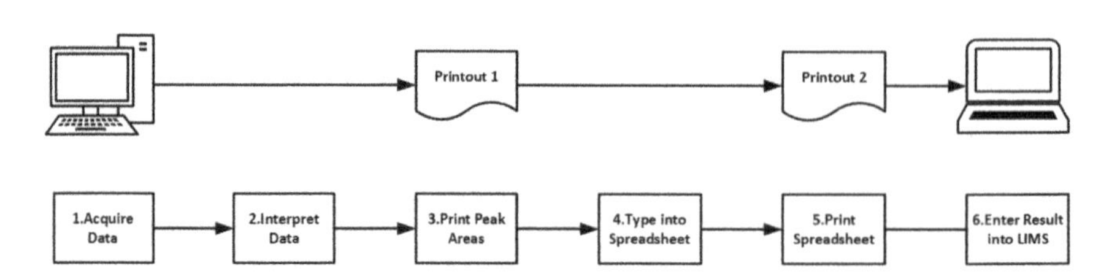

Figure 10.3 Process flow of instrument data system *via* spreadsheet calculations with the reportable result types into a LIMS.

User access to the system clock, recycle bin and data files.
Shared user identities and no user types defined – most probably all users are administrators
The password is written down.
Data can be stored in different locations by different users.
Sequence files, instrument and data acquisition methods can be changed but there is no knowing who created or changed a record.

- Activities 4 and 5: Spreadsheet
  Error prone manual entry into the spreadsheet.
  Completed spreadsheet file is not saved.
  Handwritten signatures and e-record are not linked.
  Error correction requires that all data must be re-entered and checked again.
- Activity 6: LIMS
  Error prone manual entry into the LIMS.

A simple process mapping methodology, such as IDEF, can help identify record vulnerability. Note the contrast between the system assessment presented in Section 10.3 and with the data process mapping outlined here in Section 10.4. The data process mapping is more powerful and reflects how work is performed in a laboratory. Figure 10.3 covers a CDS, spreadsheet and finishes with data entry into a LIMS, this is a better approach to data integrity assessments – process (including systems) not systems in isolation. Issues that may be identified in a system may actually occur prior to being entered into an application for example, in the sampling or sample preparation stages of an analysis.

10.4.4 Spreadsheets Used for GMP Calculations Are High Risk

Many laboratories use spreadsheets for calculation of analytical data generated by analytical instruments. Data are transcribed from a paper printout and entered into the spreadsheet and the final calculations printed out and the file may or may not be saved. There are several regulatory compliance issues with this approach:

- The spreadsheet is a hybrid system.
- The master template must be validated.
- The completed spreadsheet must be saved to a secure location and backed up as it is part of the complete data for the analysis.
- The resulting printout must be linked to the saved spreadsheet electronic record.

Facta Farmaceutici S.p.A., FDA Warning Letter, January 2017:[9]
1. Your firm failed to ensure that laboratory records included complete data derived from all tests necessary to assure compliance with established specifications and standards (21 CFR 211.194(a)).

You stored original data in an "unofficial" and uncontrolled electronic spreadsheet on a shared computer network drive. Your analyst stated that original data was first recorded in this "unofficial" spreadsheet and transcribed later to an "official" form. This spreadsheet showed failing results above the limits you established in your procedure, PCH 035 Visible Particle Determination in use prior to September 1, 2014.

Sandoz Private Ltd, FDA Warning Letter, October 2015:[10]

1. Your firm failed to prepare batch production and control records for each batch of drug product that include documentation of the accomplishment of each significant step in the manufacture, processing, packing, or holding of the batch (21 CFR 211.188(b)).

On August 28, 2014, FDA investigators identified instances of non-contemporaneous documentation of batch production activities. Two uncontrolled Excel spreadsheets were used to record discrepancies and certain in-process drug quality data. This data was initially missing in the batch manufacturing record. Your firm later entered this data into batch records and backdated them.
For example, according to a March 2, 2013 entry in one spreadsheet, you did not perform ..... testing as required after .... operations of ...... mg batch X. Despite this notation, the associated "In Process Sample Analysis Sheet" documents .... testing results from February 22, 2013.

Although spreadsheets can be very useful in the laboratory, their misuse can bring serious regulatory consequences. As part of any data integrity remediation, spreadsheets when possible should be eliminated from a process. This may not be possible in all cases due to the functionality offered by an application or the need to review data across several batches.

10.4.5 Critical Activities in a Process

From the discussion above, you can see that there are critical activities in a process where both data integrity and data quality can be compromised. Listed below are some of these, so that when data process mapping is undertaken some red flags can be raised:

- Access to computerised systems: are there unique user identities?
- Are roles segregated so that administrators are not users?
- If an analytical instrument is used to measure a property, is there objective evidence available, *e.g.* printout or a permanent e-record?
- Is the process activity manual that generates paper records?
- Are the paper records used uniquely numbered?
- Are data input manually into a computerised system?

Where these, and other critical activities occur, then the mapping team can delve deeper into the detail.

10.4.6 Fix and Forget *versus* Delivering Business Benefits?

As we plan the resolution of the record vulnerabilities following a process or system assessment, there are two main options to consider, as shown in Figure 10.4:

1. Short-term remediation – quick fixes to ensure data integrity of the process or system. This may involve some technical controls but typically involves procedural controls where further records are written to cover the process holes.
2. Long term solution – the redesign of the process to streamline it and remove paper. Typically, this could involve the implementation of functionality in an existing software application or implement a new application or even a new analytical technique (*e.g.* replacement of a traditional loss on drying analysis with a moisture analyser (analytical balance coupled with a heat source)).

However, in planning the process improvement, often we meet the immovable object that is management with or without the hindence of the finance department. Many senior managers just have an approach to remediation that requires a solution of quick fixes and get analytical staff back to work. Whilst this is an admirable approach for impressing their boss and "protecting" shareholder value, it is less than impressive in the laboratory. There is little or no thought about the adverse impact of these short-term remediations on the laboratory working practices.

Many remedial actions involve procedural controls, the use of paper records and even the introduction of more log books to paper (sorry!) over the cracks and ensure data integrity. Although the data integrity vulnerabilities may be resolved, what is the impact of this short-sighted approach on

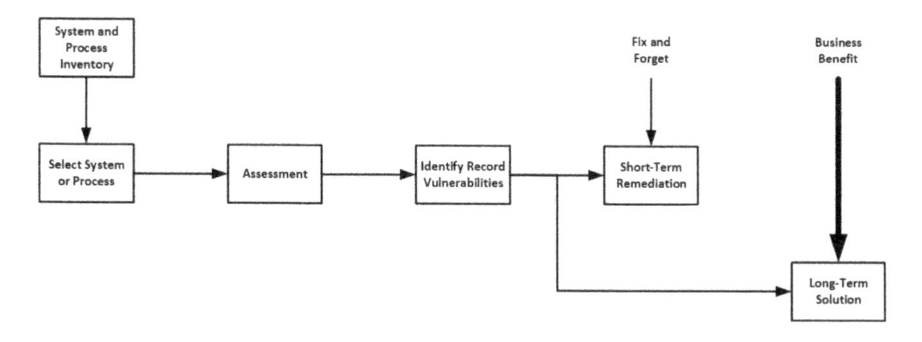

Figure 10.4 Process for assessment, remediation and long-term solution.

laboratory efficiency? Completing paper records during analysis will be slow and error prone. This will have a knock-on effect on second person review, where the review could take longer than the actual analysis. If this short-sighted approach is taken, management should never expect that laboratory to work at the same efficiency as before the remediation. Management must ensure that a laboratory's concerns are heard and provide the funding for an effective resolution by eliminating paper as much as possible. Gemba walks are an ideal opportunity for laboratory staff to feedback to senior management these problems and explain what is required. The problem is that the hidden cost of the procedural remediation is increased labour in the process not less.

10.4.7 Short Term Remediation Leading to Long Term Solution

From the list of issues listed above in Section 10.4.3, there will need to be remediation to bring the process back under control and ensure data integrity. Not everything can be achieved immediately so that there needs to be a prioritisation for short term remediation to fix critical items quickly and plans made for long term solutions, as shown in Figure 10.4. Going back to paper does not reflect the *current* in cGMP[5] that is discussed in more detail in Section 13.3.3. In Chapter 13 we discuss plans for long term solutions resulting in business benefits and process improvements that ensure data integrity.

From the data process mapping analysis in Section 10.4.3, we can develop several short-term solutions to ensure traceability and integrity of the results as well as some process improvement as shown in Table 10.4.

- The only capital item is the network card to put into the back of the CDS workstation to transfer data automatically, *via* a script or backup tool, to a secure and resilient network server to protect electronic records.
- Users will have their own log-on and user types can be defined and allocated to the appropriate users so that users can only access CDS functions appropriate to their user type.
- Of immediate business benefit is the elimination of the spreadsheet by incorporating the calculations performed into the CDS application. This eliminates one printout and one set of manual data entry into the spreadsheet.
- More problematical is the procedural controls for ensuring that data are only stored in specific locations. Procedural controls are ineffective and require more extensive second person review as well as quality oversight during data integrity audits.

The improved process can be visualised at the top of Figure 10.5.

The long-term solution to the problem and the one that provides the best business benefit is replacement of the CDS application by one with a

Table 10.4 Short term remediation and long-term solutions for a CDS, spreadsheet, LIMS manual process.

Time for resolution	Data integrity issue	Data integrity remediation
Short term remediation	• Data stored on a single hard drive local laboratory backup	• Install a network card and a script/utility for automated backup of data to a networked drive
	• User access to the system clock, recycle bin and data files	• Implement Windows security to restrict user access to the operating system
	• Shared user identities and no user types defined	• Implement individual user identities for all users
	• Sequence files, instrument and data acquisition methods can be changed with no knowledge who made the change	• Devise and implement user types suitable for role, *e.g.* analyst, supervisor, power user, IT admin • Remove sticky note with generic logon credentials
	• Data can be stored in different locations by different users	• Procedural controls enforced by detailed second person review
	• Error prone manual entry into the spreadsheet Completed spreadsheet not saved	• Eliminate spreadsheet and one printout by incorporating calculations into CDS
Long term solution	• Data stored on a single hard drive	• Replace CDS with a networked solution with a database
	• Manual data entry to LIMS	• Implement and validate CDS to LIMS interface to eliminate manual data entry between the two systems

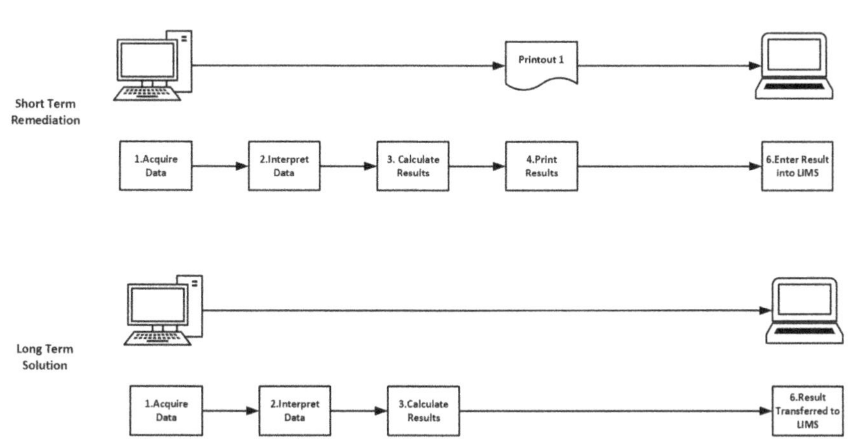

Figure 10.5 Process maps following short term remediation and long term solution implementation.

database that can be networked, as presented in Table 10.4. This will provide better enforcement of data integrity as well as provide the most business benefit with the elimination of the second printout and manual entry of the reportable result into the LIMS. Interfacing can go further with the download of sample information into the CDS from the LIMS reducing, but not eliminate, data entry to the sequence file. The simplified process can be seen at the bottom of Figure 10.5 and this approach is consistent with that advocated by WHO[1] and presented in Section 10.1.1.

10.5 Data Integrity Issues with Analysis by Observation

In a regulated analytical development or quality control laboratory there are several analyses that are manual and results are recorded by observation, *e.g.*:

- appearance;
- colour;
- odour.

The problem is that how can you ensure the integrity of the observation?

10.5.1 Potential Problems with Analysis by Observation

When an analytical procedure calls for analysis by observation, an analyst will take the sample and perform the test and typically write the result down in either a laboratory notebook or onto a controlled analytical sheet. These tests are non-destructive and are either subjective based on the observation by the analyst, *e.g.* appearance or by comparison, *e.g.* colour by using a colour chart.

The data integrity issue is that when a second person reviewer looks at the laboratory notebook or analytical sheet there are no data to look at, only the written observation.

In some cases, laboratories have taken the extreme approach of having a witness standing next to the tester to agree that what is written is correct. This is a problem. From a regulatory perspective there is no GXP requirement for a witness when conducting analysis and from a business perspective, this is a waste of resource.

However, these tests by observation are non-destructive and can easily be repeated. Let us examine an alternative and risk-based approach to this area.

10.5.2 A Risk Based Approach to Analysis by Observation

You will recall the PIC/S PI-041 Section 5.3.4 (ref. 2) quoted in Section 10.1.2 that stated: *Not all data or processing steps have the same importance to product quality and patient safety.* Take this statement a stage further; not all specifications carry the same weight when it comes to product release, which is more important:

- Impurity profile or appearance?
- Label claim or colour?

The point is that not many of the tests performed by observation carry the same weight as instrumental tests, and as said before the observational tests are non-destructive. Therefore, a risk based approach would be sensible and pragmatic.

Therefore, these tests should be carried out by an analyst as usual and without a witness. When the test is reviewed by a second person they should simply repeat the test and write down that the result is conformed in the notebook or analytical sheet with minimal impact on business efficiency. There will always be somebody who would argue that witness testing is the only way to do this but simply ask for the regulatory requirement for this and there will be no response to the question. Do not waste scarce resources as there are much more important data integrity issues to manage as we shall now see.

10.5.3 Melting Point Determination

Another test that is traditionally performed by observation is that of melting point determination. A sample is placed in a capillary tube and the temperature when the solid melts is recorded by the analyst. As with other tests by observation, there is no objective evidence for the second person review, merely a written record by the analyst. This raises some questions, *e.g.*:

- Did the test actually take place?
- Was the temperature recorded accurately?
- Did the analyst make a transcription error when recording the temperature?

During the assessment of the process – what is the impact of the test result? What risk is there to a submission, patient or product if the result is wrong? Here, remediation and long term solution could be the same: the purchase of a new melting point instrument with a video recording of the process. Now we have objective evidence of the testing. The instrument could also be controlled by a networked data system to ensure that the instrument was qualified, and have audit trails of the analysis and reporting of the results.

10.6 Data Integrity Issues with Paper Records

The current situation in many GMP laboratories is that many are paper based and working practices have hardly changed over last 2–3 decades. In fact, the FDA Guide to Inspection of Pharmaceutical Quality Control Laboratories issued in 1993 is still relevant a quarter of a century after publication as many working practices have not changed. The key issue is one of the master templates and blank paper forms.

Do not take from me that blank forms are a high regulatory risk, here is the FDA perspective on the problems with blank forms with two regulatory citations from warning letters.

Facta Farmaceutici S.p.A., FDA Warning Letter, January 2017 (ref. 9) Citation 2 states:

> Your firm failed to establish an adequate quality control unit with the responsibility and authority to approve or reject all components, drug product containers, closures, in-process materials, packaging materials, labelling, and drug products (21 CFR 211.22(a)).
> Our investigator observed many copies of uncontrolled blank and partially-completed CGMP forms … without any accountability or oversight of your quality unit.
> For example, a supervisor said he photocopied a blank OOS form and transcribed the information because he had made mistakes in the original document. Although your procedures required correcting mistakes on the original form, he made a new copy of a blank OOS form and rewrote the data.
> Our investigator documented that your employees used paper shredders to destroy critical laboratory and production records without the appropriate controls and procedures.

Abbott AMO (Hangzhou) Co, 483 Observation, January 2017,[11] citation 3 states that document control procedures are not established:

> Specifically, your firm's QC raw data used within the microbiology and analytical laboratories are accessible electronically by all employees. There are no procedures established to control the issuance, use and reconciliation of laboratory raw data worksheets.

10.6.1 Blank Forms Must be Controlled with Accountability

Since 2016, there have been increasing regulatory requirements for the control of master templates and blank forms, *i.e.* forms that are used to record data to demonstrate that a procedure has been executed correctly. The FDA, PIC/S and EMA all have stringent requirements for the control of both the master templates (*e.g.* uniquely identified, controlled, linked to a procedure, reviewed and approved) as well as the blank forms themselves (*e.g.* uniquely numbered, printed so that copying is easily identified, blank forms must be allocated to a department or individual and accounted for when work is completed and damaged forms must be kept). We will look at this subject next in Chapter 11.

References

1. *WHO Technical Report Series No.996 Annex 5 Guidance on Good Data and Records Management Practices*, World Health Organisation, Geneva. 2016.
2. *PIC/S PI-041 Draft Good Practices for Data Management and Integrity in Regulated GMP/GDP Environments*, Pharmaceutical Inspection Convention/Pharmaceutical Inspection Co-Operation Scheme, Geneva, 2016.
3. *MHRA Gxp Data Integrity Guidance and Definitions*, Medicines and Healthcare products Regulatory Agency, London, 2018.
4. A. Shanley, *Will the Pharmaceutical Industry Ever Get to Six Sigma?*, 2017, Available from: http://www.pharmtech.com/will-pharmaceutical-industry-ever-get-six-sigma.
5. Part 211-Current Good Manufacturing Practice for Finished Pharmaceuticals, *Fed. Regist.*, 1978, **43**(190), 45014–45089.
6. European Parliament, Council of the European Union, Directive 2001/83/EC of the European Parliament and of the Council of 6 November 2001 on the Community code relating to medicinal products for human use *Off. J. Eur. Communities*, 2001, **311**, 67.
7. *FDA Compliance Program Guide CPG 7346.832 Pre-approval Inspections*, Food and Drug Adminsitration, Silver Springs MD, 2010.
8. R. D. McDowall, *Validation of Chromatography Data Systems: Ensuring Data Integrity, Meeting Business and Regulatory Requirements*, Royal Society of Chemistry, Cambridge, 2nd edn, 2017.
9. *FDA Warning Letter: FACTA Farmaceutici S.p.A. (Warning Letter: 320-17-17)*, Food and Drug Administration, Silver Spring, MD, 2017.
10. *FDA Warning Letter: Sandoz Private Limited (Warning Letter 320-16-01)*, Food and Drug Administration, Silver Spring, MD, 2015.
11. *FDA 483 Observations: Abbott AMO (Hangzhou) Company, January 2017*, Food and Drug Administration, Silver Spring, MD, 2017.

CHAPTER 11

Data Integrity and Paper Records: Blank Forms and Instrument Log Books

Document control is a fundamental tenant of GXP regulations. This is easily understood and can be implemented for paper and electronic records relatively easily. However, a major problem is found at the back of many SOPs in analytical laboratories and this is a blank form. This has been thoughtfully designed by the subject matter expert who wrote the procedure to help people record the execution of the SOP. The problem is that blank forms can be filled in, then discarded if the analyst is not happy with the result, photocopied and filled in again and again. Blank forms are high on every regulator's hit list, as we will see in this chapter.

Another paper document is the log book for analytical instruments and chromatographic columns that should be used to record the use and maintenance of each one. The logs need to be controlled but if used correctly can be a valuable resource in ensuring data integrity in the second person review presented in Chapter 17.

11.1 What Do the Regulators Want? – Blank Forms

11.1.1 Focus on the Key Data Integrity Issues with Paper Records

Documentation, as the Principle of EU GMP Chapter 4[1] states, is key to being in compliance with Good Manufacturing Practice or indeed any Good Practice regulation within the pharmaceutical industry. Chapter 4 also states that

Data Integrity and Data Governance: Practical Implementation in Regulated Laboratories
By R. D. McDowall

Published by the Royal Society of Chemistry, www.rsc.org

documentation can exist in many forms. However, the purpose of this chapter is to focus on the main data integrity issues with paper records and look at two specific areas:

1. master templates and blank forms;
2. instrument log books and their role in data integrity.

As such, I will not discuss document management or document control.

We will start our discourse by addressing the regulatory concerns around master templates and blank forms. Of necessity, this section will contain much regulatory guidance as this is an area where falsification is easy and blank forms must be controlled to ensure the integrity of any paper records. We can also see how the regulatory approach has changed from a single sentence to much more extensive and stringent requirements, especially over the past few years.

11.1.2 FDA Guide to Inspection of Quality Control Laboratories

We will go back to the future to discover what the FDA said about control of blank forms over a quarter of a century ago. In 1993 the Agency published a Guide to Inspection of Quality Control Laboratories[2] that was written after the Barr Laboratories court case, discussed in Chapter 2, and written to help FDA inspectors in the aftermath of the case. Why mention an old publication when it is nearly a quarter of a century old? Quite simply, it is still relevant now as it was then as many regulated laboratories work the same way now as they did them.

In Section 13 on the topic of laboratory records and documentation, it states:[2]

> We expect raw laboratory data to be maintained in bound, (not loose or scrap sheets of paper), books or on analytical sheets for which there is accountability, such as prenumbered sheets.

11.1.3 MHRA GMP Data Integrity Guidance

In the section on Designing Systems to Assure Data Quality and Integrity of the MHRA Data Integrity Guidance for Industry, 2015[3] it states:

Systems should be designed in a way that encourages compliance with the principles of data integrity. Examples include:

- Control over blank paper templates for data recording

11.1.4 MHRA Draft GXP Data Integrity Guidance

In July 2016, MHRA issued an update of the GMP guidance document where the scope was extended to cover GCP and GLP as well as GMP[4] where the section on blank forms was expanded to state:

'Free access' to blank paper proformae for raw/source data recording should be controlled where this is appropriate. Reconciliation may be necessary to prevent recreation of a record.

11.1.5 MHRA GXP Data Integrity Guidance and Definitions

In the 2018 publication of the final version of the GXP Data Integrity Guidance,[5] there are two sections on blank forms and their control; MHRA has expanded their guidance from 8 words in 2015 simply asking for "control", into two paragraphs with much more detail three years later.

Section 5.1
Access to blank paper proformas for raw/source data recording should be appropriately controlled. Reconciliation, or the use of controlled books with numbered pages, may be necessary to prevent recreation of a record.

Section 6.7
When used, blank forms (including, but not limited to, worksheets, laboratory notebooks, and master production and control records) should be controlled. For example, numbered sets of blank forms may be issued and reconciled upon completion.

11.1.6 WHO Guidance on Good Data and Record Management Practices

In the final guidance from the WHO, there is a brief requirement for blank forms:[6]

Good document design, which encourages good practice: documents should be appropriately designed and the availability of blank forms/documents in which the activities are recorded should be ensured;

However, this can be difficult to interpret for blank forms as there is little further information available.

11.1.7 FDA Data Integrity and Compliance with cGMP

A draft data integrity guidance was issued by the FDA in 2016 (ref. 7) in the form of a question and answer approach. Question 6 asks, "How should blank forms be controlled?" The answer is reproduced below:

There must be document controls in place to assure product quality (see §§ 211.100, 211.160(a)).

FDA recommends that, if used, blank forms (including, but not limited to, worksheets, laboratory notebooks) be controlled by the quality unit or by another document control method.

For example, numbered sets of blank forms may be issued as appropriate and should be reconciled upon completion of all issued forms. Incomplete or erroneous forms should be kept as part of the permanent record along with written justification for their replacement (for example, see §§ 211.192, 211.194).

Similarly, bound paginated notebooks, stamped for official use by a document control group, allow detection of unofficial notebooks as well as of any gaps in notebook pages.

This is a reiteration of the 1993 guidance[2] but it also goes into more detail in that blank forms must be prenumbered, they are issued to individuals. After a form has been completed there needs to be a reconciliation process to show that each one has been returned. You must count the forms out and count them back again. If one form is incomplete or errors have been made, there needs to be a written justification for issuance of a new form. The returned form must be retained. In addition, there need to be document controls such as design, approval and secure storage of the master template.

11.1.8 PIC/S PI-041 Good Practices for Data Management and Integrity in Regulated GMP/GDP Environments

Section 8 of the PIC/S guidance[8] focuses on the QMS and the control of master templates, blank forms and records.

8.3.1 Why is managing and controlling master records necessary?

Managing and controlling master records is necessary to ensure that the risk of someone inappropriately using and/or falsifying a record 'by ordinary means' (*i.e.* not requiring the use of specialist fraud skills) is reduced to an acceptable level. The following expectations should be implemented using a quality risk management approach, considering the risk and criticality of data recorded (see sections 5.4, 5.5).

8.4.1 An index of all documents and the template records should be maintained by QA organisation within the pharmaceutical quality system. This index should mention for each type of template record at least the following information: title, reference number including version number, location (*e.g.*, documentation data base, effective date, next review date, *etc.*)

8.5.1 Records should be available to operators at the point-of-use and appropriate controls should be in place to manage these records. These controls should be carried out to minimize the risk of damage or loss of the records and ensure data integrity. Where necessary, measures must be taken to protect records from being soiled (*e.g.* getting wet or stained by materials, *etc.*).

There are tables outlining the further regulatory expectation that will be discussed in Sections 11.2.2–11.2.4.

11.1.9 EMA GMP Questions and Answers on Data Integrity

EMA has issued a question and answer on data integrity for GMP[9] and question 14 focuses on:

> How should the company design and control their paper documentation system to prevent the unauthorised re-creation of GMP data?
>
> The template (blank) forms used for manual recordings may be created in an electronic system (Word, Excel, *etc.*). The corresponding master documents should be approved and controlled electronically or in paper versions. The following expectations should be considered for the template (blank) form:
>
> - Have a unique reference number (including version number) and include reference to corresponding SOP number
> - Should be stored in a manner which ensures appropriate version control
> - If signed electronically, should use a secure e-signature
>
> The distribution of template records (*e.g.* 'blank' forms) should be controlled. The following expectations should be considered where appropriate, based on data risk and criticality:
>
> - Enable traceability for the issuance of the blank form by using a bound logbook with numbered pages or another appropriate system. For loose leaf template forms, the distribution date, a sequential issuing number, the number of the copies distributed, the department name where the blank forms are distributed, *etc.* should be known
> - Distributed copies should be designed to avoid photocopying either by using a secure stamp, or by the use of paper colour code not available in the working areas or another appropriate system.

11.1.10 Regulatory Guidance Summary

Regulatory concerns about blank forms have been evident for a quarter of a century. As can be seen by reading through this section the guidance from regulatory authorities has expanded to include the master templates used to create the blank forms as well as stringent control of the use of the forms themselves with accountability. Why is this you may ask? The simple answer is that the industry listens but does not act. In the absence of global movement by the pharmaceutical industry, the regulators are making industry's mind up for them.

What does the guidance boil down to?

- Master templates for blank forms need to be designed correctly, controlled, approved and protected.
- Blank forms for routine use need to be uniquely numbered and secure from copying, are issued to specific individuals or groups and reconciled after use. Damaged forms must be retained.
- Uniquely numbered blank forms must be completed correctly and be reviewed.

If there is any doubt about the problems with blank forms, see the FDA regulatory citations presented in Chapter 10.6 with the Facta Farmaceutici warning letter[10] and the 483 Observation for Abbott AMO (Hangzhou) Co.[11] These regulatory citations show that problems with blank forms persist over 25 years after the FDA first mentioned that the area needs control.[2] It is a good example of the pharmaceutical industry not keeping current and still relying on paper.

11.2 Control of Master Templates and Blank Forms

11.2.1 Understanding Master Templates and Blank Forms

Before we go much further into this discussion it is important to have a clear understanding of what is meant by a master template and a blank form. This is shown in Figure 11.1.

At the top left is a written procedure or recording form, *e.g.* standard operating procedure, work instruction, analytical procedure or analytical batch record. This will go through a process of drafting, review, revision, approval, update and finally retirement that is common throughout the pharmaceutical industry. An analytical batch record is a blank form. Also, at the back of many laboratory procedures will be a form designed to be completed by an operator as they perform the procedure to demonstrate compliance with the procedure and forms a record of the work. This will go through the document cycle as part of the procedure and not as a separate document.

The master template is the approved electronic form and a copy that is the blank form attached to the approved SOP as an appendix or the template of the analytical batch record used instead of as the master templates. The copies of either are the blank forms. PIC/S Guidance PI-041 (ref. 8) has in Section 8 of the document three tables about the master templates and blank forms. In the next sections, the tables presented here have been adapted rather than quoted verbatim, please refer to the guidance document to see the original version.

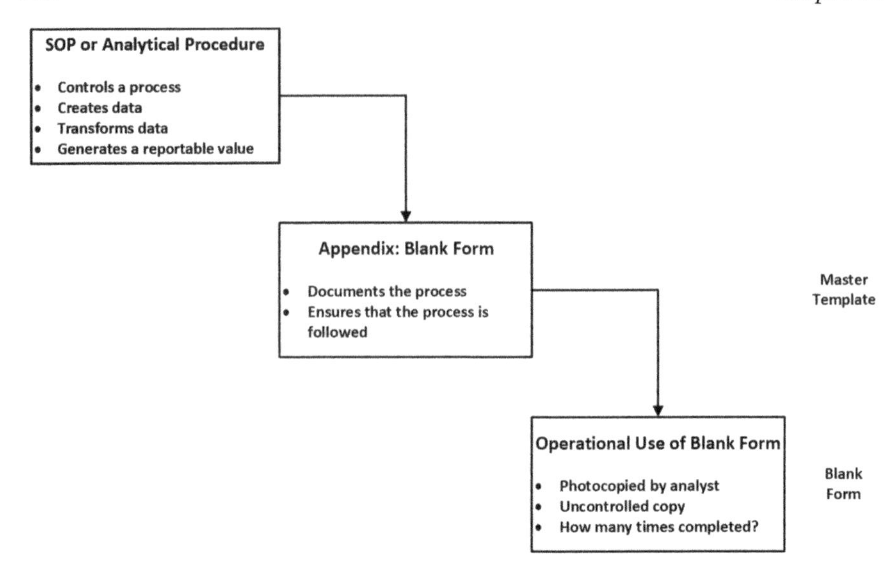

Figure 11.1 Understanding master templates and blank forms.

11.2.2 Requirements for the Design, Approval and Storage of Master Templates

There will be a procedure for the writing of procedures (typically an SOP on SOPs) that will define the process within a company for the design, drafting, reviewing, updating, approving, regular review and update and finally retiring a procedure. This overarching SOP will also detail the document controls and be linked to a document template so that all procedures have the same look and feel within an organisation.

Given the regulatory focus, each the SOP on SOPs needs to be reviewed to assess if there are adequate controls for the master templates and blank forms. Table 11.1 shows the PIC/S regulatory expectations for the generation and control of master templated that has been adapted from PIC/S PI-041 (ref. 8) in the preparation of this book.

11.2.3 Process for Generation, Review and Approval of a Master Template

The process for designing, reviewing and controlling the master template is shown in Figure 11.2. This is a cross functional process flow diagram that shows the interaction between the writer and reviewer (analytical laboratory), approver (Quality Assurance) and master template secure storage (IT).

- Each master template should have an owner allocated to it. This will typically be the subject matter expert (SME) responsible for writing and maintaining the standard operating procedure, analytical procedure or work instruction that controls the use of the form.

Table 11.1 PIC/S PI-041 expectations for the generation and control of master templates. (Adapted from ref. 8.).

Regulatory expectations	Data integrity issue
1. All documents should have a unique identification number (including the version number) and should be checked, approved, signed and dated.	• An analytical batch record must be linked to the product/material that it relates to. • The master template needs to be linked to the procedure or work instruction it relates to. • Each form needs to be version controlled, signed, approved and dated. There is a risk of using superseded forms if there is no version control or controls for issuance. • If records can be created and accessed without control, it is possible that the work may not have been recorded at the time the event occurred.
2. The use of uncontrolled documents should be prohibited by local procedures.	• Uncontrolled documents increase the potential for omission or loss of critical data as these documents may be discarded or destroyed without traceability. • Uncontrolled records may not be designed to correctly record critical data. • It is easy to falsify uncontrolled records – just take another form and repeat the work.
3. The use of temporary recording practices, *e.g.* scraps of paper should be prohibited.	• Use of temporary recording practices may lead to data omission, and these temporary original records are not specified for retention.
4. The document design should provide sufficient space for manual data entries.	• Handwritten data may not be clear and legible if the spaces provided for data entry are not sufficiently sized. • Documents should be designed to provide sufficient space for comments, *e.g.* in case of a transcription error, there should be sufficient space for the operator to cross out, initial and date the error, and record any explanation required. • If additional pages of the documents are added to allow complete documentation, the number of, and reference to any pages added should be clearly documented on the main record page and signed. • There should be no writing on the reverse (unused side) of existing pages as this would typically be omitted when copied.
5. The document design should make it clear the data that is to be entered into each area on the form.	• Ambiguous instructions may lead to inconsistent/incorrect recording of data. • Ensures all critical data are recorded. • Ensures clear, contemporaneous and indelible/durable completion of entries. • The document should also be structured in such a way as to record information in the same order as the operational process and related SOP, to minimise the risk of inadvertently omitting critical data.

(*continued*)

Table 11.1 (*continued*)

Regulatory expectations	Data integrity issue
6. Documents should be stored in a manner that ensures appropriate version control.	• Inappropriate storage conditions can allow unauthorised modification, use of expired and/or draft documents or cause the loss of master documents. • The processes of implementation and the effective communication, by way of appropriate training prior to implementation when applicable, are just as important as the document. • Master copy (in soft copy) should be prevented from unauthorised or inadvertent changes. For example: For the template records stored electronically, the following precautions should be in place: • Process controls for creating and updating versions should be clear and practically applied/verified;
7. Master copies should contain distinctive marking so as to distinguish the master from a copy, *e.g.* use of coloured papers or inks so as to prevent inadvertent use.	• Access to master templates should be controlled; • Master documents should be stored in a manner that prevents unauthorised changes;

- This form can be designed using a word processor or even a spreadsheet, however, the template must have the name and the version number of the form embedded into it as well as the procedure number to which the form relates.
- The form will be designed to accommodate the work and collect records according to the applicable process and/or procedure. One point that must be ensured is that there is sufficient space for an analytical chemist to enter a value or result in each place on the form.
- When complete the form needs to be reviewed by a different person to ensure that the form is complete, correct and accurate and matches the requirements of the controlling procedure. If changes are required, the form is returned to the author for an update.
- When technical reviews are complete, there needs to be a quality function review and approval and when approved, the date of approval is added to the master form.
- The form should be signed either handwritten or by electronic signature.
- Assuming that no changes are required, the template master now needs to be stored securely either in an IT environment on the network or in an electronic document management system with restricted access to it.

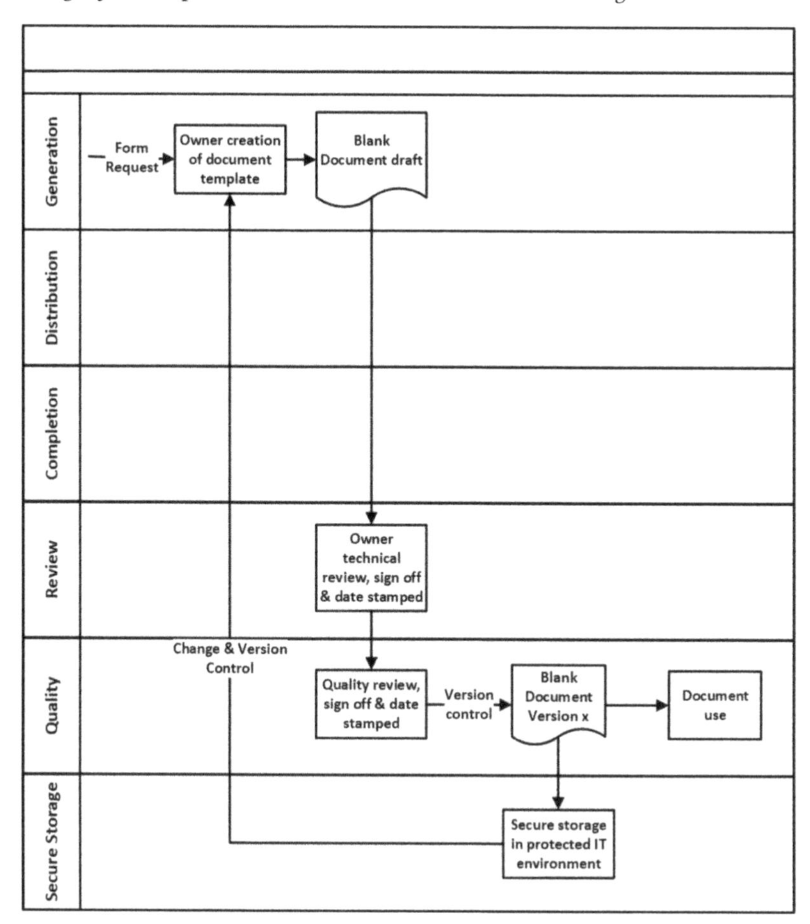

Figure 11.2 A process for generating and controlling a master template for a blank form.

At the end of this first process, we now have a blank template that is version controlled and stored securely.

11.2.4 Requirements for the Issue and Reconciliation of Blank Forms

The regulatory expectations from PIC/S PI-041 for the issue and reconciliation of blank forms are presented in Table 11.2. There is a very bureaucratic process for issuing and reconciling the forms as shown in Table 11.2. Note that there is a difference between PIC/S[8] and the FDA[7] guidance documents for what to do with superfluous and damaged forms. PIC/S guidance permits the destruction of the forms or secure storage, however, the FDA guidance[7] requires all forms to be securely retained and a justification for reissue of damaged forms.

Table 11.2 PIC/S PI-041 expectations for the issue and reconciliation of blank forms. (Adapted from ref. 8.).

Regulatory expectations	Data integrity issue
1. Updated versions should be distributed in a timely manner.	• Obsolete master documents and files should be archived and their access restricted. • There may be a risk that obsolete versions can be used by mistake if available for use. • The obsolete version can be used intentionally or by error. • Any issued and unused physical documents should be retrieved, reconciled, and where permissible, destroyed accordingly.
2. The issue of blank forms must be controlled by a written procedure.	• Ensuring that only the current approved version of each form is available for use. • Allocating a unique identifier to each blank document issued and recording the issue of each document in a register. • Name of the person and date the copies were issued. • Name of the individual to whom the controlled copies were issued. • Use of a secure stamp, metal seal to bind pages together, or paper colour code not available in the laboratory areas to ensure that pages cannot be copied. • Without the use of security measures, there is a risk that rewriting or falsification of data may be made after photocopying or scanning the template record (which gives the user another a template to copy). • A filled record with an anomalous data entry could be replaced by a new rewritten template. • All unused forms should be accounted for, and either defaced and destroyed or returned for secure filing. • *NOTE*: There is a difference between the PIC/S and FDA guidance documents, the FDA[7] requires all forms including unused and damaged forms to be retained and reconciled and PIC/S allows reconciliation and then an option to destroy damaged forms.[8]
3. The unique numbering of every distributed blank form.	• Given the number of forms in an analytical laboratory, there needs to be a blank form numbering convention, *e.g.* SOP123/01/00001 for the SOP and version number plus a sequential numbering for each individual form • Where the re-issue of additional copies of the blank form is necessary, a controlled process regarding re-issue should be followed. • All issued forms should be reconciled following use to ensure the accuracy and completeness of records.

A prudent approach would be to retain all copies of both damaged and unused forms.

The sane approach would be to eliminate blank forms wherever possible and automate the process.

11.2.5 Process for Issue and Reconciliation of Blank Forms

Now we come to the issue of how to request and reconcile after use a blank form. Instead of printing or photocopying a blank form when you want to do some work, you now must have a formal process for issuing a controlled and numbered version of the document and this is shown in Figure 11.3.

- The process starts with a request made to the person or group who manages the issue and reconciliation of blank forms. This function should be outside of the laboratory, typically this would be a Quality Assurance

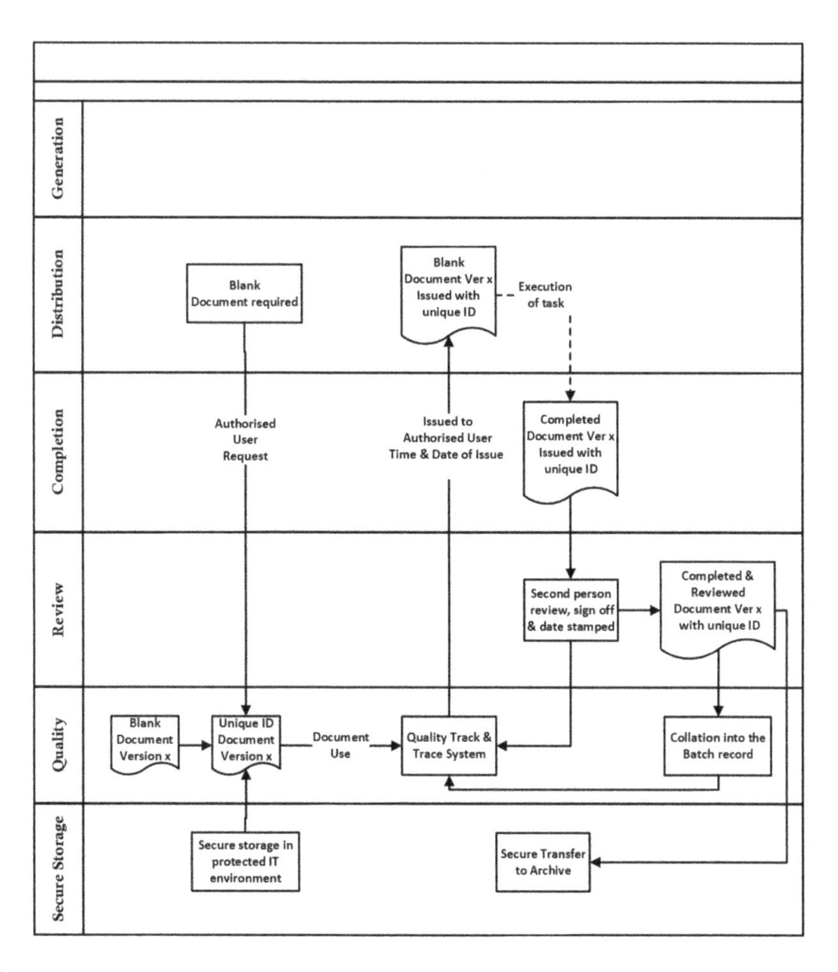

Figure 11.3 Process for the issue and reconciliation of controlled blank forms.

role to ensure the independence of operation. An authorised analyst will request a specific form from the document controller who will issue a uniquely numbered version of the blank form created from the current version of the template.

- The unique number is entered into a register or a track and trace system and the name of the requester is entered along with the date/time of issue. This track and trace system in its simplest form is a bound logbook with numbered pages with the entries handwritten by the document controller with information such as date of issue, unique form number, person/department to whom the form was issued to. However, consider the practicalities of a paper based register. How many procedures are there in laboratories that are there with blank forms? 10, 50, 100, 500? Regardless of the actual number, managing this paper will be an administrative nightmare. You will need a file numbering convention as outlined in Table 11.2.
- The distributed copies of these blank forms should be designed to avoid photocopying either by the use of a secure stamp and ink colour or by the use of coloured paper not available in the laboratory.
- An electronic system that can issue forms with a unique number, as well as copy number (controlled copy printing), may be an alternative but stringent validation needs to ensure that this process is secure and only one copy is allowed to be printed. Otherwise, a controlled copy will only be issued on paper as an electronic version could be reprinted (the same as blank forms today).

Do you really want to do this? You can see the compliance overhead by just reading the text above or looking at Figure 11.3.

11.2.6 Process for Issue and Reconciliation of Blank Forms

To visualise the process for the issuance and reconciliation of blank forms, see Figure 11.3:

- An authorised user makes a requires for a controlled blank form. A uniquely numbered copy is made from the working master template, logged in the register or a track and trace system and distributed to the requester.
- If everybody did this for every blank form the laboratory or QA would grind to a halt, so a batch of controlled and numbered blank forms may be prepared instead. This needs careful handling to ensure that reconciliation and accountability of the forms are ensured during their working life, therefore the batch would be for a short-term period rather than, say, annually.
- The form is completed as an analysis is undertaken as described in Section 11.2.7.
- If there is a mistake and a new form is needed, then the form needs to be returned to the document controller but before a new form is issued

there needs to be a documented rationale for why the form needs to be replaced. Get your excuses and grovel pads at the ready for the document controller! The old form must be retained and a new form issued.
- At the completion of the work, a reviewer needs to check that work is completed correctly and if any calculations are included on the form, they need to be verified as correct including rounding.
- The form along with other documented evidence is collated into the analytical batch or study record and the form is reconciled with the track and trace system. It is important to know the location of every form in the register or track and trace system.

11.2.7 Completing Blank Forms and Creating GXP Records

In addition to the processes for the control of master templates and issue and reconciliation of blank forms, one more data integrity issue needs to be considered: the completion of the form itself. Completion of the forms should follow good documentation practices as shown in Table 11.3 and the reader is referred to Appendix 1 of the WHO guidance[6] for further information on this subject.

11.3 What Do the Regulators Want? – Instrument Log Books

We will look at the regulatory requirements for instrument log books and by implication any other log books used in an analytical laboratory such as chromatography column logs, sample management logs, chromatographic sequence logs, *etc.*

11.3.1 EU GMP Chapter 4 on Documentation

Clause 4.31 of EU GMP Chapter 4[1] states:

> Logbooks should be kept for major or critical analytical testing, production equipment, and areas where product has been processed.
>
> They should be used to record in chronological order, as appropriate, any use of the area, equipment/method, calibrations, maintenance, cleaning or repair operations, including the dates and identity of people who carried these operations out.

11.3.2 FDA GMP 21 CFR 211

The US GMP regulations have requirements for equipment maintenance and use logs in 21 CFR 211.182:[12]

> A written record of major equipment cleaning, maintenance (except routine maintenance such as lubrication and adjustments), and use shall be

Table 11.3 PIC/S PI-041 expectations for the completion blank forms. (Adapted from ref. 8.).

Regulatory expectations	Data integrity issue
1. Handwritten entries must be made by the person who executed the task.	• Check that handwriting is consistent with entries made by the same person.
2. Correct completion of the form according to GXP documentation principles.	• Check that written entries are in ink, which is not erasable and/or will not smudge or fade over the retention period. • Check that the records were not filled out using pencil prior to being overwritten by pen. • Unused, blank fields within documents should be crossed-out, dated and signed. • All entries must be legible, clear and unambiguous. • Does not include the use of unknown symbols or abbreviations. • Ditto (") marks should not be used. • Write the date and do not use a date stamp (Note: Informal discussions with individual GMP inspectors indicate that many are not concerned about the use of date stamps). • Backdating and postdating are not allowed. • All corrections to the original entries should be initialled and dated with an explanation for the change.
3. Form completeness.	• Performer of the test to check that the form is correct and entries are complete. • Second person reviewer to verify completeness and accuracy of data recorded. • Check correct pagination of the records and are all pages present. • The completion of date fields should be done in the format that is unambiguous, *e.g.* DD MMM YYYY or format defined for the organisation
4. The completion of records of operations should be contemporaneous.	• Verify that forms and applicable procedures are available within the immediate areas in which they are used, *i.e.* Inspectors should expect that sequential recording can be performed at the site of operations. • If the form is not available at the point of use, this will not allow operators to fill in records at the time of occurrence.
5. Original records and true copies.	• If thermal paper is used, ensure that a true copy has been taken, signed indelibly and dated. • Both the true copy and the original should be retained.
6. Records should be signed and dated using a unique identifier that is attributable to the author.	• Check that there are signature and initials logs that are controlled and current and that demonstrate the use of unique examples, not just standardized printed letters. • Ensure that all key entries are signed and dated, particularly if steps occur over time, *i.e.* not just signed at the end of the page and/or process.

> included in individual equipment logs that show the date, time, product, and lot number of each batch processed. ....
>
> The persons performing and double-checking the cleaning and maintenance (or, if the cleaning and maintenance is performed using automated equipment under 211.68, just the person verifying the cleaning and maintenance done by the automated equipment) shall date and sign or initial the log indicating that the work was performed.
>
> Entries in the log shall be in chronological order.

11.3.3 FDA Good Laboratory Practice 21 CFR 58

The US GLP regulations in 21 CFR 58.63(c) also have a requirement for equipment records:[13]

> Written records shall be maintained of all inspection, maintenance, testing, calibrating and/or standardizing operations.
>
> These records, containing the date of the operation, shall describe whether the maintenance operations were routine and followed the written standard operating procedures.
>
> Written records shall be kept of nonroutine repairs performed on equipment as a result of failure and malfunction. Such records shall document the nature of the defect, how and when the defect was discovered, and any remedial action taken in response to the defect.

11.3.4 OECD GLP Regulations

Instrumentation and equipment used in the analytical laboratory are referred to as apparatus and in Section 4.2 of the OECD GLP regulations there is the following requirement:[14]

> Apparatus used in a study should be periodically inspected, cleaned, maintained, and calibrated according to Standard Operating Procedures.
>
> Records of these activities should be maintained.
>
> Calibration should, where appropriate, be traceable to national or international standards of measurement.

11.3.5 Summary of Regulatory Requirements for an Instrument Log Book

There is the need for an instrument log book that covers the use of the instrument, calibration, maintenance and repairs. What is not included in all the regulatory quotes above is that instruments also must be calibrated, qualified and requalified but this is contained in other clauses of the relevant regulations or pharmacopoeias. These activities also must be included in the log book. The log book must be a sequential record of all work performed on the instrument and not divided up into sections. The regulatory requirement is that a log book is controlled and bound with sequentially numbered pages.

11.4 The Role of an Instrument Log Book for Ensuring Data Integrity

From the data integrity model discussed in Chapter 5 the importance of a correctly qualified and calibrated analytical instrument is immediately apparent. Level 1 of the model is the right instrument for the right job and is placed immediately above the Foundation layer.

11.4.1 Why is an Instrument Log Book Important?

A good question to begin with. One answer is that one is required by either the Good Laboratory Practice (GLP) or Good Manufacturing Practice (GMP) regulations as seen above. Another answer is that it is good analytical science as provides a history of the instrument, *e.g.*:

- initial set up and qualification of the instrument;
- use of the instrument;
- calibration or point of use checks to demonstrate that the instrument is working correctly;
- preventative maintenance visits from a service provider;
- repairs and maintenance carried out in-house and/or by a service provider;
- periodic re-qualification of the instrument.

As such, a log book is essential to demonstrate that an analytical instrument is under control, works correctly and the results generated can be relied upon.

An instrument log book is also one of the essential documents for data integrity and this is typically a bound and paginated paper book that is completed manually. Not loose-leaf sheets or use of a ring bound notebook purchased from a local supermarket but a real book. However, you would be surprised by the number of laboratories that cannot get even this basic element right. The log book is issued by a document control function, typically the Quality Assurance department, and when completed the log is returned to document control and then archived. To ensure that it is completed correctly there needs to be a written procedure with training for making entries in an instrument log book but also for reviewing them. There need to be a means of referencing the reports from service visits and any qualification of the instrument both need to be accommodated to ensure a complete record for each instrument. The other item that is often neglected is who is going to review the entries, when and how often?

11.4.2 What Needs to be Entered in the Log Book?

As we all work to the same regulations you would think that the contents of an instrument log book would be easy and straightforward. Unfortunately, this is not the case. We follow the same regulations but everyone interprets the words differently. Summarising these GMP and GLP regulations in Section 11.3, the log book entries must be:

- Written.
- Each entry must be linked to the use of the instrument, *e.g.* use, calibration, maintenance, *etc.*
- Chronological, which means that the entries must be dated and must be made contemporaneously in the sequence they are performed on a day.
- Do not separate the tasks of use, calibration, maintenance, *etc.* but list them all chronologically in a single section.
- Checks for correct operation of the instrument.
- Point of use checks or system suitability test samples.
- Description of the analytical work undertaken by the instrument *e.g.* batch or study number, analytical procedure used, number of samples or sample identities, as appropriate.

In addition to the elements listed above, you also need to consider any pharmacopoeial requirements such as system suitability test results or sound science such as point of use checks made on the day to show that the system is working correctly, *e.g.* calibration of a balance or pH meter or running a reference standard spectrum for a spectrometer.

You would think that from this list it would be simple, right? Wrong! For example, how will use of the instrument be interpreted? Does each use mean just the work package performed or individual runs within that package? It depends on the interpretation by an individual organisation of the regulations. BUT.... We now have additional requirements for the log book that potentially cloud the issue:

- The FDA's Part 11 Scope and Application Guidance[15] allows laboratories to record audit trail information if an application does not have such functionality.
- Data integrity requirements mean that the log book takes on an increased importance for recording the work potentially resulting in increased granularity for the entries. This may include verification by a second person of an instrument reading where the output is not captured electronically or by an attached printer. Or entries *in lieu* of an audit trail for corrections to data.

And yes, you have guessed – all entries are handwritten!

11.4.3 Inspectors Know the Importance of an Instrument Log

The importance of the instrument log book has been known by regulatory agencies for a long time. Enter stage left my North American advertising agency .... The Food and Drug Administration. The FDA guidance for inspectors on the Inspection of Pharmaceutical QC Laboratories was issued in 1993, following the Barr Laboratories case, and has an interesting approach that is also very useful for second person reviews and data integrity audits:[2]

> Laboratory records and logs represent a vital source of information that allows a complete overview of the technical ability of the staff and of overall quality control procedures.

Note the three key phrases:

- a vital source of information;
- the technical ability of staff;
- overall QC procedures.

Why does the FDA think this about a humble instrument log book? The reason is that a log book is not humble but a vital link between the analytical instrument plus any data system and the work carried out, as you can see in Figure 11.4. The log book links the analytical instrument with the work performed by or on the instrument and the data system such as qualification and requalification work, method development, method validation, calibration checks, system suitability test, routine work, repairs and preventative

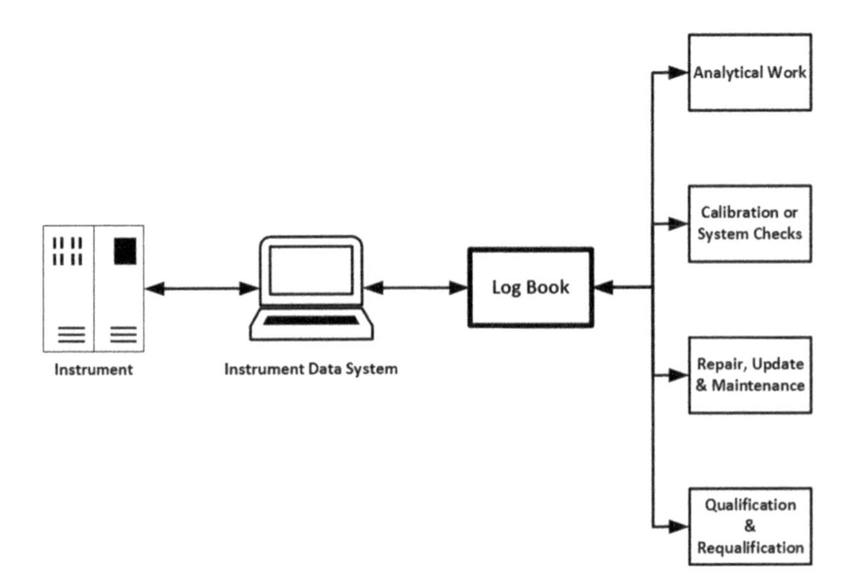

Figure 11.4 Relationship between the instrument log book, the analytical instrument and the work performed.

maintenance. The aim is to have a complete record to identify issues and be an aid in troubleshooting issues as well as trending calibration or system suitability test results.

We will now compare the theory with the practice as found in inspections.

11.4.4 FDA Citations for Laboratory Log Books

In May 2016 Akorn AG received a 483 citation for discrepancies in the sample log book entries used to document storage conditions as either ambient or 2–8 °C:[16]

- Between 17th February and 18th May, there were 24 instances where a single sample was noted as being stored at both temperatures so it was not known if samples were correctly stored prior to analysis.
- Write overs of entries where dashes were changed to X's.
- Review of the log book by a supervisor only consists of checks for empty spaces. Discrepant entries (*e.g.* write overs) had not been identified or investigated for their impact on stored samples or the corresponding analytical data.

In the Aarti warning letter of July 2013 there was the following citation:[17]

- On October 27, 2012, our investigator noticed that a QC analyst was performing a Loss on Drying (LOD) analysis for <redacted> and had recorded the completion time as <redacted> and total time as <redacted> in the usage log book for the LOD oven usage logbook although the step was not yet completed.

How do these citations compare with the criteria identified in a quarter of a century old FDA guidance document?

- What do you think of the technical ability of the staff or should that be the inability of staff?
- Are the written procedures adequate or even followed?
- If you were an auditor or an inspector what would be your view of the technical competence of this laboratory?

Just from these two citations, you can start to see the importance of a correctly completed and reviewed instrument log book. The instrument log book is also very important in second person review, data integrity audits and data integrity investigations.

11.4.5 Instrument Log Books in Practice

Typically, the majority of instrument log books are paper based using bound books with sequentially numbered pages to prevent pages being removed and replaced. You would be surprised that I still find instrument log books that fail to meet these simple requirements, *e.g.*:

- bound books but without numbered pages;
- ring files containing uncontrolled loose pages for documenting instrument use;
- spiral bound books with unnumbered pages where pages can be removed without any record.

An instrument log book requires all users to complete the entries chronologically, manually, consistently and contemporaneously. The way the use of an instrument is recorded will depend on the interpretation of the regulations in Section 11.3 by each individual organisation. However, as the log book is paper it is always difficult to know if an entry has been made contemporaneously or before or after the event.

As an auditor, one item that is usually overlooked is that each instrument log book in any analysis should be subject to second person review to ensure that the work is correctly documented. As can be seen in the 483 observations earlier, some reviews can be superficial and ineffective. The second person review may look at a single entry or connected entries for a study or batch release but there is also the need for the instrument owner to check periodically all entries to confirm that over time the instrument is operating correctly.

11.5 Role of the Instrument Log Book in the Second Person Review

One of the important stages of analysis is the second person review – the third and fourth eyes of the four eyes principle. Primarily, it is good analytical science to have a check of results including have procedures been followed, have samples been analysed correctly, data interpreted correctly, have calculations are right and is the reportable result accurate. The second person review is also a means of ensuring data integrity and preventing poor data management practices or falsification of data. An outline of a second person review is shown in Figure 11.5 and Chapter 17 is devoted to this subject. Here, the role of the log book is to verify what has been performed in the data system as evidenced by all electronic records, the paper printouts and the log book entries are congruent and are consistent. Omissions or discrepancies with anything between the three parts need to be investigated thoroughly – is a discrepancy due to a mistake or is it something more sinister?

11.5.1 Is an Instrument Performing OK?

An instrument log book is often a repository of data about the day to day performance against a reference substance, calibration standard or system suitability parameters. Over time it can tell you if an instrument is continuing to

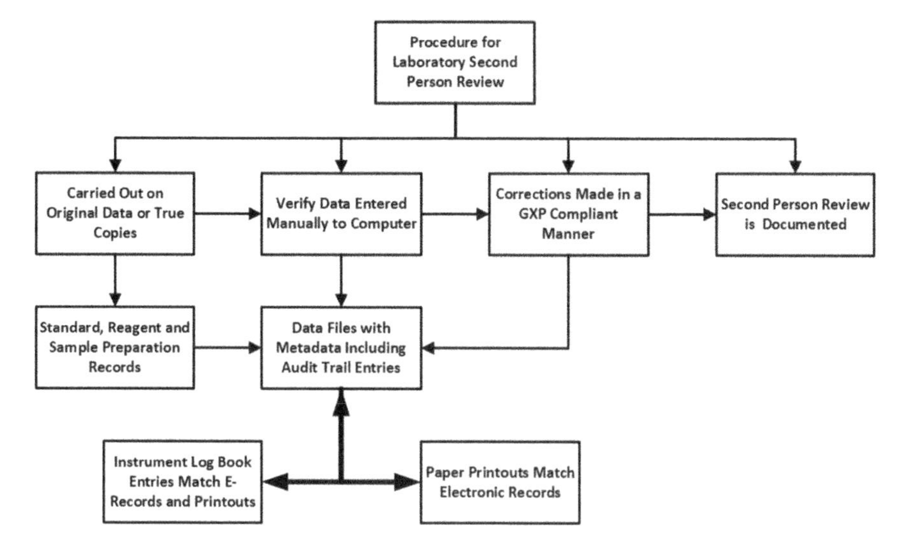

Figure 11.5 Role of the instrument log in a second person review.

perform acceptably. However, as this information is stored on paper there is often little attempt made to trend the data to see if there is a potential problem and stop it getting worse either though service or repair. This is a shame as often out of specification situations could be mitigated or even prevented if these data were plotted and trended – even if only the results were visualised over time against acceptance and warning limits. This is one of the roles of the instrument owner – to confirm that the instrument is working as expected. In addition, the individual may also be responsible for co-ordinating service and repair of the instrument.

11.6 Automating Blank Forms and Instrument Log Books

In this section, the way to overcome data integrity problems paper records is to eliminate paper and work in a secure electronic environment. For the two topics covered in this chapter, we discuss how this may be achieved. Please remember, just because you have always worked this way does not mean it is not incredibly stupid and that you should continue to work this way.

11.6.1 Automating Master Templates and Blank Forms

As you can see, the processes outlined in Figure 11.2 and 11.3 are far more complex than just using a blank form as we have done previously. However, the whole pharmaceutical industry is now picking up the bill for many organisations' laxity, mistakes and falsification. We now come to our earlier question – is paper the best way to record GMP data? Looking at these two figures the answer is an unequivocal no.

BUT.....

The problem is that many of the software applications that are used in regulated laboratories today are ill-equipped to take over many of the functions currently performed on paper today. This is due either to inability to expand from their core functionality or poor compliance features such as records stored in operating system files, inadequate audit trail functions including review and electronic signatures not on the records that are signed. Suppliers and users need to work together to ensure adequate functionality and compliance features and this will take time.

BUT......

Even with the best software, some activities in laboratories may still need to be recorded on paper such as any dilutions made during sample preparation. This work will need to be recorded on paper and manually entered into an application such as a chromatography data system. Alternatively, some applications such as one or two instrument data systems and electronic laboratory notebooks can automate the whole process and eliminate the paper entirely. For an overview of the principles of automating a regulated laboratory, the reader is referred to a paper[18] and a book chapter[19] on the subject.

11.6.2 Instrument Log Book

When we move our attention to the instrument log books for analytical instruments and laboratory computerised systems we return to the mediaeval age. We have the ridiculous situation where an amazing amount of money has been spent on the instrument with amazing capabilities complete with (mostly) an amazing software application to control it. What about the instrument log book? Paper! The application software has the sequence file, the work performed, a record of the samples, the analysis method, *etc.* but what does the poor analyst have to do? They must write all this information into the instrument log book complete with typos and mistakes. What an amazing waste of time and effort when the information – in probably more detail than is required by local procedures – already exists in the instrument data system.

Why can we not have an electronic instrument log book that is generated automatically by the instrument control software? It would be protected and secure like an audit trail and would be able to be viewed as part of a second person review with a function to document when selected entries have been reviewed. An automated instrument log function would not be able to be turned off so that like an audit trail will be a true record of the work performed by and on the instrument, generated and maintained automatically. The log would be an additional source of information to ensure the integrity of the work performed by an instrument. In addition, there should be the ability to cross reference between the audit trail and log book entries and drill down into the underlying data and metadata.

Qualification and maintenance work may be more difficult to document automatically but this should be surmountable by instrument companies as many instrument data systems can identify an instrument down to the model and serial number. Therefore, some linkage between qualification work and the on-line instrument log book would also be required and to any paper records of service visits outside of the log. This may be *via* a controlled user input to the automated instrument log in the data system.

Acknowledgements

I would like to thank Chris Burgess who contributed to the section on blank forms including the diagrams.

References

1. *EudraLex - Volume 4 Good Manufacturing Practice (GMP) Guidelines, Chapter 4 Documentation*, E. Commission, Editor, Brussels, 2011.
2. *Inspection of Pharmaceutical Quality Control Laboratories*, Food and Drug Administration, Rockville, MD, 1993.
3. *MHRA GMP Data Integrity Definitions and Guidance for Industry*, Medicines and Healthcare products Regulatory Agency, London, 2nd edn, 2015.
4. *MHRA GXP Data Integrity Definitions and Guidance for Industry, Draft Version for Consultation July 2016*, Medicines and Healthcare products and Regulatory Agency, London, 2016.
5. *MHRA GXP Data Integrity Guidance and Definitions*, Medicines and Healthcare products Regulatory Agency, London, 2018.
6. *WHO Technical Report Series No.996 Annex 5 Guidance on Good Data and Records Management Practices*, World Health Organisation, Geneva, 2016.
7. *FDA Draft Guidance for Industry Data Integrity and Compliance with CGMP*, Silver Spring, MD, USA. 2016.
8. *PIC/S PI-041 Draft Good Practices for Data Management and Integrity in Regulated GMP/GDP Environments*, Pharmaceutical Inspection Convention/Pharmaceutical Inspection Co-Operation Scheme, Geneva, 2016.
9. *EMA Questions and Answers: Good Manufacturing Practice: Data Integrity*, 2016, Available from: http://www.ema.europa.eu/ema/index.jsp?curl=pages/regulation/general/gmp_q_a.jsp&mid=WC0b01ac058006e06c#section9.
10. *FDA Warning Letter: FACTA Farmaceutici S.p.A. (Warning Letter: 320-17-17)*, Food and Drug Administration, Silver Spring, MD, 2017.
11. *FDA 483 Observations: Abbott AMO (Hangzhou) Company, January 2017*, Food and Drug Administration, Silver Spring, MD, 2017.
12. *21 CFR 211 Current Good Manufacturing Practice for Finished Pharmaceutical Products*, Food and Drug Administration, Sliver Spring, MD, 2008.
13. *21 CFR 58 Good Laboratory Practice for Non-clinical Laboratory Studies*, Food and Drug Administration, Washington, DC, 1978.

14. *OECD Series on Principles of Good Laboratory Practice and Compliance Monitoring Number 1, OECD Principles on Good Laboratory Practice*, Organisation for Economic Co-operation and Development, Paris, 1998.
15. *FDA Guidance for Industry, Part 11 Scope and Application*, Food and Drug Administration, Rockville, MD, 2003.
16. *FDA Form 483 for Akorn AG Switzerland*, Food and Drug Administration, Silver Spring, MD, 2016.
17. *FDA Warning Letter Aarti Drugs Limited July 2013*, Food and Drug Administration, Silver Spring, MD, 2013.
18. R. D. McDowall, Developing a Strategy for a Regulated Electronic Bioanalytical Laboratory, *Bioanalysis*, 2014, **6**(2), 1–20.
19. R. D. McDowall, *Validation of Chromatography Data Systems: Ensuring Data Integrity, Meeting Business and Regulatory Requirements*, Royal Society of Chemistry, Cambridge, 2nd edn, 2017.

CHAPTER 12

The Hybrid System Problem

Hybrid systems are computerised systems that generate electronic records with signed paper printouts. They are very common in regulated analytical laboratories as suppliers do not design adequate systems and users do not press for better solutions. Their use presents several problems such as traditional reliance on the paper printouts as the raw data, most hybrid systems are standalone workstations unconnected to a network, allowing deletion of records and time travelling as well as the difficulty in synchronising the electronic and paper records throughout the record retention period. Even systems that can work electronically are often operated in a hybrid mode as laboratories are very conservative and unwilling to change working practices.

In short, hybrid systems are THE WORST possible solution for ensuring data integrity in a regulated laboratory. The simple answer is to replace them with electronic solutions wherever possible but this requires suppliers to deliver better applications for controlling analytical instruments. As the market is driven by user requests will users of these systems see the light and demand changes to the architecture and way hybrid systems work?

12.1 What Do the Regulators Want?

12.1.1 Electronic Records and Electronic Signatures Regulations (21 CFR 11)

Section 11.70 of the regulation, on record/signature linking, states:[1]

> Electronic signatures and handwritten signatures executed to electronic records shall be linked to their respective electronic records to ensure that

Data Integrity and Data Governance: Practical Implementation in Regulated Laboratories
By R. D. McDowall

Published by the Royal Society of Chemistry, www.rsc.org

> the signatures cannot be excised, copied, or otherwise transferred to falsify an electronic record by ordinary means.

Hybrid systems are covered by this clause of Part 11 in that signed paper printouts from a laboratory computerised system must be linked to the relevant underlying electronic records (*handwritten signatures executed to electronic records …*). This includes spreadsheets and reinforced to answers to questions 10, 12 and 14 from the FDA guidance on data integrity.[2]

12.1.2 WHO Guidance on Good Data and Record Management Practices

Clause 11.3 of this guidance states:[3]

> Data integrity risks are likely to occur and to be highest when data processes or specific data process steps are inconsistent, subjective, open to bias, unsecured, unnecessarily complex or redundant, duplicated, undefined, not well understood, hybrid, based upon unproven assumptions and/or do not adhere to GDRP.

Furthermore, in Appendix 1, under special risk factors in the Attributable section:[3]

> The use of hybrid systems is discouraged, but where legacy systems are awaiting replacement, mitigating controls should be in place. In such cases, original records generated during the course of GXP activities must be complete and must be maintained throughout the records retention period in a manner that allows the full reconstruction of the GXP activities.
>
> A hybrid approach might exceptionally be used to sign electronic records when the system lacks features for electronic signatures, provided adequate security can be maintained. The hybrid approach is likely to be more burdensome than a fully-electronic approach; therefore, utilizing electronic signatures, whenever available, is recommended. For example, the execution and attribution of an electronic record by attachment of a handwritten signature may be performed through a simple means that would create a single page controlled form associated with the written procedures for system use and data review. The document should list the electronic dataset reviewed and any metadata subject to review and would provide fields for the author, reviewer and/or approver of the dataset to insert a handwritten signature. This paper record with the handwritten signatures should then be securely and traceably linked to the electronic dataset, either through procedural means, such as use of detailed archives indexes, or technical means, such as embedding a true-copy scanned image of the signature page into the electronic dataset.

Replacement of hybrid systems should be a priority.

The WHO guidance goes about as far as a regulatory guidance can go by stating that hybrid systems are discouraged and they should be replaced as a matter of priority. It also provides a way of simplifying paper printouts from hybrid systems that will be discussed in Section 12.3.4.

12.1.3 PIC/S PI-041 Good Practices for Data Management and Integrity in Regulated GMP/GDP Environments

In Section 9.3 of this guidance document we have the following statement:[4]

> Increased data review is likely to be required for hybrid systems.

This is true as a second person reviewing hybrid records must review both paper and electronic records plus the linkages between the two types of records as shown in Figure 12.1 and this makes the review more labour intensive and slow.

12.1.4 EU GMP Chapter 4 on Documentation

This chapter on documentation has the following requirements for records and hybrid systems:[5]

Principle:

> Suitable controls should be implemented to ensure the accuracy, integrity, availability and legibility of documents. The term 'written' means recorded, or documented on media from which data may be rendered in a human readable form.

In Clause 4.1 there is the following requirement:

> Many documents (instructions and/or records) may exist in hybrid forms, *i.e.* some elements as electronic and others as paper based.
>
> Relationships and control measures for ... records need to be stated for ... hybrid ... systems.
>
> Appropriate controls should be in place to ensure the integrity of the record throughout the retention period.

12.1.5 FDA Guidance for Industry Data Integrity and cGMP Compliance

In this guidance for Industry[6] there are two questions pertinent to the discussion of hybrid systems

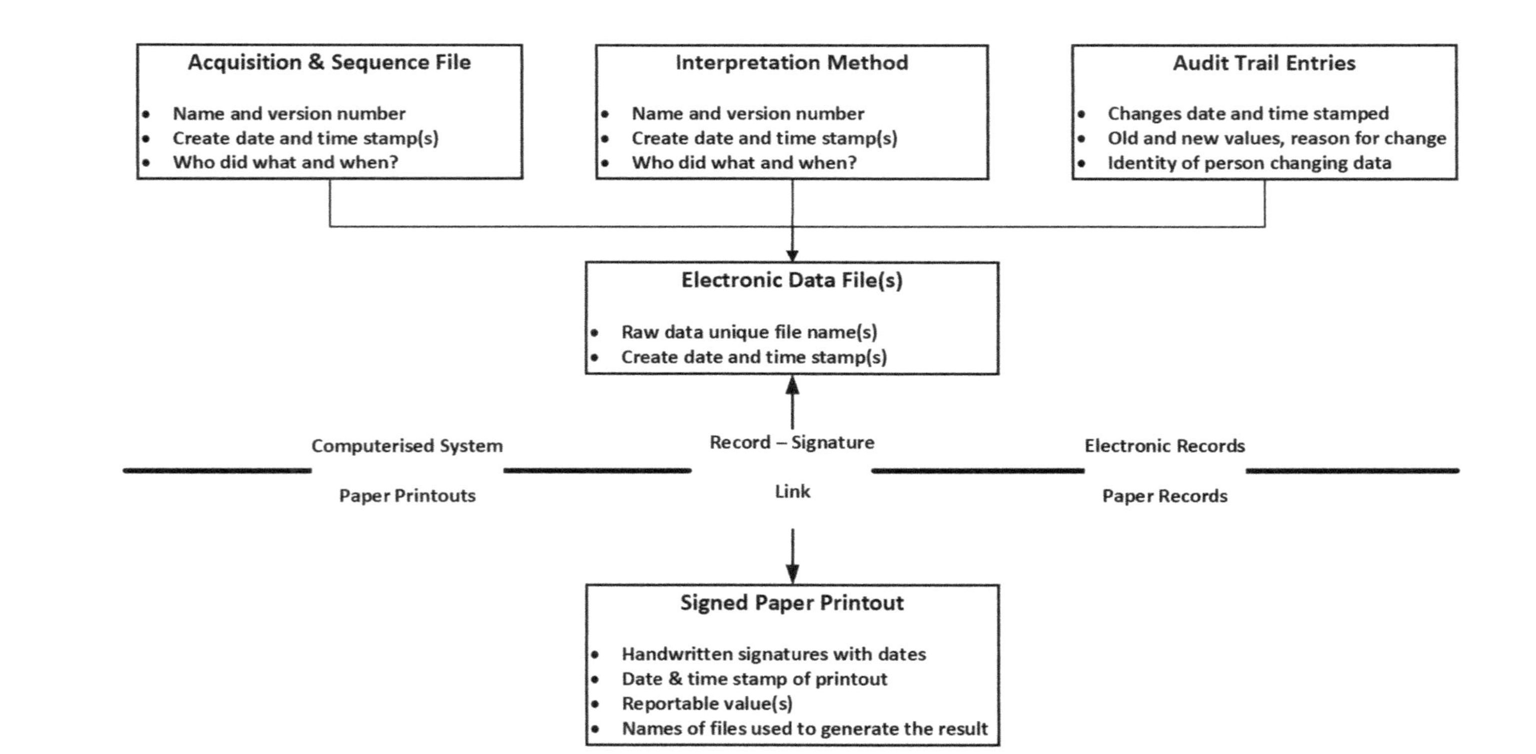

Figure 12.1 Diagrammatic representation of a hybrid laboratory computerised system.

Question 10. Is it acceptable to retain paper printouts or static records instead of original electronic records from stand-alone computerized laboratory instruments, such as an FT-IR instrument?

Answer: NO

Question 12. When does electronic data become a cGMP record?

Answer: When generated to satisfy a cGMP requirement, all data become a cGMP record.

Therefore, a laboratory cannot rely on paper. All electronic records must be saved including all supporting contextual metadata. Printing electronic records to PDF is not acceptable as the underlying electronic files are part of complete data (21 CFR 211.194(a))[7] and both FDA[2] and MHRA[8] have stated that dynamic data should not be converted into static data.

12.1.6 FDA Level 2 Guidance for Records and Reports

A better discussion on this issue will be found in the Level 2 guidance on the FDA web site: Questions and Answers on Current Good Manufacturing Practices, Good Guidance Practices, Level 2 Guidance - Records and Reports:[9]

Q3: How do the Part 11 regulations and "predicate rule requirements" (in 21 CFR Part 211) apply to the electronic records created by computerized laboratory systems and the associated printed chromatograms that are used in drug manufacturing and testing? (Posted 2010)

Some in industry misinterpret the following text from "The Guidance for Industry – Part 11, Electronic Records; Electronic Signatures – Scope and Application" (Part 11 Guidance; lines 164 to 171) to mean that in all cases paper printouts of electronic records satisfy predicate rule requirements in 21 CFR Part 211.

On the other hand, when persons use computers to generate paper printouts of electronic records, and those paper records meet all the requirements of the applicable predicate rules and persons rely on the paper records to perform their regulated activities, FDA would generally not consider persons to be 'using electronic records *in lieu* of paper records' under §§ 11.2(a) and 11.2(b). In these instances, the use of computer systems in the generation of paper records would not trigger part 11

BUT "...persons must comply with applicable predicate rules, and records that are required to be maintained or submitted must remain secure and reliable in accordance with the predicate rules."

21 CFR 211.180(d) requires records to be retained "either as original records or true copies"

21 CFR 211.68 further states that: "backup data are exact and complete".

> The printed paper copy of the chromatogram would not be considered a "true copy" of the entire electronic raw data used to create that chromatogram, as required by 21 CFR 211.180(d).
>
> The printed chromatogram would also not be considered an "exact and complete" copy of the electronic raw data used to create the chromatogram, as required by 21 CFR 211.68(b). The chromatogram does not generally include, for example, the injection sequence, instrument method, integration method, or the audit trail, of which all were used to create the chromatogram or are associated with its validity.
>
> Therefore, the printed chromatograms used in drug manufacturing and testing do not satisfy the predicate rule requirements in 21 CFR Part 211. The electronic records created by the computerized laboratory systems must be maintained under these requirements. However, the electronic record must be maintained and readily available for review by, for example, QC/QA personnel or the FDA investigator.

This principle of the electronic record being part of complete data applies to all GMP records generated in the laboratory. Electronic records must not be deleted unless a written procedure is followed by management authorisation.

12.1.7 Regulatory Summary

Hybrid systems are used in many laboratories but there are now more stringent controls placed on them, especially if the system is run on a standalone workstation not connected to the IT network. Regulatory guidance is that, if used, this will result in more work compared with an electronic process. The WHO guidance is the most radical by stating that hybrid systems should be discouraged and replaced as a matter of priority.

12.2 What Is a Hybrid System?

In this section, we will explore the definition and key features of a hybrid laboratory computerised system.

12.2.1 WHO Definition of a Hybrid System

In the definitions section of the WHO Good Records Management guidance[3] there is the following definition of a hybrid system:

> This refers to the use of a computerized system in which there is a combination of original electronic records and paper records that comprise the total record set that should be reviewed and retained.

This can be explained in the next section.

12.2.2 Key Features of a Hybrid System

Figure 12.1 is a diagrammatic representation of a hybrid system with the focus on the electronic and paper records created during normal operation, as such, there is no mention of interfacing with other computerised systems or manual input of data by users.

The main features of a hybrid system are:

- Control of an analytical instrument, *e.g.* chromatograph or a spectrometer (this is a USP <1058> Group C instrument) consisting of an instrument controlled by application software on a separate workstation. The software controls and acquires data from the instrument, processes the files and generates the result of the analysis. All data are stored by the software, usually on the workstation.
- The application will create data files from the instrument with associated contextual metadata files for controlling the acquisition and processing of data. There should also be an audit trail that records all changes to the data and contextual metadata.
- Running a configurable report generates the print out that is signed by hand.
- The print out should be linked to the instrument data files that have been used to generate the reportable results as well as the metadata files used to control the instrument, acquire the data and interpret the files to generate the results. This information is essential for record–signature linking.
- The print out is signed by the tester and reviewed by a second analyst who signs to indicate that these data have been performed.

The WHO guidance[3] notes that when a hybrid approach is used, laboratory analysts use computerised instrument systems that create original electronic records and then print a summary of the results. The hybrid approach requires a secure link between all record types, including paper and electronic, throughout the records retention period. Where hybrid approaches are used, appropriate controls for electronic documents, such as templates, forms and master documents, that may be printed, should be available (see Chapter 11 for controls for master templates and blank forms).

12.3 The Core Problems of Hybrid Systems

12.3.1 A Typical Hybrid System Configuration

A typical hybrid system is shown in Figure 12.2 and consists of three components:

- analytical instrument;
- controlling workstation where the instrument control software is loaded and where data are stored;
- printer.

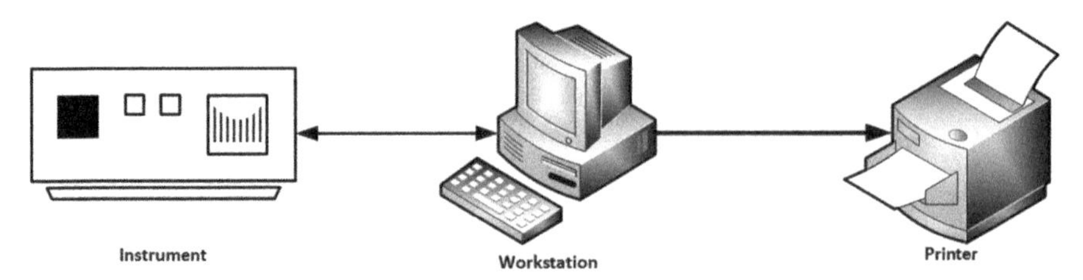

Figure 12.2 A typical hybrid laboratory computerised system.

The workstation is not connected to a network; therefore, backup of the electronic records is either performed by laboratory staff or not backed up at all.

We will look at this configuration to examine the problems of this approach that tend to focus on the standalone workstation:

- Workstation Resilience
  There is a single hard drive and controller on most workstations representing single points of failure with the potential loss of electronic records.
- Keyboard Contention
  Only one person can use the system at any one time. If data are being interpreted by one analyst, the instrument is idle and cannot be set up for the next analysis. This contrasts with a networked system where data are acquired to a network server and the previous night's data can be processed while the instrument is being set up for the next analysis.
- Backup and Recovery
  Protection of the electronic records is the responsibility of the laboratory. Starting with good intentions the laboratory will initiate a backup procedure but it will not be resilient or followed as shown by the FDA warning letter for Ohm Laboratories.[10] The backup will typically be weekly, meaning in the event of a disk crash up to a week's data will be lost. The media for backup may not be appropriate, *e.g.* CD or USB media and the backup may be manually not automatic resulting in a potential regulatory citation as found by BBT Biotech[11] and Zhejiang Hisun Pharmaceutical Co.[12]
- From the previous bullet point, electronic records are often ignored and even deleted without procedure or management approval and this will inevitably attract a regulatory citation.[13]
- Raw Data Definition
  Even today, raw data are defined as the signed paper printouts and not the original electronic records and the associated contextual metadata including audit trail entries.
- Media Formats
  There are two incompatible media formats: paper printouts and electronic records that must be managed and synchronised. Effective

linking of the two can be difficult if the supplier has not designed the system correctly or if ignored by the analytical staff.

In addition, there are further issues that can arise as the user access is not controlled by an IT administrator:

- All users share the same user identity making it impossible to identify who has performed a specific task in the system (breaking the ALCOA principle for Attributable).
- All users have administration privileges which enable then to change the configuration of the software, *e.g.* turn the audit trail on and off. As such there is a conflict of interest, *e.g.* users with administrator privileges as well as user rights can be tempted to change the configuration of the application without change control. For example, there are numerous warning letters and 483 citations where audit trails have been turned off and then on again some time later.

There is often a lack of technical controls for most standalone hybrid systems such as:

- Restricting access to the system clock to authorised individuals to prevent time travelling.
- Where data are stored in folders or directories of the operating system, they can be accessed outside of the application using Windows Explorer allowing files and directories to be deleted or files copied from one directory to another. Access to directories *via* the operating system must be restricted to authorised individuals only – ideally an IT administrator.
- Users can typically access the recycle bin that would allow them to remove deleted files from the system, access to this feature must be restricted to administrators only.
- USB sticks or flask drives should not be used as removeable media used to transfer files may also introduce malware. Use of USB sticks should be prevented by technical controls.

12.3.2 File Organisation and Printing Results

The files generated by a hybrid system need to be organised and ordered so that they are easy to retrieve in the case of a review, audit, inspection or compliant. Therefore, the following is strongly suggested for all hybrid systems, especially those that are based on file storage using operating system directories:

- There must be a naming convention for directories created and electronic records generated by the system so that it is easy to retrieve data for review, audit or inspection.

- There needs to be consistent application of data and time stamps for all systems in the laboratory, *e.g.*:
  Time format: HH:MM:SS using a 12 or 24 hour clock.
  Unambiguous date format, *e.g.*: DD MMM YYYY.
- Printouts must be linked to the underlying e-records used in the analysis, *e.g.*:
  Data files.
  Acquisition method.
  Instrument control file.
  Interpretation/processing/calculations performed.
  Spectral library files.
  Audit trail entries (entries should only be printed out upon specific request *e.g.* inspection).
  Printouts need to paginated in the format page *x* of *y* as well as dated and timed on each page.

Regulated laboratories print out anything that lives, breathes or moves and each of these printouts are signed. Some unfortunate analysts and reviewers must initial and date every single page of the printout from a run including the various control files. This slows the testing and the review portions of a procedure. However, the main problem of working with hybrid systems comes with synchronising the printouts and electronic records over time.

12.3.3 Synchronising Paper Printouts and Electronic Records

One of the major issues with any hybrid system is synchronising the paper printouts and electronic records as the two media are incompatible: physical and electronic, respectively. To illustrate the issue, see Figure 12.3.

- An analyst performs the work and the data are printed and this is Printout 1 in Figure 12.3.
- The records are reviewed and a supervisor finds an error that results in the data being returned to the analyst to correct. As the first printout is part of the complete data, this now results in Printout 2. The review is completed and Printouts 1 and 2 are stored together. The problem in the data system is that there is rarely any record of a second printout and only the audit trail may identify the changes in electronic records required by the reviewer to correct an error.
- In the future, there may be a requirement to review the data such as product complaint or new impurity or metabolite being identified. The electronic records are reviewed and result in a reinterpretation that requires a new printout – Printout 3.

What is apparent is that within the data system there is usually the ability to track the changes to electronic records *via* the audit trail but the problem

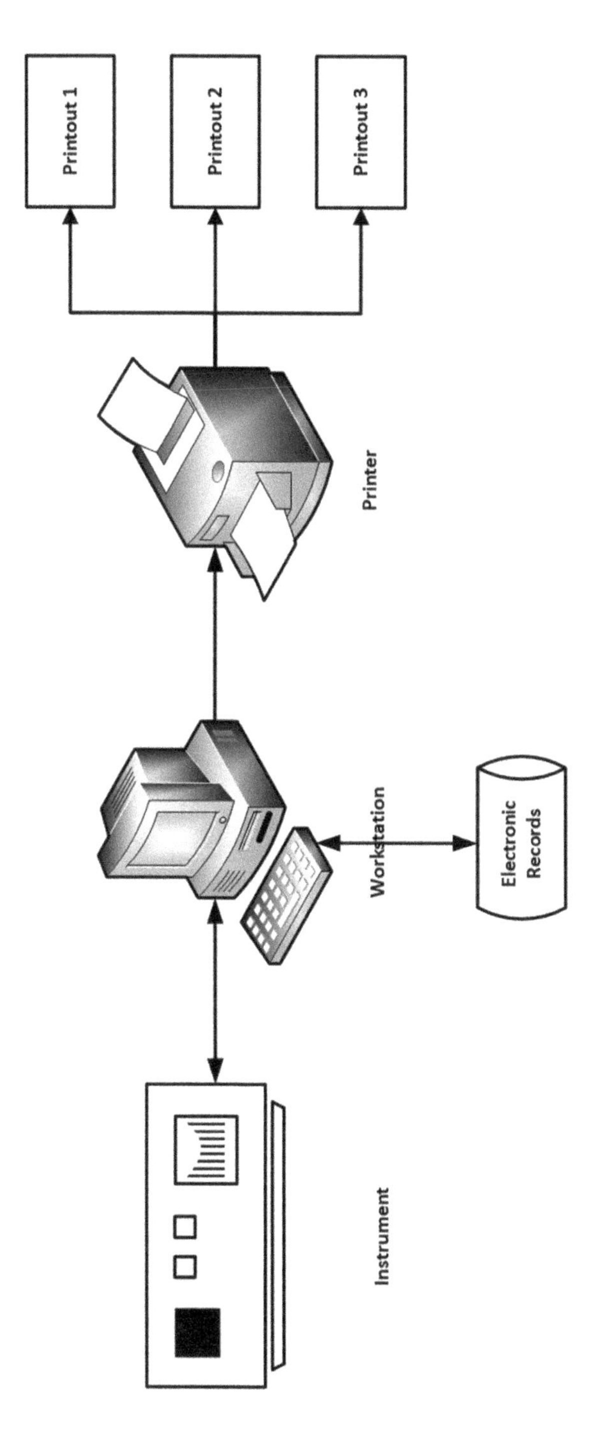

Figure 12.3 The hybrid system problem: synchronising paper printouts and electronic records.

is linking the external printouts to the internal electronic record changes. This is one of the reasons that hybrid systems are the worst solution for managing regulatory records.

12.3.4 A Simple Way to Reduce Paper with Hybrid Systems

Most hybrid systems print paper as if it is going out of fashion. Moreover, often each page must be initialled by the tester and again by the reviewer, which is an error prone and tedious task. The printouts must also be checked against the electronic records that were created or used in the analysis.

There is a simple way to reduce the amount of paper printed from a hybrid system outlined in Appendix 1 of the WHO data integrity guidance[3] and quoted completely in Section 12.1.2. Shown in Figure 12.4, is a controlled review form for the creation and review of the records created during an analysis using a hybrid system. The form is linked to the analytical procedure and SOP for review of laboratory records.

At the start of the analysis, a uniquely numbered version of the form is issued to the tester who documents the start and end data and time of the analysis (to help the reviewer search audit trail entries) as well as document the records created, modified and the location where they are stored. The only printout is the test summary of the reportable results and analysis information. The results printout and review form is signed by the tester. When records are ready for review, a second analyst reviews files electronically on screen with no printouts and this makes the task simpler and quicker. The reviewer checks the data files and contextual metadata files generated by the performer of the test and documents these on the review form. The applicable audit trail entries are also reviewed and documented. Checks for falsification of data are included in the form as this will be checked during data integrity audits and, if applicable, data integrity investigations. If changes are required to be made these are documented and sent back to the tester to update. When the review is complete the reviewer then signs the form. If required by laboratory procedures, there may be space on the form for an approval or QA signature.

12.4 Eliminate Hybrid Systems

Working with hybrid systems – computerised systems that generate electronic records with signed paper printouts – is the WORST possible world to be in. The laboratory must manage two incompatible media formats: paper and electronic records. The best advice is to eliminate these systems by using electronic systems to ensure both regulatory compliance and business efficiencies, as we will discuss in Chapter 13.

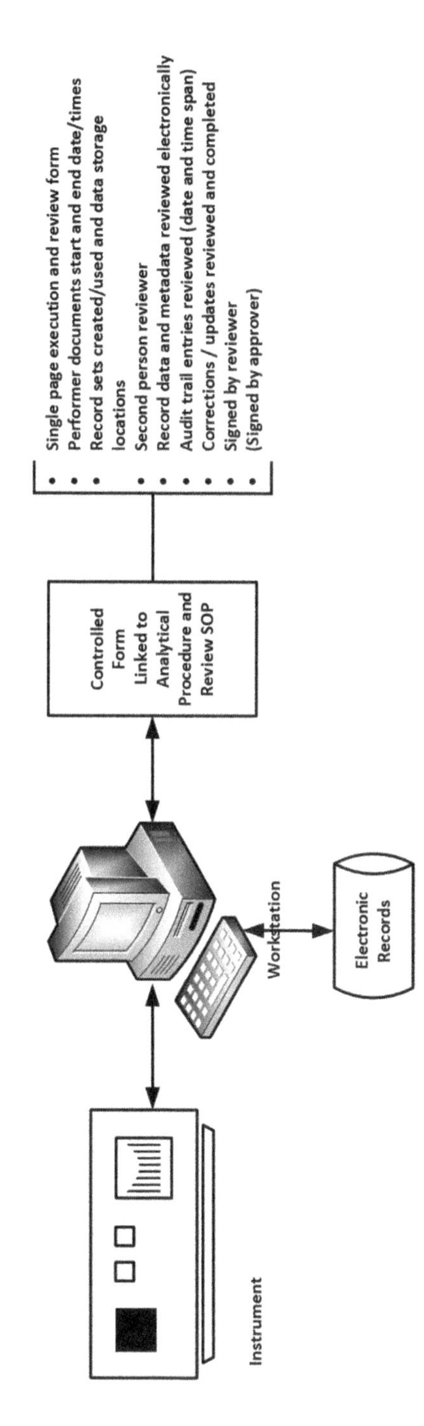

Figure 12.4 Generation and review of hybrid records.

References

1. *21 CFR 11 Electronic Records; Electronic Signatures, Final Rule, in Title 21*, Food and Drug Administration, Washington, DC, 1997.
2. *FDA Draft Guidance for Industry Data Integrity and Compliance with cGMP*, Silver Spring, MD, USA, 2016.
3. *WHO Technical Report Series No.996 Annex 5 Guidance on Good Data and Records Management Practices*, World Health Organisation, Geneva, 2016.
4. *PIC/S PI-041 Draft Good Practices for Data Management and Integrity in Regulated GMP/GDP Environments*, Pharmaceutical Inspection Convention/Pharmaceutical Inspection Co-Operation Scheme, Geneva, 2016.
5. *EudraLex - Volume 4 Good Manufacturing Practice (GMP) Guidelines, Chapter 4 Documentation*, E. Commission, Brussels, 2011.
6. *FDA Guidance for Industry Submission of Quality Metrics Data, Revision 1*, Food and Drug Administration, Rockville, MD, 2016.
7. *21 CFR 211 Current Good Manufacturing Practice for Finished Pharmaceutical Products*, Food and Drug Administration, Sliver Spring, MD, 2008.
8. *MHRA GXP Data Integrity Guidance and Definitions*, Medicines and Healthcare products Regulatory Agency, London, 2018.
9. *FDA Questions and Answers on Current Good Manufacturing Practices, Good Guidance Practices, Level 2 Guidance - Records and Reports. 2010*, 27 May 2016, Available from: http://www.fda.gov/Drugs/GuidanceCompliance-RegulatoryInformation/Guidances/ucm124787.htm.
10. *FDA Warning Letter Ohm Laboratories*, Food and Drug Administration, Rockville, MD, 2009.
11. *FDA Warning Letter BBT Biotech Gmbh (Warning Letter 320-16-12)*, Food and Drug Administration, Sliver Spring, MD, 2016.
12. *FDA Warning Letter: Zhejiang Hisun Pharmaceutical Co., Ltd (Warning Letter: 320-16-06)*, Food and Drug Administration, Silver Spring, MD, 2015.
13. *Inspection of Pharmaceutical Quality Control Laboratories*, Food and Drug Administration, Rockville, MD, 1993.

CHAPTER 13

Get Rid of Paper: Why Electronic Processes are Better for Data Integrity

Following on from the problems with hybrid systems that we discussed in Chapter 12, I now want to focus on how to resolve these problems by their replacement. In Chapter 1, we discussed that data integrity is a bigger programme of work compared with the Year 2000 and 21 CFR 11 remediation projects combined. The issue with these two earlier projects is that typically there was little or no business benefit obtained when systems were remediated. In this chapter, we consider the longer-term solutions of systems and processes to deliver business improvement and benefits for the time and money spent. This is essentially a move away from paper records to automated laboratory processes designed to ensure data integrity with transparent data capture and processing and with minimal manual data input from laboratory staff. This approach could be viewed as a second attempt to get the intent of 21 CFR 11 regulations right. Not all processes can move to electronic working either because there is not an application available or it is not cost effective to do so. However, most work within a regulated analytical laboratory can be automated – are organisations willing to invest and make their laboratories more efficient and effective?

13.1 What Do the Regulators Want?

13.1.1 WHO Guidance on Good Data and Record Management Practices

The WHO data integrity guidance[1] has several clauses noting that risk assessment of process should be undertaken and improvements made:

Data Integrity and Data Governance: Practical Implementation in Regulated Laboratories
By R. D. McDowall

Published by the Royal Society of Chemistry, www.rsc.org

11.1 Data processes should be designed to adequately mitigate and control and continuously review the data integrity risks associated with the steps of acquiring, processing, reviewing and reporting data, as well as the physical flow of the data and associated metadata during this process through storage and retrieval.

11.2 QRM (quality risk management) of the data life cycle requires understanding the science and technology of the data process and their inherent limitations. Good data process design, based upon process understanding and the application of sound scientific principles, including QRM, would be expected to increase the assurance of data integrity and to result in an effective and efficient business process.

11.3 Data integrity risks are likely to occur and to be highest when data processes or specific data process steps are inconsistent, subjective, open to bias, unsecured, unnecessarily complex or redundant, duplicated, undefined, not well understood, hybrid, based upon unproven assumptions and/or do not adhere to GDRP.

11.4 Good data process design should consider, for each step of the data process, ensuring and enhancing controls, whenever possible, so that each step is:

- consistent;
- objective, independent and secure;
- simple and streamlined;
- well-defined and understood;
- automated;
- scientifically and statistically sound;
- properly documented according to GDRP (Good Documentation Records Practice).

The above clauses in this section should also be read in conjunction with Chapter 12 on the problem with hybrid systems to obtain a full picture of the way WHO are trying to change the thought processes and working practices within the pharmaceutical industry.

13.1.2 PIC/S PI-041 Good Practices for Data Management and Integrity in Regulated GMP/GDP Environments

The PIC/S guidance document[2] notes that:

5.5.4 For computerised systems, manual interfaces with IT systems should be considered in the risk assessment process. Computerised system validation in isolation may not result in low data integrity risk, in particular, when the user is able to influence the reporting of data from the validated system.

A fully automated validated process together with a configuration that does not allow human intervention, or reduces human intervention to a minimum, is preferable as this design lowers the data integrity risk.

Procedural controls should be installed and verified where integrated controls are not possible for technical reasons.

13.1.3 EU GMP Annex 11 Computerised Systems

Clause 6 of Annex 11 is concerned with accuracy checks:[3]

> For critical data entered manually, there should be an additional check on the accuracy of the data. This check may be done by a second operator or by validated electronic means. The criticality and the potential consequences of erroneous or incorrectly entered data to a system should be covered by risk management.

13.1.4 Regulatory Summary

The take home messages from the regulatory guidance documents above are that an ideal laboratory process should be:

- Designed to be automated where possible to enforce data integrity so that human intervention is reduced to a minimum: analytical staff should focus on the interpretation of the data not the falsification of it.
- Critical data entered manually to a computerised system is high risk and needs to be checked or eliminated by automation.
- Software must be configured to ensure that records are protected throughout the analytical data life cycle.
- Data from acquisition through interpretation to reporting must be transparent.
- Computerised systems must be validated to demonstrate that a laboratory process does what it is intended to do including protecting the data and metadata.

Where this approach is not possible, then procedural controls with training can be used, however, this increases process risk and involves more work for second person review. In the author's opinion, manual processes or manual portions of processes will have increased regulatory scrutiny in the future as these will be seen by auditors and inspectors has having a greater risk to records generated compared with correctly designed, implemented and validated automated laboratory processes.

13.2 Why Bother with Paper?

13.2.1 Tradition – Why Change Our Approach?

The GXP regulations are over 40 years old and at the time that they were written the recording medium was paper. From the 1970's to the late 1980's this was appropriate as there were few means of capturing and storing any electronic records created. For example, in a chromatographic integrator, the records of a single run were transient and stored in volatile memory until the next record was created when the first set of data were overwritten.[4]

As time has progressed to the current day and the issues with data integrity have surfaced with many warning letters and import bans from the FDA perspective, electronic records and not paper printouts have become important. FDA, MHRA, PIC/S and EMA[1,2,5–8] have all stated that if a computerised system is involved, then the electronic records including all associated metadata must be captured and maintained as paper printouts are incidental. However, this message has not gotten through to many laboratories as there are still 483 citations, even now, such as:

> Your firm considers the original record to be the printed chromatograms and therefore does not perform a review of the electronic data and associated metadata.[9]

The problem is that the pharmaceutical industry is ultra conservative and if paper records were good in the past – why can we not continue to use them? This is where we need to consider the "current" in cGMP.

13.2.2 Back to the Future 2: Understanding the Current in cGMP

Chapter 1 introduced the fact that we need to understand the meaning of the word "current" in the context of cGMP from 21 CFR 211.[10] To understand the intent of the regulation this is where we need to go back to the future. To begin the discussion of understanding the current in cGMP, it begins with the title of the regulation 21 CFR 211:

> Current Good Manufacturing Practice for Finished Pharmaceuticals.[11]

It is built into the GMP regulation from the beginning not slipped in as an afterthought. In the scope section of the regulation (21 CFR 211.1), it states that:

> (a) The regulations in this part contain the minimum current good manufacturing practice for preparation of drug products for administration to humans or animals.

The 21 CFR 211 regulations are stated to be minimum. The problem with the pharmaceutical industry is this is typically interpreted as this is all we will do. This divergence from the intent of the regulation is the start of some data integrity problems. We can delve further into an understanding of "current" in comment 17 of the 1978 preamble to the GMP regulations,[11] a discussion on the use and meaning of the word "current" is as follows:

> One comment recommended that the word "current" be deleted since it is obvious that the latest regulations to be published are current, and therefore the use of the word "current" is superfluous. ....
>
> Several of these comments reflect, the Commissioner believes, a misunderstanding regarding the use of the word "current." ......
>
> The Congress intended that the phrase itself have a unique meaning and that the good manufacturing practice regulations represent sound current methods, facilities and controls for the production of drugs to assure safety...
>
> Although the practices must be "current" in the industry, they need not be widely prevalent. Congress did not require that a majority or any percentage of manufacturers already be following the proposed mandated practices, as long as it was a current good manufacturing practice in the industry, *i.e.* that it had been shown to be both feasible and valuable in assuring drug quality.[11]

The intent of the US GMP regulations, since they became effective in 1978, has been that as science, analytical instrumentation, technologies, techniques and software applications advance, then so too must the pharmaceutical industry. Even if an advance is not widely used in the industry but it can demonstrate an improvement in drug quality, then it comes under the purview of the "c" in cGMP.

However, the problem is that reality in regulated laboratories does not match regulatory intent. Once a pharmaceutical organisation has developed an interpretation of the regulation they are reluctant to change due to several reasons, such as the cost of new technology, validation costs and perceived or actual or perceived inflexibility of the inspectorate to accept such advances. The industry has preferred to remain with older processes that do not reflect the most effective and efficient due to inertia, fear of spending money or the inability to deliver a more data integrity compliant operation.

13.2.3 The Pharmaceutical Industry is a Two Sigma Industry

One of the triggers for the FDA to introduce the GMPs for the 21st century[12] in 2002 were reports that the pharmaceutical industry was a two-sigma industry. Motorola had started the move towards six-sigma where the error rate of 3.4 defects per million compared in a process or system with a two-sigma industry with 308 000 defects per million, as shown in Table 13.1.

Table 13.1 Defects per million *versus* sigma level.

Sigma level	Error rate per million	Accuracy
1	690 000	31.0%
2	308 000	69.0%
3	66 800	93.3%
4	6200	99.4%
5	233	99.97%
6	3.4	99.999997%

The main reasons for pharmaceutical industry lagging were:

- The pharmaceutical industry is ultra-conservative and not willing to implement new technology (counter to the intent of the GMP regulations which anticipated industry moving forward over time as discussed in Section 13.2.2).
- Conservative regulatory inspectors who preferred paper based processes and were unwilling to accept or understand automated workflows.

Following the adoption by the FDA of risk based approaches from 2002 and the training of inspectors globally in data integrity, the move towards six-sigma can begin in the pharmaceutical industry. However, it will be slow progress in many organisations. A 2017 publication has raised the question will the pharmaceutical industry ever get to six sigma as it appears to be stuck in a rut between two and three sigma.[13]

13.2.4 Are the Regulations Part of the Data Integrity Problem?

When a drug product is registered, part of the regulation are details of the analytical methods used by Quality Control to test the product for release against the specifications prior to release. As these methods are part of the registration dossier or marketing authorisation, they are difficult to change as the change must go through the different regulatory authorities where the product is registered. This takes time and money and pharmaceutical companies are reluctant to do this.

Herein lies part of the data integrity problem, the regulations themselves. If the industry must be "current" should there not be an easier process for changing an analytical procedure?

This is especially true for older products where the analytical procedures were developed perhaps 25–30 years ago before the publication of ICH Q2 and Q2(R1).[14] Such a registered procedure may not be robust and would not be validated to the same standard as now. For example, a chromatographic method could require extensive manual integration of peaks but due to the time and cost of registration companies stay with the *status quo*.

This is a major factor in updating ways of working in any Quality Control or GMP regulated laboratory. In GLP the issue is not so highly controlled as the methods are further away from the patient and have an indirect impact on product quality and patient safety.

13.2.5 Is Paper a Realistic Record Medium Now?

Although paper as an archive medium is well established, there are several drawbacks, as we shall discover. Let us forget all the hyperbole about data integrity and assume for a moment that paper is still the ideal option for laboratory records. Unrealistic I know, but stay with me on this discussion. We generate paper records in the laboratory and use them for GLP and GMP decisions within an organisation. An activity on its own may be acceptable to manage on paper, but there are other areas where those records will be used for other regulated activities, *e.g.*:

- Development
  Preparation of regulatory submissions such as IND/CTX and NDA/PLA. These submissions are only accepted electronically now by regulatory authorities. How grotesque would it be to take paper records from the laboratory and scan them to produce electronic copies for inclusion into a regulatory submission? The scanning process would need to be validated to demonstrate intended use. This would be followed by quality control checks on each and every page to ensure that the scan was the correct orientation and of acceptable quality.
- Quality Control
  Annual Quality Reviews and Product Quality Reviews are requirements of the FDA and EU GMP regulations respectively.[15,16] These regulations require the assessment of some or all batches of a product. If paper were the only record medium think of the time taken to read and abstract the data and information for the review. Does this make sense? Of course, it does not.

The only pragmatic and realistic approach is to use electronic records to generate regulatory submissions and quality reviews. Paper is not an option.

13.3 Design Principles for Electronic Working

Before looking at the next automation workflow for an analytical balance it is important to list and understand the three design principles for automating laboratory processes.[17] These are:

1. Capture Data at the Point of Origin
   If you are going to work electronically, then data must be electronic from when it is first generated or observed. However, there is a wide range of

data types that include observational data (*e.g.* haemolysed sample), instrument data (*e.g.*, pH, balance, LC-MS), and computer data (*e.g.* data used for statistical or trending results). Interfacing systems and applications should be a major goal but this must be balanced with a cost-effective approach. For example, are the sample numbers and volumes sufficient to justify the cost and validation of interfacing?

2. Eliminate Transcription Error Checks
   The second principle is to ensure that once data are captured electronically, they are never printed and re-entered manually into another computer system. All data transfers must be electronic using validated processes with appropriate checks, *e.g.* audit trails, checksums for file integrity, to ensure that data are not corrupted. Ideally, only networked systems should be used to minimise data loss and for sharing data and information effectively. As a corollary to this principle, raw data must be defined as electronic records. Paper should only be a by-product of an electronic process that is used for information only and is therefore not defined as raw data. The electronic records produced in these workflows should be designated as records or raw data for archiving purposes.
3. Know Where the Data Will Go
   Data storage repositories need to be designed for an electronic workflow that may be distributed across several applications, *e.g.* LIMS, CDS, or an SDMS or simply secure network drives. To help size the storage, an estimate of the data volumes will be necessary. In addition, a pre-determined plan of how data will be stored, *e.g.* by batch/lot number, drug project or product, must be developed before the repositories are implemented. This data storage plan will also include any file naming conventions to ensure that an individual file, analytical run or study can be retrieved quickly and with the minimum of effort. To ensure no data loss in this electronic environment it is essential that key hardware for storing data must be resilient and fault tolerant and that the backup and recovery processes must be robust and validated.

Implicitly required with this operating principle of the electronic laboratory is that standalone data systems are not adequate for data storage. In the view of the author, all laboratory data must be stored on networked drives that have sufficient hardware resilience to prevent data loss from the failure of a single drive as an absolute minimum. In addition, rather than use file based data systems, applications that use adequately designed databases for acquiring and managing data as the audit trails are within the database rather than incorporated in the data file in contravention of 21 CFR 11 requirements for audit trails.[18] Backup strategies also need to be developed to ensure that data are not lost if a backup fails: consider differential rather than incremental backups or full backups that are executed each working day.

The key message when designing electronic workflows is to ensure that once data are acquired at the point of origin they are not printed out or transcribed again but transferred electronically between systems using validated routines. Paper is not to be used as the transfer mechanism and should only be printed when required, *e.g.* audits, inspections, *etc.* Data storage must be networked and robust and, where necessary, sufficient thought must be given to file naming conventions to identify data uniquely and be stored under projects or studies as appropriate.

We will apply these three principles in the following sections of this chapter to demonstrate how a manual or hybrid approach can be migrated to an electronic one.

13.4 Designing Data Workflows 1 – Analytical Balances

In this section, we get to the nub of the issue: how to design simple, automated and transparent laboratory workflows that ensure data integrity. Please note that not all analytical procedures can be automated due to ether cost (there is not sufficient justification for automation) or there is not an application available to allow effective automation. First, we will consider a simple workflow that can be automated relatively easily but first, we see some regulatory citations for analytical balances.

The MHRA GMP Data Integrity Guidance for Industry[6] mandates that balances must have a printer attached to document the work or the weighing value is captured by an automated data capture solution, *e.g.* LIMS or ELN. The first thing to note is that under no circumstances will a balance weighing by observation be accepted as the data are too critical. There must be independent evidence to corroborate the activity. As a minimum, an analytical balance must have a printer attached. However, even a printer has the potential to be manipulated, as evidenced by two regulatory citations:

USV Ltd, FDA Warning Letter, February 2014:[19]

1. Your firm failed to follow and document at the time of performance required laboratory control mechanisms (21 C.F.R. §211.160(a)).

> Our investigators found that laboratory analysts did not document the balance weights at the time of sample weighing. Specifically, sample weights used in calculations were created after the chromatographic runs. The analyst admitted that the sample weights that were represented as raw data from the analysis actually were backdated balance weight printouts produced after the analysis and generated for the notebooks. These sample weights were used to calculate related compounds and impurities used in support of method validations submitted in FDA drug applications.

Hetero Laboratories, FDA 483 Observation, December 2016:[20]
On 12th December 2016, we observed the scrap area behind the production area of Buildings X and Y to contain controlled documents that had been discarded:

a) A balance printout with drug product B dated 14-Dec-2016. After discussing this finding with your firm, you failed to explain why the balance printout was post-dated by two days and therefore indicating an alternation to dates on balances. Your firm's VP of Operations explained that not all balances are password protected.

As analytical balances are so critical that access to the clock must be restricted to authorised individuals to prevent time travel. However, post run fabrication of data can be difficult to detect, especially when spreadsheets are used to calculate reportable results and not a secure instrument data system.

13.4.1 Weighing a Reference Standard or Sample

The main workflow that we will consider in this section is weighing samples, analytical reference standards or materials for buffers and mobile phases which are universal operations in regulated analytical laboratories globally. We will look at various options:

- recording a weight by observation;
- using a printer to record the sample weight;
- automating the process with software applications.

The reason for selecting this workflow, apart from its universal applicability, is the impact that an incorrect weight analytical reference standard can have on an analysis. If it is a weight of a sample, an incorrect weight can bias the analysis of that sample. In comparison, a wrong reference standard weight can have a deleterious impact on several analytical batches – more if the reference solution is very stable – and result in out of specification samples being passed for the market or passing batches being rejected. Hence, the need to ensure that weighing using an analytical balance is rigorously controlled and, where possible, enforced.

13.4.2 Recording a Weight by Observation

Do not try this in your regulated laboratory. The cost of a printer attached to an analytical balance will repay itself many times over and save you a compliance citation for several reasons:

- First, regulatory authorities do not trust an individual to record a balance weight by observation. This is not unreasonable as an analytical balance weighing can have a far-reaching impact if it is wrong, as discussed above, hence, the need to see documented evidence of the actual weighing process.

- Secondly, there is no independent evidence available for a second person review. Even if a second person witnesses the process and verifies that the value recorded is correct, this is not sufficient to ensure data integrity of any weights performed in this manner. There must be objective evidence in one form or other.
- Thirdly, the authorities will not mandate a specific way of working only to say that they need evidence of the work being carried out as long as it is not by observation.
- Fourthly, I would strongly suggest that such an approach is not "current" from the discussion outlined in Section 13.2.2 and is also not "scientifically sound" as required by US GMP and ICH Q7.[15,21]

Consequently, we will not consider recording weighing by observation further.

13.4.3 Recording Balance Weights with a Printer

From the problems highlighted in the previous section, the minimum balance configuration that is required in a regulated laboratory to ensure data integrity is an analytical balance with an attached printer. In this configuration, the balance and printer will be standalone and not connected to a network or any software application. Typically, the process to weigh a material (sample or reference standard) will contain the following steps:

- Check manually that the balance is the correct one or type for the weighing you are going to carry out.
- Check manually that the balance is qualified.
- Write up the balance information in your lab notebook or on a controlled worksheet (see Chapter 11 for the issues with uncontrolled blank forms and worksheets).
- Perform a point of use check (typically with reference masses over the calibration range or using the internal calibration of the balance).
- Write the information in the instrument log book and check the external mass result *versus* the acceptance criteria.
- Take reference material and write the information into your laboratory notebook.

Just writing this makes me breathless and we have not done any real work yet!

- Take a weighing vessel and weigh it.
- Tare the balance.
- Weigh the reference standard.
- Print the weighing sequence (or set up the printer so that each accepted weight is automatically printed).

- Complete the instrument log book for the weighing.
- Stick the printout into your laboratory notebook or attach to a controlled worksheet.
- Transfer the vessel to a volumetric flask.
- Dissolve the material and make up to volume.
- Label the flask with the solution identity, storage conditions and expiry date, *etc.*
- Record the information in your laboratory notebook or controlled worksheet.

Finished!? Not quite, as there may be some additional activities to consider, such as:

- There might be come calculations performed on the weight to calculate the solution concentration, adjust the weight taken from the salt form that was weighed to calculate the base concentration or to account for purity or water content. This may involve entering the weight into a validated spreadsheet or even using a hand-held calculator. These steps involve more writing and entry of critical data into the spreadsheet and the additional records must be available for second person review to ensure that there are no transcription errors. This is a slow and error prone process as even the second person reviewer can miss errors.
- If the reference solution is sufficiently stable, it can be used in several analyses. Each time it is used the analyst will need to write the solution identity and other information into the analytical run documentation each time.
- Printing is not intelligent (*i.e.* controlled copy printing) and therefore if the analysis is not within specification the balance printout could be heading towards the nearest paper shredder or waste bin and a new weighing sequence could be performed.

To remove the suspicion of possible data integrity violation, a better approach would be to link the analytical balance to an instrument data system, ELN or LIMS that will offer significant data integrity advantages. It also meets the requirements of the WHO and PIC/S data integrity guidance documents[1,2] in automating a critical process and removing much of the human involvement. Indeed, new top end analytical balances from one supplier can only be operated *via* their instrument control software that automates the whole process and leads us to the third option to present and discuss.

13.4.4 Connecting the Balance to an Instrument Data System

Let me ask you a very personal question. Are you lazy? The answer to this question should be an unequivocal yes. The reason is that you do not want to go through the palaver outlined in the previous section each time you weigh something. Be lazy and automate the process – even the regulators are with you on this, as you can see from the guidance presented in Section 13.1. If a process is automated, validated and transparent, then it becomes more inspection friendly. If the instrument data system is networked and the IT infrastructure has a trusted time source to synchronise with, then time travelling can be greatly reduced if not eliminated. The balance is now connected on-line, the date and time stamp is automatically applied to each weighing as the value is transferred to the database.

Let us look at how we can automate the process using an instrument data system, typically available from the supplier of the analytical balance or an ELN or LIMS. Most analytical balances have the facility to be used as a terminal so the data system can be operated from the balance screen but data are acquired directly to the database on the networked server. This eliminates two potential sources of data integrity error or falsification: access to the computer clock and not automating the whole acquisition process, meaning that the overall procedure is enforced technically, as can be seen in Figure 13.2. You will notice the difference between the two process flows: the electronic one is much simpler.

Some of the benefits of this approach are:

- It ensures that the correct type or, if required, a specific balance is selected for use and that instrument is qualified. Attempts to select a different or unqualified balance will be blocked by the instrument data system.
- The calibration weights, if using external masses, can be captured by the instrument data system to determine if the balance meets predetermined acceptance criteria. In addition, the cumulative calibration data can be analysed to see if there are any trends apparent over a user defined time period and this meets the EU GMP requirements 6.09 and 6.16.[22]
- Time and date stamps will be applied to the data by the instrument data system server, which is linked to the network time server that itself is linked to a trusted time source external to the organisation such as a national time source, GPS system or a network time protocol (NTP) server.
- Typically, the analytical reference substance is entered into the instrument data system database and this has all requisite information such as identity, lot number, acquisition date, purity, *etc.* The container will be labelled both in human readable text and also bar coded. If the balance has a bar code reader interfaced, then the identity of the standard can be verified as the weighing of the material begins.
- The whole of the balance weighing can be captured within the instrument data system without the need to record anything on paper and eliminating the need for controlled blank forms.

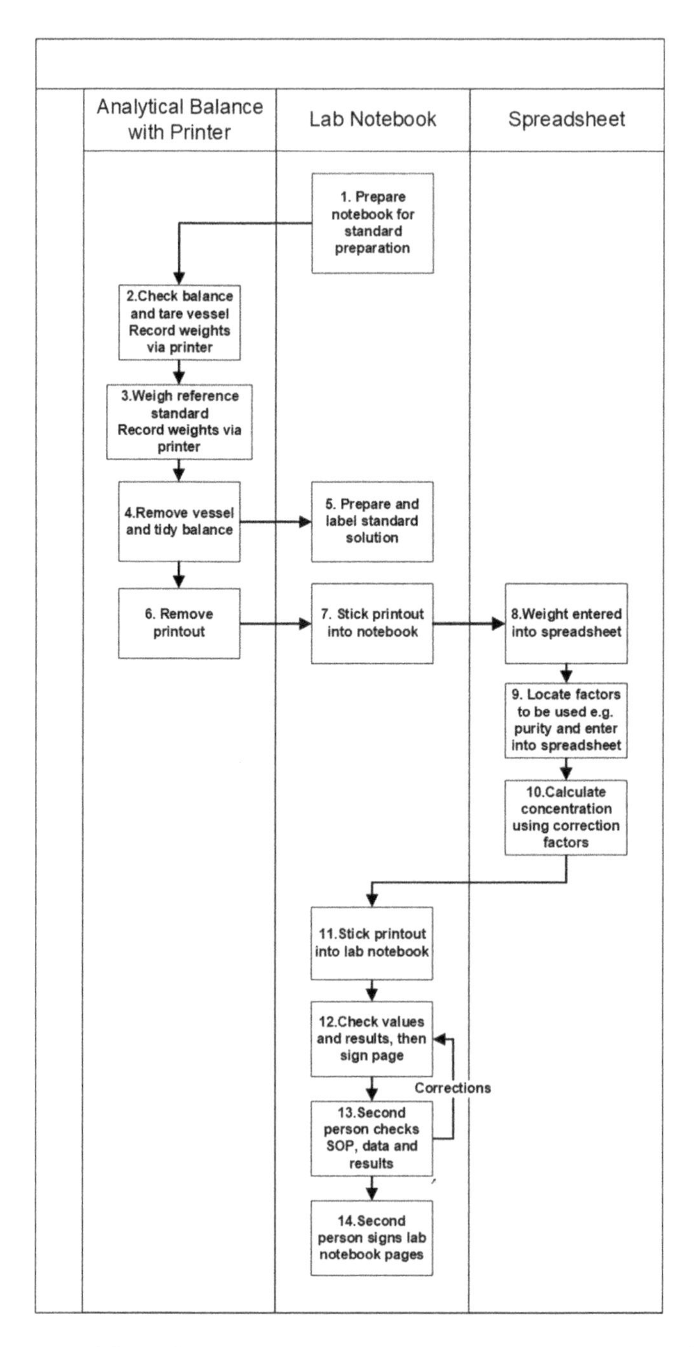

Figure 13.1 Workflow of weighing a reference standard with an analytical balance with an attached printer.

- Any calculations regarding purity or salt form to base conversion can be performed in the data system during the weighing obviating the need to use a spreadsheet, manual entry of data and the subsequent transcription error checks required in the second person review. Of course, these calculations must be specified and verified that they are mathematically correct.
- The resulting reference solution can have a label printed by the data system and the use of this up to its expiry date can be tracked by the data system as well.
- The use of the reference standard material over time can also be seen and if required trended easily.

Virtually the whole of the operation can be controlled and data captured by the instrument data system or any suitable laboratory informatics application that can perform these tasks. There is little that needs to be recorded outside of the system.

When you compare the two processes you wonder why analytical scientists still work in the old-fashioned paper based way. Just look at the simplicity of the electronic process in Figure 13.2 in comparison to the complexity of the manual process in Figure 13.1. This raises the two questions:

- Are analytical scientists masochists?
- Or is it just the sadists in Management, Quality Assurance and Finance that perpetuate these inefficient working practices?

13.5 Designing Data Workflows 2 – Chromatography Data Systems and LIMS

13.5.1 Options for Interfacing

In this section, we will consider the options for interfacing a Chromatography Data System with a Laboratory Information Management System. The latter could be a laboratory focused LIMS or a quality module from an Enterprise Management System but in either case, the term LIMS will be used to describe the work undertaken by the application. The options for interfacing we will consider are:

- Manual data entry and transfer between the two applications and is the electronic typewriter option.
- Single interface for transfer of data from the CDS to the LIMS. All laboratory data such as sample identities, study/batch number and other data such as dilutions, factors, weights are all entered manually into the CDS.
- Bidirectional interfacing between the CDS and LIMS. Note that there will still need to be some laboratory data entered manually into the CDS

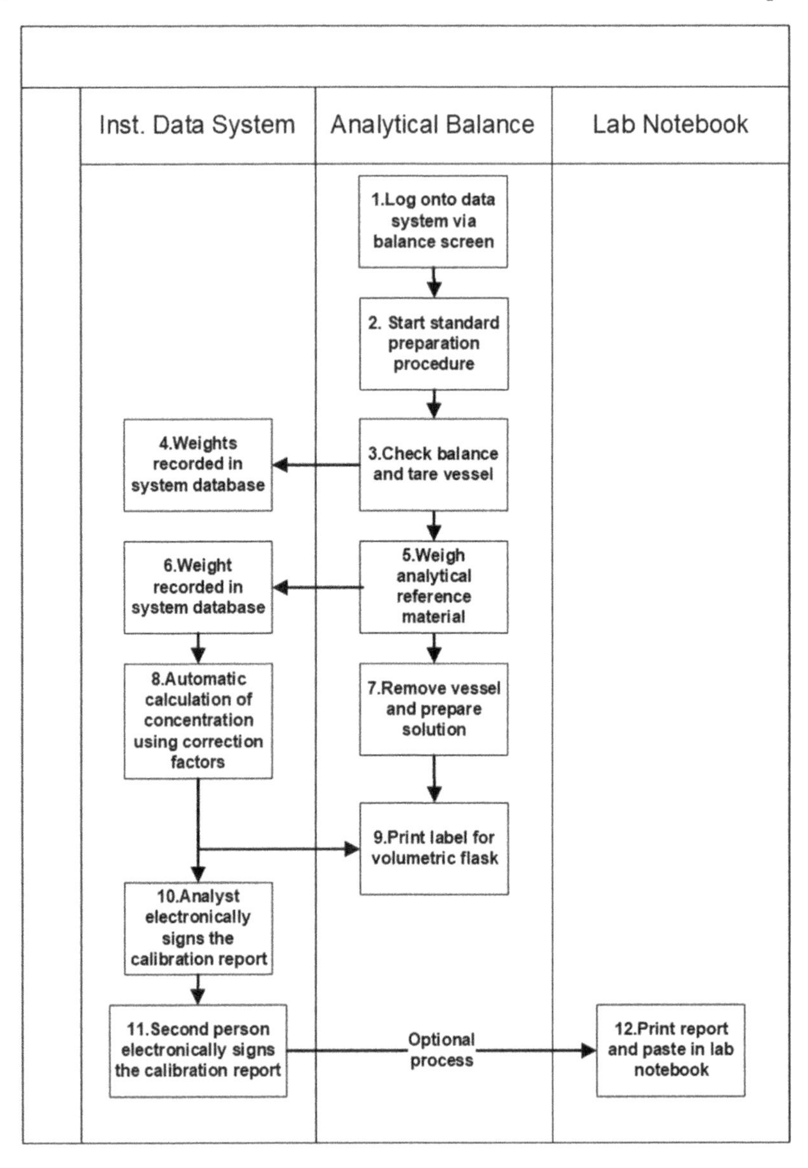

Figure 13.2 Electronic workflow with a balance connected to an instrument data system.

but this tends to be limited to dilutions or other information from the sample preparation stage.

We will look at the three interfacing options from a data integrity perspective rather than a technical interfacing discussion. A further discussion on interfacing CDS to LIMS can be found in Chapter 3 of my book on Validation of Chromatography Data Systems.[23]

It is important to understand that the process flow described in this section is one of several possibilities. The selection of which option to implement and the detail of the data to be transferred will depend on several factors such as the technical feasibility and ease of the two applications to be interfaced, skills in-house and from the application suppliers, *etc.*

13.5.2 Manual Data Transfer Between CDS and LIMS

Having spent a large sum of money and used many resources to implement and validate these two applications, the laboratory is the proud owner of two standalone electronic typewriters. Most of the business benefit is lost but, more importantly, the process is riven with data integrity issues. Depending on the work performed by the laboratory, each item of data entered into either system must be subject to detailed second person review, itself an error prone process. If the laboratory is working to GMP, then much of the data being entered into each application needs to be formally assessed to see which elements are critical and therefore are subject to a second person check as mandated by EU GMP Annex 11.[3]

Given the stupidity of implementing such a manual process between two systems from a data integrity perspective, we will not consider this as a viable option. However, many laboratories do take this approach.

13.5.3 Unidirectional Interfacing from CDS to LIMS

In this example, there is an interface between the CDS to the LIMS but all data entered into the CDS either from the LIMS or from the analytical process in the laboratory must be checked for transcription errors. On the bright side, the CDS has been configured and validated so that the audit trail is turned on and that there is functionality to identify manual changes to data and manual integration easily. In addition, any calculations performed on the data have been implemented so that the analysis can proceed from injection to the calculation of the results and occurs within the boundaries of the CDS. Only the final results are electronically transferred from the CDS to the LIMS using a validated transfer. The workflow of the process is shown in Figure 13.3 and described below:

- The samples for a specific chromatographic analysis are selected in the LIMS and the sample numbers manually entered into the sequence file of the CDS. This data input from the LIMS can be either *via* typing into the CDS the information on LIMS printouts or potentially copying and pasting the sample information from one application to another if the two applications permit this operation. Other metadata can also be entered at this time such as the number of sample replicates, the number of injections from each vial and the standard, samples and blanks

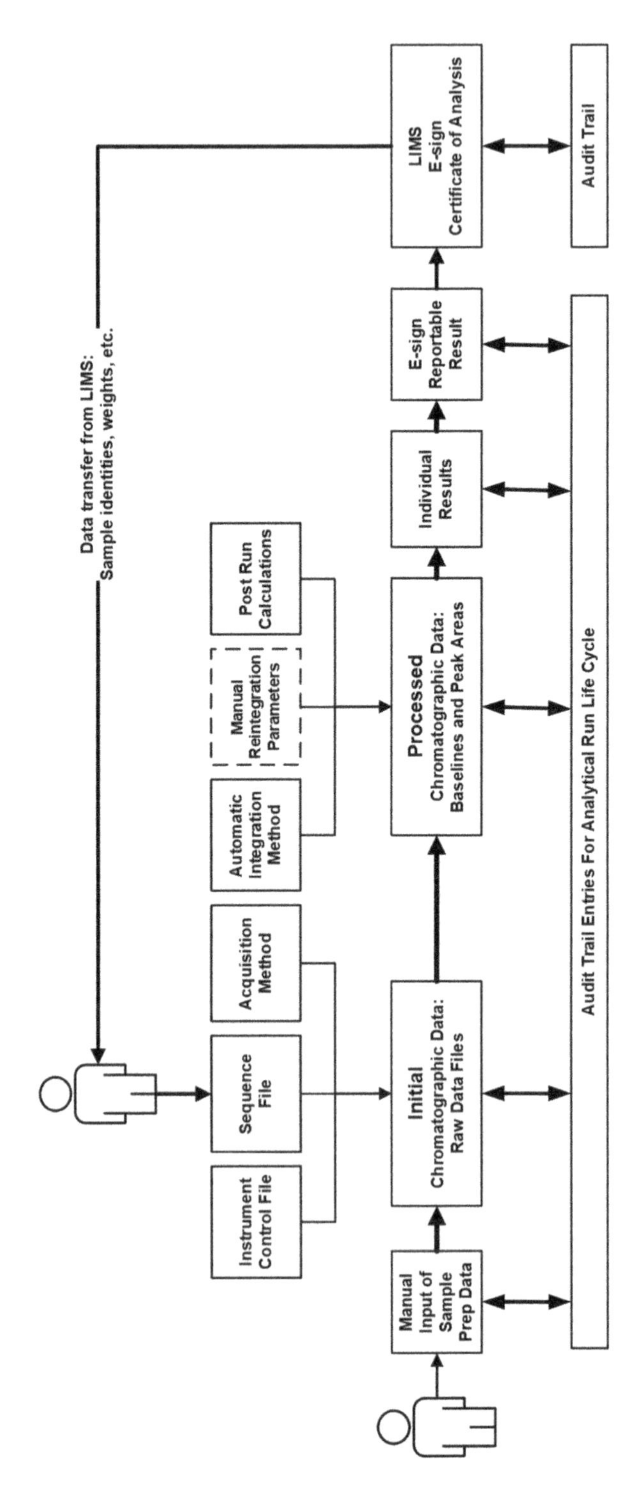

Figure 13.3 Unidirectional interfacing CDS to LIMS.

to be used along with the method to be used and the study or batch number. Manual data entry is error prone, as discussed above, and is a data integrity weak point.
- Additional manual input is required from the sample preparation phase of the analysis, *e.g.* weights, dilutions, water content, standard purity, *etc.* Again, another data integrity weak point that requires extensive second person checks.
- The analytical run proceeds and data files are acquired automatically until the run sequence is complete. The analyst then interprets the data and checks the integration is correct and that the peaks are correctly identified and any integration parameters can be adjusted as necessary (manual intervention). The type of analysis and laboratory procedures will determine if peaks can be manually integrated or not (*e.g.* manual repositioning of the baselines). Once any manual intervention and, if allowed, manual integration is complete the results will be calculated and the results electronically signed by the analyst performing the work.
- Before transfer to the LIMS, the second person review of the data must take place. This is performed by a peer in the laboratory and is not a quality assurance role. In this example, there are parts of the process where the review will proceed quickly and other portions of the work that will be slow and labour intensive. The focus will be on the data integrity weak points, *e.g.* the manual data entries to ensure that they are correct. Integration will be reviewed, although the features of the CDS should highlight where data have been changed or manual integration has been performed, as will the audit trail review. Any errors or corrections will be returned to the analyst, executed and then resubmitted for peer review. When the data are correct for the run, the second person electronically signs the results and the results can be transferred to the LIMS.
- An alternative for the timing of the second person review, could be after the input of the reportable results into the LIMS. In this case, the review would also include a check that the reportable result had been transferred correctly and that the audit trail entries in the LIMS corroborated this.

There are some advantages with a unidirectional CDS–LIMS interface but there are major data integrity weak points that require extensive second person checks to ensure that the data are correct, accurate and complete.

13.5.4 Bidirectional Interfacing Between CDS and LIMS

The process for bidirectional interfacing between CDS and LIMS is shown in Figure 13.4. Both applications have data integrity features enabled, specified and tested including electronic signatures. The process is simpler as there is little manual data input and it should be faster as we can see when we look at the details:

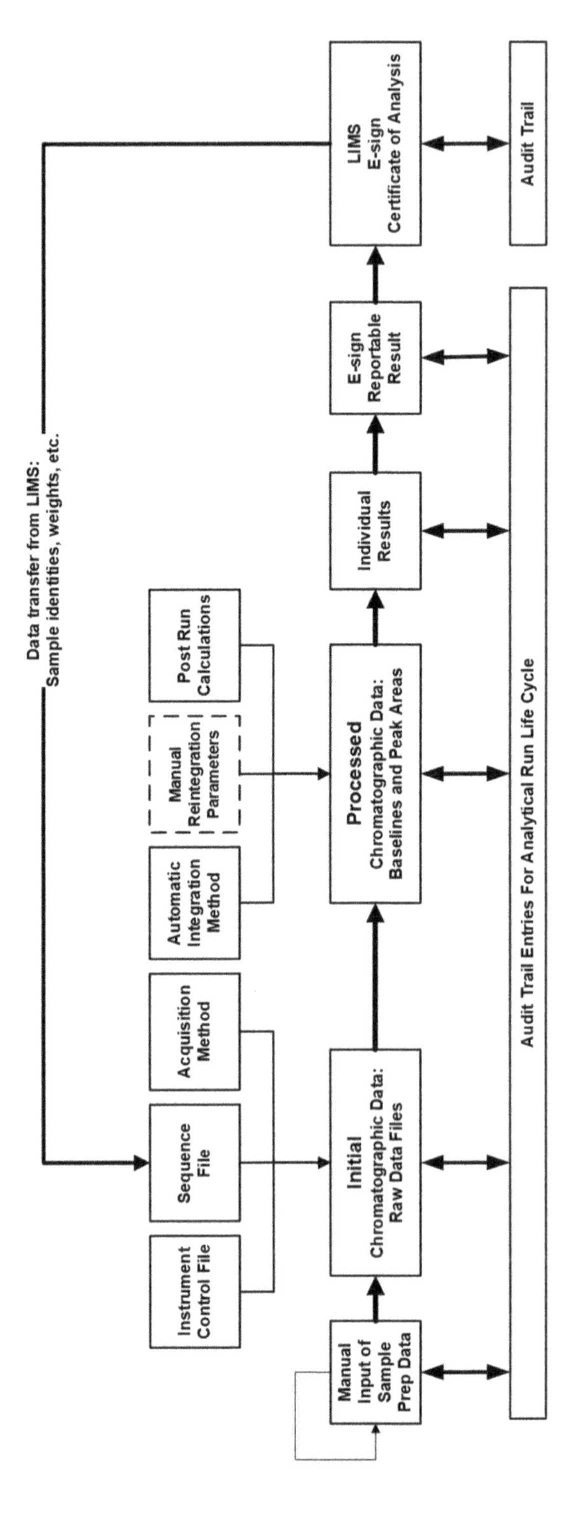

Figure 13.4 Process flow of a bidirectional interfacing between CDS and LIMS.

- The process starts in the LIMS with a download of the sample information for analysis to the CDS. This is a tested and validated step and therefore it needs no transcription checking by a second individual. The download could be just the sample identities, reference standard information with purity and study/batch information. The number of replicate injections and standard vials could be predefined in a sequence template. Alternatively, all required information could be downloaded into the sequence file for the analysis.
- Further data for input to the sequence file in the CDS will be generated during the sample preparation phase of the analysis, *e.g.* reference weight, sample weights, dilutions, *etc.* Unless this data can be captured electronically in a mobile worksheet in an ELN or LIMS for electronic transfer to the CDS, it is likely that manual data input cannot be eliminated entirely from the process. However, manual input is greatly reduced and so is the work of the second person reviewer to check it.
- After the chromatographic analysis run is complete, the interpretation of chromatograms, any manual intervention, manual integration and calculation of the final results proceeds as described above as does the second person review. Overall, the optimised process is simpler to use and quicker to operate.

Therefore, I suggest that instead of implementing point informatics solutions in your laboratory an electronic environment should be designed. This should eliminate transcription error checks and make transfer between instruments and software applications efficient and rapid. Developing an electronic environment is not a one-off process but a continuous project with different phases where single applications are implemented typically one at a time. The aim should be that succeeding implementations should bring their own advantages but must also leverage the benefits of existing and operational applications.[23]

13.6 Impact on Data Integrity and Second Person Review

13.6.1 Ensuring Data Integrity with Electronic Working

As can be seen from the two examples presented in Sections 13.4.4 and 13.5.4, each electronic process is simpler with automatic data capture and transfer between the analytical balance and instrument data system and between CDS and LIMS, respectively.

If the two processes are analysed using data process mapping as described in Chapter 10, many of the data integrity problems and poor data management practices can be eliminated by using interfaced applications. This requires investment in applications, validation and training to use them but the business benefits and ensuring data integrity will outweigh the cost.

13.6.2 Impact on Second Person Review

Second person review can be transformed as well from a slow labour-intensive and paper-based process into a streamlined one where the reviewer knows the weak data integrity points and can focus their efforts there. In addition, the data integrity of the process is enforced as much as possible by technical controls that are underpinned by computerised system validation to demonstrate intended use of the configured applications.

13.6.3 Summary of an Approach for Electronic Working that Ensures Data Integrity

To summarise the approach for electronic working that ensures data integrity consider the following when designing and implementing applications:

- Capture data electronically where it is generated.
- Never transcribe data to eliminate transcription error checks whenever possible.
- Avoid manual data input to avoid second person transcription error checking.
- Work electronically so that records only consist of a single medium.
- Enable and validate all technical controls to ensure data integrity, to enforce security and locations where data can be stored.
- Ideally, all applications should have a database to manage records and will automatically name individual files to ensure quick and easy retrieval of data and the associated metadata. Manual file naming is a slow and labour intensive process.
- Use electronic signatures where appropriate.
- Store data on secure and fault tolerance networked devices.
- Keep paper printouts to a minimum.

Once the process for electronic working is defined, the instruments involved and the associated application software must be qualified and validated, respectively. We now move to Level 1 of the Data Integrity Model to look at analytical qualification and computerised system validation that is for intended use but also to ensure data integrity. This will be covered in Chapter 14.

References

1. *WHO Technical Report Series No.996 Annex 5 Guidance on Good Data and Records Management Practices*, World Health Organisation, Geneva, 2016.

2. *PIC/S PI-041 Draft Good Practices for Data Management and Integrity in Regulated GMP/GDP Environments*, Pharmaceutical Inspection Convention/Pharmaceutical Inspection Co-Operation Scheme, Geneva, 2016.
3. *EudraLex - Volume 4 Good Manufacturing Practice (GMP) Guidelines, Annex 11 Computerised Systems*, European Commission, Brussels, 2011.
4. R. D. McDowall, An evolutionary view of chromatography data systems used in bioanalysis, *Bioanalysis*, 2010, **2**(2), 157–165.
5. *FDA Draft Guidance for Industry Data Integrity and Compliance with CGMP*, Silver Spring, MD, USA, 2016.
6. *MHRA GMP Data Integrity Definitions and Guidance for Industry*, Medicines and Healthcare products Regulatory Agency, London, 2nd edn, 2015.
7. *MHRA GXP Data Integrity Guidance and Definitions*, Medicines and Healthcare products Regulatory Agency, London, 2018.
8. *EMA Questions and Answers: Good Manufacturing Practice: Data Integrity*, 2016, Available from: http://www.ema.europa.eu/ema/index.jsp?curl=pages/regulation/general/gmp_q_a.jsp&mid=WC0b01ac058006e06c#section9.
9. *FDA 483 Observations: Abbott AMO (Hangzhou) Company, January 2017*, Food and Drug Administration, Silver Spring, MD, 2017.
10. Part 211-Current Good Manufacturing Practice for Finished Pharmaceuticals, *Fed. Regist.*, 1978. **43**(190), 45014–45089.
11. *21 CFR 58 Good Laboratory Practice for Non-clinical Laboratory Studies*, Food and Drug Administration, Washington, DC, 1978.
12. *FDA Pharmaceutical CGMPs for the 21st Century: A Risk-based Approach*, Food and Drug Administration, Rockville, MD, 2002.
13. A. Shanley, *Will the Pharmaceutical Industry Ever Get to Six Sigma?*, 2017, Available from: http://www.pharmtech.com/will-pharmaceutical-industry-ever-get-six-sigma.
14. *ICH Q2(R1) Validation of Analytical Procedures: Text and Methodology*, International Conference on Harmonisation, Geneva, 2005.
15. *21 CFR 211 Current Good Manufacturing Practice for Finished Pharmaceutical Products*, Food and Drug Administration, Sliver Spring, MD, 2008.
16. *EudraLex - Volume 4 Good Manufacturing Practice (GMP) Guidelines, Chapter 1 Pharmaceutical Quality System*, European Commission, Brussels, 2013.
17. S. Jenkins, Presentation at a Seminar on the Paperless Laboratory, in *Pittsburgh Conference on Analytical Chemistry and Applied Spectroscopy*, Chicago, 2004.
18. *21 CFR 11 Electronic Records; Electronic Signatures, Final Rule, in Title 21*, Food and Drug Administration, Washington, DC, 1997.
19. *FDA Warning Letter, USV Limited (WL: 320-14-03)*, Food and Drug Administration, Silver Spring, MD, 2014.
20. *FDA 483 Observations: Hetero Labs Limited, Telangana, India*, Food and Drug Administration, Silver Spring, MD, 2016.

21. *ICH Q7-basic Requirements for Active Substances Used as Starting Materials*, International Conference on Harmonisation, Geneva, 2000.
22. *EudraLex - Volume 4 Good Manufacturing Practice (GMP) Guidelines, Chapter 6 Quality Control*, European Commission, Brussels, 2014.
23. R. D. McDowall, *Validation of Chromatography Data Systems: Ensuring Data Integrity, Meeting Business and Regulatory Requirements*, Royal Society of Chemistry, Cambridge, 2nd edn, 2017.

CHAPTER 14

Data Integrity Centric Analytical Instrument Qualification and Computerised System Validation

Typically, computerised system validation and analytical instrument qualification have been separate tasks, however, with the publication of the updated version of United States Pharmacopoeia <1058> on AIQ,[1] the two tasks are now integrated. From Level 1 of the data integrity model presented in Chapter 5, AIQ and CSV are essential tools for ensuring data integrity. It is essential that software applications are configured to ensure the protection of electronic records and tested to demonstrate that they work as intended. This requires a system architecture that protects records and a flexible approach to computerised system validation. This chapter is not the intended to be a detailed discussion of computerised system validation of laboratory systems as this can be found in other references.[2–4] Instead, this chapter outlines a general approach for ensuring data integrity for laboratory systems when they undergo analytical instrument qualification with or without computerised system validation. As the new version of USP <1058> contains some major differences these are outlined in this chapter as they will have an impact on the way you now must approach AIQ.

Although Performance Qualification (PQ) is conducted during sample analysis to show that the instrument is functioning as intended, the subject is presented and discussed here as PQ is intimately involved with the other

Data Integrity and Data Governance: Practical Implementation in Regulated Laboratories
By R. D. McDowall

Published by the Royal Society of Chemistry, www.rsc.org

phases of the 4Qs model. Furthermore, as PQ requires the user requirements specification for an instrument, the obvious place for discussion is here rather than split the discussion.

14.1 What the Regulators Want

14.1.1 21 CFR 211 Current GMP for Finished Pharmaceutical Products

The US GMP regulations for equipment are found in Section 211.63 and these state[5]:

> Equipment used .... shall be of appropriate design, adequate size, and suitably located to facilitate operations for its intended use and for its cleaning and maintenance.

US GMP regulations for the control of analytical instrumentation is relatively short and succinct. This is also true of the corresponding FDA GLP regulations in the next section.

14.1.2 21 CFR 58 GLP for Non-clinical Studies

The corresponding regulations for GLP are in Section 58.61[6] and are as follows:

> Equipment used .... shall be of appropriate design and adequate capacity to function according to the protocol and shall be suitability located for operation, inspection, cleaning and maintenance.

The OECD GLP regulations[7] refer to apparatus but the gist of both regulations is the same.

14.1.3 United States Pharmacopoeia <1058> on Analytical Instrument Qualification

Under the software section of the revision of this USP General chapter there is the following statement[1]:

> There is an increasing inability to separate the hardware and software parts of modern analytical instruments. In many instances, the software is needed to qualify the instrument, and the instrument operation is essential when validating the software. Therefore, to avoid overlapping and potential duplication, software validation and instrument qualification can be integrated into a single activity.

An integrated approach to analytical instrument qualification and computerised system validation of laboratory systems is advocated by USP <1058>.

14.1.4 EU GMP Annex 11

In Clause 1 of EU GMP Annex 11 on computerised systems[8] there is the following requirement for risk management:

> Risk management should be applied throughout the lifecycle of the computerised system taking into account patient safety, data integrity and product quality. As part of a risk management system, decisions on the extent of validation and data integrity controls should be based on a justified and documented risk assessment of the computerised system.

The first requirement of Annex 11 is to use risk management to determine the extent of validation to ensure, amongst other things, data integrity.

14.1.5 ICH Q7 and EU GMP Part 2: Good Manufacturing Practice Guide for Active Pharmaceutical Ingredients

ICH Q7 and EU GMP part 2 [9,10] recognise the important of qualifying and, by implication validating the software, before validating an analytical method or carrying out analytical work:

> 12.82 Appropriate qualification of analytical equipment should be considered before starting validation of analytical methods.

This approach is consistent with both the USP <1058> data quality triangle[1] and the level 1 data integrity model presented and discussed in Chapter 5.

14.1.6 WHO Guidance on Good Data and Record Management Practices

Section 10.2 of this WHO guidance states[11]:

> To assure the integrity of electronic data, computerized systems should be validated at a level appropriate for their use and application.
>
> Validation should address the necessary controls to ensure the integrity of data, including original electronic data and any printouts or PDF reports from the system.
>
> In particular, the approach should ensure that GDocP will be implemented and that data integrity risks will be properly managed throughout the data life cycle.

Appendix 1 under the Attributable section, states[11]:

> The use of hybrid systems is discouraged, but where legacy systems are awaiting replacement, mitigating controls should be in place.
>
> The hybrid approach is likely to be more burdensome than a fully-electronic approach; therefore, utilizing electronic signatures, whenever available, is recommended.
>
> Replacement of hybrid systems should be a priority.

14.1.7 PIC/S PI-041 Good Practices for Data Management and Integrity in Regulated GMP/GDP Environments

In this guidance there are the following statements in Section 5.5.4 [12]:

> 5.5.4 For computerised systems, manual interfaces with IT systems should be considered in the risk assessment process.
>
> Computerised system validation in isolation may not result in low data integrity risk, in particular, when the user is able to influence the reporting of data from the validated system.
>
> A fully automated validated process together with a configuration that does not allow human intervention, or reduces human intervention to a minimum, is preferable as this design lowers the data integrity risk.
>
> Procedural controls should be installed and verified where integrated controls are not possible for technical reasons.

14.1.8 Regulatory Summary

From the sections of the various guidance documents above, there is a regulatory expectation that:

- Integrated instrument qualification and computerised validation is advocated in the new version of USP <1058>.[1]
- Systems must be validated not just for the intended purpose but also to ensure data integrity.
- Laboratories are expected to apply risk management to their validation approach, especially to ensure data integrity, and the approach taken must be documented and justified.
- Hybrid systems are not recommended due to the problem of managing and synchronising two incompatible record formats, as was discussed in Chapters 12 and 13.
- Technical controls including electronic signatures and electronic working to enforce integrity, compliance and are automatic are preferred over procedural controls that can be inconsistently applied by users.

- However, when technical controls are implemented, they must be specified and then tested to demonstrate that they work as intended as part of the overall validation of a system.

All the regulatory guidance documents do not mention analytical instrument qualification, yet this is a vital quality and data integrity building block, as stated in ICH Q7[10] and shown in the Data Integrity Model presented in Chapter 5.

14.2 GMP Regulations and the Pharmacopoeias

14.2.1 Relationship Between GMP and the Pharmacopoeias

In the regulated world of Good Manufacturing Practice (GMP) we have regulations that define what should be done but leaves the interpretation to the individual organisation of how to do it. There are some general regulations for the laboratory but there is not much detail to aid interpretation. However, when we come to the GMP regulated analytical development or quality control laboratory we also have the pharmacopoeias such as the European Pharmacopoeia (EP), Japanese Pharmacopoeia (JP) and United States Pharmacopoeia (USP), to provide further information to help interpret the regulations. These publications can have monographs for active pharmaceutical ingredients, finished products and general chapters that provide requirements for how to apply various analytical techniques such as chromatography or spectroscopy.

Figure 14.1 shows the relationship between the US GMP regulations in 21 CFR 211[5] and the various parts of the United States Pharmacopoeia. The aim of any pharmacopoeia is to provide more specific information on

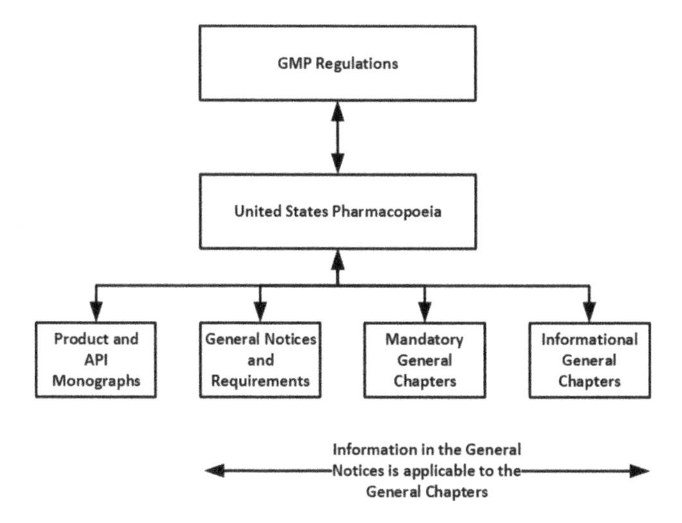

Figure 14.1 Relationships between US GMP regulations and the USP main sections.

specific topics such as the specifications of active pharmaceutical ingredients and finished products. From an analytical perspective the main elements are:

- Monographs for Active Pharmaceutical Ingredients and Products provide the basis for the analytical methods.
- General Notices and Requirements

This section presents the basic assumptions, definitions, and default conditions for the interpretation and application of the USP. A requirement in the General Notices section applies to all articles in the USP and to all general chapters unless specifically stated otherwise.

- Mandatory General Chapters

  These are numbered between <1> and <999> and cover the main analytical techniques used in the laboratory such as balances, loss on drying, dissolution, chromatography and spectroscopy. Recent general chapters are published in pairs with a mandatory and informational general chapter.

- Informational General Chapters

  These are numbered between <1000> and <1999>. You will notice that USP <1058> is an informational general chapter describing the overall approach to integrated AIQ and where appropriate CSV. Nowhere in the chapter are the parameters to qualify a specific analytical instrument, these are found in the mandatory general chapter.

14.2.2 Importance of USP <1058>

Of the major pharmacopoeias, only the USP has a general chapter on analytical instrument qualification (AIQ).[1] This came about with a 2003 conference organised by the American Association of Pharmaceutical Scientists (AAPS) on Analytical Instrument Validation. The first decision of the conference was that the name was wrong and it should be Analytical Instrument Qualification. From the conference came a white paper[13] that after review and revision became USP <1058> on AIQ effective in 2008.[14] This described a data quality triangle, general principles of instrument qualification and a general risk classification of analytical instruments into one of three groups: A, B or C. These groups are characterised as follows:

- Group A comprises instruments with no measurement capability and proper function is ensured by observation.
- Group B instruments that control a physical parameter or measure a parameter and are calibrated.

- Group C instruments are controlled by a workstation with instrument software to control the instrument, acquire and process data and report results. These systems require qualification of the instrument coupled with validation of the application software.

14.2.3 Use the USP <1058> Principles for GLP Instruments and Systems

USP <1058> is intended for laboratories working to Good Manufacturing Practice, *e.g.* analytical development and quality control laboratories. However, as instrument qualification and computer validation are also applicable to Good Laboratory Practice accredited laboratories, it would be prudent to follow the same approach as outlined in USP <1058>.

14.3 Why Is Instrument Qualification Important?

The simplest answer to this question is that you know that the instrument is functioning correctly and that you can trust the results it produces when used to analyse samples. This is based on the data quality triangle in USP <1058>.[1,14]

14.3.1 Data Quality Triangle

The USP <1058> data quality triangle is shown in Figure 14.2, this is adapted to show the 4Qs model in the lower AIQ layer along with the role of the

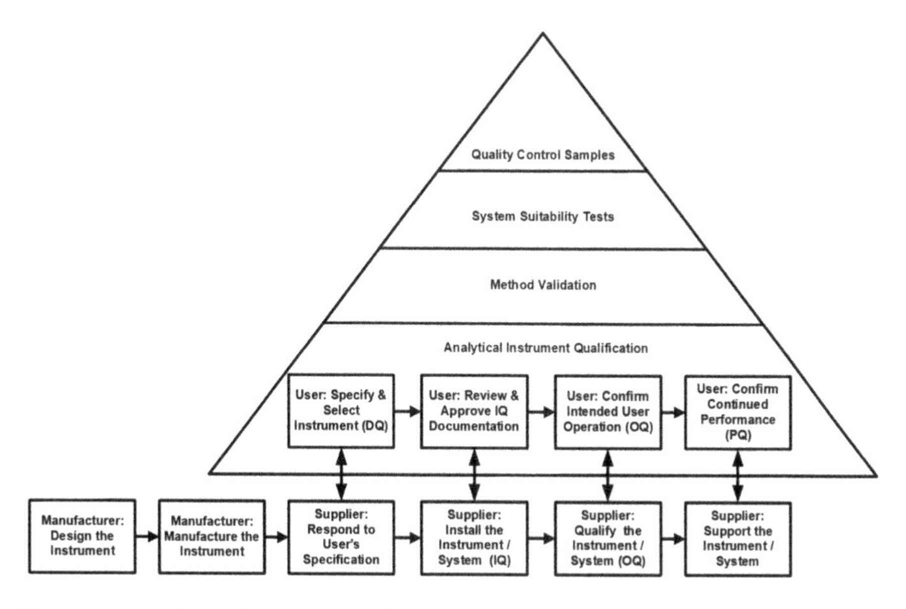

Figure 14.2 Adapted USP <1058> data quality triangle showing the 4Qs model and supplier involvement.

supplier in designing and developing the instrument and any software. The figure also shows the interaction between the laboratory users and the supplier throughout the instrument life cycle.

One point to note is that AIQ is instrument based and all other layers, *e.g.* method development and validation, system suitability tests and QC checks, are method based. It is the AIQ layer that challenges the instrument operation against reference standards and calibrated instruments.

14.3.2 Data Integrity Model

However, there is a more important reason in today's world of data integrity – integrated analytical instrument qualification and computer system validation is an essential component of a data integrity model. The Data Integrity Model works from the Foundation up with each layer providing input to the next. As seen from Chapter 5, AIQ and CSV at Level 1 come after the Foundation layer and illustrate that if the instrument is not qualified and that any software is not validated, the two layers above (method validation and sample analysis) will not be effective and can compromise data integrity. Interestingly, certainly AIQ and sometimes CSV are omitted from many of the data integrity guidance documents but you can see the importance in the overall data integrity framework of the model. Without qualifying the instrument and validating any associated software, how do you know that the system is working correctly?

14.4 Why a New Revision of USP <1058>?

14.4.1 Problems with the 2008 Version

The 2008 version of USP <1058> had several issues that were discussed by the author at a session co-organised by Paul Smith at an AAPS conference in 2010 and published in Spectroscopy.[15] The three main problems with the first version were:

- Problem 1: The true role of the supplier is missing.

  The supplier is responsible for the instrument specification, detailed design and manufacture of the instrument but this is not mentioned in <1058>, as shown in Figure 14.2. The reason is that the section on design qualification (DQ) mentions that a user can use the supplier's specification. However, a user needs to understand the conditions under which the specification was measured and how relevant this is to a laboratory's use of an instrument. A supplier cannot do this.

- Problem 2: Users are responsible for DQ.

  USP <1058> places great emphasis on the fact that in the design qualification stage it is the responsibility of the supplier: *Design qualification (DQ) is most suitably performed by the instrument developer or manufacturer*. This is wrong. Only a user can define their instrument needs and must do so to define their intended use of the instrument and comply with GMP regulations (§211.63).[5]

- Problem 3: Poor software validation guidance.

  USP<1058> describes software qualification and validation is the poorest part of this general chapter as software is pervasive throughout Group B instruments and Group C systems.
  Although the approach to handling embedded software in Group B instruments where the firmware is implicitly or indirectly validated during the instrument qualification is fine, there are omissions. Users need to be aware that both calculations and user defined programs must be verified to comply with GMP requirements in 211.68(b).[5] Note that qualification of firmware, which is a simple and practical approach, is now inconsistent with GAMP 5 that has dropped Category 2 (firmware).

Software for Group C systems is the weakest area of the whole of <1058>. The responsibility for software validation was pushed onto the supplier:

> The manufacturer should perform DQ, validate this software, and provide users with a summary of validation. At the user site, holistic qualification, which involves the entire instrument and software system, is more efficient than modular validation of the software alone.

In the days of data integrity, this approach is completely untenable. The FDA Guidance on Software Validation[16] quoted by <1058>, was written for medical device software that is not configured, unlike much of the laboratory software used today.

14.4.2 Revision Path of USP <1058>

To try and rectify some of these issues, the revision process of USP <1058> started in 2012 with the publication of a Stimulus to the Revision process article published in Pharmacopoeial Forum written by Chris Burgess and Bob McDowall.[17] This proposed two items:

1. An integrated approach to analytical instrument qualification and computerised system validation (AIQ-CSV).
2. More granularity for Group B instruments and Group C computerised systems to ensure that calculations and user defined programs were captured in the first group and an appropriate amount of validation was performed by the users for the second group.

The feedback from this article was used by the authors to draft a new version of <1058> in the summer of 2013 that was circulated to a few individuals in industry and suppliers for review before submission to the USP. Proposed drafts of the new version were published for public comment in Pharmacopoeial Forum in 2015 and 2016[18,19] and comments incorporated in the updated versions. The approved USP <1058> final version was published in January 2017 in the First Supplement to USP 40 and became effective from 1st August 2017. However, an erratum was issued clarifying the role of the URS in the <1058> and the final version is effective from Volume 41 of the USP in 2018.[1]

14.5 What Has Changed with USP <1058>?

14.5.1 Differences Between the Old and New Versions of USP <1058>

First, let us look at the overall scope of changes between the old and new versions of USP <1058>, this is shown in Table 14.1.

14.5.2 Omitted Sections in the New Version

Omitted from the new version of <1058> are the following items:

- Differences between qualification and validation. This is omitted as qualification and validation activities are integrated into the new version, so why describe the differences? You need to control the instrument and any software and if you can demonstrate this through the 4Qs process described in the new <1058>, why bother with what the activity is called?
- Table 1 in the old version of <1058> describes the timing, applicability and activities of each phase of AIQ is dropped from the new version. Rather than give a fixed and rigid approach to AIQ as the table does, there is more flexibility in the new version of <1058> and omitting this table reinforces the new approach.
- Standalone software: as standalone software has nothing to do with AIQ not surprisingly this section has been omitted from the new version. The GAMP 5 and the accompanying Good Practice Guide for Laboratory Computerised Systems[3,4] or my CDS validation book[2] will be adequate for this task.
- Examples of instruments in the three categories: As the instrument classification depends on the intended use there is no need to give a list of instruments or systems in Groups A, B and C. It is the intended use of the instrument that defines the Group and providing a list produces anomalies. For example, in the old version of <1058>, a dissolution bath is listed in Group C when it should be in Group B if there is only firmware and the instrument is calibrated for use but in Group C if the bath has a standalone software to control it. To avoid these arguments the list of examples has been dropped from the new version.

14.5.3 Additions and Changes to USP <1058>

Of greater interest to readers will be the changes and additions to the new general chapter, again these can be seen in Table 14.1. I will discuss the following three areas that reflect the main changes to the general chapter:

- roles and responsibilities;
- changes to DQ, IQ and OQ phases and how this impacts your approaches to AIQ;
- software validation.

14.5.4 Roles and Responsibilities

The USP <1058> update to the Roles and Responsibilities section makes users ultimately responsible for specifying their needs, ensuring that a selected instrument meets them and that data quality and integrity are maintained.[1] The manufacturer section now includes suppliers, service agents and consultants to reflect the real world of instrument qualification. One new responsibility is for the supplier or manufacturer to develop meaningful specifications for users to compare with their needs. Incumbent on both users and suppliers is the need to understand and state respectively the conditions under which specifications are measured to ensure that laboratory requirements can be met. We will discuss this further under the 4Qs model in the next section.

Finally, there is a requirement for a technical agreement between users and suppliers for the support and maintenance of any Group B instrument and Group C system. The agreement may take the form of a contract that both parties need to understand the contents of such an agreement and the responsibilities of each one.

14.5.5 An Updated 4Qs Model

At first sight, the new version of USP <1058> uses the same 4Qs model as the 2008 version. However, there are some significant differences. Figure 14.3 presents the 4Qs model in the form of a V model rather that a linear flow, published by Paul Smith and myself.[20] However, Figure 14.3 has now been updated to reflect the changes in the new version of USP <1058> and the erratum and we will discuss the main changes.

Design Qualification now has two phases associated with it:

- The first phase is for the users to define the intended use of the instrument in a user requirements specification: *users must define functional and operations specifications and intended use.*[1] Although the new <1058> notes that this is *expected to be minimal for commercially available instruments* it does not mean slavishly copying supplier specifications – especially if you do not know how any of the parameters have been measured.

Table 14.1 Comparison of the old and new versions of USP <1058> on analytical instrument qualification.

Section	USP <1058> 2008 version	USP <1058> 2017 version
Introduction		• Expanded introduction • Can merge activities, *e.g.* IQ & OQ • Description of Groups A, B & C moved earlier in chapter • Classification of an instrument depends on the intended use
Validation *versus* qualification	• Outline of the differences between the two terms	
Components of data quality	• Data quality triangle unchanged • Essentially the same in the two versions	
AIQ process	**Design qualification**	
	• Emphasis on the supplier to perform this task • Little if any involvement by the user	• Users must define functional and operations specifications and intended use • Expected to be minimal for commercially available instruments • Users demonstrate fitness for use • Supplier robust design, development and testing documentation • Change of use triggers review/update of specifications
	Installation qualification	
	• IQ needed for pre-owned instruments	• Extension of the section to include software installation and IT involvement for interface to a network • Risk assessment for non-qualified instruments
	Operational qualification	
		• Can be merged with IQ • New section on software functions • New section on software configuration and/or customisation • Configure software before OQ testing • Users must review Supplier qualification materials • OQ tests refer to instrument specific general chapters
	Performance qualification	
		• Expanded section on practices for PQ, change control and periodic review

Table 14.1 *(continued)*

Section	USP <1058> 2008 version	USP <1058> 2017 version
Table 1	• Timing, applicability and activities for each phase of AIQ	
Roles and responsibilities		• Expansion of section on Manufacturers to include suppliers, service agents and consultants • The requirement for a technical agreement between user and supplier
Software validation	• Standalone software	• Expanded introduction • Firmware now includes control of calculations and user defined programs • Instrument control software expanded section
Change control		• Slimmer and more concise approach to managing change
AIQ documentation	• Essentially the same in the two versions	
Instrument categories	• Description of groups A, B and C • Examples of each group	
Glossary		• Definition of 7 terms

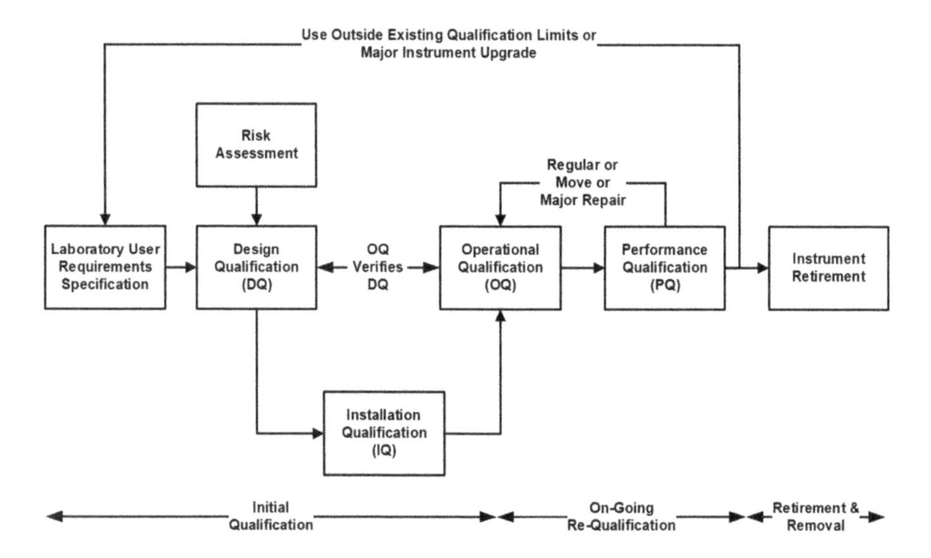

Figure 14.3 Modified 4Qs model for analytical instrument qualification. Figure adapted from ref. 20.

The output from this process is a user requirements specification (URS) or your design document.
- The second phase that is the qualification of the instrument design. This means that you confirm that the selected instrument meets your design specification or intended use. If looking outside of the analytical laboratory, medical device manufacturers call this Design Verification – assuring that the user's current requirements are in the system and any omissions are mitigated, where appropriate.

These two sections are where most laboratories get it wrong for reasons such as we know what we want (therefore why bother to document it?) or we believed the supplier's literature. This is where most qualifications fail as there is no specification upon which to base the testing in the OQ phase of the process as shown in Figure 14.3. Executing an OQ without a corresponding URS or design document is planning to fail any qualification.

- Note that when the new <1058> talks about minimal specifications for commercial instruments it does not include minimal specifications for software used to control it for Group C systems. Here, you not only need to consider the control of the instrument but also the acquisition, transformation, storage and reporting of data and results that include how data integrity and data quality are assured. This is not an activity that is expected to be minimal – especially in today's regulatory environment.
- There is also the possibility, for Group B instruments, of merging the laboratory user requirements and the DQ: the decision to purchase being made on the URS (with the proviso that the contents are adequate).
- Risk management is implicit in the <1058> classification of the instrument groups and the sub-groups of B and C instruments and systems but more needs to be done in the specification and configuration of the software. For example, access controls, data acquisition and transformation are key areas for managing data integrity risks.
- What is also shown in Figure 14.3 is that if there is any change of use during the operation of the instrument or system, this must trigger a review of the current specifications with an update of them, if appropriate.

Installation Qualification in the new version of USP <1058> now includes these new areas:

- The installation of software and the involvement of the IT function to interface an instrument to a network.
- The requirement for conducting an IQ for non-qualified instruments is replaced with a requirement to gather available information and conduct a risk assessment to determine if an IQ should be conducted. In many cases, if an instrument has been installed and maintained by a supplier with records of these activities but has not been formally qualified the risk assessment may determine that no IQ should be performed,

placing increased emphasis on the OQ phase to demonstrate fitness for intended use.
- In the introduction to <1058>, there is the mention that activities can be merged, *e.g.* IQ & OQ,[1] harmonising the approach with that in Clause 2.5 of EU GMP Annex 15[21] discussed below in more detail. I would see suppliers taking advantage of this option to have a single IQ/OQ document for ease of working as both phases of qualification are typically conducted by the same service engineer. The combined protocol must be pre-approved by the laboratory and then reviewed by them post-execution.

Operational qualification has also been extended to include:

- A new section on software functions and the differences between software configuration and customisation for Group C systems.
- It is important to configure software (and document the settings) before an OQ is conducted, otherwise, you will be repeating some tests. In practice, however, there may be a differentiation of duties as a supplier may only perform a basic qualification of the unconfigured software, leaving the laboratory to configure the application and then conduct further verification of the whole system. This is a critical point as data integrity gaps are usually closed through the configuration of the application. However, unless the application is known or a copy is already installed and configured, it is unlikely that the OQ will be performed on a configured version as the laboratory may not know the process to be automated (*e.g.* hybrid operation, electronic operation or incorporate existing spreadsheet calculations into the software).

14.5.6 Harmonisation of Qualification Approaches

There is harmonisation between USP <1058>[1] and EU GMP Annex 15 on Qualification and Validation[21] in two areas:

1. USP <1058> states that the laboratory should write a URS that is then followed by selecting an instrument or system that meets those requirements (Design Qualification or DQ).[1] This is mirrored by EU GMP Annex 15 with Clauses 3.2 and 3.3 labelled User Requirements Specification and Design Qualification, respectively[21]:

   3.2. The specification for equipment ...... or systems should be defined in a URS and/or a functional specification. The essential elements of quality need to be built in at this stage and any GMP risks mitigated to an acceptable level. The URS should be a point of reference throughout the validation life cycle.

   3.3. The next element in the qualification of equipment .... or systems is DQ where the compliance of the design with GMP should be demonstrated and documented. The requirements of the user requirements specification should be verified during the design qualification.

2. USP <1058> allows the merging of documents[1] and this is also mirrored in Annex 15 Clause 2.5 in the section on documentation[21] as discussed above:

   2.5. Qualification documents may be combined together, where appropriate, *e.g.* installation qualification (IQ) and operational qualification (OQ).

The first point contains two issues that are very important:

- The first is that the URS is a point of reference throughout the life cycle.[21] This is reflected in USP <1058> with the testing performed during OQ and PQ being linked back to the laboratory URS.
- The second issue is that in the DQ requirements in the URS are verified.[21] This is important as the URS is a living document and if there is a requirement that is not met by the proposed instrument or system, then the URS should be updated to reflect what is being purchased or select another one that is a better fit. Equally so if the proposed system has functions that will be used but are not contained in the URS, then update the specification to include them.

The second point is a way to solve a practical issue in many regulated laboratories when IQ and OQ are performed by the supplier or their agent. Two documents need to be reviewed and approved by the laboratory, typically the IQ should be reviewed and approved post execution before the OQ starts. This is rarely the case in practice as the service engineer is limited by the available time on site and when the IQ is completed, the OQ starts. Where applicable and practicable, merging the IQ and OQ documents means that there is a single document for pre-execution approval. The combined IQ/OQ can proceed and then the laboratory can review the combined document and release the instrument. There could be a potential problem when the laboratory review occurs of the executed document, if the service engineer has omitted a signature, did not complete an applicable section or there is an unresolved problem it may be an issue to get them back on site to complete the document.

14.6 Importance of the Laboratory URS for Analytical Instruments

14.6.1 Role of the URS

One area that is critical is a current laboratory URS for any analytical instrument. Typically, this is the weakest area in the analytical instrument life cycle as few analysts bother to write one or simply use the supplier's specification. USP <1058> on Analytical Instrument Qualification[1] has the following statements:

> The first activity is the generation of a user requirements specification (URS), which defines the laboratory's particular needs and technical and operational requirements that are to be met.

> The subsequent qualification activities necessary to establish fitness for purpose may be grouped into four phases: design qualification (DQ), installation qualification (IQ), operational qualification (OQ), and performance qualification (PQ).

It is clear that writing the URS for an analytical instrument is a totally separate activity from the Design Qualification (DQ) phase, as noted in Section 14.5.6. DQ is an activity that confirms that the instrument or instruments you propose purchasing meet the requirements in your URS, as shown in Figure 14.3. You test the requirements contained in the URS in both the OQ and PQ phases.

The problem now comes with a further statement in the new USP <1058>:

> It is expected that DQ requirements will be minimal for commercial, off-the-shelf instruments.

Note these words well. Requirements will be minimal. Not zero, not virtual, not imaginary, not verbal but minimal. Minimal = written down, documented, recorded and formal (approved).

14.6.2 Understand Your Intended Use

If you know how you will use the instrument writing the laboratory URS is much easier. For example, do you want a research instrument for detailed spectroscopic work or a simpler instrument that just measures the absorbance of materials at specified wavelengths? Thus, a minimal specification for the instrument is based on what you want the instrument to do and not the supplier specification. From the understanding of intended use will flow the actual specifications based on that usage. Whilst you may use the supplier's specification as a basis for your URS do not slavishly copy the specification but understand how each parameter was measured and adapt accordingly for use. Suppliers have a responsibility here to write meaningful specifications to help all users purchase the right instrument for their needs.

14.6.3 A Role of the Supplier: Write Meaningful Specifications

The supplier has an equally important role in the selection process as the USP <1058> also requires suppliers[1]:

> To aid the user, suppliers are responsible for developing meaningful specifications for the users to compare with their needs and aid selection.

Note well – meaningful specifications.

To illustrate this point, a bench top centrifuge that operated at room temperature with a fixed rotation speed and a timer needed qualification. What instrument could be simpler to qualify? The rotor speed was specified as 3500 ± 1 rpm. However, the specification was measured from the pulse train to the stepper motor when no rotor was attached to the spindle. How many laboratories centrifuge samples without a rotor? This is a simple instrument but with a specification that has no connection with the intended use. The key issue is that you need to understand how the specification was measured and not take it at face value.

14.6.4 How Minimal Is Minimal?

User requirements for commercially available instruments are expected to be minimal. But requirements must also be testable in the OQ using calibrated instruments and traceable reference standards according to the updated version of USP <1058>.[1] As mentioned above, minimal does not mean anything. We also need our instrument specifications to be scientifically sound as the FDA have put this into the regulations in 21 CFR 211.160(b).[5]

Let us assume that we want to purchase a UV-Vis spectrophotometer, there are many available that come in a range of specifications and budgets. Looking at a high-end instrument we could have one that can operate over a wavelength range between 175 nm and 900 nm. Are you really going to measure down to 175 nm (not forgetting the nitrogen purge)? Step back and think that although the instrument specification is wide, what are you going to use the instrument for? What is a realistic wavelength range for your work? Many laboratories may only use a UV-Vis instrument over a far narrower range say between 220 nm and 290 nm. Think about what you would have to do about qualifying the wavelength accuracy. Holmium perchlorate solution can be used over the range from 240 nm to 640 nm with 14 traceable peaks. But what will you do between 175–240 nm and 650 to 900 nm? The key message is to specify what you actually want rather than what the supplier can offer you.

Associated with this is the "just in case" argument. We currently work in the range 240–280 nm but we might want someday to widen the range, how will this impact the specification and how you will verify that in the OQ? The best advice is to write the specification for how you work now. When you change the intended use of the instrument, it is relatively easy to update the URS and perform an extension qualification under an approved change control request.

This approach also applies to other parts of the instrument specification, you need to specify what you need rather than what is on offer, *e.g.*:

- Single beam or double beam instrument?
- Monochromator or diode array?
- Bandwidth?
- Absorbance units?
- Photometric linearity?

14.6.5 Do Not Forget the Software!

Although USP <1058> says that a URS for an instrument is expected to be minimal, what about any associated software used to control the instrument as well as acquire, process, store and report data? Minimal? No. We must also include in the URS data integrity the software functionality requirements for the intended use of the instrument and this is not an exercise in minimalism.

The configuration of software is important for computerised laboratory systems as the intended use of the software including data integrity can only be tested when the application settings have been enabled and documented. For example: segregation of duties, enabling record protection and turning the audit trail on. A major problem is that instruments and software can be ordered without involving the QA data integrity function whose role should be to see if an application is suitable for use in a regulated environment. Some software applications will not save data automatically and will only do so when a user hits the save button. Implementing procedural controls to work around the data integrity omissions in software is not the way forward.

14.6.6 Purchasing a Second Instrument

What is the situation if you want to buy the second or third instrument of the same make and model? Do you need to write a new specification? Here you should be practical. First, review the current URS for the first instrument and ask yourself the questions:

- Are the requirements still applicable to the current instrument or do they need to be modified?
- Are the current or updated requirements applicable to the proposed instrument?

  If yes, then order and qualify the new instrument using the existing URS.

  If no, then use the current URS as a basis to modify or write additional requirements that reflect the intended use of the new instrument and qualify it on this basis.

Inertia and fear of the unknown are the factors that stop analytical scientists writing a specification for their instruments. There is also the attitude of management – why waste time, just buy the instrument....

14.6.7 It's all About Investment Protection

Although we have discussed the GXP regulations and the need to specify our instrument correctly to meet them, the bottom line is less about interpreting clauses in regulations and standards but buying the right instrument for the right job. It is investment protection: resources are scarce and you must spend money wisely. Get the specification right for business reasons and compliance comes as a bonus. You know it makes sense but does management?

14.7 Software Validation Changes to USP <1058>

The major changes to USP <1058> occur in the section on software validation. This is shown diagrammatically in Figure 14.4. As the instrument examples have been removed in the new version of <1058> and replaced with the need to determine the group based on intended use, a formal risk assessment now needs to be performed and documented. A risk assessment, based on Figure 14.4 to classify instruments based on their intended use has been published by Burgess and McDowall[22] and is based on the updated classification used in the new version of USP <1058>.[1] As can be seen from Figure 14.3, the risk assessment should be conducted at the start of the process in the DQ phase of work as the outcome of the risk assessment can influence the extent of work in the OQ phase.

Rather than classify an item as either Group B or Group C, there is now more granularity for both Groups with three sub-options in each of these two

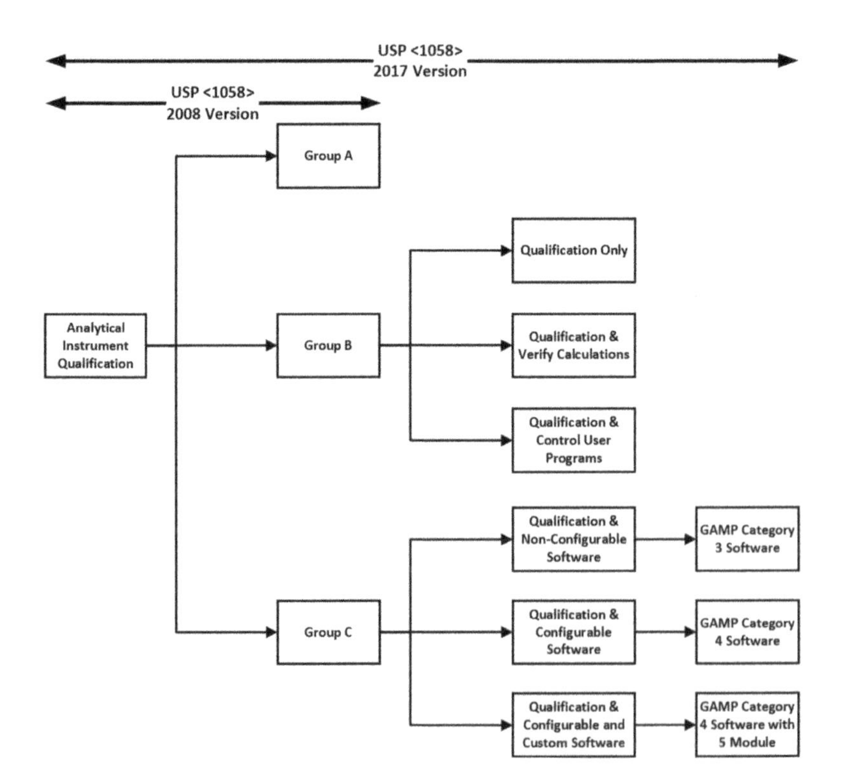

Figure 14.4 Software validation and verification options with the new USP <1058>.

groups. This allows laboratories more flexibility in qualification/validation approach but also fills the holes from the first version of <1058>.

Group B instruments now just require either qualification of the instrument plus either verification of any embedded calculations, if used, or specification, build and test of any user defined programs.

For Group C systems, the new USP <1058> divides software into three types:

- non-configurable software;
- configurable software;
- configurable software with custom additions.

As can be seen from Figure 14.4, these three sub-types can be mapped to GAMP software category 3, 4 and 4 plus category 5 modules. These changes now align USP <1058> closer but not identically with GAMP 5. The main difference is how firmware in Group B instruments is validated – directly with GAMP 5 or indirectly when qualifying the instrument with USP <1058>. Mapping of GAMP 5 software categories to the new USP <1058> groups has been published if more detail is required.[23]

However, the bottom line is that software validation of Group C systems under the new USP <1058> should be the same as any GXP system following GAMP 5 principles. One item that is not mentioned in the new <1058> is a traceability matrix. For Group B instruments, it will be self-evident that the operating range of each parameter will be tested in the OQ. However, this changes with Group C systems, especially as software and possibly networking are involved, then a traceability matrix will be mandatory.

14.7.1 Improving the Analytical Process

One of the items not covered in USP <1058> is an understanding of how an instrument, specifically one from Group C, fits into the analytical process and how it can be used to make the business process more efficient. Typically, this would be undertaken by implementing electronic working with electronic signatures with the elimination (or perhaps extermination would be a better word) of all those horrible spreadsheets that slow down the process. An organisation can spend a large amount of money on a shiny system that is capable of amazing things, only for idiots in the laboratory to print out piles of paper then enter data manually into a spreadsheet and carefully check the entries. Perhaps if Dante were to rewrite his Inferno and set it in modern times, this would be his vision of analytical hell. Endless manual data entry and transcription error checks carried out for ever in an ocean of paper. This would be coupled with the devils from Hell's QA department poking those analytical miscreants who did not spot a transcription error with sharpened poles into places poles should not be poked into. Perhaps this is a description of your laboratory?

In an ideal world we would be working electronically, the way this is achieved is to redesign the process as shown in Figure 14.5. The process is redesigned and the application software configured to match it. The focus as shown in Figure 14.5 is on a top down approach aimed at the process efficiency. Therefore, the validation of a laboratory computerised systems integrates the life cycle tasks outlined in my CDS validation book[2] with the process redesign shown in Figure 14.5. Now the system is validated! We can now relax safe in the knowledge that things are under control.

14.7.2 A Validated System with Vulnerable Records Means Data Integrity Problems

This has been the way most validation work has been carried out. However, there are three potential problems that may arise in this approach, as shown in Figure 14.5:

1. Process level

   A problem arises if the system is used as a hybrid and paper is defined as the raw data. The FDA refuted this argument in 2010 with a Level 2 guidance where they stated that paper was neither a true copy nor an exact and complete copy of the underlying electronic records[24] and reiterated this in the Draft FDA Guidance on Data Integrity.[25]

2. Application level

   The application configuration settings are not set to protect the electronic records generated by the system (*e.g.* no security, audit trail turned off) or if configured, they are not documented. This is unwise as inspectors have been trained to request this documentation and assess the accuracy of it by comparing the document with the software settings.

3. Record level

   Protection of electronic records created and managed by the application. However, if your electronic records are stored in directories in the operating system – be worried, as the application is validated but the electronic records could be unprotected if stored in directories that are open to users. This is a feature that is at the heart of many warning letters and 483 citations.

Shown in Figure 14.5 underneath the application are the data and metadata produced from the analyses performed in the laboratory. With laboratory applications there are two options for storing the data either in directories in the operating system file structure or in a database. McDowall and Burgess have published a four-part discussion that looked at the ideal chromatography data system for a regulated laboratory,[26–29] in the paper on

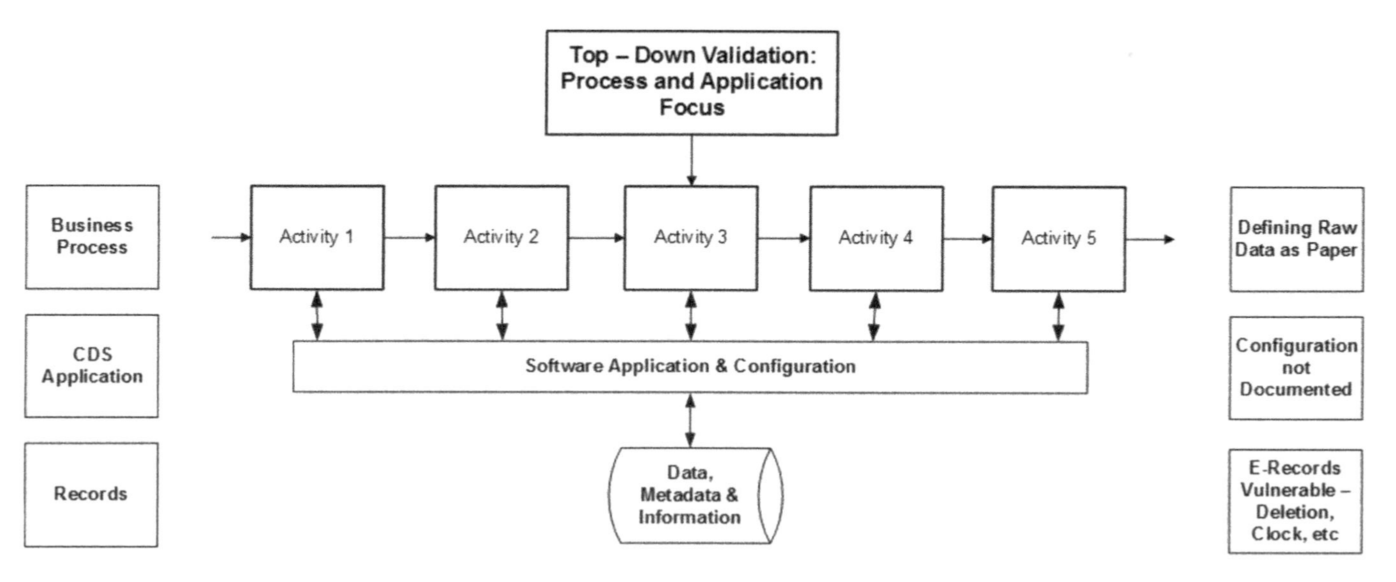

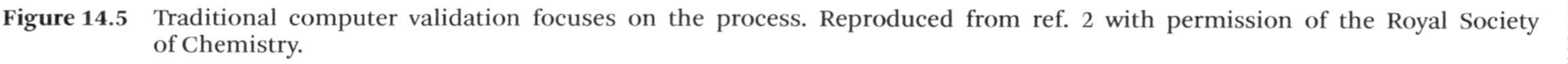

Figure 14.5 Traditional computer validation focuses on the process. Reproduced from ref. 2 with permission of the Royal Society of Chemistry.

system architecture we recommendation that standalone workstations are not fit for purpose and that a CDS must store the data and contextual metadata in an adequately designed database.[27]

Records stored in directories are vulnerable to deletion and unrestricted access to the system clock enabled time traveling on a standalone workstation. To be secure, data must be stored on a fault tolerant network drive where the clock source was a time server linked to a trusted time source, with effective and regular backup performed by the IT department.

14.7.3 Change the Validation Approach to Ensure Data Integrity

To go forward, let us go back in time. In 2005 the GAMP Forum published a Good Practice Guide (GPG) on Compliant Part 11 Electronic Records and Signatures.[30] The approach was rather different from the way I have described validation in Figure 14.5. Instead of the top-down validation approach, they took a bottom-up approach and focused on the electronic records and signatures created and used within the system. In overview, the process was to identify the records created in the system, evaluate their regulatory impact and as a result determine the controls that were necessary to control and protect them. However, there was little apparent impact of this guidance.

The problem is that this bottom up or record centric validation approach does not create process efficiencies that the top-down approach does. Simply, a focus on records creates protected records but you can still have an inefficient process. However, it is time to reconsider the bottom-up approach considering the data integrity issues that are plaguing the pharmaceutical industry.

14.7.4 Brave New CSV World?

The brave new CSV world for ensuring the integrity of our electronic records I would suggest is a hybrid of both top down and bottom up approaches to get the best of both worlds. With little additional effort but with great compliance and data integrity benefits, the vulnerability of the electronic records should be managed by controls specifically implemented that are based on the record's regulatory impact. This is shown in Figure 14.6 and would proceed in several stages.

1. The start of the project would be a focus on process improvement and efficiency gains.
2. As the selected application is prototyped and configuration settings of the CDS examined, all applicable electronic records generated during analysis (data and metadata including audit trail entries) would be identified.

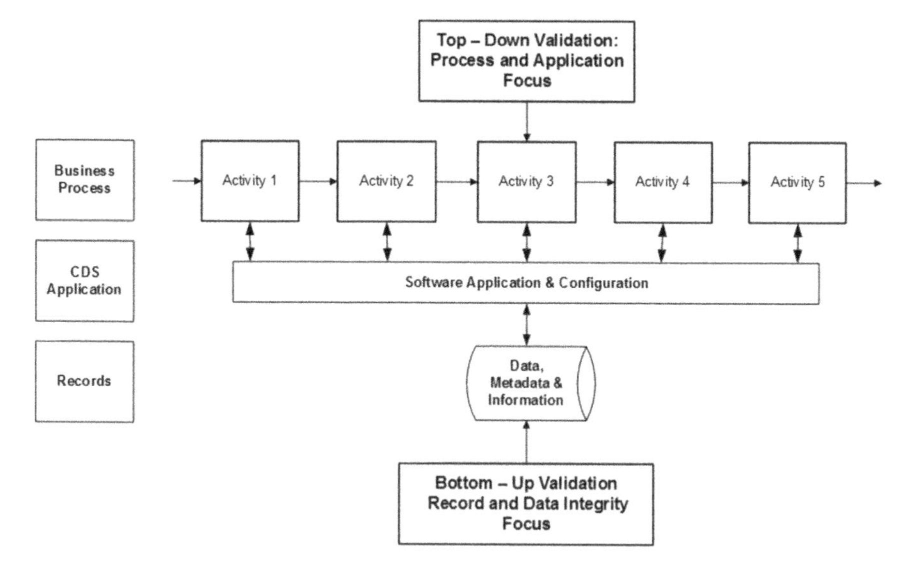

Figure 14.6 Computerised system validation using a combination of top-down and bottom-up approaches. Reproduced from ref. 2 with permission of the Royal Society of Chemistry.

3. The regulatory impact of the records would be assessed depending on their function, *e.g.* method development, method validation, batch release or protocol analysis, stability testing, *etc.*
4. The vulnerability of the electronic records would be assessed and appropriate controls to protect these records would be added to the specification documents for implementation in later stages of the validation project.
5. As the system is being configured then controls for the electronic records and signatures would be implemented at the same time as application configuration. These controls can be either technical or procedural.
6. During the performance qualification (PQ) or user acceptance testing (UAT) phase of the validation the additional controls for the records and signature would be integrated into the overall testing of the intended use of the CDS application.

14.7.5 Turning Principles into Practice

Whilst these may be the principles, you may be thinking that a few diagrams do not give sufficient detail to explain the approach. A fair point. Let us take the principles above and turn them into practice here. We join the validation of a new laboratory system at the prototyping phase where the application is being configured and the Part 11 controls are being evaluated. The

CDS is being installed in a regulated Quality Control laboratory undertaking verification of compendial methods, analysis of active ingredients, in-process materials and finished goods. Stability testing is also performed. The project team decided that electronic signatures and the 21 CFR 11 controls offered by the application will be implemented. Although the application is networked, all data are stored in directories in the operating system and not in a database.

The process for bottom-up or records based validation is outlined in Figure 14.6 and each stage will be described.

- The first task is to identify the electronic records and signatures generated and maintained in the system.
- The next task is to assess the regulatory impact of the identified records/signatures. The GAMP Part 11 Good Practice Guide classifies records into high, medium and low impact categories as shown in Figure 14.7 and Table 14.2.[30] From the descriptions of the use of the system and the

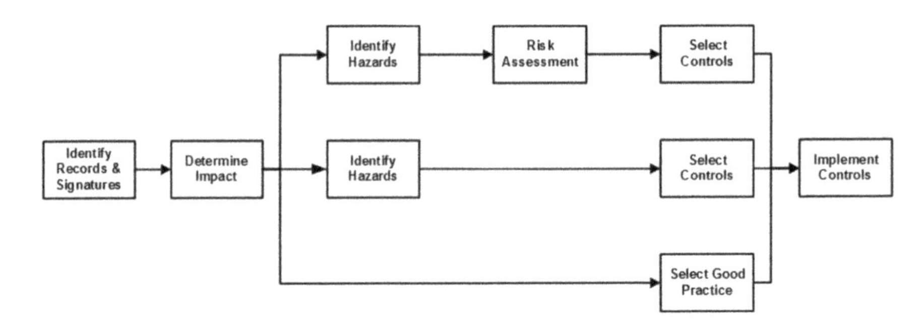

Figure 14.7 Identification of controls for high, medium and low impact regulatory records.

Table 14.2 Classification of high, medium and low impact regulatory records.

Record category	Regulatory impact
High	Direct impact: • Product quality (batch release) • Electronic signatures • Records submitted to a regulatory agency • Records required by predicate rule, *e.g.* master schedule, GLP or GCP study protocols
Medium	Indirect impact: • Records used to support product quality, *e.g.* CSV and method validation and calibration records • SOPs • Training records
Low	Negligible impact: • Calibration and maintenance plans • Project plans

Table, the records generated during operation fall into the high impact category as they are involved in product release.

- The identification of hazards that the records face, followed by a risk assessment that is documented is now performed. To cut a long story short we will assume this has been done and the highest risk is the records on the server hard drive in operating system directories because they can be deleted outside of the application, without leaving any evidence of their deletion.
- Controls need to be selected to protect these high-risk records, *e.g.* records can only be accessed to authorised users *via* the application, restriction of access to directories by a shell program, the drive could be hidden on the network, monitoring access to the drive *via* the operating system, restrict copying of the records. These controls need to be documented in the specification(s) for the system, *e.g.*
  Restricting access to the system clocks to prevent time travelling.
  Restrict access to the server *via* the operating system by implementing OS security on the drive.
  Each user has a unique identity and access privileges.
  Data can only be acquired and processed *via* the application.
  The audit trail is turned on.
  Only an IT administrator can configure the application *via* change control.
  If and how are data interpreted.
- As the validation progresses the controls will be implemented and later tested as part of the user acceptance tests for the system.

Further technical controls should also be considered such as where acquired data must be stored so that unofficial testing can be restricted if not eliminated.

14.7.6 Qualified, Validated and Released for Operational Use

Now that the instrument is qualified and any software controlling it has been validated, it can be released for operational use. In our data integrity journey, we now progress to Level 2 of the Data Integrity Model and discuss validation of analytical procedures in Chapter 15.

14.8 Performance Qualification

The last stage of the 4Qs model is Performance Qualification (PQ), as shown in Figure 14.3. To ensure the integrity and quality of data it is important to understand the role of PQ within the Data Integrity Model discussed in Chapter 5. As AIQ and CSV are found in Level 1 of the Model, their importance for ensuring data integrity is immediately obvious, especially during the analysis of samples at Level 3. However, as PQ is linked to the rest of the 4Qs model, the discussion of the role of PQ and what it consists of is best explained here.

14.8.1 Changes to USP <1058> and the Impact on Understanding of PQ

The definition of PQ in the updated version of USP <1058>[1] is:

> PQ is the documented collection of activities necessary to demonstrate that an instrument consistently performs according to the specifications defined by the user, and is appropriate for the intended use.

This definition has changed from that in the original 2008 USP <1058> PQ, as there is now alignment of the PQ with the instrument requirements documented in the laboratory URS, as shown in Figure 14.3. One of the problems associated with PQ is that few people know what the term encompasses. For example, most analytical scientists associate PQ with System Suitability Tests (SSTs) for chromatography instruments. The reason it is incorrect to define PQ as an SST is that AIQ is instrument-specific and SSTs are method-specific:

> The PQ verifies the fitness for purpose of the instrument under actual conditions of use. After IQ and OQ have been performed, the instrument's continued suitability for its intended use is demonstrated through continued PQ.

The fundamental question many laboratories have for Group 3 instruments under the updated USP <1058> is this: are SSTs alone sufficient for a PQ? This question will be discussed in Section 14.8.5. Of necessity, there will be a presentation of the USP <1058> requirements for PQ before discussing the impact on each of the three Groups.

The scope of PQ is defined in the next paragraph of USP <1058>[1] as:

> The user must define the PQ plans, including test procedures, acceptance criteria, and frequency. Preventive maintenance plans and documentation of repairs and other changes are also a necessary part of the overall instrument qualification.

USP <1058> also mentions that PQ may include[1]:

- Performance Checks: A test or series of tests to verify the acceptable performance of the instrument for its intended use between regular OQ must be undertaken. PQ tests are usually based on the instrument's typical on-site applications and may consist of analysing known components or standards. The tests should be based on good science and reflect the general intended use of the instrument. User specifications for PQ tests should demonstrate trouble-free instrument operation for the intended applications. As is the case with OQ testing, PQ tests may be modular or holistic.

Table 14.3 PQ approaches for USP <1058> Groups A, B and C.

PQ activity	Group A	Group B	Group C
Test plans *versus* URS requirements	—	✓	✓
Procedures, frequency and acceptance criteria	—	✓	✓
Repairs and requalification	—	✓	✓
Preventative maintenance	—	✓	✓
Instrument log book		✓	✓
Change control	—	✓ (firmware)	✓
Periodic review (critical systems only)	—	—	✓

- Preventive Maintenance and Repairs: Periodic preventive maintenance activities are required for many instruments. This may include calibration. Document the preventive maintenance plans, including procedures and frequency as part of the AIQ package. When an instrument fails to meet PQ criteria or otherwise malfunctions, the cause of the failure must be investigated and documented. The instrument may require maintenance or repair. The relevant OQ or PQ test(s) should be repeated after the needed maintenance or repair to ensure that the instrument remains qualified.
- Change Control and Periodic Review: All PQ, maintenance, and calibration activities must be documented. There should also be a change control procedure to ensure changes to the instrument configuration, including firmware and software, are controlled. For instruments that are critical, there should be a periodic review to ensure that the system is still under control and remains qualified/validated. This harmonises USP <1058> with EU GMP Annex 11.[8]

What does this mean in practice? First and foremost is that PQ is not a single activity as the overall approach to performance qualification for Groups A, B and C can be seen in Table 14.3, and this will be discussed in more detail in the following sections. Note that the table does not include periodic repetition of the OQ that is typically performed annually following a preventative maintenance service by a supplier or service agent.

14.8.2 Linking the URS, OQ, and PQ

The 2017 USP <1058> PQ definition relates PQ testing to user requirements. The problem is that an OQ typically tests the user requirements directly through traceable standards, metrology measurements using calibrated equipment, and use of appropriate reference materials, designed to test the instrument performance and range of use. In contrast, a PQ is usually application- or method-based (*for example, OQ and PQ test different attributes of system performance, which is why both are required*):

- OQ is related to testing the instrument performance under standardised conditions, so that the correct operation of the instrument in the laboratory against the URS can be demonstrated. For example, for HPLC, flow rate accuracy, and reproducibility can be measured directly as metrology measurements, using a calibrated and traceable digital flow meter. The range of use (for example, maximum and minimum settings) is measured in the OQ phase.
- PQ addresses the suitability of the instrument under actual conditions of use in between repetition of the OQ/PQ cycle. A PQ will indirectly measure the laboratory user requirements. For example, flow rate accuracy, and reproducibility can be measured indirectly in a PQ using retention time windows and % RSD of retention time. Because the range of use is measured in the OQ phase, it does not need to be measured in the PQ.

The key issue is that there must be a laboratory URS upon which both PQ and OQ tests should be based.

The earliest regulatory publication on Analytical Instrument Qualification is the 1994 paper by Furman *et al.*,[31] which includes a discussion of modular *versus* holistic qualification of chromatographic instruments. The argument was that if the performance of each module or component was within acceptance limits, in principle, the system could potentially fail due to the addition of errors or the system components not working correctly together. The authors therefore proposed the inclusion of an overall system, or holistic test, in the qualification of chromatography instruments, as module testing alone would not detect this.

A holistic PQ test executed after an OQ, or even as part of the OQ, would provide a link between the functional and operational-based OQ and the method-based PQ, as shown in Figure 14.9.

14.8.3 PQ for Group A Instruments

The simplest PQ is for instruments that are classified by their intended use into Group A. As shown in Table 14.3, the PQ is performed by observation, as is the OQ, and simply confirms that an instrument is functioning correctly. No records are required for this function as there is no measurement capability of the instrument and therefore the operation is low risk.

14.8.4 PQ for Group B Instruments

For Group B instruments, the focus in this section will be on the PQ rather than verification of calculations or specification and validation of user defined programs. The OQ is typically performed using calibrated instruments and standards or reference materials that are traceable to national or international standards. Here, the user requirements can be directly tested by metrology, as shown in Figure 14.8, and the work can be performed either

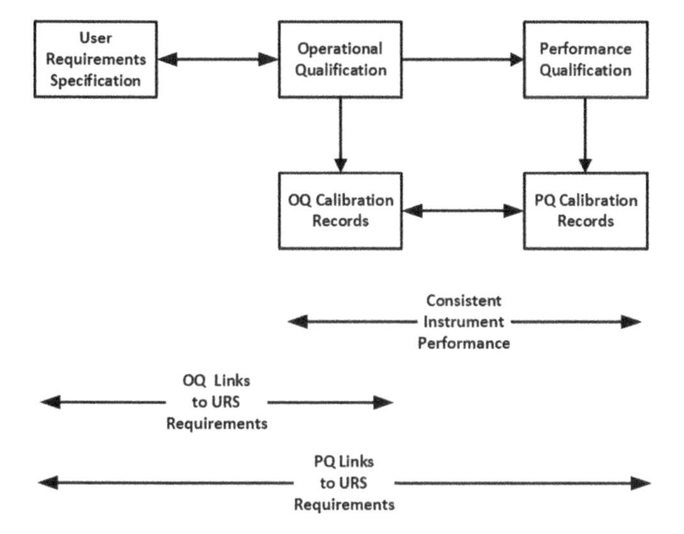

Figure 14.8 Relationship between the URS, OQ and PQ for Groups B instruments.

by a service agent or an in-house metrology group. The advantage for the PQ is that the laboratory can also perform calibration checks using reference materials that can be directly related to the requirements in the URS, also as shown in Figure 14.8.

Some examples of this approach are:

- Analytical balances where the supplier or service agent uses calibrated masses and the laboratory has their own set of calibrated masses and the operating range specified in the URS can be shown to be performing as intended.
- pH meters can take a similar approach by using reference buffers over the limits specified in the laboratory URS both during the OQ and PQ.
- Pipettes can be sent for external calibration at regular intervals but also checked by laboratory staff monthly by pipetting and weighing water over the operating range of each pipette. This will confirm that each pipette remains within the operating parameters specified in the URS for each model and make of pipette.

14.8.5 PQ for Group C Instruments

Group C instruments consist of a wide range of analytical techniques, *e.g.* chromatography, spectroscopy, thermal analysis. There are two main types of PQ test that could be applied, shown in Figure 14.9 and discussed below:

- Direct Measurement Using Reference Materials and Standards: The performance of some instruments can be directly measured and equated to specific user requirements by using traceable standards and reference

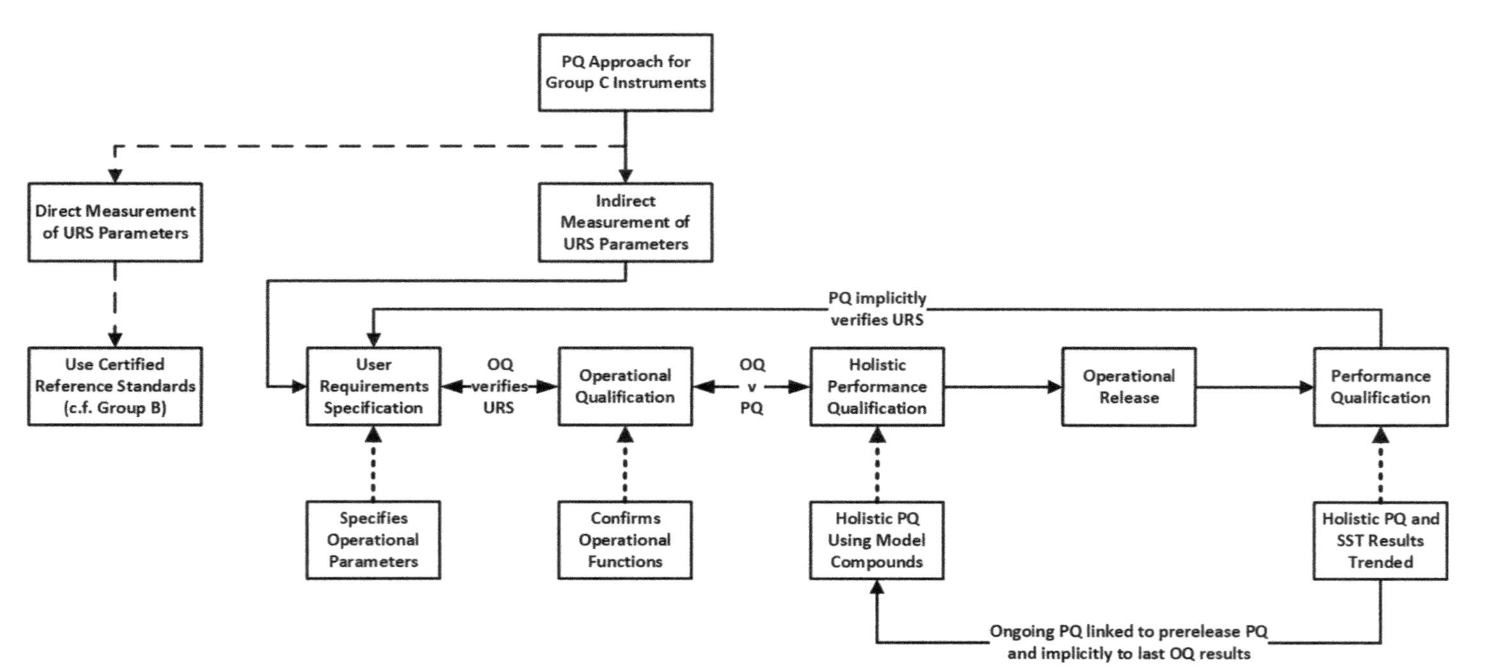

Figure 14.9 Performance qualification options for Group C instruments.

Table 14.4 PQ test approaches for chromatography instruments.

Approach to PQ instrument testing	Comments and observations
1. Only Use System Suitability Tests (SSTs)	• SSTs are method-specific • The validity of SSTs included in each method? • How are SST parameters related to user requirements? • An auditor may not accept this interpretation
2. Use a Universal Holistic PQ Method in Conjunction with SSTs	• Holistic PQ implicitly tests user requirements • Simplifies PQ requirements • Builds on SST during use • Justifies that PQ testing is representative of use
3. Use a Specific PQ Method per Instrument/Use plus SSTs	• Complicates PQ requirements across a laboratory (*e.g.* could have specific and different PQ requirements for each chromatograph) • Builds on SST during use • Ideal where instruments are dedicated to specific methods

materials. For example, the wavelength accuracy of spectrometers can be assessed using such materials, *e.g.* holmium perchlorate solution in sealed cuvettes for UV detectors, polystyrene wavelength and transflectance standards for near infra-red spectrometers and UV photometric linearity at 257 nm can be measured using propyl parabenz solutions. Thermal analysis instruments can use polymorphs in known mixtures for differential scanning calorimetry.
- However, some techniques such a chromatography do not have this luxury and have method based PQ tests and the options for this are presented in Table 14.4.

14.8.6 System Suitability Tests as Part of a PQ

Option 1 in Table 14.4 uses SST samples alone for PQ testing. Designed to satisfy pharmacopeia requirements, such as USP <621>[32] or EP 2.2.46,[33] SSTs play a pivotal role in documenting the performance of the chromatography system at the analytical run level. Therefore, a natural evolution of this is to consider how SSTs can contribute to an overall PQ test plan. If SST results are to be used as part of PQ testing of a chromatograph, it is important that:

- Method Development: SST parameters and acceptance criteria must be defined during development of the method using a qualified instrument.
- Method Validation: During the validation the suitability of the selected SSTs for performance monitoring will be confirmed, moreover, there is now traceability between the method validation and the use of SSTs and provides a baseline for monitoring during operational use. The baseline

permits monitoring that the instrument is meeting its user requirements during operational use.
- Trending SST Parameters: As required by EU GMP 6.9 and 6.16[34] and the draft USP <1220>[35] the summaries of each method test will be part of the overall PQ acceptance criteria.

14.8.7 Keep It as Simple as Possible – But No Simpler

Many laboratories have implemented "Lean Initiatives" to reduce potentially unnecessary work. However, this must be balanced with scientific soundness as required under the GMP regulations, such as 21 CFR 211.160(b),[5,36] for example:

- Blank Injection Removal: To save time a blank injection of mobile phase is often removed from the injection sequence. The problem arises when troubleshooting of a problem, as there will be no chromatogram of the injection of mobile phase any investigation can be severely limited. A blank injection can determine if there is any carryover from the autosampler and the level of baseline flatness/noise in the detector response. These can be traced to user requirements for the injector and detector, respectively.
- Removal of a Control Sample: Similar considerations need to be made for the inclusion of an approved and well-characterised control sample, particularly for impurity characterization. For example, it is not uncommon in some laboratory for chromatographic methods not to include a standard to serve as a comparison for the run.

  Given the fact that chromatography is a comparative analytical technique, this is scientifically indefensible and not compliant with regulations.

The problem is that relying on lean designed SST tests means there could be a higher risk of PQ failure and an inability to investigate Out of Specification (OOS) results adequately, or provide scientific evidence that an instrument failure did not affect analytical results (because there is no evidence, depending on what is performed in SST).

If time is a major factor, then serious consideration should be given to developing and validating a UPLC method where is a bottleneck is moved from the analysis to the data processing phase of the analytical procedure.

14.8.8 Holistic HPLC PQ Test

As part of an overall approach to PQ as listed in Table 14.4 and on the right-hand side of Figure 14.9, there should be a holistic test performed after the OQ and predefined periodic thereafter that can show the user requirements are still being met. The principle of a good holistic PQ test is to design an analytical procedure that is sensitive to instrument performance but is also robust and reliable. Ideally, the procedure must use stable model compounds, with simple and stable chromatography to minimize analytical variance from the reference compounds and the column. The performance of the procedure is therefore dependent on instrument performance. Some of the characteristics of an HPLC holistic test are:

- Two Stable Model Compounds: Well-separated model compounds with good peak shape and a short overall run time.
- Analytical Column: Use a relatively short easily available analytical column to reduce run time for the overall PQ test.
- Simple Mobile Phase: Use a simple organic–aqueous mobile phase that is easy to prepare.
- Prepare Standards in the mobile phase to minimise disturbance when injecting.
- Prepare Standard Solutions Gravimetrically to avoid pipetting errors and minimize overall method variance.
- Use Four Solution Concentrations – (25, 50, 75 and 100%) containing the two compounds to test that autosampler and detector reproducibility and linearity are prepared.
- Run Sequence – consists of a blank, injected once at the beginning of the sequence and at the end of each standard set to assess carryover and six injections of each standard.

These overall holistic standards allow limits to be set for:

- detector reproducibility and linearity;
- autosampler precision;
- the combination of the pumping system and thermostatic control of the column.

All parameters will be measured with the instrument under actual conditions of system use.

Acknowledgement

I would like to thank Paul Smith for input and collaboration in the writing of the USP <1058> portions of this chapter.

References

1. *USP 41 General Chapter <1058> Analytical Instrument Qualification*, United States Pharmacopoeia Convention, Rockville, MD, 2018.
2. R. D. McDowall, *Validation of Chromatography Data Systems: Ensuring Data Integrity, Meeting Business and Regulatory Requirements Second Edition*, Royal Society of Chemistry, Cambridge, 2017.

3. *Good Automated Manufacturing Practice (GAMP) Guide Version 5*, International Society for Pharmaceutical Engineering, Tampa, FL, 2008.
4. *GAMP Good Practice Guide a Risk Based Approach to GXP Compliant Laboratory Computerised Systems, Second Edition*, International Society for Pharmaceutical Engineering, Tampa, FL, 2012.
5. *21 CFR 211 Current Good Manufacturing Practice for Finished Pharmaceutical Products*, Food and Drug Administration, Sliver Spring, MD, 2008.
6. *21 CFR 58 Good Laboratory Practice for Non-clinical Laboratory Studies*, Food and Drug Administration, Washington, DC, 1978.
7. *OECD Series on Principles of Good Laboratory Practice and Compliance Monitoring Number 1, OECD Principles on Good Laboratory Practice*, Organisation for Economic Co-Operation and Development, Paris, 1998.
8. *EudraLex – Volume 4 Good Manufacturing Practice (GMP) Guidelines, Annex 11 Computerised Systems*, European Commission, Brussels, 2011.
9. *ICH Q7-Basic Requirements for Active Substances Used as Starting Materials*, International Conference on Harmonisation, Geneva, 2000.
10. *EudraLex – Volume 4 Good Manufacturing Practice (GMP) Guidelines, Part 2-Basic Requirements for Active Substances Used as Starting Materials*, European Commission, Brussels, 2014.
11. *WHO Technical Report Series No. 996 Annex 5 Guidance on Good Data and Records Management Practices*, World Health Organisation, Geneva, 2016.
12. *PIC/S PI-041 Draft Good Practices for Data Management and Integrity in Regulated GMP/GDP Environments*, Pharmaceutical Inspection Convention/Pharmaceutical Inspection Co-Operation Scheme, Geneva, 2016.
13. *AAPS White Paper on Analytical Instrument Qualification*, American Association of Pharmaceutical Scientists, Arlington, VA, 2004.
14. *USP 31 General Chapter <1058> Analytical Instrument Qualification*, United States Pharmacopoeial Convention, Rockville, MD, 2008.
15. R. D. McDowall, Focus on quality: where are we now with USP <1058>? *Spectroscopy*, 2010, **25**(11), 24–31.
16. *FDA Guidance for Industry General Principles of Software Validation*, Food and Drug Administration, Rockville, MD, 2002.
17. C. Burgess and R. D. McDowall, Stimulus to the Revision Process: An Integrated Risk Assessment for Analytical Instruments and Systems, *Pharmacop. Forum*, 2012, **38**(1).
18. USP <1058> Analytical Instrument Qualification in process revision, *Pharmacop. Forum*, 2015, **41**(3).
19. USP <1058> Analytical Instrument Qualification in process revision, *Pharmacop. Forum*, 2016, **42**(3).
20. P. Smith and R. D. McDowall, Life Cycle Risk Assessment of HPLC Instruments, *LCGC Eur*, 2015, **28**(2), 110–117.
21. *EudraLex – Volume 4 Good Manufacturing Practice (GMP) Guidelines, Annex 15 Qualification and Validation*, European Commission, Brussels, 2015.
22. C. Burgess and R. D. McDowall, An integrated risk assessment for analytical instruments and computerised laboratory systems, *Spectroscopy*, 2013, **28**(11), 21–26.

23. L. Vuolo-Schuessler, *et al.*, Harmonizing USP <1058> and GAMP for analytical instrument qualification, *Pharm. Eng.*, 2014, **34**(1), 46–56.
24. *FDA Questions and Answers on Current Good Manufacturing Practices, Good Guidance Practices, Level 2 Guidance – Records and Reports*, 2010 [27 May 2016], Available from: http://www.fda.gov/Drugs/GuidanceCompliance-RegulatoryInformation/Guidances/ucm124787.htm.
25. *FDA Draft Guidance for Industry Data Integrity and Compliance with cGMP*, Silver Spring, MD, USA, 2016.
26. R. D. McDowall and C. Burgess, The Ideal Chromatography Data System for a Regulated Laboratory, Part 1: The Compliant Analytical Process, *LCGC North Am.*, 2015, **33**(8), 554–557.
27. R. D. McDowall and C. Burgess, The Ideal Chromatography Data System for a Regulated Laboratory, Part 2: System Architecture Requirements, *LCGC North Am.*, 2015, **33**(10), 782–785.
28. R. D. McDowall and C. Burgess, The Ideal Chromatography Data System for a Regulated Laboratory, Part 3: Essential Chromatographic Functions for Electronic Ways of Working, *LCGC North Am.*, 2015, **33**(12), 914–917.
29. R. D. McDowall and C. Burgess, The Ideal Chromatography Data System for a Regulated Laboratory, Part 4: Assuring Regulatory Compliance, *LCGC North Am.*, 2016, **34**(2), 144–149.
30. *GAMP Good Practice Guide a Risk-based Approach to Compliant Electronic Records and Signatures*, International Society for Pharmaceutical Engineering, Tampa, FL, 2005.
31. W. B. Furman, T. P. Layloff and R. Tetzlaff, Validation of computerised liquid chromatographic systems, *J. OAC Int.*, 1994, **77**, 1314–1317.
32. *USP General Chapter <621> Chromatography*, United States Pharmacopoeia Commission Inc, Rockville, MD.
33. *EP 2.2.46 Chromatographic Separation Techniques*, European Pharmacopoeia, Strasbourg.
34. *EudraLex – Volume 4 Good Manufacturing Practice (GMP) Guidelines, Chapter 6 Quality Control*, European Commission, Brussels, 2014.
35. G. P. Martin, *et al.*, Stimuli to the Revision Process: Proposed New USP General Chapter: The Analytical Procedure Lifecycle ⟨1220⟩, *Pharmacop. Forum*, 2017, **43**(1).
36. Amendments to the current good manufacturing practice regulations for finished pharmaceuticals, *Fed. Regist.*, 2008, **73**(174). 51919–51933.

CHAPTER 15

Validating Analytical Procedures

Validation of Analytical Procedures is to ensure that there is the right analytical procedure for the right job and represents Level 2 of the Data Integrity Model described in Chapter 5. The correct operation of Level 2 is predicated upon the correct implementation of Foundation (Data Governance and Data Integrity) and Level 1 (Analytical Instrument Qualification and Computerised System Validation).

Currently ICH Q2(R1) is used for validation of analytical procedures in a GMP context but this does not take a life cycle approach as advocated by ICH Q8, the latter guidance takes a holistic approach to controlling pharmaceutical processes. In several stimuli to the revision process articles, the United States Pharmacopoeia Validation and Verification expert committee have proposed a life cycle approach to the validation of analytical methods that has culminated in publication of a draft general chapter USP <1220>. An updated FDA guidance for bioanalytical method validation guidance in 2018 only now has started to consider method development but not in the same degree of depth as the USP approach. In light of the USP update, ICH Q2(R1) will be updated to a life cycle approach.

Regardless of the approach to validation, it is imperative that analytical procedures are validated, under control and that a laboratory knows the impact of changing key variables in the procedure.

Data Integrity and Data Governance: Practical Implementation in Regulated Laboratories
By R. D. McDowall

Published by the Royal Society of Chemistry, www.rsc.org

15.1 What the Regulators Want

15.1.1 US GMP 21 CFR 211

In Section 211.194(a) there is the following requirement for analytical methods[1]:

> a) Laboratory records shall include complete data derived from all tests necessary to assure compliance with established specifications and standards, including examinations and assays, as follows:
>
> (2) A statement of each method used in the testing of the sample.
>
> The statement shall indicate the location of data that establish that the methods used in the testing of the sample meet proper standards of accuracy and reliability as applied to the product tested. ....
>
> The suitability of all testing methods used shall be verified under actual conditions of use.

15.1.2 EU GMP Chapter 6 on Quality Control

Chapter 6 of EU GMP requires the following[2]:

> 6.15 Testing methods should be validated.
>
> A laboratory that is using a testing method and which did not perform the original validation, should verify the appropriateness of the testing method.
>
> All testing operations described in the marketing authorisation or technical dossier should be carried out according to the approved methods.

15.1.3 EU GMP Annex 15: Qualification and Validation

Section 9 of EU GMP Annex 15 covers validation of test methods[3]:

> 9.1 All analytical test methods used in qualification, validation or cleaning exercises should be validated with an appropriate detection and quantification limit, where necessary, as defined in Chapter 6 of the EudraLex, Volume 4, Part I.

15.1.4 Bioanalytical Method Validation Guidances

Section 4.1 of the European Medicines Agency Guideline on Bioanalytical methods Validation, dated 2011, notes[4]:

> A full method validation should be performed for any analytical method whether new or based upon literature.

> The main objective of method validation is to demonstrate the reliability of a particular method for the determination of an analyte concentration in a specific biological matrix, such as blood, serum, plasma, urine, or saliva. Moreover, if an anticoagulant is used, validation should be performed using the same anticoagulant as for the study samples. Generally, a full validation should be performed for each species and matrix concerned.

The draft FDA Bioanalytical Methods Validation guidance for industry (2013) notes in the introduction section[5]:

> For pivotal studies that require regulatory action for approval or labeling, such as Bioequivalence or Pharmacokinetic studies, the bioanalytical methods should be fully validated.
>
> For exploratory methods used for the sponsor's internal decision making, less validation may be sufficient.
>
> When changes are made to a previously validated method, additional validation may be needed.

In the section on Guiding Principles of the final and updated version of the FDA Bioanalytical Method Validation guidance for chromatographic and ligand binding assays 2018[6] there are the following statements:

> The purpose of bioanalytical method development is to define the design, operating conditions, limitations, and suitability of the method for its intended purpose and to ensure that the method is optimized for validation.
>
> Before the development of a bioanalytical method, the sponsor should understand the analyte of interest (*e.g.*, determine the physicochemical properties of the drug, *in vitro* and *in vivo* metabolism, and protein binding) and consider aspects of any prior analytical methods that may be applicable.
>
> Method development involves optimizing the procedures and conditions involved with extracting and detecting the analyte.
>
> Bioanalytical method development does not require extensive record keeping or notation. However, the sponsor should record the changes to procedures as well as any issues and their resolutions during development of the bioanalytical method to provide a rationale for any changes during the development of the method.
>
> Bioanalytical method validation proves that the optimized method is suited to the analysis of the study samples.

Whilst the latest version of the FDA guidance for bioanalytical methods has started to include method development, the importance of this phase of an analytical life cycle is not fully recognised as it is in the GMP world. The principles outlined in the draft USP <1220> reflect more accurately a Quality by Design (QbD) approach to method development and validation[7] that are intended to deliver more robust analytical procedures.

15.1.5 Regulatory Requirements Summary

The regulations and guidance documents require that all analytical procedures used in GMP or GLP/GCP analysis must be validated or verified under actual conditions of use. This is simply a requirement that any procedure should be fit for its intended use. However, the regulations do not state how that is to be achieved. In a GMP context that is the role of the pharmacopoeias and ICH guidance documents and a GLP/GCP role the role of guidance documents. Within a GMP context there is a movement towards QbD (Quality by Design) with a focus on understanding what parameters influence a procedure and how each can be controlled to reduce variability of the method. Ultimately, better methods should reduce OOS results and the subsequent laboratory investigations.

15.1.6 Outsource Analytical Work with Care

In an FDA warning letter to Quali-Controle, a contract testing laboratory, there is the following citation[8]:

> 1. Your firm failed to establish and document the accuracy, sensitivity, specificity, and reproducibility of its test methods (21 CFR 211.165(e)).
>
> For example, your non-compendial test methods used to analyze your customers' drug products had not been validated.
>
> In your response, you indicated that a comment has been added to your Certificate of Analysis (CoA) to inform your customer that the analyses are conducted using unvalidated test methods. You also committed to inform your customers that all future test methods will need to be properly transferred or validated to legitimize the results obtained in your laboratory.

This response by the laboratory is lazy and inadequate, which sponsor wants work conducted using unvalidated methods? The utilisation of validated analytical methods is one of the basic requirements for laboratories in any regulated environment and is a key requirement of the Data Integrity Model at Level 2. If a laboratory works without proof of reliability and reproducibility of their analytical procedures, it violates these requirements. If this happens at a contract laboratory, which performs analyses for various customers, the consequences are even more extensive and troublesome for all parties involved. Quali-Controle also broke the obligations laid down in the quality agreement with the customer. Instead of the utilisation of pharmacopoeial procedures as described, alternative and unvalidated methods had been used. The customers had neither been informed about this beforehand, nor had their consent been obtained.[8]

However, when outsourcing any regulated laboratory work it is important that sufficient due diligence is performed not just during the qualification phase but also when performing the contracted analytical work. This is to ensure not only the data integrity but also the data quality. We will return to outsourcing analytical work in Chapter 23.

15.2 Current Method Validation Guidance

15.2.1 Terminology: Analytical Method or Analytical Procedure?

The title of this chapter is validating analytical procedures and this section is entitled method validation guidance, is there a difference? As can be seen in Section 15.1, the terminology used varies between GXP disciplines.

It is important to understand what is meant by analytical procedure and whether there is any difference if you used the term analytical method. Analytical procedure refers to the whole process from sampling through to reporting the result, as shown in Figure 15.1. In contrast, the term analytical method usually only refers to the instrument portion of the analytical procedure, also as shown in Figure 15.1. Therefore, the term analytical procedure is preferred when discussing validation as there may be issues in the sampling (*e.g.* light sensitive analyte) or during transport or storage (*e.g.* unstable analyte) that may not be considered if just focused on the analytical method portion of the procedure.

For GLP/GCP bioanalytical studies, the study protocol defines how samples are to be taken, preserved and stored and the analytical method refers to the sample preparation, instrumental analysis, interpretation of the data and generation of the reportable results. Within the context of this book, the term analytical procedure will be used and this can be interpreted for GLP analyses relatively easily.

15.2.2 Business Rationale for Procedure Validation/Verification

Today's pharmaceutical industry is characterised by long, convoluted or even virtual supply chains. Globalisation and outsourcing are the key words used within regulated organisations. The pharmaceutical industry outsources anything, *e.g.* analytical development work, stability testing, finished product testing, manufacturing and if the industry made kitchen sinks, production of these would be outsourced as well.

In the context of outsourcing, are validated analytical procedures important? Oh, yes! If you doubt the issue see the Quali-Controle warning letter[8] described in Section 15.1.6. The rationale is that an analytical procedure must be described in sufficient detail that allows any laboratory (contract

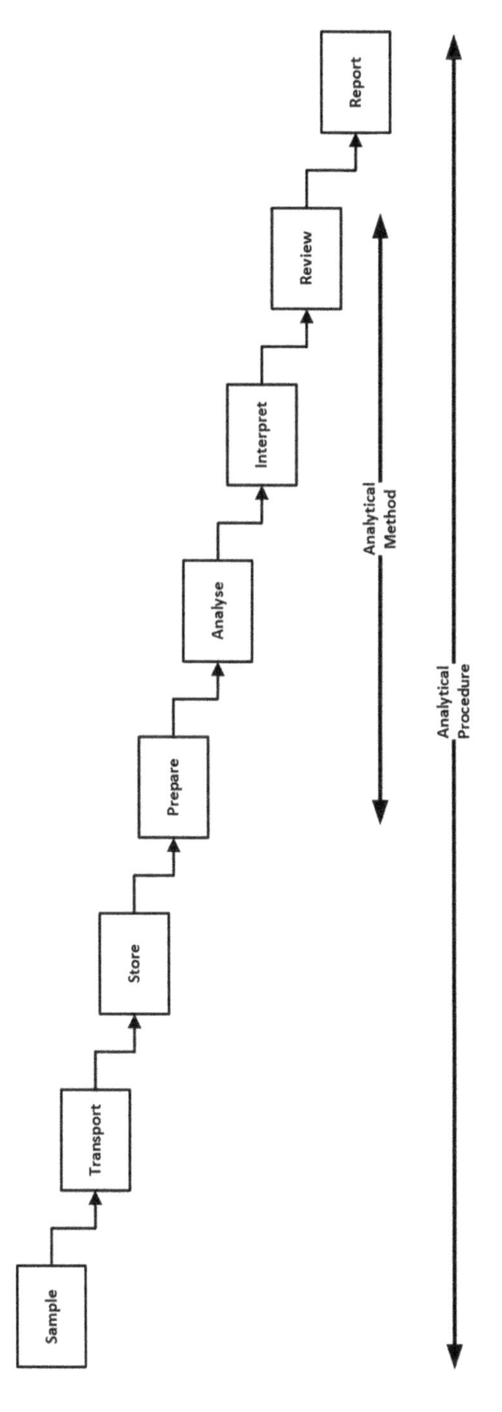

Figure 15.1 Differences between analytical procedure and analytical method.

acceptor) performing work on behalf of a sponsoring company (contract giver) to establish and verify that the procedure works more easily than if this were not the case. This is a key pre-requisite for speedy and efficient technology transfer.

15.2.3 ICH Q2(R1) Validation of Analytical Procedures: Text and Methodology

One of the first tripartite ICH guidance documents, Q2(R1) entitled Guideline on Validation of Analytical Procedures: Text and Methodology[9] is the basis of method validation for GMP laboratories and has been incorporated into the pharmacopoeias of USP, EP and JP, *e.g.* USP <1225>.[10] The problem with this guidance is that analysts treat it like a bible to be followed without thinking, *e.g.* if measuring an active pharmaceutical ingredient between 80% and 120% of label claim, many procedures also include an experiment to measure the limit of quantification (LOQ). Why? This is a meaningless parameter for such a method. The rationale that many would offer is that this parameter is written in ICH Q2(R1)!

There needs to be a different approach to developing robust methods. It is essential to know which parameters have the greatest impact on performance of the method, resulting in a lower likelihood of generating out of specification results as a direct result of method variability.

15.2.4 FDA Guidance for Industry on Analytical Procedure Validation

In 2000, the FDA issued a draft guidance for industry on Analytical Procedures and Methods Validation[11] that outlined the FDA expectations for validation. The main problem is that this guidance did not address one of the most critical stages of the whole process: method development. In 2015, the FDA replaced the 2000 draft guidance with yet another draft guidance entitled Analytical Procedures and Methods Validation for Drugs and Biologics.[12] There was some improvement in the content with some reference to method development but the emphasis in this chapter will be with the work of the USP and the draft general chapter <1220> on a life cycle for analytical procedures[7] that will be described in Section 15.5.

15.2.5 Update of ICH Q2(R1) to a Life Cycle Approach

Following the publication of the draft USP <1220>[7] there is an ICH project to update Q2(R1) as there is no corresponding ICH guideline on Analytical Procedure Development[13] as when companies submit a regulatory dossier, the validation report for an analytical method is rarely accompanied by an evaluation of the development of the procedure and identifying the critical parameters that need to be controlled. This is especially so for real-time release testing using multi-variate spectroscopic methods, *e.g.* NIR as the current guidance focuses on chromatographic methods. If a validation report for non-chromatographic analytical techniques does not contain enough information, this can lead to a sequence of questions and answers between the regulator and the company. Therefore, the current approach of ICH Q2(R1) is not sufficient to establish suitability of multivariate methods where they are used within a pharmaceutical company. More importantly, if the lack of guidance does not allow a submission to contain the scientific basis then the regulators cannot respond with a flexible approach to post-approval changes in any analytical procedure.

There is now a project to update ICH Q2(R1)[13] and to:

- Address an analytical procedure life cycle that includes method development.
- Define the validation elements for specific procedures, such as NIR, nuclear magnetic resonance spectroscopy (NMR), and hyphenated techniques, including CE-MS, CE-ICP-MS, LC-NMR, GC-MS, LC-MS.
- Define the validation characteristics applicable to multivariate methods that may differ with the area of application (*e.g.*, identification *vs.* quantitation, batch *vs.* continuous process, dosage form assay *vs.* blending monitoring).
- Identify the important method parameters established during method development.
- Develop quantitative measures for robustness.
- Discuss the statistical issues in method validation.

This would result in a new version of ICH Q2 with an expanded scope Q2 that would cover validation issues outside HPLC, and specifically address validation of some of the newer analytical procedures, address lifecycle issues, and include a discussion of statistical issues in validation. The outcome of the revised ICH Q2 document:

> will facilitate selecting or identifying development approaches that will reduce risk of post-approval change to procedures discussed in ICH Q12,[14] and enable more efficient and science-based change management. Applying the enhanced approach for analytical procedures will contribute to the resource-efficient drug development and post-approval CMC changes.[13]

15.2.6 Pharmacopoeial Methods Do Not Work as Written

One specific instance of the need to verify that an analytical procedure works is the case of a pharmacopoeial method. Imagine an ethical pharmaceutical company that develops an analytical procedure for a new molecular entity (NME). Say, this uses a C18 HPLC column from a specific manufacturer, due to the non-commercial nature of a pharmacopoeia, the column listed in the monograph for the API or finished product is not the specific column used by the ethical company but more of a HPLC column.

This is the reason that most HPLC analytical methods do not work straight from the pages of a pharmacopoeia. Let us look at an example for the HPLC column for analysing cimetidine USP:

> HPLC analysis uses a column that is packed with octadecylsilane (ODS or C18 silica), particles 5 to 10 μm in diameter packed in a column 25 cm long with a 4.6 mm internal diameter.

How many LC supplier's columns fit this vague description? Unfortunately, quite a few! Given the fact that 211.160(b)[1] requires anything that is done in the regulated GMP laboratory is required to be scientifically sound, how scientifically sound is this description in the USP? As a bonus, cimetidine is a relatively small molecule and that could be able to be analysed on many different C18 columns. However, consider a more polar compound also quantified on a C18 column where a specific carbon loading with narrow range of end capping is required and this is only met by a relatively few or even only one column manufacturer. This name of the column supplier would not be published in a compendial monograph as it is considered commercial information. Looked at objectively, it is amazing that given such a brief description laboratories can get an analytical procedure to operate. The monograph should just be considered a starting point for developing a method.

Hence, when considering qualification of suppliers, as discussed in Chapter 23, laboratories that only rely on a method as written in a pharmacopoeia should be treated with grave suspicion as you know they do not work as written. Analytical procedures from a pharmacopoeial monograph MUST be interpreted and verified under actual conditions of use. You must also realise that what is written in the monograph is only a part of the requirements of a pharmacopeia, the section on General Notices is applicable as well as some of the general chapters, *e.g.*:

- USP <1058> Analytical Instrument Qualification;
- USP <41> Balances;
- USP <1251> Weighing on an Analytical Balance.

All of these ancillary sections and general chapters can have a direct impact on the validation of an analytical procedure as well as the application of them, as we shall see in Chapter 16 entitled Performing an Analysis.

15.3 Role of Analytical Procedure Validation in Data Integrity

15.3.1 Method Validation in the Data Integrity Model

From the data integrity model described in Chapter 5, Analytical Procedure Validation is at Level 2. To be successful, it requires all the elements of the Foundation (management leadership, policies and procedures, staff training, open culture, *etc.*) and Level 1 (analytical instrument qualification and computerised system validation to be performed adequately). As stated in ICH Q7/EU GMP Part 2[15]:

> 12.82 *Appropriate qualification of analytical equipment should be considered before starting validation of analytical methods.*

As you can see, the dependencies of the data integrity model are acknowledged in ICH Q7/EU GMP Part 2 regulations: ensure that the bottom layers of the model are in place and ensure that they are effective before starting to develop and validate analytical procedures.

15.3.2 Equating the Data Integrity Model with the USP <1058> Data Quality Triangle

As noted in Chapter 14 on AIQ and CSV, USP <1058> has a data quality triangle that shows the relationship between analytical instrument qualification, analytical procedure validation, system suitability tests and quality control samples. Similarly, the data integrity model has at Level 1 AIQ and CSV, Level 2 analytical procedure validation and Level 3 Sample analysis. Figure 15.2 maps the data integrity model and the data quality triangle together[16] and this shows how close the two approaches are. Only the Foundation layer is omitted from the USP data quality triangle.

15.4 Current Approaches to Validation and Verification of Procedures

15.4.1 Good Manufacturing Practice

In the GMP regulated environment there is ICH Q2(R1) on Validation of Analytical Procedures: Text and Methodology[9] that has been incorporated into the United States, European and Japanese Pharmacopoeias, *e.g.*[10] In addition, the USP has two additional general chapters <1224> on Transfer of Analytical Procedures[10] and <1226> on Verification of Compendial Procedures.[17] There is not much detail in these documents about the experimental design and in any case, most analysts validating procedures tend to take a tick-box approach – if it is listed in ICH Q2(R1) it is included in a validation – regardless if it is relevant or not, as we will discuss in Sections 15.4.3 and 15.4.4.

15.4.2 Bioanalytical Method Validation

Unlike the GMP regulations, the bioanalytical method validation guidance documents from the EMA[4] and FDA[5,6,18] contain much more information about how to validate and operate a bioanalytical method rather than focus on method development. The latest FDA guidance[6] has a section on method

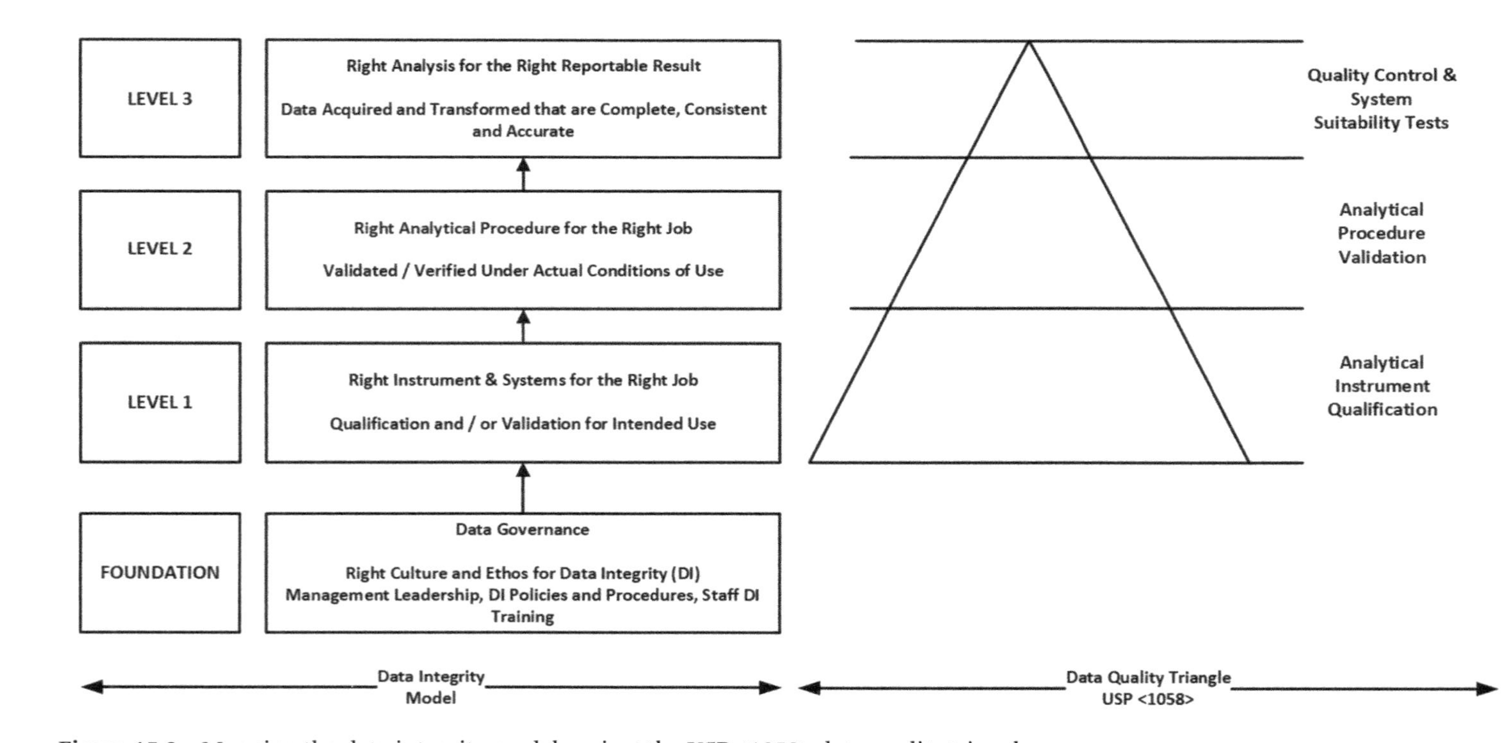

Figure 15.2 Mapping the data integrity model against the USP <1058> data quality triangle.

development, quoted in Section 15.1.4, but does not go into details. This guidance includes detail of how the regulators require control of integration of the resultant chromatograms.[19]

15.4.3 Validation Documentation for Analytical Procedures

ICH Q7 and EU GMP Part 2 in Section 12.2 Validation Documentation require the following[15]:

12.20 A written validation protocol should be established that specifies how validation of a particular process will be conducted. The protocol should be reviewed and approved by the quality unit(s) and other designated units.
12.21 The validation protocol should specify critical process steps and acceptance criteria as well as the type of validation to be conducted (*e.g.* retrospective, prospective, concurrent) and the number of process runs.
12.22 A validation report that cross-references the validation protocol should be prepared, summarising the results obtained, commenting on any deviations observed, and drawing the appropriate conclusions, including recommending changes to correct deficiencies.
12.23 Any variations from the validation protocol should be documented with appropriate justification.

For bioanalytical method validation, there are the following expectations from the FDA[6] for planning and reporting work:

> Establish a detailed, written description (*e.g.*, protocol, study plan, and/or standard operating procedure) for the bioanalytical method before initiating validation. The description should identify procedures that control critical parameters in the method (*e.g.*, environmental, matrix, procedural variables) from the time of collection of the samples to the time of analysis to minimize their effects on the measurement of the analyte in the matrix.
> Document and report (in the method validation report) all experiments used to make claims or draw conclusions about the validity of the method.
> Validate the measurement of each analyte in the biological matrix. The specific requirements and acceptance criteria for each bioanalytical parameter are listed in Table 1 of the guidance.[6]

Table 1 in the guidance document consists of 8 pages of requirements with predefined acceptance criteria for each validation parameter for inclusion in the validation report and is very proscriptive.[6] Interestingly, there is no repeat analysis of standards or quality control samples allowed during a validation study. In Table 2 there is the requirement to include in the validation report the reasons for and mode of reintegration of chromatograms.[6]

15.4.4 Validation Parameters

Depending on the type of analytical procedure being assessed one or more of the following parameters will be determined during the validation:

- Specificity or selectivity (depending on the analytical technique used).
- Linearity.
- Range.
- Accuracy and Precision (reproducibility and intermediate precision). Note that accuracy and precision are being replaced by uncertainty in the proposed USP <1220>[7] that brings the pharmaceutical industry into line with ISO 17025 where uncertainty has been used since the standard was issued in 1999. There is also a Stimulus to the Revision Process discussing the use of uncertainty measurement in the pharmaceutical industry.[20]
- Limit of Quantification (LOQ). This can be LLOQ and ULOQ for lower and upper limits of quantification, respectively, for bioanalytical methods.
- Limit of Detection (LOD).
- Stability: reference standard solutions, extracts following sample preparation, autosampler conditions, *etc.*
- Robustness (although this may be defined as a method development parameter but has no measure).
- System Suitability Test parameters verified, as appropriate.

It is important to ensure that when validating a method a tick-box approach is NOT used. For example, if a method is being developed for assay of an active pharmaceutical ingredient in a finished product between 80% and 120%, then why would limits of quantification and detection be measured? Each analytical procedure validation needs to have an appropriate experimental design to measure only those parameters pertinent to the needs. Slavishly following all criteria of ICH Q2(R1) simply as it has a list is not acceptable. Perhaps engaging the brain in logical thought might be a more useful exercise?

15.5 Overview of the Life Cycle of Analytical Procedures

15.5.1 USP <1220> and Stimuli to the Revision Process

The approach outlined in this section is based upon the on-going work of the Validation and Verification Expert Committee of the United States Pharmacopoeia. Although focused on GMP and compendial analytical procedures, the sound scientific principles outlined here are applicable to bioanalytical methods as well. The intention of the rest of this chapter, albeit derived from a GMP perspective, is also intended to be applicable to any analytical procedure used in any regulated laboratory.

Since 2012, the USP Validation and Verification Expert Panel have been publishing several stimuli to the revision process articles that has culminated with the proposed USP general chapter <1220> on The Analytical Procedure Lifecycle.[7] This work is on-going. With the update of ICH Q2(R1) and alignment with ICH Q12 on Technical and Regulatory Considerations for Pharmaceutical Product Lifecycle Management,[14] will mean that life cycles for analytical procedures will become a regulatory expectation over time, although there will be a transition period where the current and life cycle approaches work in parallel.

15.5.2 Life Cycle of Analytical Procedures

The life cycle of analytical procedure advocated by USP <1220> that is shown in Figure 15.3 consists of three stages:

1. Procedure Design and Development (or method development) that is based on the Analytical Target Profile (ATP) for the procedure.
2. Procedure Performance Qualification (or method validation).
3. Procedure Performance Verification (Ongoing assessment of the procedure).

The life cycle approach is based on the FDA guidance for industry for Process Validation as well as ICH Q8 on Pharmaceutical Development.[21]

Shown also in the figure are the feedback loops from stage 3 to 2 and stage 2 to stage 1 as well as to the ATP, representing continual improvement of the procedure. The key is continual as the pharmaceutical industry is regulated and some procedures are part of a registration dossier might need to be modified under change control. The three stages will be discussed in more detail in Sections 15.6 to 15.9. This chapter will only provide an overview of the life cycle process but for a more detailed understanding of the USP <1220> process and best practices in procedure validation the reader is referred to the book by Ermer and Nethercote.[22]

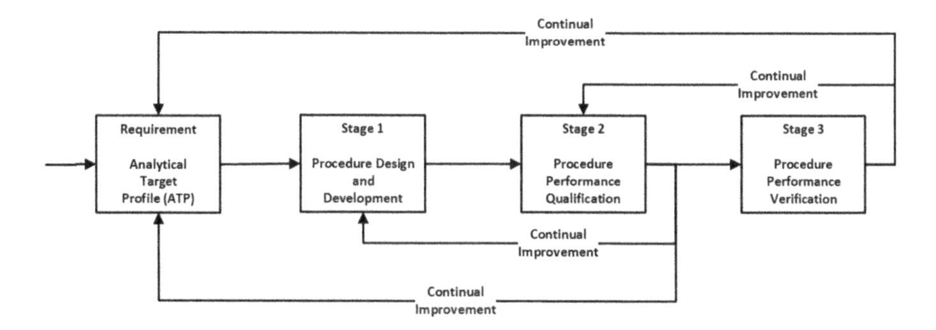

Figure 15.3 Life cycle of an analytical procedure.

15.6 Defining the Analytical Target Profile (ATP)

15.6.1 Specification for an Analytical Procedure

The first step in developing an analytical procedure is writing the Analytical Target Profile or ATP, that can be considered as the specification or intended use for the procedure. This term was developed by an EFPIA/PhRMA working group on Analytical Design Space and Quality by Design of analytical procedures and has been incorporated by the USP Validation and Verification Expert Panel into two Stimuli to the Revision Process on ATP itself and the proposal for a new USP general chapter <1220> on the Analytical Procedure Lifecycle.[7]

The Analytical Target Profile (ATP) for an analytical procedure is a pre-defined objective of a method that encapsulates the overall quality attributes required of the method, *e.g.*:

- sample to be tested;
- matrix that the analyte will be measured in;
- analyte(s) to be measured;
- range over which the analyte(s) are to be measured for the reportable result;
- quality attributes such as selectivity and precision/accuracy of the whole procedure or total measurement uncertainty (TMU).

From the above, this is the core of the life cycle approach as it defines the high-level objectives with no mention of any analytical technique used to meet the ATP as this could bias the analytical approach. An example ATP could be:

> To quantify analyte *X* over a range between *a*% and *b*% (or whatever units are appropriate) with *X*% RSD precision and *Y*% bias in a matrix of *Z* (or in the presence of *Z*).

This means that the requirements for an analytical procedure are defined before any practical work begins or even the analytical technique to be used has been selected. It provides the method developer with an explicit statement of what the procedure should achieve. This is a documented definition and can be referred to during development of the procedure or revised as knowledge is gained.

15.6.2 Advantages and Limitations of an Analytical Target Profile

From the example of an ATP in Section 15.6.1, you can see that it is easy to understand and is relatively straightforward. The accuracy and precision required of the reportable result are known before the analytical technique is selected and may be used to select it. Most importantly, defining the ATP allows an understanding and subsequent control of the variables in the procedure and will allow a defined and risk based control strategy.

As the pharmaceutical industry still works with separate assessments of precision and accuracy the total measurement uncertainty (TMU) is not actually defined. One of the problems with the pharmaceutical industry is that criteria for accuracy (bias) and precision are often based on generally accepted industry practices using default criteria. However, in a Quality by Design (QbD) approach, these criteria are aligned with the specification and product and process needs, and the criteria focus on the reportable value.[7]

15.7 Stage 1: Procedure Design and Development

15.7.1 Overview

The most important part of an analytical procedure life cycle is missing from the current regulatory and pharmacopoeial documents, namely method development. Knowing how sampling, transport, storage, instrumental analysis parameters and interpretation of data impact the reportable value is vitally important to reducing analysis variability and hence out of specification (OOS) results. The aim of a Quality by Design (QbD) approach is a well understood and characterised analytical procedure and this begins with the design and development of the procedure.

Procedure design and development consists of a number of stages, as shown in Figure 15.4 and will be discussed in the following sections.

15.7.2 Information Gathering and Initial Procedure Design

Once the ATP has been defined, the first stage of the life cycle begins and this is shown in Figure 15.4. The procedure design and development stage begins with knowledge and information gathering including:

- chemical information about the analytes of interest, *e.g.* structure, solubility and stability (if known);
- literature search if a known analyte or discussions with medicinal chemists if a new molecular entity (NME).

From this knowledge coupled with the ATP the most appropriate procedure including the measurement technology can be derived, *e.g.*:

- Type of procedure, *e.g.* assay or impurity in an active pharmaceutical product or determination of a new molecular entity in animal or human plasma.
- Sampling strategy such as the sample amount or volume required, how the sample will be taken, any precautions required to stabilise the analyte in the sample, *etc.*

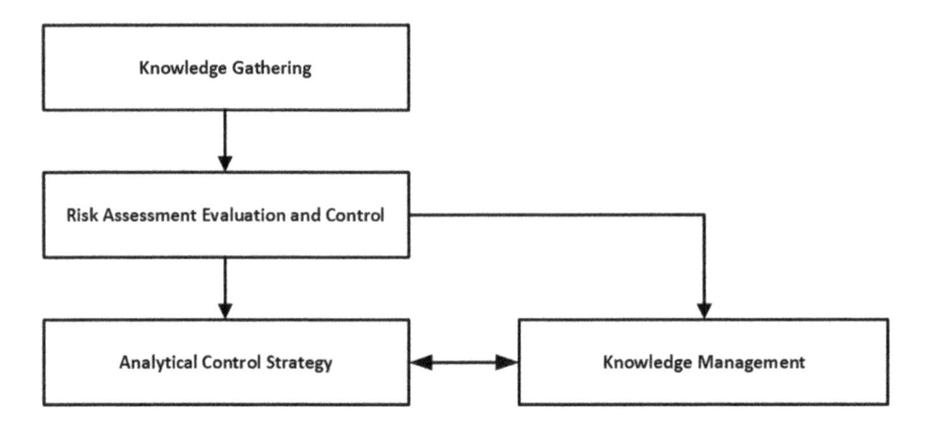

Figure 15.4 Method development workflow overview.

- Design of the sample preparation process to present the sample to the instrument
- Is there any need to derivatise the analyte to enhance detection characteristics?
- Appropriate analytical technique to use, *e.g.* LC-MS, LC-UV, GC-FID, UV, titration, *etc.*, based upon the ATP and the chemical structure of the analyte.
- Outline separation needs based on previous analytical methods with analytes of similar chemical structure, if appropriate,
- In addition, business factors such as time for the analysis and cost should be considered when developing a method. Quicker is better, provided that the ATP is met, and UPLC may be a better alternative to conventional HPLC.

15.7.3 Iterative Method Development and Method Optimisation

Assume that we are dealing with a liquid chromatographic analysis, as this is one of the most common analytical techniques used in the pharmaceutical industry, the sample needs to be prepared so that it can be introduced into the chromatograph to separate and quantify the analytes. Development of the sampling, sample preparation and separation should proceed in tandem and iteratively, as shown in Figure 15.5. Some considerations for this phase of the development, covering all sample types and concentration or amount ranges defined in the ATP are:

- Physicochemical properties of the analyte such as solubility and stability and impact on the preparation method that may vary from dilution

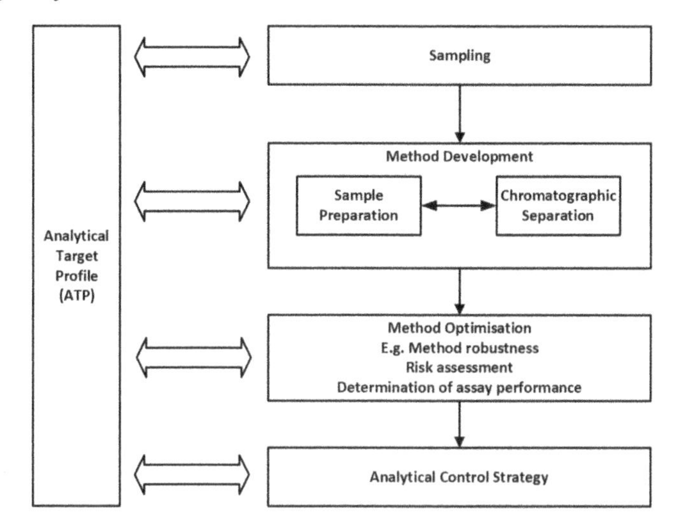

Figure 15.5 Method development workflow for an HPLC procedure. Reproduced from ref. 22 with permission from John Wiley & Sons, Copyright © 2015 Wiley-VCH Verlag GmbH & Co. KGaA.

to liquid–liquid or solid phase extraction. In addition, the pK_a values of any ionisable groups and lipophilicity and matching these to potential column packings.

- How much sample is required to achieve the ATP?
- Does the sample have to be dissolved, homogenised, sonicated, crushed, *etc.* before sample preparation can begin?
- Does the analyte require derivatisation either to stabilise the compound or to enhance limits of detection or quantification?
- Screening experiments are run to see how the analyte(s) run on a variety of columns (*e.g.* C18, C8, *etc.*) and mobile phases of varying composition of organic modified and pH value of the aqueous buffers. It is important to note here that the KISS (Keep It Simple, Stupid) principle applies here. Do not over complicate a method as it will usually need to be established in one or more other laboratories and makes method transfer more difficult.
- A better approach for screening is to automate using method development software to design and execute experiments using a statistical design, *e.g.* factorial design such as Plackett–Burman. This is a more expensive option but it will produce design space maps for optimum separation faster than a manual approach. These design space maps provide the basis for a robust separation as the factors controlling the separation can be more easily identified and the optimum separation to meet the ATP can be predicted and then confirmed by experiment.

The overarching principle in the method development and method optimisation is to keep the method as simple as possible to achieve the ATP requirements. For example, a commonly available column and simple mobile phase preparation should be the starting point for most separations, depending, of course, on the type of analyte, *e.g.* simple heterocyclic, chiral determination, or macromolecule, *etc.*

15.7.4 Risk Assessment and Management

Management of risk is a key element in the analytical procedure life cycle approach. This involves identifying and then controlling factors that can have a significant impact on the performance of the separation. Such factors may be:

- pH value of the aqueous buffer or proportion of organic modifier used in an LC mobile phase;
- type or dimensions of the column used;
- autosampler temperature;
- impact of light during sampling or sample preparation.

A formal risk assessment can be undertaken such as Failure Mode Effects Analysis (FMEA) to identify the risk with the highest impact.[22] The aim of risk assessment is to either mitigate or eliminate the risk posed by variables in the sample preparation, instrumental analysis or operating practices. Method variables can be classified as controlled, noise or experimental (C, N or X), as shown in Table 15.1. A discussion of how these variables are investigated is outside the scope of this book and the reader is referred to Ermer and Nethercote's book for more details.[22]

When key variables have been identified, then robustness studies can be started to understand the impact of each one on the overall analytical procedure. There will be a study design for robustness experiments and the results will be examined statistically. The aim is to identify the acceptable range of each key variable – the greater the range the more flexible the method, again please see Ermer and Nethercote for more information about this approach.[22]

Table 15.1 C, N, X classification of method variables.

Method variable	Classification
Controlled (C)	Variable controlled *via*: • Explicit instructions in the analytical procedure • Fixing the value of a parameter in an instrument
Noise (N)	Variable difficult to control or predict and may vary randomly: • Common variations are investigated experimentally (*e.g.* precision)
Experimental (X)	Variable that can varied deliberately • Impact is investigated by experiment (*e.g.* stability and robustness) • Establish acceptable ranges of performance

15.7.5 Analytical Control Strategy: Identifying and Controlling Risk Parameters

The analytical control strategy for each analytical procedure is based upon the outcome of the risk assessment and, where appropriate, in combination with the robustness studies. This should provide a list of method parameters and variables that have significant impact on the method and its performance as well as what to avoid when executing the analytical procedure. The outcome is the establishment of controls for critical parameters such as how to perform a specific task with sufficient detail to ensure consistent performance, the type of integration, define conditions in the procedure that have significant effects or to avoid certain situations where, say, light can impact the stability of the analyte.

The outcome of the analytical control strategy is to have a set of instructions that are explicit and unambiguous when executing the procedure such as:

- How to sample and the required sample size.
- Specification of sample containers, transport conditions to the laboratory, storage conditions.
- Preparation of the sample for analysis.
- Preparation of reference standard solutions and mobile phases.
- Performance of the analysis, as well as integration and interpretation of data.
- Identify the system suitability test parameters to be used and determine the acceptance criteria for each one.

15.7.6 Procedure Development Report

The outcome of Stage 1 should be a comprehensive method development report describing the optimised procedure. It should also contain practical details for the procedure including the robustness of the analytical procedure, the analytical control strategy and the SST parameters to use and their acceptance criteria.

This is in stark contrast with the FDA bioanalytical method guidance that notes that *bioanalytical method development does not require extensive record keeping or notation.*[6] In response, if you do not have any understanding of how critical parameters impact the performance of an analytical procedure then how can you control them? In my opinion, method development needs a report that highlights those key parameters and how they impact performance of the procedure. This report, along with a validation report for the procedure, will allow a faster transfer of a procedure from one laboratory to another.

15.8 Stage 2: Procedure Performance Qualification

15.8.1 Planning the Validation

Method validation should be simply confirmation of good method development and demonstrates that the analytical procedure is fit for purpose [C. Burgess, personal communication]. Procedure Performance

Qualification demonstrates that the developed analytical procedure meets the ATP quality attributes and that the performance is appropriate for the intended use.

To control the work, there will be a validation plan or protocol describing the experiments to be performed with predefined acceptance criteria, as appropriate to the procedure, *e.g.* API, impurities or measurement in biological fluids. The various experiments will depend on the criteria described in the ATP and on the intended use of the procedure, *e.g.*:

- Linearity experiments should be used to support the use of the specific calibration model used in the procedure (the calculations for which have been verified in Level 1 in the computerised system validation of the data system used for this work).
- Specificity or selectivity (depending if the instrumental technique is absolute or comparative respectively) *e.g.* resolution for impurities, peak purity assessment for stability indicating methods.
- Precision: injection precision, repeatability and intermediate precision. The minimum number of runs could be two but four or more provides better understanding of the intermediate precision for routine use.
- Accuracy can be run in the same experiments as precision.
- Determination of analyte stability under storage, laboratory and/or instrument conditions, *etc.*
- System suitability test parameters and their acceptance criteria will be verified during this work.

It is important that the acceptance criteria are defined in the validation plan and are based on outcomes from Stage 1, the procedure design and development. The plan will also define how the data from the various experiments will be evaluated statistically against the acceptance criteria.

15.8.2 Validation Report

Once the work is completed, a report is written that describes the outcome of the validation experiments and how the procedure meets the requirements of the ATP.

As the draft USP <1220>[7] notes,

> *the analytical control strategy may be refined and updated as a consequence of any learning from the qualification study. For example, further controls may be added to reduce sources of variability that are identified in the routine operating environment in an analytical laboratory, or replication levels (multiple preparations, multiple injections, etc.) may be modified based on the uncertainty in the reportable value.*

The scope and the various parameters with the acceptance criteria for a bioanalytical method validation report are defined in the updated FDA Guidance for Industry on Bioanalytical Method Validation.[6]

15.8.3 Analytical Procedure Transfer

Analytical method transfer is not easy as there are always items that are not well described or even omitted altogether from analytical procedures. However, the work must be planned and a protocol developed between the originating and receiving laboratories that includes pre-defined ways that the data will be interpreted with acceptance criteria. A report should be produced summarising the transfer results against the data generated by the receiving laboratory. The extent of the work required is outlined in Table 15.2.

To reduce the effort required when establishing an analytical procedure in another laboratory, a subject matter expert could travel to the receiving laboratory to provide help and advice. Alternatively, an analyst from the receiving laboratory could go to the originating laboratory to learn the procedure. Management often looks at the up-front cost of this but dismisses the hidden cost in time wasted in establishing the method without help from the originating laboratory.

Table 15.2 Draft USP <1220> recommendations for qualifying changes to an analytical procedure.

Change	Action
A change to a procedure variable to a value within the range that was previously qualified	• Document the change • No additional experimentation required
A change to a procedure variable to a value outside the range that was previously qualified to produce fit-for-purpose data	• Perform a risk assessment to determine parameters impacted by the change • Perform an appropriate procedure performance qualification study to confirm that the change does not impact the method's ability to meet the ATP
A change to a new laboratory	• Review the risk assessment • Determine an appropriate qualification study (which might include comparability testing or a reduced or full requalification)
A change to a new procedure or technique	• Perform the appropriate development and qualification activities (Stages 1 and 2) to demonstrate conformance of the new procedure to the ATP
A change impacting the ATP, *e.g.*, a specification limit change or need to measure levels of analytes not considered in the original ATP	• Update to the ATP • Review the existing procedure design and qualification data (Stages 1 and 2) to determine if the procedure still meets the new ATP • The level of activities required to confirm that a changed analytical procedure is producing fit-for-purpose data will depend on an assessment of 1) the risk associated with the change, 2) the knowledge available about the procedure, and 3) the effectiveness of the control strategy. • Conduct a risk assessment to determine the appropriate level of activities required to provide confidence that the modified method will produce results that meet the ATP

When considering method transfer, one of the issues when using a CRO laboratory is the quality of the written procedure used for method transfer. Often the originating laboratory (sponsor) may make a minimal effort at validation before passing the procedure to a CRO.

15.9 Stage 3: Procedure Performance Verification

15.9.1 Routine Monitoring of Analytical Performance

Verification of an analytical procedure's on-going performance is equivalent to the PQ phase in Analytical Instrument Qualification discussed in Chapter 14. It provides assurance that the analytical procedure remains in a state of control throughout its life cycle and provides a proactive assessment of a procedure's performance. The aim of verification is that the reportable result is fit for purpose and can be used to make a decision.

Part of this verification can be trending of SST and sample replicates over time. However, there is a note of caution that SST results can also be used to measure directly (Group B and some Group C instruments) or indirectly (some Group C instruments) instrument performance.

Data that could be collected and tracked are:

- system suitability results including failures;
- trending individual results and the reportable result including OOS and outputs from investigations.

These data should be monitored against limits so that when there is a trend indicating a parameter is out of control, an investigation can be started early before the situation gets out of hand. When a root cause is identified in an investigation, it may be appropriate to update the analytical control strategy or to update the analytical procedure.

15.9.2 Changes to an Analytical Procedure

Change is inevitable over time with an analytical procedure either due to modifications of the process generating the samples, availability of chemicals and columns, retirement of analytical instruments and changes due to continual improvement of the procedure. The question is what is the impact of these changes? The draft USP <1220>[7] provides some help in determining what if anything needs to be done to qualify the change, and this is shown in Table 15.2.

15.9.3 Validated Analytical Procedure

Now that the analytical procedure is validated, it can be used for the analysis of samples from R&D or manufacturing and this will be covered in Chapter 16 for the analysis and Chapter 17 for the second person review.

References

1. *21 CFR 211 Current Good Manufacturing Practice for Finished Pharmaceutical Products*, Food and Drug Administration, Sliver Spring, MD, 2008.
2. *EudraLex – Volume 4 Good Manufacturing Practice (GMP) Guidelines, Chapter 6 Quality Control*, European Commission, Brussels, 2014.
3. *EudraLex – Volume 4 Good Manufacturing Practice (GMP) Guidelines, Annex 15 Qualification and Validation*, European Commission, Brussels, 2015.
4. *EMA Guideline on Bioanalytical Method Validation*, European Medicines Agency, London, 2011.
5. *FDA Draft Guidance for Industry, Bioanalytical Methods Validation*, Food and Drug Administration, Rockville, MD, 2013.
6. *FDA Guidance for Industry: Bioanalytical Methods Validation*, Food and Drug Administration, Silver Spring, MD, 2018.
7. G. P. Martin, *et al.*, Stimuli to the Revision Process: Proposed New USP General Chapter: The Analytical Procedure Lifecycle ⟨1220⟩, *Pharmacop. Forum*, 2017, **43**(1).
8. *FDA Warning Letter Quali Controle and Qual-Controle-C.E.Bac (Warning Letter 320-18-38)*, Food and Drug Administration, Silver Spring, MD, 2018.
9. *ICH Q2(R1) Validation of Analytical Procedures: Text and Methodology*, International Conference on Harmonisation, Geneva, 2005.
10. *USP General Chapter <1225> Validation of Compendial Procedures*, United States Pharmacopoeia, Rockville, MD.
11. *FDA Draft Guidance for Industry: Analytical Procedures and Methods Validation*, Food and Drug Administration, Rockville, MD, 2000.
12. *FDA Guidance for Industry: Analytical Procedures and Methods Validation for Drugs and Biologics*, Food and Drug Administration, Silver Springs, MD, 2015.
13. *Concept Paper: Analytical Procedure Development and Revision of ICH Q2(R1) Analytical Validation*, International Council on Harmonisation, Geneva, 2018.
14. *ICH Q12: Technical and Regulatory Considerations for Pharmaceutical Product Lifecycle Management, Step 2*, International Council for Harmonisation, Geneva, 2017.
15. *EudraLex – Volume 4 Good Manufacturing Practice (GMP) Guidelines, Part 2-Basic Requirements for Active Substances Used as Starting Materials*, European Commission, Brussels, 2014.
16. *USP 41 General Chapter <1058> Analytical Instrument Qualification*, United States Pharmacopoeia Convention, Rockville, MD, 2018.
17. *USP General Chapter <1226> Verification of Compendial Procedures*, United States Pharmacopoeia, Rockville, MD.
18. *FDA Guidance for Industry, Bioanalytical Methods Validation*, Food and Drug Administration, Rockville, MD, 2001.
19. R. D. McDowall, *Validation of Chromatography Data Systems: Ensuring Data Integrity, Meeting Business and Regulatory Requirements*, Royal Society of Chemistry, Cambridge, 2nd edn, 2017.

20. M. L. J. Weitzel, *et al.*, Stimuli to the Revision Process: Measurement Uncertainty for the Pharmaceutical Industry, *Pharmacop. Forum*, 2018, **44**(1).
21. *ICH Q8: Pharmaceutical Development*, International Conference on Harmonisation, Geneva, 2008.
22. J. Ermer and P. Nethercote, *Method Validation in Pharmaceutical Analysis, a Guide to Best Practice*, Wiley-VCH, Weinheim, Germany, 2nd edn, 2015.

CHAPTER 16

Performing an Analysis

We have waited until Chapter 16 to get to the heart of data integrity with the focus on sample analysis. However, the preceding chapters are essential to understanding, *via* the Data Integrity Model, of what an organisation and regulated laboratory needs to have in place to ensure data integrity before any analysis occurs, *e.g.* management leadership, an open culture, data integrity policies and procedures, trained staff, qualified analytical instruments with validated computerised systems and validated analytical procedures. These elements are all applied to the analysis of samples that we will discuss in this chapter.

This chapter focuses on Level 3 of the Data Integrity Model – the execution of a validated analytical procedure to analyse samples and generate reportable results. As discussed in Chapter 9, the analytical life cycle can encompass sampling, sample management, sample preparation, instrumental analysis, interpretation of analytical data and the generation of the reportable result in part or in whole. This work may be performed by a single analyst or by a group of analysts but excludes the second person review, which is presented in Chapter 17. Sample analysis is where most of poor data management practices or data violations are seen but often the root causes of these problems are found in the underlying layers of the Data Integrity Model.

In regulatory and industry data integrity guidance documents there is no mention of sampling or sample preparation, yet these two processes that are typically manual operations, can hide a multitude of data integrity problems even before instrumental analysis occurs.

Data Integrity and Data Governance: Practical Implementation in Regulated Laboratories
By R. D. McDowall

Published by the Royal Society of Chemistry, www.rsc.org

16.1 What the Regulators Want

In this section, the high level regulatory requirements for sample analysis are presented. In addition, as we discuss the various sections of the analytical process, additional regulatory requirements will be presented so that readers can understand the regulatory emphasis on individual stages of an analysis.

16.1.1 EU GMP Chapter 1 Pharmaceutical Quality System

In Chapter 1 of EU GMP Section 1.9 covers Quality Control[1]:

> 1.9 Quality Control is that part of Good Manufacturing Practice which is concerned with sampling, specifications and testing, and with the organisation, documentation and release procedures which ensure that the necessary and relevant tests are actually carried out and that materials are not released for use, nor products released for sale or supply, until their quality has been judged to be satisfactory. The basic requirements of Quality Control are that: ....
> (ii) Samples of starting materials, packaging materials, intermediate products, bulk products and finished products are taken by approved personnel and methods;
> (iv) Records are made, manually and/or by recording instruments, which demonstrate that all the required sampling, inspecting and testing procedures were actually carried out. Any deviations are fully recorded and investigated;

Note in the quotations above there is a phrase that is repeated twice: "actually carried out" as opposed to fabricated or falsified.

In addition, there are further and more detailed requirements for GMP laboratory operations in:

- EU GMP Chapter 6 on Quality Control[2];
- Annex 8 for the Sampling of Starting and Packaging Materials.[3]

16.1.2 US GMP 21 CFR 211 GMP for Finished Pharmaceutical Products

In the US GMP regulations there are the two sections related to testing[4]:

Section 211.84 Testing and approval or rejection of components, drug product containers, and closures.

> (a) Each lot of components, drug product containers, and closures shall be withheld from use until the lot has been sampled, tested, or examined, as appropriate, and released for use by the quality control unit.
> (b) Representative samples of each shipment of each lot shall be collected for testing or examination. The number of containers to be sampled, and

the amount of material to be taken from each container, shall be based upon appropriate criteria such as statistical criteria for component variability, confidence levels, and degree of precision desired, the past quality history of the supplier, and the quantity needed for analysis and reserve where required by 211.170.

Section 211.160 Laboratory Controls – General requirements.

(a) The establishment of any specifications, standards, sampling plans, test procedures, or other laboratory control mechanisms required by this subpart, including any change in such specifications, standards, sampling plans, test procedures, or other laboratory control mechanisms, shall be drafted by the appropriate organizational unit and reviewed and approved by the quality control unit.

The requirements in this subpart shall be followed and shall be documented at the time of performance.

Any deviation from the written specifications, standards, sampling plans, test procedures, or other laboratory control mechanisms shall be recorded and justified.

(b) Laboratory controls shall include the establishment of scientifically sound and appropriate specifications, standards, sampling plans, and test procedures designed to assure that components, drug product containers, closures, in-process materials, labeling, and drug products conform to appropriate standards of identity, strength, quality, and purity.

16.1.3 FDA Guide for Inspection of Pharmaceutical Quality Control Laboratories

Section 13 of this Guidance document[5] is quoted at length as it focuses on Laboratory Records and Documentation and has several items for consideration during inspections. Knowing the focus of the inspection areas means that laboratories can ensure that the records are complete:

Review personal analytical notebooks kept by the analysts in the laboratory and compare them with the worksheets and general lab notebooks and records. Be prepared to examine all records and worksheets for accuracy and authenticity and to verify that raw data are retained to support the conclusions found in laboratory results.

Review laboratory logs for the sequence of analysis *versus* the sequence of manufacturing dates. Test dates should correspond to the dates when the sample should have been in the laboratory. If there is a computer data base, determine the protocols for making changes to the data. There should be an audit trail for changes to data.

Carefully examine and evaluate laboratory logs, worksheets and other records containing the raw data such as weighings, dilutions, the condition of instruments, and calculations. Note whether raw data are missing, if records have been rewritten, or if correction fluid has been used to conceal errors. Results should not be changed without explanation. Cross reference the data that has been corrected to authenticate it. Products cannot be "tested into compliance" by arbitrarily labeling out-of-specification lab results as "laboratory errors" without an investigation resulting in scientifically valid criteria.

Test results should not have been transcribed without retention of the original records, nor should test results be recorded selectively. For example, investigations have uncovered the use of loose sheets of paper with subsequent selective transcriptions of good data to analyst worksheets and/or workbooks. Absorbance values and calculations have even been found on desk calendars.

Cut charts with injections missing, deletion of files in direct data entry systems, indirect data entry without verification, and changes to computerized programs to override program features should be carefully examined. These practices raise questions about the overall quality of data.

The firm should have a written explanation when injections, particularly from a series are missing from the official work-sheets or from files and are included among the raw data. Multiple injections recorded should be in consecutive files with consecutive injection times recorded. Expect to see written justification for the deletion of all files.

Determine the adequacy of the firm's procedures to ensure that all valid laboratory data are considered by the firm in their determination of the acceptability of components, in-process, finished product, and retained stability samples. Laboratory logs and documents when cross referenced may show that data has been discarded by company officials who decided to release the product without a satisfactory explanation of the results showing the product fails to meet the specifications. Evaluate the justification for disregarding test results that show the product failed to meet specifications.

16.2 The Analytical Process

16.2.1 Linking the Data Integrity Model to the Analytical Process

Chapter 5 presents and discusses a Data Integrity Model, Figure 16.1 shows the four layers of the Data Integrity Model in a column down the left-hand side against which the various tasks in an analytical process. The model in this context is dynamic and shows the interaction between all layers of the model.

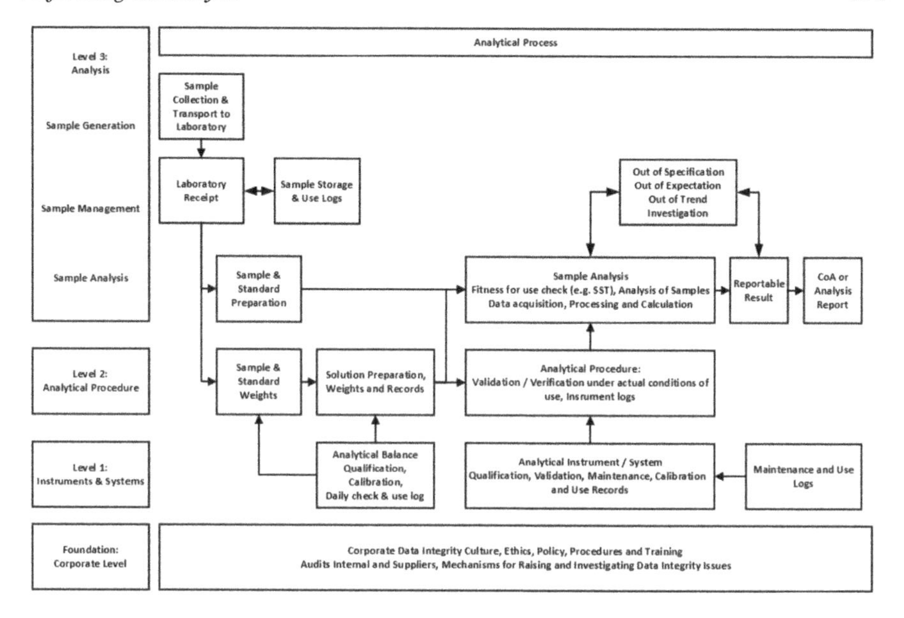

Figure 16.1 The analytical process linked to the Data Integrity Model. Reproduced from ref. 6 with permission of the Royal Society of Chemistry.

- The foundation layer shows in outline what is required with management leadership, culture, ethos and data integrity policies, procedures and planning.
  For an actual analysis, the requirements are trained analytical staff that know the data integrity policies of the company and who know that if they make a mistake they can raise the issue without fear of retaliation as well as understanding that if they falsify data, then disciplinary action will be taken.
- Level 1 shows qualification of an analytical balance as well as analytical instrument such as a chromatograph coupled with the validation of the controlling chromatography data system. When used for analysis there are system suitability tests or point of use checks made of the instrument, *e.g.* SST samples for a chromatograph and checks for the analytical balance.
- Level 2 is the validation of analytical procedures.
  In the analysis, there is the preparation of reference standard solutions, sample preparations, and execution of the analysis using a validated analytical procedure.
- Level 3 The qualified instruments and validated instrument data system (*e.g.* CDS) with a validated analytical procedure are applied to the analysis of a sample.
  The process flow starts with the sampling, transporting the sample to the laboratory, sample management, application of the analytical procedure using the qualified chromatograph and validated CDS, calculation of the reportable result as well as any out of specification investigation.

16.2.2 Process Overview

The analytical process, from a different perspective, is shown in Figure 16.2. Here, we can see the full scope of an analysis from sampling to the generation of a reportable result. At the top of the figure are the six stages of an analysis leading to the reportable result of 98.3 on the right. Underneath each stage are the main tasks to be performed correctly to ensure data integrity.

The first two stages of the analytical process, sampling and sample preparation are performed manually in most laboratories and have all the right pre-requisites for data integrity issues such as:

- actions are executed by people (sometimes interacting with instruments or systems);
- recording text and numbers and labelling items are involved;
- where many steps of the operation are not second-person reviewed at the time of data entry/labelling, and;
- where errors can lead to incorrect business decisions.

Sampling and preparation represent a hidden problem: it is *NOT* covered in any data integrity guidance from any regulatory agency. However, if sampling and sample preparation is not controlled adequately, business decisions can be made on inaccurate or falsified analytical information.

16.2.3 Analytical Instruments Are Qualified and/or Validated

Level 1 of the Data Integrity Model involves the qualification of analytical instruments and the validation of the application software. In this chapter, it is assumed that the instruments involved are adequately qualified with validated software, as appropriate. Therefore, these two aspects will not be discussed and the reader is referred to Chapter 14 for these aspects in more detail.

16.2.4 System Suitability Tests and Point of Use Checks

Even if analytical instruments are qualified and any software validated, there still needs to be a check on the day to see if an instrument or analytical system can perform the analysis, as discussed in Chapter 14 under Performance Qualification. This is good analytical science that is backed up by GXP regulations and, where appropriate, pharmacopoeial requirements. When using an analytical instrument during the execution of an analytical procedure there are pharmacopoeial requirements for point of use checks or system suitability tests. Such tests are typically performed before using the instrument, *e.g.*:

- Analytical balance checks using calibrated masses.
- pH meters are calibrated before use with reference solutions provided by the instrument supplier.

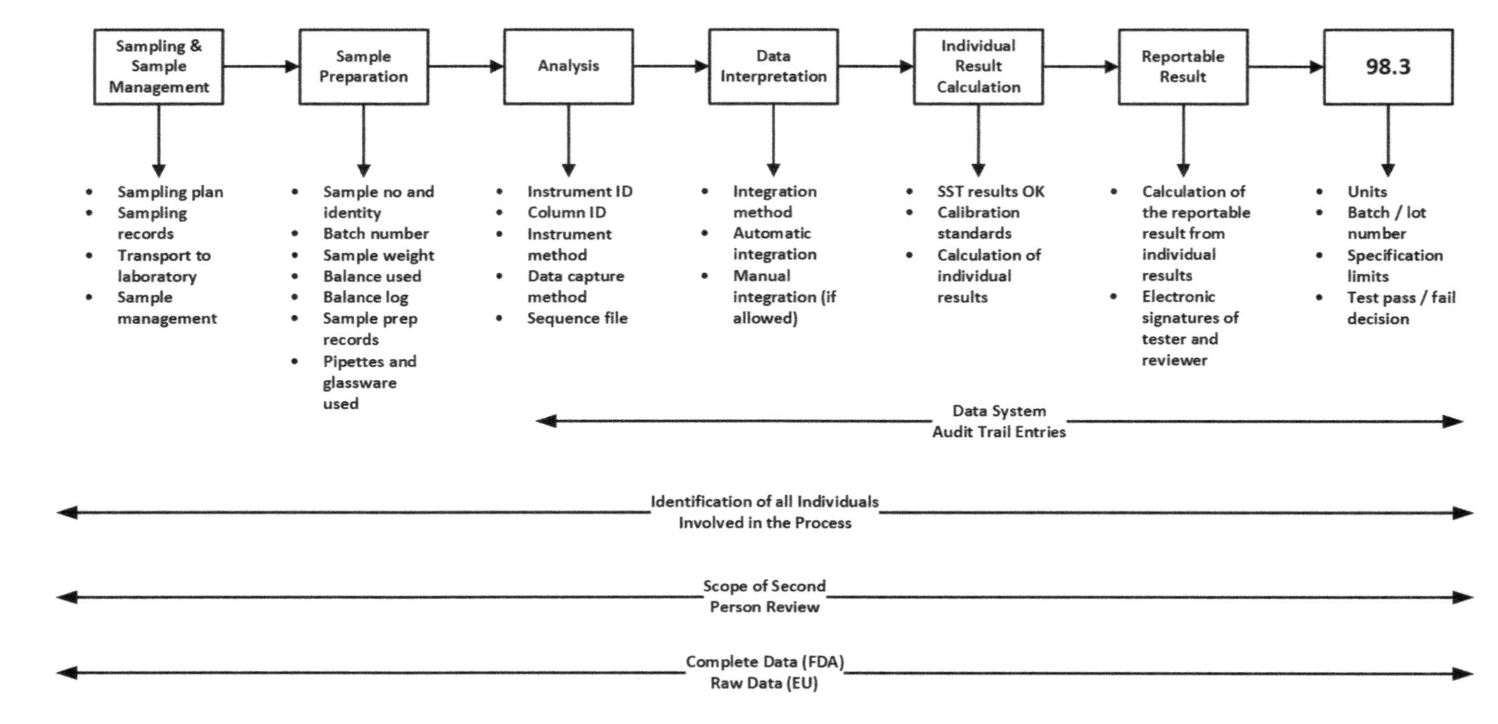

Figure 16.2 The analytical process from sampling to reportable result.

- Spectrometers will be checked for wavelength accuracy using a traceable standard appropriate to the use of the instrument, *e.g.* holmium in the range 241 nm to 650 nm, quinine sulfate for fluorescence detectors in the wavelength range 375 nm to 675 nm or polystyrene references used for wavelength checks in the near-and mid-infrared regions (1.1 μm to 18.5 μm, 9000 cm^{-1} to 540 cm^{-1}, respectively).
- Chromatographs have system suitability tests using parameters defined in the United States Pharmacopoeia <621>[7] or European Pharmacopoeia 2.2.46.[8]
- Mass spectrometers used as LC detectors use a polyethylene glycol (PEG) solution to calibrate the MS.

In this chapter, the instrument calibration or running SST will be assumed to occur unless there is a specific data integrity reason for a discussion on this subject.

16.3 The Scope of Analytical Procedures

In this chapter we will consider the four following types of analytical procedure, as shown in Figure 16.3, these follow the analytical data life cycle discussed in Chapter 9 in this book:

- Observation Tests: the sample requires no sample preparation and tests are carried out by looking at the sample and writing down the reportable result. There is no instrumental analysis involved.
  Under this subject, we will also consider titration tests using an indicator for pharmacopoeial content analysis, although there is some sample preparation involved, the discussion will focus on the titration phase of the assay alone.

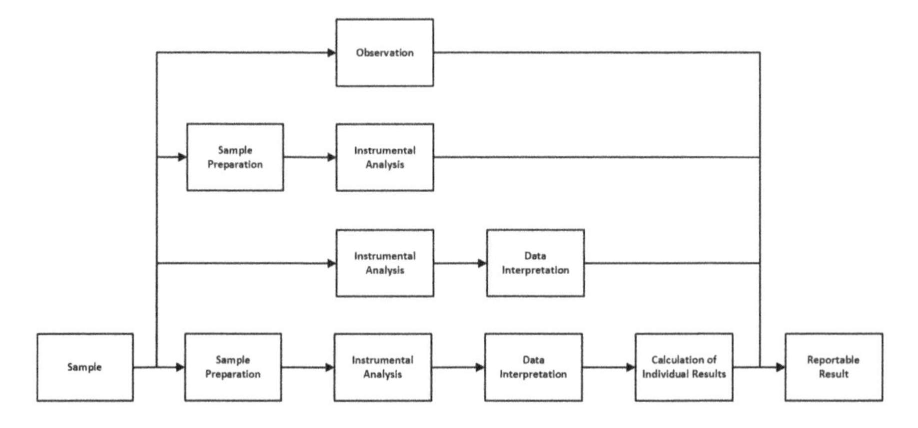

Figure 16.3 Scope of analytical procedures.

- Simple Instrumental Analysis: involving a sample preparation phase followed by instrumental analysis. Here, the instrumental analysis can be as simple as a pH value determination or a loss on drying analysis.
- Instrumental Analysis with Data Interpretation: Here, we can consider a spectroscopic analysis such as Near Infra-Red analysis (NIR) for identity confirmation where a probe is placed at or in a sample, a spectrum is taken and compared with a library. There is no sample preparation involved but the emphasis of the analytical procedure is on the interpretation of the spectrum to confirm the identity of the sample.
- Sample preparation, instrumental analysis coupled with data interpretation: Represented by a chromatographic analysis with the instrument linked to a chromatography data system. Here, the whole process is automated by the CDS including SST parameter calculations, calibration curves and calculation of the reportable result. A report is generated and signed electronically.

These are not all options that could be considered but covers most analyses in regulated laboratories. Regardless of the scope of the analytical procedures, they all have the same starting point: sampling and sample management.

16.4 Sampling and Sample Management

16.4.1 What the Regulators Want

ICH Q7, GMP for Active Pharmaceutical Ingredients[9] and EU GMP Part 2[10] has in Clause 7.33 the following statements:

> Samples should be representative of the batch of material from which they are taken. Sampling methods should specify the number of containers to be sampled, which part of the container to sample, and the amount of material to be taken from each container.
>
> The number of containers to sample and the sample size should be based upon a sampling plan that takes into consideration the criticality of the material, material variability, past quality history of the supplier, and the quantity needed for analysis.

EU GMP Chapter 4 states in Clause 4.25 regarding sampling[11]:

> There should be written procedures for sampling, which include the methods and equipment to be used, the amounts to be taken and any precautions to be observed to avoid contamination of the material or any deterioration in its quality.

Note, that in Section 211.160(b) of US GMP there is the requirement for *scientifically sound* sampling plans as well as the following requirements for sampling in Section 211.84[4]:

(4) If it is necessary to sample a component from the top, middle, and bottom of its container, such sample subdivisions shall not be composited for testing.

(5) Sample containers shall be identified so that the following information can be determined: name of the material sampled, the lot number, the container from which the sample was taken, the date on which the sample was taken, and the name of the person who collected the sample.

(6) Containers from which samples have been taken shall be marked to show that samples have been removed from them. ......

(d) Samples shall be examined and tested as follows:

(1) At least one test shall be conducted to verify the identity of each component of a drug product. Specific identity tests, if they exist, shall be used.

(2) Each component shall be tested for conformity with all appropriate written specifications for purity, strength, and quality. *In lieu* of such testing by the manufacturer, a report of analysis may be accepted from the supplier of a component, provided that at least one specific identity test is conducted on such component by the manufacturer, and provided that the manufacturer establishes the reliability of the supplier's analyses through appropriate validation of the supplier's test results at appropriate intervals.

The requirements for sampling are that samples must be homogeneous and at least an identity test is carried out on APIs and excipients used for drug product manufacturing. The assumption is that the supplier of such materials has been adequately qualified and has data integrity approaches that are satisfactory, the latter aspect is covered in Chapter 23.

16.4.2 Sampling Is Critical

Sampling is the most critical part of the analytical process. If the sample is not representative or homogeneous, then the rest of the laboratory work is worthless. Given its criticality in analysis and ensuring both data integrity and data quality, it is surprising that there is no mention of sampling in any regulatory guidance. In most cases, sampling is manual and can be an area where data integrity can be compromised, *e.g.* substitution of a sample with material from a batch known to be within specification with very little chance of detection.

Table 16.1 Required information when taking a representative GMP sample.

Sampling information
• Site/Location from where the sample was taken
• Sample name If the sample is a product then the product code, strength and dosage form
• Batch number and date of manufacture Expiry date if a product
• Date of sampling (time may only be required when testing for an intermediate from continuous production)
• Identification of the person or persons taking the sample
• A visual description of the material

The sample collection container must be robust enough to withstand poor handling during transport to the laboratory, of adequate size for the analysis including any repeat work and made of a material that does not interact with the analyte or degrade it.

16.4.3 GMP Sample Plan and Sampling

The regulations in Section 16.4.1 make it clear that sampling plans must be scientifically sound. To help develop a sampling plan there is USP <1097> on bulk powder sampling strategies[12] that consists of two steps:

1. Primary sample taken from the material. The sampling plan can be based on either a statistical or non-statistical basis but the aim is to take a representative sample consistent with the analytical procedures and the material specifications.
2. Secondary sample, reducing the primary sample to a size suitable for analysis in the laboratory.

It is important that there is the right sampling equipment available for the material under test, *e.g.* solid, liquid, powder, *etc.* The equipment to be used must be defined in each sampling plan. In addition, it is essential to know the physico-chemical properties of the material being sampled, *e.g.* light sensitivity, thermal and chemical stability as this will influence the selection of the sample container, storage conditions as well as transport from the sampling location to the laboratory.

For finished goods there are additional sampling requirements, *e.g.* a minimum of 20 units for table and capsule dosage forms, that must be sufficient for at least two full tests[3] as shown in Table 16.1.

16.4.4 GLP Protocol and Sampling

In US GLP regulations, 21 CFR 58, there are the requirements in Clause 120 for the protocol.[13]

(a) Each study shall have an approved written protocol that clearly indicates the objectives and all methods for the conduct of the study. The protocol shall contain, as applicable, the following information:

(9) The type and frequency of tests, analyses, and measurements to be made.

There will be a section in the study protocol for taking samples from the study animals. This will define the sample time and sampling site as well as the volume of the biological fluid to be sampled. The protocol will also define:

- How the sample will be treated, *e.g.* blood will be placed in a tube with a defined anticoagulant and then centrifuged at 4 °C to harvest plasma.
- The resultant plasma will be transferred to a plain tube.
- The plasma will be frozen using dry ice before storage at −20 °C, −40 °C or −80 °C and the storage conditions to be used during shipment to the bioanalytical laboratory.

Dependent on analyte stability, there may be additional controls required, *e.g.* for light sensitive compounds work may be carried out under yellow light, catecholamine analysis requires a preservative such as citric acid or EDTA to prevent oxidation of the compounds prior to analysis.

A similar process will be used for samples taken during a clinical study involving either human volunteers or patients.

16.4.5 Ensure Correct Sample Labelling

Proper sample collection and identification is critical for test result values that accurately characterise the process or study under test. Therefore, regardless of the regulatory discipline that the samples are collected under, all samples must be correctly labelled with enough information to enable correct identification. The labels must be of adequate size for the sample container and have the correct tack adhesive to ensure that they do not become detached from the container during the sample life cycle. Label and the printing must be robust enough to withstand the environmental conditions that the sample will be stored under, *e.g.* a bioanalytical sample could be stored from −80 °C to +60 °C, the latter condition is used in the sample preparation for deactivation of viruses; thus, a thermally printed label would not be appropriate.

Until now we have discussed the front of the label, there is also a serious consideration for the back of the label: the tack of adhesive used. Adhesive tack must be sufficiently strong to withstand the conditions that the sample and label will go through without coming off the containers likely to be used during analysis but not so strong that it cannot be removed from reusable apparatus such as volumetric flasks when they are cleaned.

Mixing up sample identities and labels can happen in one of several ways:

1. Batch printing identification labels from LIMS, ELN or Lab Execution Systems offers the opportunity to select the wrong label from the sheet and affix it to the wrong sample.
2. If tubes are already labelled, a sample could be placed in the wrong container.
3. Incorrectly transcribing the sampling location to the sample container while reading the location from a paper list (or report).
4. If labels have come unstuck from containers during transport or storage they could be incorrectly re-applied to the wrong sample.

Once sample identity is compromised, data integrity is also lost, and it becomes impossible to reconstruct data with confidence.

In addition to human readable information, each sample label should have a bar code or, when the cost is acceptable, a Radio Frequency Identification (RFID) tag. The use of such labels is discussed in Section 16.4.7 on sample receipt and storage.

16.4.6 Transport to the Laboratory

Some samples require timely transport under specific conditions (*e.g.* temperature and/or humidity) to assure the accuracy of test results. For such unstable samples, it is imperative to assure compliant transport to the test facility and this can be a weak link in the process, especially when third-parties are involved in transport. Situations where a single temperature/humidity is required for all samples may represent a low risk, but when multiple humidity and temperatures are required for delivery of samples, risks rapidly rise. Remember that the integrity and quality of test results are tied to the correct storage and transport of these samples, which could be in the hands of a person with little training, and little understanding of the importance of proper storage and timely transport.

In cases where transport conditions are critical, automated environmental recording of temperature and/or relative humidity using calibrated portable data loggers is the preferred solution. At a minimum, at least one and preferably two calibrated data loggers should be transported with the samples to collect data, providing evidence that the samples remained within the required temperature range prior throughout shipment to receipt in the laboratory. In addition, an investment in transportable data loggers and the associated software application enables the laboratory to review the actual sample transport conditions.

Environmental data from the transport may hold the key to understanding an unusual laboratory test results from what seemed to be a process in control. Additionally, data logger results can be reviewed and analysed (*e.g.* plotting against the required temperature conditions for storage or by performing a Kinetic Mean Temperature (MKT) calculation in the case of a temperature excursion) to verify the adequacy of the sample/transport/receipt process. In the case of an MKT calculation it is important that the calculation itself is not used to hide unacceptable temperature excursions by using an extended time axis as to make the excursion meaningless.

In situations where data loggers are not feasible, sample stability studies can provide evidence that temperature excursions will not significantly impact the reportable test results for specific sample/temperature/time ranges.

Upon arrival at the testing facility, critical samples should be removed from the transport containers by a laboratory analyst to verify that correct transport containers and conditions were used while delivering samples. A delivery person should never remove samples from transport containers and place them in a receipt storage area as essential information about sample transport conditions could be lost. Only use couriers or transport companies that have a good understanding of GXP regulations and if refreezing with dry ice is required for biological samples, that they are trained to do so and are reliable.

16.4.7 Sample Receipt and Storage

Sample receipt begins when the samples arrive at the laboratory. The best scenario is that a trained person immediately receives the samples, records them in the laboratory data system or logbook, and places them in correct storage until testing begins. Reality can be far from ideal: busy labs may permit samples to sit for several hours before processing and storing the samples, unless they are sensitive samples that require immediate attention.

Once again, manual data entry provides an opportunity for errors in sample locations, sample identity mix-ups, or omission of sampling comments. In addition, there might be original data attached to the samples (*e.g.* paper records) that must be either retained, entered into a computer system, or scanned and retained as an image (verified true copy). It is important to assure that all paper records, including information on sample labels, are retained. When shipping large numbers of samples, a spreadsheet can be used instead of paper to generate the shipping list of samples but ensure that a paper copy accompanies the samples as well as an e-mailed electronic copy.

Best practice sample receipt is one that does not require the receiving analyst to make any entry from the keyboard: rather, samples are identified with scanned barcode labels or RFID tags, and comments about the sample are entered directly into the lab system by the person who receives them and observes their condition. Unfortunately, this robust approach to data is difficult when external parties are involved in sample collection, transport or receipt.

Sample management will also cover when samples are removed from storage for analysis and returned, moved to long term storage and finally disposed of. Paper records are difficult to manage and the best way of ensuring accurate records is using bar code labels in combination with a laboratory informatics application, *e.g.* LIMS or ELN.

In addition to the above errors, consider who will manage them: it is not your senior analyst, who understands the importance of the right sample location and the right container/transport system to preserve sample integrity. Instead, these errors are managed by a contract person or a new junior analyst who has a few months on the job, with a minimal amount of training to collect and deliver samples. Faced with the fact that eight sample labels, attached with rubber bands, are now shuffled inside a box, he/she will attempt to correctly affix the labels, rather than report it and risk disciplinary action. As a result of this hidden knowledge, the lab might have a set of atypical results with no assignable cause known to them.

Once compromised, a new sample can sometimes be collected, while specific point-in-time samples will simply be lost, as the process has moved past that process point. So how can we overcome these potential data integrity issues?

16.4.8 Sample Collection Best Practice

From a data integrity perspective, a sampling process should be designed to minimise incorrect data entries and to assure sample identity throughout the life cycle until sample disposal. It should not permit recording of original data (raw data) on the sample/label, and should prevent the loss of sample information (including identity) from handling/transport.

Best practice sample collection acquires and processes sample data using an electronic system:

- Users access the system to view sample information, typically with mobile devices.
- Sample labels include barcoded (or RFID) number/character strings that can be affixed to containers at any time, because they have no meaning until associated with a specific sample.

This approach allows a wide range of possible label materials and sampling data is collected immediately in electronic form using non-contact readers in a mobile computer or a workstation in a sample receipt booth. Attribution of action is supplied by the user identity of the person logged into the system account, rather than manual entries and each entry is date and time stamped with associated audit trail entries. When sampling locations (and/or other routine sample information) are also barcoded for rapid entry, human data entry is minimised, and the process requires less time.

16.5 Reference Standards and Reagents

Much of the work in laboratories is involved with the preparation of reference standard solutions, buffers, mobile phases, *etc.* and these processes are manual. It provides an opportunity to make mistakes or to falsify solution data that can change the outcome of an analysis: preparation data can be generated after an analysis and simple mistakes are easy to hide. Both actions can impact the integrity of several analyses if it involves a stable analytical reference solution.

16.5.1 What the Regulators Want

The requirements in EU GMP Chapter 6[2] for reagents and reference standards are:

> 6.20. Reference standards should be established as suitable for their intended use. Their qualification and certification as such should be clearly stated and documented. Whenever compendial reference standards from an officially recognised source exist, these should preferably be used as primary reference standards unless fully justified (the use of secondary standards is permitted once their traceability to primary standards has been demonstrated and is documented). These compendial materials should be used for the purpose described in the appropriate monograph unless otherwise authorised by the National Competent Authority.
>
> 6.21. Laboratory reagents, solutions, reference standards and culture media should be marked with the preparation and opening date and the signature of the person who prepared them. The expiry date of reagents and culture media should be indicated on the label, together with specific storage conditions. In addition, for volumetric solutions, the last date of standardisation and the last current factor should be indicated.
>
> 6.22. Where necessary, the date of receipt of any substance used for testing operations (*e.g.* reagents, solutions and reference standards) should be indicated on the container. Instructions for use and storage should be followed. In certain cases it may be necessary to carry out an identification test and/or other testing of reagent materials upon receipt or before use.

US GMP in 21 CFR 211.194(c) has the following requirement[4]:

> Complete records shall be maintained of any testing and standardization of laboratory reference standards, reagents, and standard solutions.

The 1993 FDA guidance on Inspection of Pharmaceutical Quality Control Laboratories[5] provides, in Section 14, a valuable insight on how to inspect the preparation of laboratory standard solutions:

> Ascertain that suitable standards are being used (*i.e.* in-date, stored properly).
>
> Check for the reuse of stock solutions without assuring their stability.
>
> Stock solutions are frequently stored in the laboratory refrigerator. Examine the laboratory refrigerators for these solutions and when found check for appropriate identification.
>
> Review records of standard solution preparation to assure complete and accurate documentation.
>
> It is highly unlikely that a firm can "accurately and consistently weigh" to the same microgram. Therefore, data showing this level of standardization or pattern is suspect and should be carefully investigated.

In summary, all reference standard solutions, buffers and mobile phases must be properly prepared with adequate documentation and be appropriately labelled and stored. Reference standards can have a major impact on both the integrity and quality of the data generated during an analysis.

16.5.2 Preparation of Reference Standards and Solutions

Some of the data integrity issues that can arise with the preparation and storage of reference standards and solutions are:

- Is the analytical balance used to weigh material suitable? For example, do not weigh milligram amounts using a 2 or 3 place balance.
- Ensure that the date and time stamp of the analytical balance has been set correctly and cannot be changed by an unauthorised individual and this will ensure contemporaneous criterion for data integrity.
- Each user must have their own user identity and password to undertake work and meet the requirement for regulated work to be attributable.
- Calibration checks of the analytical balance used to weigh the compounds must use the appropriate traceable masses. It is possible to use the internal calibration function of an analytical balance so that there are fewer external calibration checks. A potential problem with this approach is that if the external check highlights an error with the balance, then all work since the previous calibration is suspect and must be investigated to determine the impact of the balance failure. For this reason, most laboratories use external masses for balance calibration.
- Best practice instructions for weighing reference standards should define a range of weights rather than an exact amount as this would trigger regulatory interest,[5] as noted in Section 16.5.1.

- Recording weights on any balance by observation is an unacceptable practice. All balances must have either a printer attached (adequate) or be linked to a data system such as ELN, LES or LIMS, *etc.* (preferred). Enforcement of the weighing sequence using the technical controls in an informatics application is the best practice and speeds up the second person review, as we shall discuss in Chapter 17.
- If recorded on paper, the whole of the weighing sequence needs to be printed in order so that it is contemporaneous, *e.g.* calibration checks for each mass used, the weight of the vessel, tare weight and sample weight. All should be recorded on a single continuous strip of paper from the balance printer to be above board and not invite any regulatory questions.
- Record of the preparation of the solution must be recorded in a laboratory notebook, solution log or authorised analytical record, as defined in the laboratory procedures including the printout of the balance weighing sequence.
- The work must also be recorded in the balance log book.
- All solutions must be correctly labelled including storage conditions and expiry date.

It is preferable that work such as this, especially analytical reference standard solutions, be automated as much as possible. Analytical standards and even general chemicals can be bar-coded and scanned when used to identify them and provide an automatic check to ensure that the batch to be used has not expired as well as the correct standard, chemical or solution is used. If calculations are required, then these should be performed in an instrument data system interfaced to a balance rather than manually or *via* a spreadsheet; the data system calculation will be validated and will avoid manual entry of data by a user. The principles for this approach were discussed in Chapter 13.

16.5.3 Sweep Under the Carpet or Own Up to a Mistake?

Even with all the automation and computerisation in the world there is still the human factor to consider. Consider the following situation. The analytical balance is qualified and has been calibrated, the reference standard is within expiry, the weight taken is within limits and the vessel is transferred to a volumetric flask. One of three things could happen:

1. The material is transferred to the flask correctly and the solution is made up to volume as required. All is well with the world.
2. During transfer some material is dropped outside the flask but the analyst still dissolves the material and makes the solution up to volume.
3. All material is transferred to the flask correctly but the flask is overfilled past the meniscus.

At this point, only the analyst preparing the reference solution stands between your organisation and a data integrity disaster. The analyst is the only person who knows that options 2 and 3 are wrong. What happens next depends on several factors:

- corporate data integrity policies and training;
- the open culture of the laboratory to allow individuals to admit mistakes;
- the honesty of the individual analyst;
- laboratory metrics, *e.g.* Turn Around Time (TAT) targets that can influence the actions of individuals;
- the attitude of the supervisor and laboratory management to such errors.

STOP! This should be the correct and only action taken by the analyst. Document the mistake and repeat the work. This is the simplest and least painful outcome. Documenting this as a deviation should be done but a CAPA is not required for such a minor issue as the deviation management procedure should have a triage process for classifying deviations.

BUT what actually happens depends on those factors described above. Preparation of a reference standard is one area where the actions of an individual analyst can compromise the integrity of data generated. If the mistake is ignored one of the possible outcomes could be an out of specification result or the release or an under or over strength batch. In the subsequent investigation unless the mistake is mentioned it may not be possible to have an assignable cause. The culture of an organisation can have a major bearing on the outcome of such a mistake. However, before forgetting about mistakes what is the FDA's view on this?

16.5.4 What Is the FDA's View of Analyst Mistakes?

Hidden in the Responsibilities of the Analyst section in the FDA's Guidance for Industry on Investigating OOS Results is the following statement[14]:

> If errors are obvious, such as the spilling of a sample solution or the incomplete transfer of a sample composite, the analyst should immediately document what happened.
>
> Analysts should not knowingly continue an analysis they expect to invalidate at a later time for an assignable cause (*i.e.*, analyses should not be completed for the sole purpose of seeing what results can be obtained when obvious errors are known).

As mentioned in Section 16.5.3, the only ethical option open to an analyst is to stop the work, document it and repeat the appropriate steps. This not only applies to the preparation or reference solutions but any analytical work. You have been informed and warned.

16.6 Sample Preparation

The purpose of sample preparation is to treat a sample so that it is in a form ready for instrumental analysis. This can take the form of a simple weighing of the sample, dissolving in a solute followed by dilution, to more complex forms of sample preparation such as homogenisation, complex extraction, solvent evaporation and reconstitution in a solvent suitable for introduction to the analytical instrument. As this phase of an analytical procedure is typically manual there are many opportunities for poor data management practices as well as falsification of records.

16.6.1 What the Regulators Want

There are no explicit GXP regulations for sample preparation.

However, the subject is implicitly included in regulations for analytical procedures that have been described earlier in this chapter (21 CFR 211.160(b)[4] and EU GMP Chapter 6.15[2]). Remember that the former regulation requires scientific soundness. For sample preparation for bioanalytical methods there are FDA and EMA guidance documents for method validation[15,16] that will implicitly include sample preparation, as discussed in Chapter 15.

16.6.2 Sample Preparation Current Practices

Sample preparation provides many opportunities to compromise the integrity of test results as the process is typically manual that can suffer from lapses in concentration to falsification. For example, analysts can make small changes in execution that can bias results in the desired direction (toward product acceptance):

- adding a little extra analyte;
- slightly under or over-filling a volumetric flask of the reference standard;
- recording a weight for a sample or reference standard but using a different one that is not recorded anywhere.

These are but a few possible risks when preparing samples and/or standards for an assay that could occur – in addition to sample preparation and data recording errors made by honest analysts.

Fortunately, some sample preparation actions can be verified later in the process: for instance, a reviewer can look at a volumetric flask to verify that 1 mL was removed for a dilution or instrument injection. Weights can be printed to a report and retained for later verification. On the other hand, a weighed powder, dissolved in a liquid, cannot be verified with any certainty unless the operation is observed contemporaneously.

To be trustworthy, manual sample preparation needs a second set of eyes viewing it as it is performed; but, this is not required in the current GXP regulations and is also economically unfeasible. Therefore, we are then forced to either[17]:

(1) Rely on the quality culture (and lack of incentives/fear) to trust analysts to act responsibly on the behalf of patients; or
(2) Develop semi-automated or fully automated sample preparation methods, where technically feasible, that remove the human from the sample preparation process. Rather than use robotics, dedicated liquid handling systems are a simpler and easier option to implement but have limitations in that only liquid samples can be handled. Bioanalysis of non-clinical and clinical studies are key areas for exploiting automated sample preparation due to the large numbers of liquid samples for extraction.

The end product of preparation is a sample extract that is ready for instrumental analysis either by manual feeding to the instrument or automatically *via* an autosampler. The key items for consideration are:

- Preserving the identity of all standards, blanks, quality controls and samples throughout the various sample preparation stages to the instrumental analysis is critical. Relying on sample position alone is a risky proposition, as a dropped sample rack or autosampler carousel or tray can destroy all positions, and errors also happen when creating a manual preparation list to correlate sample position with sample identity.
- The recommended solution is barcoding if the label can be attached to the vessel containing the sample extract. Some supplier's instruments and associated data systems can read barcodes as each sample is processed to assure the identity of each sample processed in sequence.
- Like the sample labels discussed earlier, the sample preparation glassware can have printed barcoded labels with unique number and text that can be affixed to each container. Each number can be recorded using a reader for inclusion in the electronic laboratory records of the work for second person review. The tack of the adhesive used is important, strong enough so that the labels will not fall off any glassware but not so strong they cannot be removed during the cleaning process.
- Where laboratory records are paper, small labels printed in duplicate can be affixed to the sample preparation container and the other affixed to a controlled blank form that is the paper record for the analysis. This creates a means to preserve preparation identity while testing, for second person review and, if required, laboratory investigations.

16.6.3 Automate Where Technically Feasible

The human option, trust us our staff are trained, has been the only option for laboratories in the past as automation was limited and expensive. Where used, it was typically standalone and not interfaced with other applications for data sharing, and automation often lacked the basic security and data integrity functions such as individual user accounts and ability to segregate roles. Consequently, beyond high sample volume throughput, there was little else to entice laboratories to move toward automated sample preparation.

Where technically feasible, automated sample preparation removes the human from dilutions and extractions, resulting in greater consistency. Additionally, it provides accurate and trusted timestamps for all actions that permit troubleshooting, should the need appear. Records of the work are generated automatically and automation has the potential to provide better data integrity and productivity benefits compared with a human and manual process.

16.7 Recording Data by Observation

16.7.1 Typical Tests Recording Results by Observation

Some tests, typically colour, odour and appearance, are non-destructive and are recorded by observation. The recording of the result may be a statement that the observed result matches the specification and an analyst writes "complies" or the actual observation in a laboratory notebook or more commonly a controlled analytical form.

In the current regulatory environment, there is a school of thought that such tests require a witness for the observation. Note that there is no regulatory requirement for a witness of a laboratory test in any GXP regulations. As these tests are simple, non-destructive and the sample is available, the second person reviewer can simply repeat the test. This is discussed in more detail in Chapter 18 as part of a second person review of analytical records.

Consider the risk with such a test. What risk is there to the product, patient or regulatory application if a result is falsified? If an analyst records the colour as white when it actually is off-white, what is the risk? Before a knee-jerk reaction to have an army of witnesses looking over the shoulder of analysts performing such tests, answer the question – what is the risk? The answer to this question should enable a sensible decision to be taken as to the approach with such tests.

16.7.2 Instruments with No Printer or Data Transfer Capability

More problematical are analytical instruments that have no recording capability either because they are old or a printer was not considered essential when purchased. Examples of these instruments could be a pH meter, melting point apparatus or a viscometer. As part of a short-term remediation many companies use a witness to verify the instrument reading. This is an expensive approach and not really a solution. Whilst there is no capital outlay, the hidden cost is lost productivity in the laboratory, a fact that is often lost on management.

The simplest solution is to purchase a printer, provided the instrument is not too old, and to have documented evidence of any work performed. However, this may mean that data from the printout may be entered manually in a laboratory informatics application with subsequent transcription error checks. Instead calculate how much additional cost for performing the analysis this way and then purchase a new instrument with a printer and throw the old one away.

A much better approach is to automate as much of the work as possible by interfacing the instrument to an instrument data system, LIMS or an electronic laboratory notebook. This approach will enforce the work using validated technical controls of the application.

16.7.3 Pharmacopoeial Indicator Tests

There are many pharmacopoeial tests for content that use titration with an indicator (*e.g.* litmus) and that the results are used for release. The test will involve two manual observations of a burette: one at the start of the titration and one at the end to calculate the volume of titrant used by difference. These tests are typically manual and only the record of the work is the analyst readings of the burette at the start and end of the experiment.

Here, there is the old argument of trust us our people are trained. However, in the current regulatory environment that trust between the industry and the regulatory agencies is being lost with each new data falsification warning letter or regulatory citation. As a long-term solution to ensure the integrity of data, automation of the process should be undertaken by using auto-titrators with automated data capture and calculation of the reportable result. Here, automation is accompanied by trusted time stamps, individual user accounts, segregation of duties and audit trails of changes and dilutions. Automation would enforce the operations, making it harder to perform the wrong activity or to falsify records.

16.8 Sample Preparation Followed by Instrumental Analysis Methods

16.8.1 An Illustrative Analytical Procedure

Some analytical procedures only involve sample preparation followed by instrumental analysis, this can be typified by a pH measurement or loss on drying analysis. The sample preparation may involve dissolving the sample or just presenting an existing solution to a pH meter. In the case of a loss on drying analysis, two representative sample aliquots need to be weighed and transferred to a drying oven set at 105 °C for 3 h, removed and cooled in a desiccator and then the sample reweighed. The difference between the two weights is calculated for the percentage loss on drying.

16.8.2 Ensuring Data Integrity

To ensure the quality and integrity of the data, the following must be performed:

- A calibration or point of use check must be carried out on the day that the instrument is used. If the instrument is an analytical balance the check can involve one or more traceable and calibrated masses against the acceptance criteria in the laboratory URS, as discussed in Chapter 14. If the instrument is a pH meter, the check will use an appropriate reference solution provided by the instrument supplier in the range used for the analytical procedure. Obviously, records will be generated during these checks, who performed them and entries made in the instrument's log book.
- The work will be performed and the results printed out for attachment in a laboratory notebook or a controlled blank form.
- The instrument log book will be completed with the work performed, by whom and the date.
- Normally time will not be recorded with a paper based process, however loss on drying is where time must be recorded to ensure that the time in the oven is 3 h.
- The calculation of the LOD will usually involve a spreadsheet resulting in transcription error checks.
- The reportable result will be collated in a data system for preparation of the Certificate of Analysis requiring yet more transcription error checks.

16.8.3 Consider Alternate Analytical Approaches

A pharmacopoeial loss on drying analytical procedure is very labour intensive and slow, the actual drying taking three hours. An alternate approach can be to use a moisture analyser (an integrated analytical balance and halogen heat source) to perform the whole analytical procedure including drying in minutes rather than hours. The instrument can also calculate the moisture content of the sample with integral calculations provided that the sample is not heat labile. If the instrument is connected to an instrument data system the whole analytical procedure can be automated, generates all audit trail entries and can have electronic signature capability.

This technique is not in any pharmacopoeia and therefore must be considered an alternate method that must be cross validated with the pharmacopoeia procedure. Although there is an additional cost for a new instrument and software, the time saving when executing the automated drying procedure will more than compensate for the capital outlay and validation costs.

16.9 Methods Involving Instrumental Analysis and Data Interpretation

In this section we will look at analytical procedures where no sample preparation is required – only instrumental analysis followed by interpretation of the acquired data. This is typified by infra-red testing (usually in the near IR region but analytical procedures can also use Raman spectroscopy as well), this discussion will focus on using NIR for identity testing of raw materials and active pharmaceutical ingredients, typically in a warehouse receiving material deliveries.

16.9.1 What the Regulators Want

Although there are no direct regulations concerning the use of NIR for pharmaceutical analysis there is regulatory guidance issued by both the European Medicines Agency (EMA) and the US FDA. The EMA guidance issued in 2012 is entitled Guideline on the use of Near Infrared Spectroscopy (NIRS) by the pharmaceutical industry and the data requirements for new submissions and variations.[18]

> Near Infrared Spectroscopy (NIRS) is a technique, usually requiring tandem chemometric statistics, with a wide and varied use in pharmaceutical, chemical, physical and process analysis. This includes identification, qualification and assay of pharmaceutical starting materials, intermediates and finished products and verification of physicochemical properties.

The 2015 FDA's draft Guidance for Industry on Development and Submission of Near Infrared Analytical Procedures[19] notes that NIR analytical procedures are increasingly being used in the pharmaceutical industry for the identification and assay of pharmaceutical starting materials, intermediates, and finished products. The guidance is focused on the generation and maintenance of one or more spectral libraries (called calibration sets in the guidance) used to identify compounds and materials noting that:

> The spectra that comprise the calibration set are acquired from calibration samples. To create a robust model, the variation built into the calibration samples should include an appropriate concentration range for the component to be analysed and other possible sources of variability (*e.g.*, process, analyser, variation in physical characteristics of materials).

Both US and European regulatory authorities have concerns about how NIR methods and libraries are developed. From a personal perspective, with the deskilling of the pharmaceutical industry there is a critical need to understand how any computerised system works and not just rely on the output.

16.9.2 Near Infra-red (NIR) Identity Testing

The aim of NIR identity testing is a rapid test to confirm the identify of materials at a delivery location to verify that what is stated to be in a container is what it purports to be. The test is based on the fact that a spectrum taken from the sample in each container is compared with a spectral library generated and maintained by the Quality Control laboratory. The identification is supported by a certificate of analysis from a supplier that has been qualified with full in-house testing carried out every 10 to 20 batches delivered.

16.9.3 Building a Spectral Library

A spectral library is typically developed after the initial instrument qualification and software validation discussed in Chapter 14. As the EMA guidance document notes[18]:

> The suitability of a NIR spectroscopic procedure is dependent upon many factors, including the instrumentation and applied chemometrics, as well as the sound understanding of the physicochemical basis of the measurement.
>
> The identification or qualification of a substance (*e.g.* drug substance, excipient, blend, drug product, intermediate) using NIRS is based on the comparison of the spectral data of the substance with the spectral data of several samples of several batches of different substances present in a spectral reference library. It may be necessary to apply chemometrics in order to compare the data and to draw conclusions (pass, no match or ambiguous). The appropriate confidence level of the conclusion should be justified.

When developing a library for identification of materials some of the considerations when building each spectral library for a single spectrometer are:

- What compounds need to be identified using the library?
- How close in chemical structure are they – what level of discrimination between homologues is required?
- Is one library enough or should others be built in parallel?
- What chemometric method will be used for the library, *e.g.* partial least squares,
- For a given compound, how many suppliers are there and how variable is their material between each batch?
- What is the degree of spectral match between the sample and the library?
- If a library is to be shared across instruments then inter-instrument variability needs to be factored in during library development.

The library needs to be maintained and will also be subject to inspection if requested. Further details on developing and verifying a qualitative spectral library can be found in the two regulatory guidance documents discussed in Section 16.9.1.[18,19]

16.9.4 Performing the Analysis

When identity testing is performed, the work can be performed by either QC analysts or, more likely, by trained warehouse staff or sometimes by a combination of the two (the QC analyst sets up and checks that the instrument is functioning correctly and warehouse staff perform the actual identity testing). It is essential that all instrument checks performed are documented in the instrument log book and all the files generated are saved. For all files generated there needs to be a file naming convention to ensure that all records are uniquely identified and can be found quickly in case of an inspection. This requires a procedure for naming each file generated so that it can be retrieved: naming may be based on the material code, compound name but also needs to include perhaps the date, lot number and even container number.

There should be the following information available on the paper or electronic report:

- the spectrum of the sample including filename, date and time of acquisition and the name of the tester;
- the composite spectrum from the library;
- the library name and version;
- the closeness of match between the sample and the library with the confidence interval.

If more than one individual is involved in identity testing at the same time, it is important that one logs out of the spectrometer software and another one logs in, otherwise user identities have been shared and one of the ALCOA principles has been broken. If the spectrometer is used as a hybrid system, the handwritten signature on the paper must be linked with the spectrum obtained from the container tested.

As most NIR spectrometers are designed as standalone instruments, a laboratory needs to determine how to back up the records to ensure that they are available in case of inspection, audit or review.

16.10 Chromatographic Analysis and CDS Data Interpretation

Here, we will focus our discussion on the analysis and interpretation of chromatographic data using a validated network CDS with electronic signatures implemented. The review of sample preparation earlier in Section

16.6 is sufficient for chromatographic analysis and thus the topic will not be included here.

16.10.1 What the Regulators Want

There are no explicit regulatory requirements for chromatographic analysis. However, as chromatography data systems have been involved in a multitude of data falsification situations and poor data management practices[6] the key requirements of chromatographic data are:

- Complete data is captured as mandated by 21 CFR 211.194(a).[4]
- Relevant tests are actually carried out as required by EU GMP Chapter 1.9.[1]

16.10.2 Setting Up the Chromatograph and Acquiring Data

Before running the injection sequence, a chromatograph needs to be set up for the run and the CDS needs to specify the instrument control and data acquisition parameters to be used. In addition, sample and standard data need to be input into the sequence file such as sample identities, weights of samples used, dilutions, standard purities, *etc.* The overall sequence of events is shown in Figure 16.4 and detailed in Table 16.2.

16.10.3 Entering Factors, Weights, and Other Assay Values into the Sequence File

Chromatographic assays usually require the input of additional data values that are combined with the chromatographic value to generate reportable results that are compared against specifications. For example, the estimated potency, based on the chromatography injection, is combined with the sample weight and moisture results to determine the anhydrous potency, which is the reportable result that determines the fitness of the product for release (specification test). Some biologics may have potency or conversion factors that are applied. Whatever the specifics, chromatography methods require the input of other (external) values to generate results that will be compared to specifications or generate final results.

These additional values are critical, directly impact the accuracy of the calculated result and if entered manually they must be checked by a second person, as required by EU GMP Annex 11 Clause 6.[20] The rationale is that any input error is likely to create a corresponding error in the reportable result or standard curve that will influence the reportable value. Therefore, these figures must be correct. In many instances, these will be entered manually from paper records generated during sample preparation or sometimes from other assays (*e.g.* water content from a Karl Fischer titration). However, the use of a second person to check and review of such data is tedious and error prone.

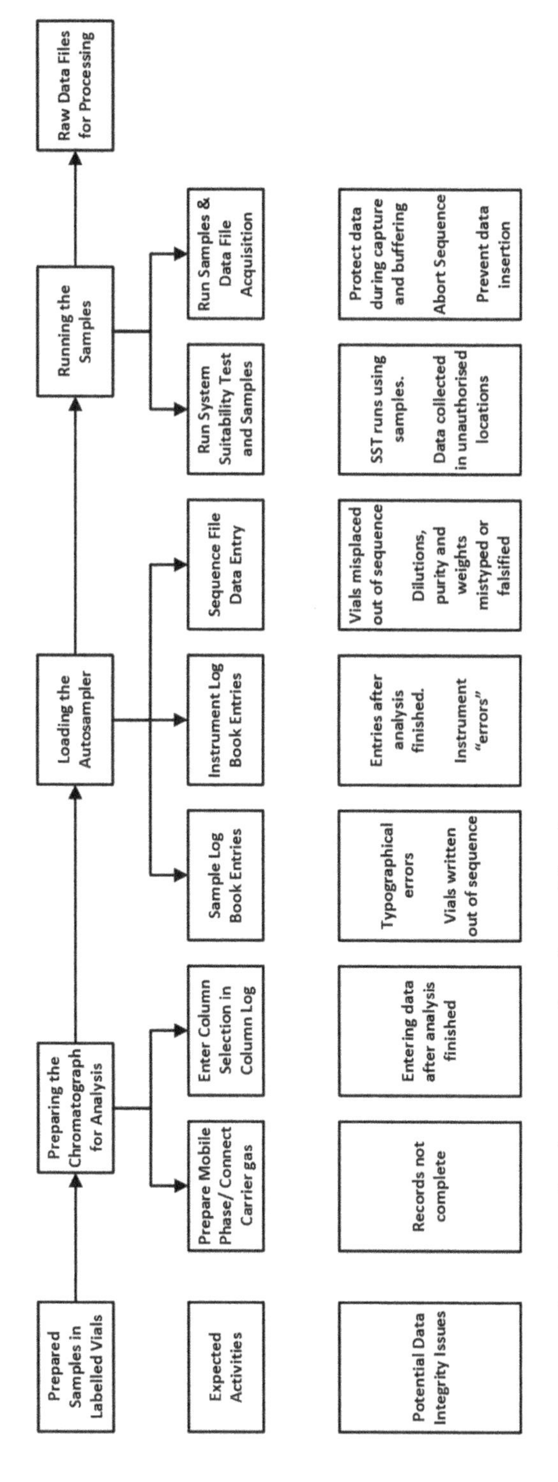

Figure 16.4 Setting up the chromatograph and acquiring data.

Table 16.2 Activities for setting up a chromatograph and acquiring data with their potential impact on data integrity.

Activity	Considerations for data integrity
Preparing the instrument for analysis	• Correct calibration model used • Injection sequences and their management • Sample log book • Preparing mobile phase/carrier • Correct column selection and management; column logbook
Loading the autosampler	• Wrong sample IDs • Dropped vials • Vials swapped positions • Correlate injection vials to sample identity • Incorrect injection sequence • Preserving injection sequence and solutions after assay for potential laboratory investigations
Temperature/humidity/light control	• Potential impact on the integrity of a sample • Data retention/review of these controls as part of the data set
Acquiring the data	• Permitted changes to acquisition and instrument control methods • Data storage location defined and alternatives limited, if possible • Naming conventions used for data generated • Aborted runs or short run sequences • Backdoor access *via* the operating system to laboratory data servers to modify data undetected by removing the network cable

Instead, consider acquiring them electronically into an instrument data system, ELN or LIMS and then transferring the results electronically to the CDS. When validated, this process can be undertaken automatically and the amount of second person review time is greatly reduced as technical controls enforce the sequence of activities and transcription checks are eliminated.

16.10.4 An Alternate Approach to Weights and Factors

An alternate approach to inputting weights, *etc.* into a CDS is available when the laboratory uses an ELN or LIMS. Rather than entering or transferring dilutions, factors or weights these additional data values into the chromatography system for calculations, do the opposite. Transfer the basic chromatographic data times, peak identities and area counts into the ELN, LES or LIMS, and perform the post-assay calculations there. This approach is common in some bioanalytical LIMS to overcome the problem of having different supplier's CDS applications by keeping the quantification algorithms constant.

Which approach is better? It probably depends on the chromatography system and the specific informatics application to be used. In general, ensuring data integrity would lead you toward calculations in a location where there is the more robust audit trail, where changes in values are most easily detected, and where re-calculations are more restrictive and controlled. For example, some ELN systems do not, or can be poorly configured not to, provide audit trail records on every action. In contrast, many LES and LIMS create audit trails for every action where any value is overwritten. Chromatography systems vary in their audit trails around calculation/factor changes. The final decision should provide the most robust, secured, and detectible environment for entry and management of external factors, weights, and assay results and will depend on the critical and in-depth assessment of the compliance features of the applications involved.

Electronic transfer of data values (weights, factors, data values) into an informatics solution will provide superior data integrity as it eliminates human data entry errors, and generates audit trail records in both the host and destination systems. However, this superiority is lost if the path from host to the destination is not secured against external manipulation due to poor security design, technical limitations, or poor administration. Finally, the second person reviewer must be trained to verify calculation values against source records and must not assume that the values are correct because the interface has been validated, this is discussed in more detail in Chapter 17.

16.10.5 System Evaluation Injections

When a chromatograph has been set up with the correct column and mobile phase, the analyst does not always know if the system has equilibrated and if the separation is adequate to meet SST requirements. There is a case for using system evaluation injections to determine if the separation is acceptable.

BUT.

Major data falsification problems occur when actual samples are used as SST injections to see if the analysis passes: this is testing into compliance. System evaluation injections can be used provided the process is handled correctly and all injections used to assess if the system is ready can be seen and are included under *complete data* as required by 21 CFR 211.194(a).[4]

The following approach is suggested:

- The use of system evaluation injections must be included in laboratory procedures.
- The solution used to evaluate the chromatograph MUST be prepared from a reference standard with traceable evidence to show that does not originate from the sample material under analysis.
- The maximum number of injections allowed under the procedure must be defined along with the acceptance criteria, that are best linked to the SST criteria of an individual analytical procedure.
- Chromatographic systems failing to meet acceptance criteria must be documented and the cause determined, documented and rectified.

The procedure for system evaluation injections must answer these questions:

- When are extra (unplanned) injections acceptable?
- If they fail to meet expectations, will they result in repeats and/or investigations?
- How many repeats are acceptable?
- When do these injections stop?
- How will the unplanned injections be reviewed to determine scientific validity?

In a regulated environment, where analytical instruments are qualified and software and analytical methods are validated and users are trained, EVERY injection must be part of the GMP record. Once the system is ready then the injection sequence can begin starting with the SST injections.

16.10.6 System Suitability Tests – What the Regulators Want

The regulatory expectation under USP <621>[7] and EP 2.2.46[8] is for system suitability samples to be run to demonstrate that the chromatographic system is fit for the analytical procedure. In the FDA Guidance for Industry on Out of Specification Results[14] there is the following requirement for system suitability results:

> Certain analytical methods have system suitability requirements, and systems not meeting these requirements should not be used.
>
> For example, in chromatographic systems, reference standard solutions may be injected at intervals throughout chromatographic runs to measure drift, noise, and repeatability. If reference standard responses indicate that the system is not functioning properly, all of the data collected during the suspect time period should be properly identified and should not be used.
>
> The cause of the malfunction should be identified and, if possible, corrected before a decision is made whether to use any data prior to the suspect period.

The message is clear that if an instrument fails a point of use check or system suitability test it must not be used, as required by 21 CFR 211.160(b).[4] Failure of SST injections is justification for invalidation of all subsequent standard and sample injections in the test run; therefore, data reviewers must look cautiously at the suitability process, especially when the test run is to be declared invalid along with test results, such as the situation where suitability, standard, and samples were placed in an autosampler and assayed overnight without human intervention.

The potential for system suitability abuse begs a question: will failures result in an investigation and corrective measure(s)? The answer to this question should be included in your procedures for chromatography and laboratory working.

To avoid generating data that will not be used, many CDS applications have a stop on fault function that allows a laboratory to define the SST acceptance criteria. After the SST injections, the system will automatically integrate the SST chromatograms and calculate the SST parameters, if one or more parameters fail acceptance criteria, the chromatograph will stop. This avoids wasting time and allows troubleshooting of the problem to begin earlier than otherwise would be the case if the function were not used.

16.10.7 Integrating Chromatograms

Integration is a key area where data can be manipulated to get results that a laboratory wants. Thus, this process must be carefully controlled. During method development the integration parameters must be defined and verified during the validation. When used for routine analysis the order of chromatographic integration is shown in Figure 16.5.

Control of chromatographic integration is a key regulatory requirement, as discussed in a book on the Validation of Chromatography Data Systems.[6] Lack of integration control has resulted in several FDA warning letters and 483 observations such as:

At Leiner Health Products, FDA investigators found[21]:

> In addition, our investigators documented many instances with extensive manipulation of data with no explanation regarding why the manipulation was conducted. This manipulation would include changing integration parameters or relabelling peaks such that previously resolved peaks would not be integrated and included in the calculation for impurities...

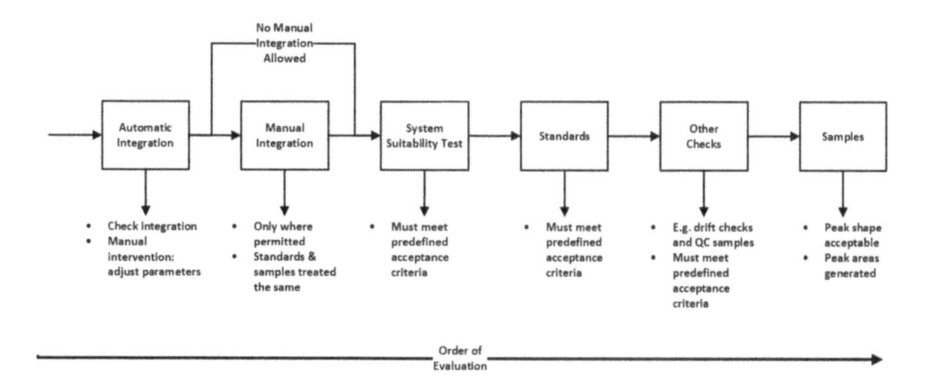

Figure 16.5 Scope and order of chromatographic integration.

> There were no Standard Operation Procedures (SOP) to describe the policy, standard practice, and circumstances under which manual integration would be allowed. .....
> There was no documentation of a justification for the manual integration.[21]

This is not an isolated case: Divi's Laboratories received a Warning Letter in 2017[22]:

> Your firm reintegrated multiple chromatograms to determine <redacted> levels; however, the parameters for the reintegration were not retained.

The problem is that regulatory agencies want to see control of integration, specifically manual integration. However, we also need to acknowledge that chromatographic systems are dynamic by their nature and separations can change during a run, so getting the right overall integration constantly throughout a run can be a balancing act. This is especially so with long overall run times comprising many samples and injections, complex separations or biomolecules where the peak shape is broad.

16.10.8 General Principles for Ensuring Good Chromatographic Integration

Before going into detail for an SOP for chromatographic integration, we need to understand the general principles for good integration:

- Know how key parameters such as peak width and threshold impact the integration of a chromatogram.[6,23] Never rely on the CDS software to do the job, understand what is happening when your chromatograms are being integrated.
- Never use a default processing method. Always develop the integration specifically for an analytical procedure.
- Having a robust and reliable analytical procedure as the sole purpose of a CDS is not to compensate for your abysmal chromatography. Understand that good integration requires good chromatography and that process begins with the development of the method, as discussed in Chapter 15.
- All integration in the first instance *MUST* be performed automatically.
- Chromatography is a comparative technique; therefore, standards and samples must be treated the same throughout an injection sequence or run.
- Given the regulatory concerns, only perform manual integrations under conditions permitted in your laboratory's chromatography SOP or analytical procedures.

The SOP for chromatographic integration will cover all the above points.

16.10.9 SOP for Integration of Chromatograms

Having established the fact that a chromatography procedure is a regulatory expectation, what should be the content of such a procedure, Newton and McDowall[24] provided a list of topics and behavioural expectations shown in Table 16.3. Depending upon your organisational approach, some topics might be moved into the data integrity training material, rather than the chromatography procedure; nevertheless, the topic should be addressed to assure consistency in practice, and avoid the appearance of testing into compliance.

This chapter uses the previously discussed terms manual intervention and manual integration to describe non-automated adjustments made to chromatograms by personnel.[6]

- Manual intervention, changing integration processing parameters but not manually repositioning baselines is an acceptable practice for all chromatograms. This can occur in any regulated analysis as the positioning of baselines is determined by the CDS and not a human operator.
- Manual integration, manually repositioning baselines and this process must be carefully defined and controlled.

Above all, the type of integration, especially manual repositioning of baselines must be scientifically justified. This will depend on several factors such as:

Table 16.3 Outline contents of an SOP on chromatographic integration.

SOP topic	Procedure content
1. Definitions	• Definitions of key words and phrases
2. Expected Behaviours	• To set the foundation of the procedure, a set of fundamental expectations must be introduced and be part of training
3. File Naming Convention	• A file and data naming convention should describe the proper label for every type of material that will be injected into a chromatography system and handled by the CDS
4. Complete Record	• The means of ensuring that all users know what is required to ensure complete data for every run
5. Handling Aborted Runs	• What is the process for managing aborted runs? • Why must all data be retained?
6. Extra Injections	• Injections not defined in any procedure – how can these be documented and justified? • Why must the files be retained?
7. Manual Intervention and Integration	• Control of integration to assure that the system is not being used to integrate into compliance

- Where in the research, development and manufacturing continuum the method is being used and if it is fully validated. For example, there is more justification for manual integration during early development before the manufacturing process and the analytical method are validated as compared with routine analysis for product release in a QC laboratory.
- The type of analysis being performed such as small heterocyclic molecules *versus* fermentation or cell culture media containing macromolecules, the latter having broader peak shape compared with the former compounds.
- Active pharmaceutical ingredient *versus* impurity profile analysis close to the limits of quantification and detection (LOQ and LOD, respectively). Manual integration during impurity profile testing can be justified scientifically as opposed to API testing (unless dealing with macromolecules).

Regardless of the analysis, any manual intervention or manual integration must be scientifically justified.

Table 16.3 outlines the contents of an SOP on integration and the individual sections[24] are detailed below and includes the following items:

1. Definitions: automatic integration, manual intervention, manual integration, analysis, method, test, sample, inhibit, peak masking, processing method – are among many terms that must be defined for clarity in interpretation among analytical staff. This need becomes more critical as the number of users and laboratory sites are increased, *e.g.* a global procedure.
2. To set the foundation of the procedure, a set of fundamental behaviours and expectations must be defined. Without these, the balance of the procedure is not achievable. These include:
   - Auto-integration of all injection peaks is the expectation. Automatic integration reduces the data integrity risk to the organisation, increases consistency in results and significantly reduces the time to review chromatograms; therefore, manual interventions or integrations are accepted only after no algorithm can be developed to process chromatograms from this material. Automatic integration must be applied to any chromatogram regardless of where the analysis is carried out (*e.g.* research, development or manufacturing) or status of the method (*e.g.* in development or validated).
   - Disabling any audit trail or adjusting a computer system clock connected with the chromatography system is unacceptable.
   - Performing any sequence of steps to avoid leaving an audit trail entry is unacceptable.
   - Audit trail reasons entered by users must provide sufficient detail to permit an inspector to reconstruct the sequence and actions performed by laboratory personnel. If the CDS application has the ability to pre-define audit trail reasons for change that are context sensitive

this is a preferable approach rather than just a free text entry, as the reasons will vary for the same situation.

- Post-injection data processing will follow a defined sequence, as shown in Figure 16.5: system suitability and criteria acceptance; reference standards and criteria acceptance; drift checks/other checks and criteria acceptance; samples and other solutions. A prescribed order reduces the temptation to reject a chromatograph run due to undesirable sample results. As mentioned earlier, these expectations must become part of training.

3. A file and data naming convention should describe the proper label for every type of material that will be injected into a chromatography system: suitability, standards, blanks, samples, control samples, drift checks, *etc.* There should be no place for analysts to use informal terms like "Test", "Wash", "Injection" or other such terms found in laboratories by inspectors. In a regulated environment every injection has a defined purpose; therefore, every injection must have a standardised name. The naming convention must enable the users, reviewers and inspectors to quickly understand its purpose. Conformance to a naming convention also makes a review of data move more quickly – once a convention is in use, non-standard names are easily seen in a list of sample identities.
4. Steps to assure a complete record for Review and Release: to avoid testing into compliance, the procedure must require personnel to include all injections made while testing, whether used to calculate a reportable result or not. Documentation must provide a discussion of data included—and excluded—and the rationale for excluded values. This includes aborted runs, runs that fail to meet suitability or method acceptance criteria, excluded replicate values or re-injections of any solution (standard, sample, control, *etc.*). In addition, a complete record requires behaviours that create transparency of actions: for example, injections should be labelled as standards, samples, wash, *etc.* prior to initiating the run. Peak names should be included in the processing method. These actions create an initial record, so changes made post-injection will create an audit trail entry. Labelling injections and peaks post-analysis provide an opportunity for analysts to mislead reviewers by labelling undesired injections or peaks to hide their true content and intent.
5. The process to manage failed (or aborted) runs: personnel need to have a clear understanding of the conditions that permit a set of injections to be stopped or excluded from further use. This includes documentation within the system, or in other official documentation that provides justification for the exclusion. There must be a means to list the excluded injections in the final review package, as the excluded data must be reviewed for scientific validity prior to releasing the reportable result(s).
6. The process to manage "extra" injections: in this context, "extra" injections (as defined in your procedure) are any injections of material for chromatographic analysis that are not explicitly addressed in

the written method. A classic example is an injection of a reference standard solution (often the middle standard in a series) to assess the response of the system prior to initiation of system suitability injections. The suitability injections are specified in the analytical method – but the injection prior to suitability is not authorised in the analytical method or any other reference document (*e.g.* USP monograph).

7. The process to manage manual adjustments to chromatographs or calculations: all manual adjustments from integration parameters to manual baselines, renaming of peaks, processing calculations and external factors included in reported results, must be transparent to the reviewer and included as part of test records. In addition, the procedure should describe a risk-based review of manual adjustments: for example, all manual adjustments may require review by a senior scientist with extensive assay experience – someone who is able to assess the scientific merit of the integrations and modified calculations.

16.10.10 Bioanalytical Guidance for Integration of Chromatograms

Bioanalytical method validation guidances from the EMA[15] and the FDA[25] require control and justification for manual integration. EMA requires that any manual integration is scientifically justified but the FDA are more stringent, stating that managerial approval is required before manual integration is performed.[25] However, in the revised 2018 Bioanalytical Method Validation guidance, the FDA has stepped back from requiring prior managerial approval but have further requirements for chromatographic integration.[16]

The revised guidance defines reintegration as *a reanalysis of the chromatographic peak*.[16] However, this definition does not differentiate between manual intervention and manual integration as discussed above in Section 16.10.8 and in Validation of Chromatography Data Systems.[6]

There is a further requirement for chromatographic integration for both method validation as well as routine analysis of samples from bioavailability and bioequivalence studies[16]:

- An SOP or guideline for sample data reintegration for chromatographic assays should be established *a priori*.
- This SOP or guideline should define the criteria for re-integration and how the re-integration will be performed.
- The rationale for the re-integration should be clearly described and documented.
- Audit trails should be maintained.
- Original and re-integrated data should be documented and reported.

The original text is a single bullet point but additional bullet points have been added here to highlight each of the requirements from the Agency. As

Table 16.4 Integration and record requirements for bioanalytical reports.[16]

Records and data	Documentation of the analysis
Analytical Runs	• 100% of run summary sheets of passed and failed runs, including: Calibration curve, regression and weighting function used, analyte and internal standard response, response ratio, integration type • 100% e-chromatograms of original and re-integrations from passed and failed runs
Chromatograms and Reintegration	• Electronic audit trail: original and re-integration • Reason for re-integration • Mode of re-integration
Validation Report	• Representative chromatograms (original and re-integration) • Reason for re-integration
Analytical Study Report	• Chromatograms from 20% of serially selected subjects for bioequivalence studies in ANDAs • Randomly selected chromatograms from 5% of studies submitted in NDAs and BLAs • Original and re-integrated chromatograms and initial and repeat integration results for bioequivalence studies • Reason for re-integration • SOP for re-integration

the definition of reintegration in the guidance document is poor and fails to distinguish between manual intervention and manual integration it places any change in integration parameter under reintegration and places a high compliance burden on bioanalytical laboratories.

There are further integration and associated analytical record requirements for both method validation and analytical study reports, as listed in Table 16.4. Many of these requirements arose because of issues of falsification of data in the laboratories found in two contract research organisations, *e.g.* Cetero[26] and Semler.[27]

16.10.11 Incomplete (Aborted) Runs

While collecting data, it may be necessary to abort a run, due to any number of issues that impact the test run, *e.g.* pump head leaking. However, there is also a dark side: like system suitability aborted runs can be used as a means of stopping the chromatography system from generating undesired (out of specification) test results. Rather than managing the test result, the run is simply aborted, preventing the creation of a result value.

For this reason, it is necessary to manage aborted runs as unplanned events, and review the system for the number of aborted runs, along with the group, person, method, instrument, and abort reason comment associated with the aborted run. The review should look for trends in aborted runs. In addition, it is important to include aborted runs with the subsequent test record, so the aborted data is reviewed along with the acceptable data prior to test result release. The management of aborted runs should be included in your Chromatography SOP.

16.10.12 Other Unplanned Injections

There are circumstances where it may be justified to perform an additional injection of a solution, such as a standard or suitability solutions. As for other solutions above, it is important to specify the circumstances that warrant these injections, the number of injections (re-injections) that is acceptable, how the results will be used, and assuring that the injection results are reviewed with all associated test records, to assure retention of the complete testing record.

16.10.13 Data Storage Locations

Once the assay has met suitability criteria and injections have been collected, the raw injection results must be stored for processing and assay calculations. One issue that has found its way into some laboratories is a diversion of test data into "alternate" directories. This is more difficult to carry out with relational database systems, but can be readily done with file-based chromatography systems. For example, test runs with failing injections are placed into a folder named "Test" to make them appear as non-production test results. This type of behaviour can be detected by creating a report to inventory test runs referenced in reported test results, looking for missing runs in the sequence (each run is given a unique ID by the system). A second detection report can look for the presence of folders or injections with suspicious names like "Test", "Practice", or other terms that merit review.

To assure data integrity, laboratory personnel must be trained to submit every injection for data review. When data is rejected for use, a reason for rejection must be documented to defend the data exclusion from the reported values.

16.10.14 Chromatography Falsification Practices 1: Peak Shaving and Enhancing

Chromatography is a comparative technique, comparing the response of known standards with unknown samples. It is imperative that standards and samples be processed consistently; otherwise there will not be a linear relationship between absorption and concentration (Beer–Lambert Law). Consistency in peak processing is a strength in automated integration, and one (of many) reasons for adopting it.

When integrating chromatograms, it is important that chromatographers are aware of bad integration practices used to falsify data such as:

- Skimming is a technique where the baselines are manually repositioned to reduce the peak area (see line 2 in Figure 16.6, where line

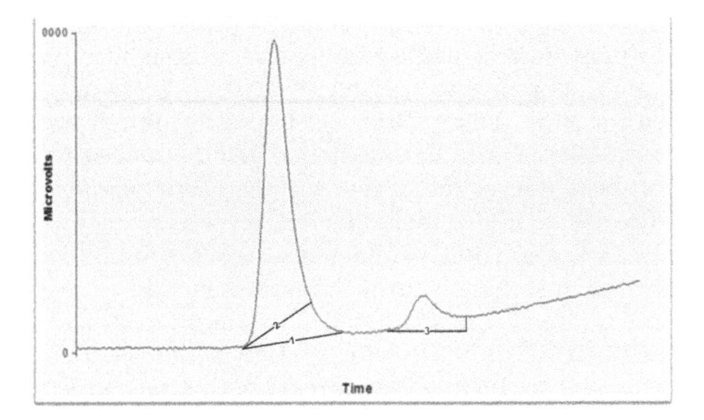

Figure 16.6 Peak skimming and enhancement. Reproduced from ref. 6 with the permission from the Royal Society of Chemistry.

1 is automatic peak placement). If performed to the standards it can increase the amount of analyte in the standards and *vice versa*.

- Enhancing is the opposite where the peak area is increased by either the standards or the samples to obtain the desired result (see line 3 in Figure 16.6).

These two practices must be eliminated in all regulated laboratories.

Keeping these out of integrations can be a challenge for biologics that have large numbers of peaks, often broad, owing to their natural origins. The first step in controlling these products is admitting their technical challenges, and therefore their risks, and devoting sufficient resources to method development to automate them. It is a tall order to expect consistency from an analyst who will spend a considerable time manually integrating peaks from a single assay, especially when proper assay development could reduce this to as little as one hour of labour. Dyson[23] recommended that method-related issues be resolved by additional method development rather than chromatographic processing.

16.10.15 Chromatography Falsification Practices 2: Inhibiting Integration

It is difficult to imagine that anyone involved in the manufacture of pharmaceuticals would consider inhibition of peak areas a good idea, but a Warning Letter issued to Divi's Laboratories in 2017[22] lists this as a practice in the firm.

1. Failure to ensure that test procedures are scientifically sound and appropriate to ensure that your API conform to established standards of quality and/or purity.
   Our investigators observed that the software you use to conduct high performance liquid chromatography (HPLC) analyses of API

> for unknown impurities is configured to permit extensive use of the "inhibit integration" function without scientific justification.
> For example, our investigator reviewed the integration parameters you used for HPLC identification of impurities in release testing for <redacted>. These parameters demonstrated that your software was set to inhibit peak integration at four different time periods throughout the analysis. Similarly, in the impurities release testing you performed for <redacted>, your HPLC parameters were set to inhibit integration at four different time periods throughout the analysis.

You have been warned: inhibiting integration is hiding data from review, just as surely as placing printed batch records in a waste can. There can be circumstances where inhibiting is reasonable, such as solvent fronts or early baseline instability; these situations will be documented in the validated analytical method. Because of the data integrity risks associated with inhibiting integration, its use should be restricted to validated circumstances within your chromatography SOP (above) and training materials.

16.10.16 Chromatography Falsification Practices 3: Integrating Samples First

One foundation of a robust chromatographic SOP requires that samples be integrated and reviewed last, as shown in Figure 16.5. When analysts are permitted to review sample results first, there is a temptation to find a way to reject (or abandon) the entire run because it is now obvious that an undesired test result will be the outcome. To assure scientific validity, system suitability should be the first set of injections to be processed and compared against acceptability criteria. A system suitability failure invalidates all subsequent injections,[14] as discussed in Section 16.10.6. Assuming the suitability criteria are met, the reference standard injections should be processed and compared against assay acceptability criteria. Only if all criteria are met should other injections be processed, then samples. Processing samples must ensure that good scientific judgement is driving the process rather than testing samples into compliance.

Once the integration is complete and acceptable, then the calculation of the reportable results can begin.

16.11 Calculation of Reportable Results

Calculation of reportable result may entail the use of calibration curves, comparison of sample results with standards, averaging of individual results from a specified number of replicate determinations. For an analytical procedure using observation, *e.g.* colour, the observation is the reportable result, for NIR tests the identity confirmation is the reportable result and these tests will not be discussed in this section. We will focus our attention on the calculation of results from chromatographic analyses.

To be specific, the term reportable result has a specific meaning that originates with the FDA OOS guidance[14]:

> The term reportable result as used in this document means a final analytical result. This result is appropriately defined in the written approved test method and derived from one full execution of that method, starting from the original sample.

16.11.1 What the Regulators Want

There are many regulations and regulatory guidance for calculation of reportable results that are presented below:

The current good manufacturing practice for finished pharmaceutical products in 21 CFR 211.68(b)[4] requires that calculations be checked for accuracy (implying that the calculations should be specified first):

> (b) ... Input to and output from the computer or related system of formulas or other records or data shall be checked for accuracy. The degree and frequency of input/output verification shall be based on the complexity and reliability of the computer or related system. ......

Furthermore, in the laboratory control section of the same regulation there is the following requirement in 211.194(a)[4] for any calculations used in the laboratory:

> (5) A record of all calculations performed in connection with the test, including units of measure, conversion factors, and equivalency factors.

There is a similar requirement in EU GMP Chapter 6[2]:

> 6.17. The tests performed should be recorded and the records should include at least the following data: ......
> iv. Test results, including observations and calculations, and reference to any certificates of analysis;

Regulations also impact active pharmaceutical ingredient manufacturing in EU GMP Part 2/ICH Q7[10]:

> 6.60: A record of all calculations performed in connection with the test, including, for example, units of measure, conversion factors, and equivalency factors;

FDA OOS guidance, Section B[14]

3. Verify that the calculations used to convert raw data values into a final test result are scientifically sound, appropriate, and correct; also determine if unauthorized or unvalidated changes have been made to automated calculation methods.

The OECD GLP regulations have requirements for documenting calculations in Section 6 on results[28]:

c) A presentation of the results, including calculations and determinations of statistical significance;

In addition, the OECD Series No 17 on the Application of GLP Principles to Computerised Systems[29] states:

4. .... All computerised systems used for the generation, measurement, calculation, assessment, transfer, processing, storage or archiving of data intended for regulatory submission or to support regulatory decisions should be validated, and operated and maintained in ways that are compliant with the GLP Principles.

Following the interpretation of chromatograms, all regulations require that any calculations performed must be scientifically sound, defined adequately and verified to ensure that they are correct, this includes computerised systems, including spreadsheets.

16.11.2 General Considerations for Calculations

It is essential that any calculations performed have the following characteristics to ensure that they are reliable, robust and compliant[30]:

1. Calculations must accurately reflect the true state of the material under test.
2. The calculations and any changes, either prior to or during result calculations, must be documented and available for review.
3. Calculated values should not be seen by the analyst until the value has been recorded in permanent medium – this includes an audit trail of the event.
4. After initial calculation, changes made to factors, dilutions and weights used in the calculation must be preserved in an audit trail.
5. All factors and values used in the calculation must be traceable from the original observations or other source data to the reportable result.

No matter where calculations happen, it must be possible to see the original data, the original calculation procedure (method), and the outcome. In addition, there must be sufficient transparency so that changes to factors, values or the calculation procedure are captured for review. A bonus would be a system that flags changes made to any of the above after initial use of the calculation procedure—this flag tells the review that audit trails should be reviewed to assess the scientific merit of the change(s).

16.11.3 Avoid Using Spreadsheets for Analytical Calculations Whenever Possible

Having mentioned spreadsheets earlier, this section is devoted to their elimination. This elimination is not because spreadsheet applications are poor, far from it, but the main problem is that a spreadsheet is a hybrid system. As the WHO guidance recommends, hybrid systems are not encouraged and should be replaced.[31] Additionally, spreadsheets are typically used with manual data input from paper printouts from analytical instruments that requires much time checking that the entries are correct following initial input and during the second person review. This slows the whole analytical process and delays the release of the product.

Instead, as discussed in Chapter 12, the calculations performed by spreadsheets should be included either in the CDS or peak areas information transferred automatically and electronically to another informatics solution, as presented in Section 16.10.4. If the calculation of the reportable result is carried out in the CDS using validated calculations and eliminating manual entry post data acquisition the overall analytical process is speeded up and data integrity is ensured.

16.11.4 Calculation of Reportable Results and Out of Specification Results

The way that reportable results are to be calculated must be defined in the applicable analytical procedure and this will be implemented in the laboratory informatics application where this will take place. The equation and how it is implemented needs to be documented and the calculation verified with data sets design both to pass and to fail. Where appropriate, truncation and rounding should only be performed on the final calculated results following the laboratory SOP on rounding.

Depending on the type of analysis performed, the reportable results may be included in a study report, *e.g.* non-clinical or clinical bioanalytical reports or compared to see that a batch of material complies with a specification. In the case of an out of specification result (OOS), this will trigger an investigation as outlined in the FDA Guidance for Industry on the subject.[14] As this subject involves the analyst's supervisor the topic is discussed in Chapter 17 on Second Person Review.

Out of specification results are one of the key areas that regulators are focusing on during inspections and one question often asked is show me your OOS investigations over the past 6 or 12 months to assess the quality of laboratory investigations. A laboratory should know their OOS rate across all tests performed. However, there can be problems, as seen from a 483 citation of a company in 2014[32]:

> Since beginning manufacturing operations in 2003, your firm has initiated a total of 0 out of specification investigations for finished products.

Out of specification results do occur and the regulatory authorities know this, if there are no OOS results or laboratory investigations only conclude that the cause was an analyst error that can be resolved by additional training, then expect an in-depth assessment of the analytical systems, electronic data and associated records.

16.11.5 Completion of Testing

The analyst responsible for the testing phase of the work must ensure that the procedure is executed correctly and that all records and data required have been acquired and interpreted correctly. Where data have been entered into a computerised system, the entries are correct and can be traced to the original record. In addition, any preparations, extracts, vials or solutions from the work are retained in case of an out of specification result.

Now, the analytical records are to be reviewed by a second person to check if they are complete and accurate and procedures have been followed, as we shall see in Chapter 17.

Acknowledgement

I would like to thank Mark Newton for input and collaboration for this chapter.

References

1. *EudraLex – Volume 4 Good Manufacturing Practice (GMP) Guidelines, Chapter 1 Pharmaceutical Quality System*, European Commission, Brussels, 2013.
2. *EudraLex – Volume 4 Good Manufacturing Practice (GMP) Guidelines, Chapter 6 Quality Control*, European Commission, Brussels, 2014.
3. *EudraLex – Volume 4 Good Manufacturing Practice (GMP) Guidelines, Annex 8 Sampling of Starting and Packaging Materials*, European Commission, Brussels, 2009.
4. *21 CFR 211 Current Good Manufacturing Practice for Finished Pharmaceutical Products*, Food and Drug Administration, Sliver Spring, MD, 2008.
5. *Inspection of Pharmaceutical Quality Control Laboratories*, Food and Drug Administration, Rockville, MD, 1993.
6. R. D. McDowall, *Validation of Chromatography Data Systems: Ensuring Data Integrity, Meeting Business and Regulatory Requirements Second Edition*, Royal Society of Chemistry, Cambridge, 2017.

7. *USP General Chapter <621> Chromatography*, United States Pharmacopoeia Commission Inc, Rockville, MD.
8. *EP 2.2.46 Chromatographic Separation Techniques*, European Pharmacopoeia, Strasbourg.
9. *ICH Q7-Basic Requirements for Active Substances Used as Starting Materials*, International Conference on Harmonisation, Geneva, 2000.
10. *EudraLex – Volume 4 Good Manufacturing Practice (GMP) Guidelines, Part 2-Basic Requirements for Active Substances Used as Starting Materials*, European Commission, Brussels, 2014.
11. *EudraLex – Volume 4 Good Manufacturing Practice (GMP) Guidelines, Chapter 4 Documentation*, E. Commission, Brussels, 2011.
12. *USP General Chapter <1097> Bulk Powder Sampling Strategies*, United States Pharmacopoeia, Rockville, MD.
13. *21 CFR 58 Good Laboratory Practice for Non-clinical Laboratory Studies*, Food and Drug Administration, Washington, DC, 1978.
14. *FDA Guidance for Industry Out of Specification Results*, Food and Drug Administration, Rockville, MD, 2006.
15. *EMA Guideline on Bioanalytical Method Validation*, European Medicines Agency, London, 2011.
16. *FDA Guidance for Industry: Bioanalytical Methods Validation*, Food and Drug Administration, Silver Spring, MD, 2018.
17. M. E. Newton and R. D. McDowall, Data Integrity in the GxP Chromatography Laboratory, Part I: Sampling and Sample Preparation, *LCGC North Am.*, 2018, **36**(1), 46–51.
18. *Guideline on the Use of Near Infrared Spectroscopy (NIRS) by the Pharmaceutical Industry and the Data Requirements for New Submissions and Variations*, European Medicines Agency, London, 2012.
19. *FDA Draft Guidance for Industry Development and Submission of Near Infrared Analytical Procedures*, Food and Drug Administration, Silver Spring, MD, 2015.
20. *EudraLex – Volume 4 Good Manufacturing Practice (GMP) Guidelines, Annex 11 Computerised Systems*, European Commission, Brussels, 2011.
21. *FDA Warning Letter Leiner Health Laboratories*, Food and Drug Administration, Rockville, MD, 2006.
22. *FDA Warning Letter: Divi's Laboratories Ltd. (Unit II) (Warning Letter 320-17-34)*, Food and Drug Administration, Silver Spring, MD, 2017.
23. N. Dyson, *Chromatographic Integration Methods*, Royal Society of Chemistry, Cambridge, 2nd edn, 1998.
24. M. E. Newton and R. D. McDowall, Data Integrity in the GxP Chromatography Laboratory, Part III: Integration and Interpretation of Data, *LCGC North Am.*, 2018, **36**(5), 330–335.
25. *FDA Draft Guidance for Industry, Bioanalytical Methods Validation*, Food and Drug Administration, Rockville, MD, 2013.
26. *Cetero Research Untitled Letter (11-HFD-45-07-02)*, Food and Drug Administration, Silver Spring, MD, 2011.
27. *FDA Untitled Letter: Semler Research Center Private Limited*, Food and Drug Administration, Silver Spring, MD, 2016.

28. *OECD Series on Principles of Good Laboratory Practice and Compliance Monitoring Number 1, OECD Principles on Good Laboratory Practice*, Organisation for Economic Co-Operation and Development, Paris, 1998.
29. *OECD Series on Principles of Good Laboratory Practice and Compliance Monitoring Number 17 on Good Laboratory Practice Application of GLP Principles to Computerised Systems*, Organisation for Economics Co-Operation and Development, Paris, 2016.
30. M. E. Newton and R. D. McDowall, Data Integrity in the GxP Chromatography Laboratory, Part IV: Calculation of Reportable Results, *LCGC North Am.*, 2018, **36**(7), 458–462.
31. *WHO Technical Report Series No. 996 Annex 5 Guidance on Good Data and Records Management Practices*, World Health Organisation, Geneva, 2016.
32. *FDA 483 Observations, Sri Krishna Pharmaceuticals Limited*, Food and Drug Administration, Silver Spring, MD, 2014.

CHAPTER 17

Second Person Review

A second person review is an essential part of the four eyes principle that is used in all GXP disciplines to ensure that work has been conducted correctly and that the results are correct. A suitably qualified peer scientist is used to check that an analysis has been carried out correctly, *e.g.* the analytical procedure and applicable SOPs have been followed, all data have been collected and interpreted and that the reportable result is correct. Furthermore, checks must be made to ensure that data have not been deleted or falsified, as required by several data integrity guidance documents. Second person review is a critical part of the analytical process, especially where the interpretation of data is involved, *e.g.* spectroscopic or chromatographic analysis. The role of the second person reviewer is to ensure that the interpretation is scientifically sound, is justified and that interpreting, testing or falsifying into compliance has not occurred.

Owing to the wide range of analytical procedures, this chapter will consider second person review from the perspectives of a test by observation, a manual process with paper records, a hybrid system and an electronic process utilising electronic signatures. When reviewing records generated by either hybrid or electronic computerised systems, this will include relevant audit trail entries. Note, even for electronic systems there may be paper records to review such as instrument log books or sample preparation records.

Care must be taken during a second person review to see if failing SST results are an excuse to invalidate OOS results. There are two possible issues: the first is that the SST results did actually fail, thereby invalidating the run and the second is that the test results failed, and now the analyst is seeking to invalidate the run by manipulating the suitability results. A wise reviewer

Data Integrity and Data Governance: Practical Implementation in Regulated Laboratories
By R. D. McDowall

Published by the Royal Society of Chemistry, www.rsc.org

should look at the potential SST values for potential failures as part of the system suitability review.

Another key part of a GMP second person review is investigation and documentation of aberrant results and out of specification (OOS) results.

17.1 What Do the Regulators Want?

17.1.1 cGMP for Finished Pharmaceutical Products (21 CFR 211)

Second person review is presented in §211.194(a)(8)[1]:

> The initials or signature of a second person showing that the original records have been reviewed for accuracy, completeness, and compliance with established standards.

17.1.2 EU GMP Chapter 6 Quality Control

Clause 6.17 of EU GMP Chapter 6 covers second person review[2]:

> The tests performed should be recorded and the records should include at least the following data: ....
> g) initials of the persons who verified the testing and the calculations, where appropriate;

17.1.3 EU GMP Annex 11

There are three clauses in Annex 11[3] that are applicable to the audit trail review or electronic systems if no audit trail is used for a second person review:

> Clause 1: Risk management should be applied throughout the lifecycle of the computerised system taking into account patient safety, data integrity and product quality.
>
> Clause 9: Consideration should be given, based on a risk assessment, to building into the system the creation of a record of all GMP-relevant changes and deletions (a system generated "audit trail"). For change or deletion of GMP-relevant data, the reason should be documented. Audit trails need to be available and convertible to a generally intelligible form and regularly reviewed.

The key requirement is that audit trails need to be reviewed regularly but the frequency for this review is determined on a risk based approach that will be discussed in Section 17.7.

If no audit trail is available or required for a computerised system, Clause 12.4 still applies:

> Management systems for data and for documents should be designed to record the identity of operators entering, changing, confirming or deleting data including date and time.

The ALCOA principles still apply to electronic data created, modified or deleted even if there is no audit trail functionality available in a computerised system.

17.1.4 MHRA GXP Data Integrity Guidance and Definitions

The 2018 MHRA GXP Data Integrity Guidance and Definitions has Section 6.15 on data review and approval[4]:

> The approach to reviewing specific record content, such as critical data and metadata, cross-outs (paper records) and audit trails (electronic records) should meet all applicable regulatory requirements and be risk-based.
>
> There should be a procedure that describes the process for review and approval of data. Data review should also include a risk-based review of relevant metadata, including relevant audit trails records. Data review should be documented and the record should include a positive statement regarding whether issues were found or not, the date that review was performed and the signature of the reviewer.
>
> A procedure should describe the actions to be taken if data review identifies an error or omission. This procedure should enable data corrections or clarifications to provide visibility of the original record, and traceability of the correction, using ALCOA principles (see 'data' definition).

17.1.5 FDA Guidance on Data Integrity and cGMP Compliance

The FDA guidance on Data Integrity for cGMP Compliance[5] has two questions that are relevant to the review by a second person including audit trail entries:

> 2. When is it permissible to exclude CGMP data from decision making?
>
> Any data created as part of a CGMP record must be evaluated by the quality unit as part of release criteria (see § 211.22) and maintained for CGMP purposes (*e.g.*, §211.180). Electronic data generated to fulfill CGMP requirements should include relevant metadata.
>
> To exclude data from the release criteria decision-making process, there must be a valid, documented, scientific justification for its exclusion.

All data created in regulated work comes under the applicable GXP regulation and there must be a scientifically sound reason that is documented and approved for any data to be excluded.

> 7. How often should audit trails be reviewed?
>
> FDA recommends that audit trails that capture changes to critical data be reviewed with each record and before final approval of the record. Audit trails subject to regular review should include, but are not limited to, the following: the change history of finished product test results, changes to sample run sequences, changes to sample identification, and changes to critical process parameters.
>
> FDA recommends routine scheduled audit trail review based on the complexity of the system and its intended use.

Audit trail reviews should be risk based depending on the system complexity and intended use, however, for critical data, the audit trail pertaining to that record must be reviewed before the record is approved.

> 8. Who should review audit trails?
>
> Audit trails are considered part of the associated records. Personnel responsible for record review under CGMP should review the audit trails that capture changes to critical data associated with the record as they review the rest of the data. For example, all production and control records, which includes audit trails, must be reviewed and approved by the quality unit. This is similar to the expectation that cross-outs on paper be assessed when reviewing data.

Audit trail entries are part of complete data under GMP[1] that are subject to second person review of analytical data and records, therefore the personnel responsible for audit trail review are analytical staff and not quality assurance.

> 10. Is it acceptable to retain paper printouts or static records instead of original electronic records from stand-alone computerized laboratory instruments, such as an FT-IR instrument?
>
> Control strategies must ensure that original laboratory records, including paper and electronic records, are subject to second-person review (§ 211.194(a)(8)) to make certain that all test results are appropriately reported.

One area for regulatory citation is the lack of a second person review. This is an essential part of any data integrity approach in a laboratory.

17.1.6 WHO Guidance on Good Data and Record Management Practices

There are several sections in this guidance document that are relevant to second person review and the main ones are presented below[6]:

> Clause 1.3 .... organizations subject to ... GXP requirements have been using validated computerized systems for many decades but many fail to adequately review and manage original electronic records and instead often only review and manage incomplete and/or inappropriate printouts.
>
> Clause 1.4: .... training of personnel who use computerized systems and review electronic data in basic understanding of how computerized systems work and how to efficiently review the electronic data, which includes metadata and audit trails
>
> 3. Definitions: Metadata necessary to evaluate the meaning of data should be securely linked to the data and subject to adequate review.
>
> Clause 6.4: adequate review of audit trails, including those reviewed as part of key decision-making steps (*e.g.* GMP batch release, issuance of a GLP study report or approval of case report forms), may reveal:
>
> > incorrect processing of data,
> >
> > help prevent incorrect results from being reported and
> >
> > identify the need for additional training of personnel;
>
> Clause 8.4: Supervisors responsible for reviewing electronic data should learn which audit trails in the system track significant data changes and how these might be most efficiently accessed as part of their review.
>
> Clause 9.3: .... all GXP activities conducted can be fully reconstructed by the people reviewing these records at any point during the records retention period.
>
> Cause 11.6: .... Additional controls may include locking critical data entries after the data are verified and review of audit trails for critical data to detect if they have been altered.
>
> Clause 11.9: The evaluation should take into consideration all data, including atypical, suspect or rejected data, together with the reported data. This includes a review of the original paper and electronic records.
>
> To ensure that the entire set of data is considered in the reported data, the review of original electronic data should include checks of all locations where data may have been stored, including locations where voided, deleted, invalid or rejected data may have been stored.

A second person reviewer needs to have detailed knowledge of the data systems used in the various analytical procedures as well as authority to access all relevant records including the audit trail records. They need to check not only where records are expected to be stored but also locations where data from unofficial testing could be stored.

17.1.7 Regulatory Compliance Summary

A second person review is mandated by both US and EU GMP regulations but with little detail of the actions required. The publication of data integrity guidance documents by regulatory bodies expands the detail and the importance of the second person review to include all paper and electronic records generated during an analysis but also to ensure that audit trails are reviewed and to check for evidence of falsification or testing into compliance.

There must be an SOP covering the process and also this must link with the procedure for data integrity investigation (covered in Chapter 22) in terms of how data integrity violations should trigger an investigation.

17.2 What the Regulators Want: Out of Specification (OOS) Results

One of the key data issues with GMP analysis is the requirement to compare the reportable result with the specification and to investigate any aberrant results, comes from the Barr Laboratories case in 1993 discussed in Chapter 2. However, one of the major issues of data integrity in GMP regulated laboratories is ignoring or deleting out of specification results or justifying their exclusion based on poor investigations or failing SST results.

17.2.1 21 CFR 211

Section 211.192 of US GMP is on the topic of production record review[1]:

> All drug product production and control records, including those for packaging and labelling, shall be reviewed and approved by the quality control unit to determine compliance with all established, approved written procedures before a batch is released or distributed.
>
> Any unexplained discrepancy .... or the failure of a batch or any of its components to meet any of its specifications shall be thoroughly investigated, whether or not the batch has already been distributed.
>
> The investigation shall extend to other batches of the same drug product and other drug products that may have been associated with the specific failure or discrepancy. A written record of the investigation shall be made and shall include the conclusions and follow-up.

17.2.2 EU GMP Chapter 6 Quality Control

There are two clauses in EU GMP Chapter 6 about trending results and out of specification investigations[2]:

> 6.16. The results obtained should be recorded. Results of parameters identified as a quality attribute or as critical should be trended and checked to make sure that they are consistent with each other.
>
> 6.35. Out of specification or significant atypical trends should be investigated. Any confirmed out of specification result, or significant negative trend, affecting product batches released on the market should be reported to the relevant competent authorities.

17.2.3 FDA Guidance for Industry on Investigating OOS Test Results

Following the Barr Laboratories judgement in 1993, the FDA produced a draft OOS guidance that was released as a final version in 2006[7] that described the various stages of an OOS investigation as shown in Figure 17.1.

- Phase 1 is the laboratory investigation, which is to determine if there is an assignable cause for the analytical failure.
- Phase 2A: If no assignable cause is found then the investigation looks to see if there is a failure in production.
- Phase 2B: then comes back to the laboratory with additional testing carried out in the laboratory.

The details of the laboratory testing in the two phases of the investigation will be discussed in Section 17.9.

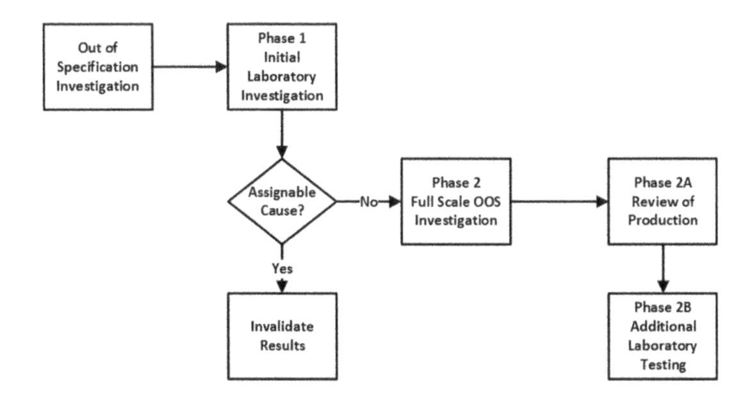

Figure 17.1 Flow chart of out of specification results investigations.

17.2.4 FDA Guidance on Quality Metrics

The FDA Draft Guidance on Quality Metrics has the key laboratory metric as the percentage of invalidated OOS rate that is defined as follows[8]:

> Invalidated Out-of-Specification (OOS) Rate (IOOSR) as an indicator of the operation of a laboratory. IOOSR = the number of OOS test results for lot release and long-term stability testing invalidated by the covered establishment due to an aberration of the measurement process divided by the total number of lot release and long-term stability OOS test results in the current reporting timeframe

The rationale for using the invalidated OOS rate can be seen in a 483 observation to a company,[9] here the IOOSR was 97% for stability testing and 71% for product testing. The reason for invalidating so many stability OOS results is that the company avoids having to send a field alert to the FDA. Quality metrics for data integrity and laboratories will be discussed in more detail in Chapter 19.

17.2.5 OOS Definitions

The following definitions are used[7]:

- Reportable Result: The term reportable result as used in this document means a final analytical result. This result is appropriately defined in the written approved test method and derived from one full execution of that method/procedure, starting from the sample. Compared with the specification to determine pass/fail of a test.[7]
- Out of Specification (OOS) Result: A reportable result outside of specification or acceptance criteria limits. As we are dealing with specifications OOS results can apply to test of raw materials, starting materials, active pharmaceutical ingredients and finished products but not for in-process testing. If a system suitability test fails this will not generate an OOS result as the whole run would be invalidated, however, there needs to be an investigation of the failure.[7]
- Out of Trend (OOT) Result: Not an out of specification result but does not fit with the expected distribution of results. This can include a single result outside of acceptance limits for a replicate result used to calculate a reportable result. If investigated the same rules as OOS should be followed.

17.2.6 OOS Regulatory Summary

Out of specification laboratory results are a key concern in GMP laboratories and need to be investigated with a scientifically sound approach and the results formally recorded. Owing to the number of instances where companies have ignored OOS results by deleting or carrying out inadequate investigations, *e.g.* "operator error" followed by retraining, this area is subject to high regulatory scrutiny during inspections.

In addition, from the data integrity model discussed in Chapter 5, the importance of analytical instrument qualification (Level 1) combined with valid and robust analytical methods (Level 2) are key items in ensuring that OOS results are reduced in a GMP regulated laboratory.

17.3 Procedures for the Second Person Review

17.3.1 Who Should Conduct a Second Person Review?

Under US GMP the whole production batch record (§211.192) should be reviewed by the quality control unit, a vague term that can encompass either quality control and/or quality assurance. However, laboratory records (§211.194(a)) should be reviewed by a second person from the laboratory.[1]

17.3.2 The Scope of the Procedure

The importance of the second person review means that it is imperative that there is an effective and comprehensive SOP for second person review along with effective training for the individuals who are authorised to perform this task. There are three main types of review that will need to be covered in this procedure and these are:

- Paper processes with written data entries made by the analyst performing the work, these are where analytical results are recorded by observation, *e.g.* colour, odour and appearance.
- Paper processes combined with a simple instrument such as a loss on drying analysis.
- Analytical procedures involving instrumental analysis using a hybrid computerised system. Note that there may also be manual sample preparation and where the calculation of the reportable result can be carried out using a validated spreadsheet.
- Instrumental analyses with electronic workflows and automatic calculation of results utilising electronic signatures with the elimination of paper. Note that although spreadsheet calculations have been eliminated from the workflow and incorporated in the data system there is still sample preparation that is manual in most laboratories that will result in paper records as well as log book entries.
- As the MHRA guidance states, presented in Section 17.1.4, *A procedure should describe the actions to be taken if data review identifies an error or omission. This procedure should enable data corrections or clarifications to provide visibility of the original record, and traceability of the correction, using ALCOA principles (see 'data' definition).*[4]

The procedure needs to be accompanied by effective training. When computerised systems are involved, it is worth reiterating that an individual must have the detailed technical knowledge of how the system works and has appropriate access privileges including viewing audit trail entries, as discussed in Section 17.3.6.

17.3.3 The Troika of Record Review

It is important to understand the relationships between the three main types of records involved in a second person review. This is especially true when dealing with any hybrid computerised system and to a lesser extent an electronic computerised system. These records are:

- paper printouts;
- electronic records;
- instrument log books.

Figure 17.2 shows the scope of a second person review and highlights the need to ensure that these three types of records, log book, printout and electronic record, correlate with each other and are consistent. For example, the work documented in the log book should match what is printed out from the instrument and be present in the instrument data system. Date and time stamps on printouts should be the same as those on the corresponding electronic records. We will look at the importance of the three types of records being consistent when we discuss data integrity investigations in Chapter 22.

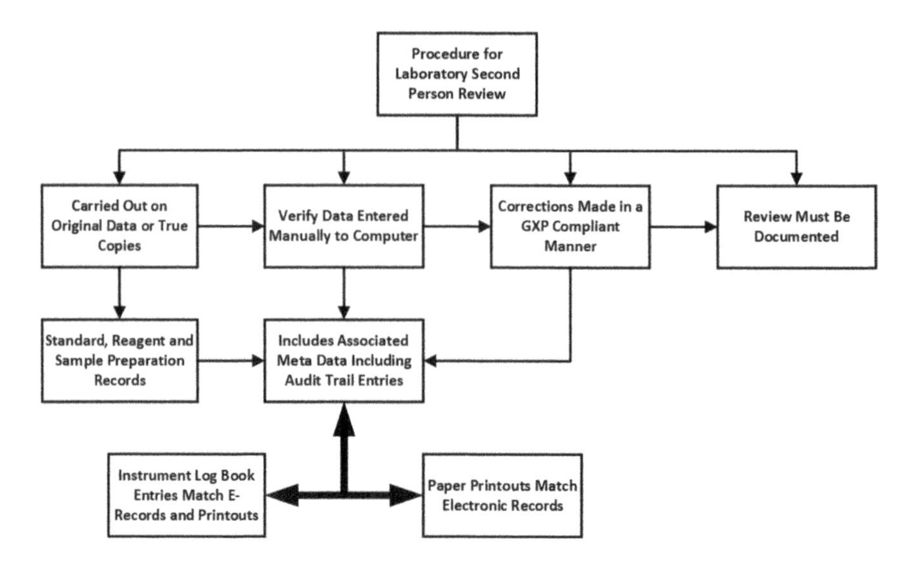

Figure 17.2 Scope of a second person review highlighting the essential correlations between printouts, electronic records and log books.

17.3.4 Timeliness of the Second Person Review

Individuals who are responsible conducting second person reviews of analytical records must ensure that the review is completed in a timely manner. This is especially the case with OOS results where the tester needs to inform their supervisor and retain all records, and solutions to investigate the work, as discussed in Section 17.9.

Time for the second person will also be a factor in many GMP laboratories where the Quality Control laboratory is at the end of the manufacturing process and a certificate of analysis is required for laboratory release of a batch of product. However, too much pressure on a reviewer to complete their work may result in a rushed review and a batch being released with errors, possibly leading to a recall when the problem is discovered.

Senior management also must realise that if procedural controls of hybrid systems are used due to a failure to invest in adequate automation then complex second person reviews will delay reporting results.

17.3.5 Documenting the Audit Trail Review

Currently, there are few laboratory informatics applications that have the functionality that will enable a reviewer to document that they have reviewed the audit trail entries pertinent to a specific analytical run. Therefore, this will have to be done procedurally until instrument and laboratory informatics suppliers provide the correct tools to do the job. This will be discussed in Section 17.6 under hybrid systems.

An alternative approach when using electronic workflows with electronic signatures is that the meaning of the reviewer's signature includes the fact that the audit trail entries relating to creation, modification and deletion of records have been reviewed. To achieve this, it must be documented in the second person review procedure.

17.3.6 Training for Second Person Review

Training to use a second person review procedure can never be read and understand. There must be practical training both in ensuring that the correct analytical procedures and associated laboratory SOPs have been followed but also each individual needs to be trained in detecting poor data management and falsification practices. The training should include scenarios where poor data management practice, data deletions and falsification are set up for the trainee to find. Where a laboratory computerised system is involved, the reviewer must have:

- Appropriate access privileges to view not only the files used to control and interpret the data generated during the analysis but also access to read and search applicable audit trail entries.

- The reviewer must be technically proficient in operating each data system for which they have review duties, otherwise, how can any individual conduct an effective second person data review?
- Refresher training is also required as more regulatory guidance becomes available and also as the software applications used in the laboratory evolve over time.

To ensure that the second person review is efficient, wherever possible, implement and validate as many technical controls in applications to restrict possible poor practice and falsification actions by testers. For example, if no deletion privileges are allowed for laboratory users and these have been documented and verified to work, then why should a reviewer be forced to check that deletions have occurred? Checks that the technical controls remain effective must be covered in data integrity audits.

17.3.7 Out of Specification (OOS) Procedure

Closely allied with the second person review SOP is a procedure for investigating the laboratory out of specification results. This will be based on the overall process shown in Figure 17.1 and the detailed discussed in Section 17.9.

17.4 Second Person Review of Analytical Procedures Involving Observation

17.4.1 What Is an Analytical Procedure Involving Observation?

Explicit mention of manual testing where the performer of the test records results by observation appears to have been forgotten in the majority regulatory data integrity guidance documents. As discussed in Chapter 16, these are analytical procedures for assessing the colour, odour and appearance of samples, manual titration as well as preparation of buffers, reagents and mobile phases. Also falling into this category is the manual colony counting of microbiological plates used in environmental testing, *etc.* Typically, many of these procedures rely on observation by the performer of the test and there may not always be objective evidence available for second person review, other than that written down in a laboratory notebook or on a controlled blank sheet (see Chapter 11).

17.4.2 Improving Manual Analytical Procedures

As part of the assessment of processes and systems described in Chapter 10, each manual procedure should be assessed to determine if there need to be changes so that execution of analytical procedures provide objective documented evidence to ensure that the second person review is simple and efficient. Such changes can be:

- Counting Microbiological Plates: Photographic evidence of each plate to be counted is a minimum expectation so that if the plate has been discarded there is objective evidence available to recount the number of colonies. However, rather than rely on an inefficient and error prone human process, implement and validate an automated colony counting system that provides digital evidence for each plate counted. This would enable not only a second person reviewer to conduct a speedier review. Photographic evidence with or without automated counting would also enable an auditor or inspector to check the work objectively.
- Direct data acquisition (best) from or printers attached to (adequate) analytical instruments such as analytical balances and pH meters will provide evidence that work was performed as claimed and enable an effective second person review.

Such is the regulatory environment that without objective evidence it can be difficult to convince a second person reviewer, auditor or inspector that work was performed as claimed. Therefore, data integrity assessments should ensure that high risk procedures are adequately documented to avoid intense regulatory scrutiny and aggressive questioning. However, there are still analytical procedures that do not generate any objective evidence during their execution. What is the position with these procedures?

17.4.3 Witness Testing or Second Person Review?

When we come to the area of tests with results recorded by observation there is a debate that has intensified since data integrity became a major issue in the pharmaceutical industry. Some organisations require that a witness is present when such a manual observation is made.

Let me be very clear, as mentioned in this topic in Chapter 16, there is no explicit regulatory requirement for witness testing in the analytical laboratory.

The applies to both GMP and GLP regulations. Therefore, there should be a push back by managers of regulated laboratories if there is a "requirement" for a witness to be present when an observation is made. What should be done is assess the risk associated with such a test, as recommended by the PIC/S data integrity guidance.[10] Ask the question, would a test failure stop the release of a batch or approval of a report?

This still leaves the question of the second person review: all that a reviewer could be presented with is a sheet with the study or batch information, the test performed and the written observation of the tester. Not an ideal situation – where is the second pair of eyes for a proper four eyes check? In most of these cases, there is not one. Leaving aside the issues of microbiological colony counting, which needs objective evidence and possible automation, as discussed above, and preparation of reference solutions and buffers that needs an instrument printout, a slightly different approach is required.

In the simplest cases, *e.g.* assessment of sample colour, odour and appearance all these procedures are non-destructive and the sample will still be available either as the original or as a retained/retention sample. For the minimal time that is required, the second person reviewer should access the sample analysed and then verify the observation made by the tester and record this fact on a controlled form or laboratory notebook. This is a much simpler process and consistent with the applicable GXP regulations that a laboratory works under. The controlled forms or laboratory notebooks used for recording the observations of tests must have space for the reviewer to verify the observation and state that they have reviewed the work. If an auditor or inspector wishes to verify what was recorded, then all that is required is for them to request the retained/retention sample and they can see for themselves.

17.5 Sample Preparation and Instrumental Analysis

17.5.1 Loss on Drying Analysis

From Chapter 16, the analytical procedure involving sample preparation and instrumental analysis was a loss on drying involving two sample aliquots weighed, dried for three hours at 105 °C, cooled and reweighed. Unfortunately, the loss on drying was calculated by a validated spreadsheet.

The documented evidence for the second person review should consist of the following items:

- Completed controlled blank form for the whole analysis containing sample information and the required analysis details and the reportable result.
- The balance printouts of the original and dried samples plus any calibration mass weights used to check that the balance worked correctly.
- Printout of the spreadsheet signed by the tester, the printout is linked to the spreadsheet file name plus create and print data and time on the printout.
- Saved spreadsheet file in a secure location.
- Completed instrument log books for:

  The analytical balance used with entries for initial calibration check of the balance, weights of the aliquots at the start and end of the drying process.

  Drying oven again with checks using a calibrated thermometer to show that the oven is performing correctly at the start and end of the drying process as well as record of the analysis.

In addition, the reviewer will need the applicable SOPs, knowledge of applicable pharmacopoeial general notices and the general chapter for loss on drying plus the loss on drying analytical procedure available to help with the review. The reviewer will also require access to the secure location where the completed spreadsheet file is stored.

17.5.2 Review of the Second Person Review of the Analytical Records

The aim of the second person review is to ensure that work has been done correctly, all the records required have been generated and are consistent with the procedure, *e.g.* samples have been analysed as required and date and time stamps of files are consistent and correct. In addition, there needs to be a focus to see if any work has been falsified. Although a loss on drying analysis is essentially a manual analytical procedure, an analytical balance and a spreadsheet are involved in the process and, as such, we start to enter the realm of computerised systems, albeit on a small scale, with instrument printouts and a saved spreadsheet electronic record that must be linked to the signed paper printout. Figure 17.3 shows the loss on drying process with the records expected to be created during the execution of it, the assumption here is that just one sample with two aliquots is taken through the procedure to calculate the reportable result.

- The reviewer would start with ensuring that the analysis form had the correct sample information entered and any corrections made by the

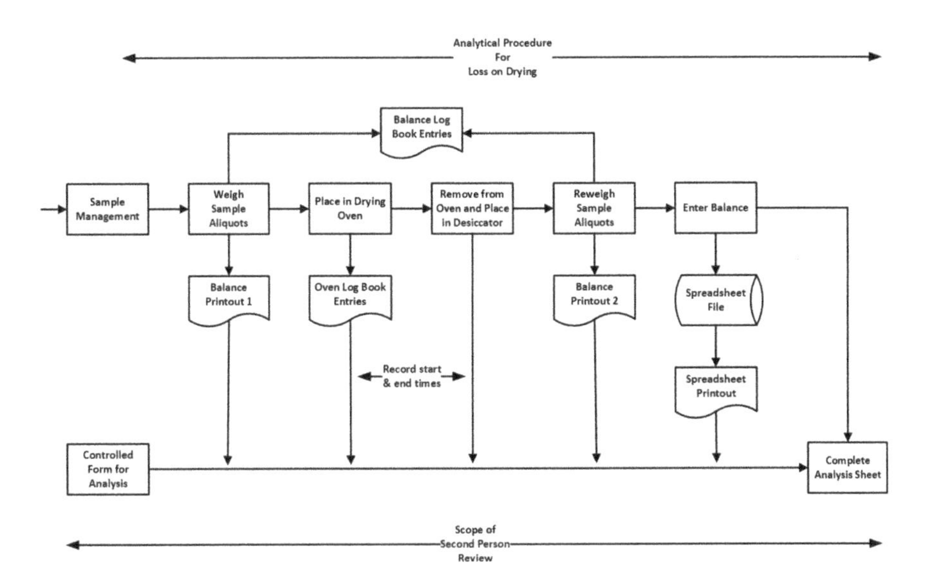

Figure 17.3 Scope of a second person review for a loss on drying analysis.

tester have been made in accordance with good documentation practice principles (the latter would apply throughout the review).

- Next, the reviewer would move to the weighing of the sample aliquots and check the balance printouts and entries in the analysis form including the date and time of the balance.
  Has the balance been calibrated, is it within acceptance criteria and has the whole sequence of the vessel, tare and sample sequence for each aliquot been correctly annotated by the tester?
  Are the weighing vessels correctly labelled?
- Do the entries in the analytical balance log book match the work sequence?
- The start and end times of the drying will be manually recorded, and the calculation of the drying time will be obtained by difference: does this time meet the 3 h criterion of the analytical procedure and the pharmacopoeia? This part of the procedure will be difficult to determine if data are falsified as there will be no independent evidence available.
- Check the drying oven log book entries to see that they correspond to the start and end of the analytical work.
- There will be a record of the sample transfer to a desiccator and then the removal of the aliquots prior to reweighing.
- The post-drying weights and balance log book entries will be checked as above.
- Date and time sequences throughout the analysis are important. The reviewer should check the date and time stamps on the two balance printouts and the spreadsheet file and printouts to see that they were consistent and there was sufficient time between weighing sessions for the drying and cooling to occur. The date and time stamp of the spreadsheet file also needs to be checked to ensure it was created after the second weighing session.
- The spreadsheet contains critical data that has been manually entered and therefore must be checked for accuracy under EU GMP Annex 11.[3] The data on the printout needs to be cross checked with the balance printouts and drying start and end times recorded on the analytical form.
- The reviewer also must check the data entries in the electronic spreadsheet file are the same as those on the printout. If data integrity controls are built into the validated spreadsheet template, as outlined in Chapter 7, Figure 7.9, then the reviewer can see the file name, file create and file print date and times both on the printout and in the electronic file are identical.
- The calculation of the loss on drying value should be correctly transferred to the analysis form and the form signed by the tester.
- Throughout the process, the reviewer needs to ensure that there are no missing signatures or initials on the documentation – including their own! Missing initials and signatures are a main finding in FDA 483 observations each year.

When this process is complete the reviewer can then sign the controlled analysis form to confirm that all work is complete, accurate and consistent. Depending on the second person review procedure, the reviewer may be required to initial or sign key stages of the work to denote that the complete review has taken place.

17.6 Second Person Review of a Hybrid System Records

17.6.1 Increased Scope of Record and Data Review

Moving from analytical records consisting mainly of paper to those that involve a hybrid computerised system brings the regulatory requirement to review the complete electronic record set. The scope of this electronic record review includes data files, contextual metadata and pertinent audit trail entries that are generated during the analysis. In this example, we will consider an analysis involving a UV-Vis spectrophotometer analysis followed by a spreadsheet used to calculate the reportable result. The spreadsheet complicates the overall process as it adds additional data entry and transcription checking. In these cases, it appears that analysts have a masochistic streak in them to work harder rather than work smarter.

17.6.2 Technical *Versus* Procedural Controls for Second Person Review

When any computer application is configured it is important that there are sufficient technical controls enabled to ensure that the correct sequence of operations takes place and data can only be stored in the correct location. However, there are several laboratory applications that are unable to allow complete technical control coverage and so procedural controls are required. Figure 17.4 shows a comparison of two types of control. In essence, technical controls are preferred over procedural controls as they are more reliable and can be validated to demonstrate that they work. Procedural controls, being operated by humans, are slow, error prone and inconsistent. As shown in Figure 17.4, each type of control can be further divided into two areas: controls to either prevent or detect falsification.

When there are full technical controls implemented and validated, *e.g.* access controls, effective audit trails, *etc.*, then the second person review becomes easier and faster. This is especially the case when there are functions to highlight if there are GMP relevant modifications or deletions in the audit trail entries. Highlighting modifications and deletions will allow a review by exception approach to be implemented: if there are no changes or deletions why review the audit trail. A risk based approach to audit trail review can go further if the system does not allow deletion then why look for such events?

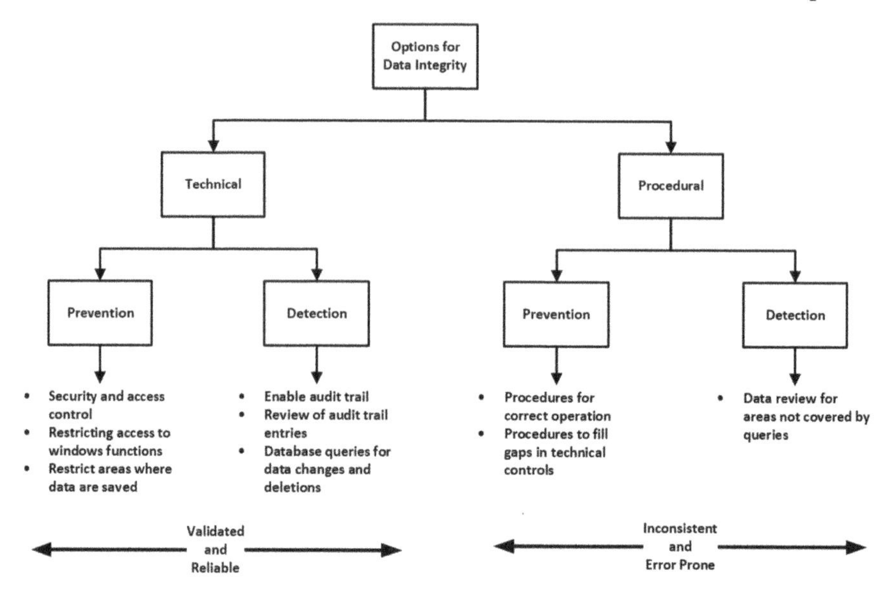

Figure 17.4 Comparison of procedural and technical controls for ensuring data integrity.

17.6.3 The Scope of an Analytical Procedure Involving a Hybrid System

The second type of second person review involves instrumental analysis controlled by a hybrid computerised system. Depending on the nature of the sample, type and complexity of the analytical procedure the scope of records generated may vary from direct analysis of the sample (*e.g.* NIR of incoming goods) to an extensive sample preparation and clean up prior to the instrumental analysis (*e.g.* chromatographic analysis of creams and lotions). In this chapter, we will discuss a more complex example involving sample preparation followed by UV-Vis instrumental analysis using a hybrid system.

An analytical procedure can consist of the following steps:

- retrieval of the sample from the storage location;
- weighing aliquots of sample and analyte reference standard on an analytical balance;
- dissolving the standard and samples in volumetric flasks and making up to volume;
- sample preparation to remove sample interferences and prepare the same for instrumental analysis;
- recording spectra of all solutions in the spectrophotometer;
- calculating the amount of analyte present in the sample aliquots;
- calculating the reportable result for the sample.

The analytical instruments and computerised systems involved in this analysis are:

- analytical balance;
- UV-Vis spectrometer connected to a standalone workstation with a validated data system for controlling the instrument, acquiring, processing, reporting and storing data;
- validated spreadsheet for calculation of the reportable result.

There will be instrument log books for the balance and spectrophotometer.

17.6.4 Technical Controls to Aid a Second Person Review

To reduce the amount of work required in a review, technical controls should be established in the instrument data system to protect electronic records. These can be:

- Setting up within the spectrophotometer application software user roles with no permissions to delete records. This function should be validated and used in conjunction with operating system security.
- Invoking e-records protection settings within the applications, these are the system policies in some laboratory informatics software.
- Establishing operating system security prevents laboratory users from accessing the operating system directories, data files, system clock, recycle bin and prevent USB sticks from being used to transfer data.
- The system clock must be set up with the correct and unambiguous date and time format that is used in the site or whole organisation, *e.g.*
  DD MMM YYYY.
  HH:MM:SS with 12 or 24 hour clock.
  If connected to the network, the workstation of the instrument should have the correct time and this will be synchronised with the network time server.
- A naming convention for files and directories has been established.
- Users have individual names and there is segregation of responsibilities, *e.g.* system administrators should be separate from the users.

In addition, there should be controls implemented in the analytical balance such as:

- identification of the users and their access level;
- correct setting of the system time and date and protection of this from unauthorised changes.

Repeating an earlier point in Chapters 12, 13 and 14, the rationale for this approach is that technical controls can be qualified (balance) or validated (instrument data system) and can be consistently applied in contrast to procedural controls with staff training.

17.6.5 Paper and Electronic Records to be Reviewed

The records generated from the analytical procedure described in Figure 17.5 are:

- sample storage location and retrieval records;
- sample preparation records and transfer to cuvettes for measurement in the spectrophotometer;
- analytical balance printout and associated instrument log entries;
- spectrophotometer printouts;
- spectrophotometer electronic records: spectra and metadata including audit trail entries;
- spectrophotometer instrument log entries;
- spreadsheet printout linked to the saved spreadsheet file.

The main records generated by a tester and that must be reviewed by a second person are shown in Figure 17.5.

17.6.6 Recording the Work Performed and the Review

As noted in Appendix 1 of the WHO data integrity guidance[6] under the Attributable section of special risk management considerations, there is an approach described to documenting the review by a second person by generating a controlled form for the analysis. This could be used for instrumental analysis involving a hybrid system. This form should be for the tester to

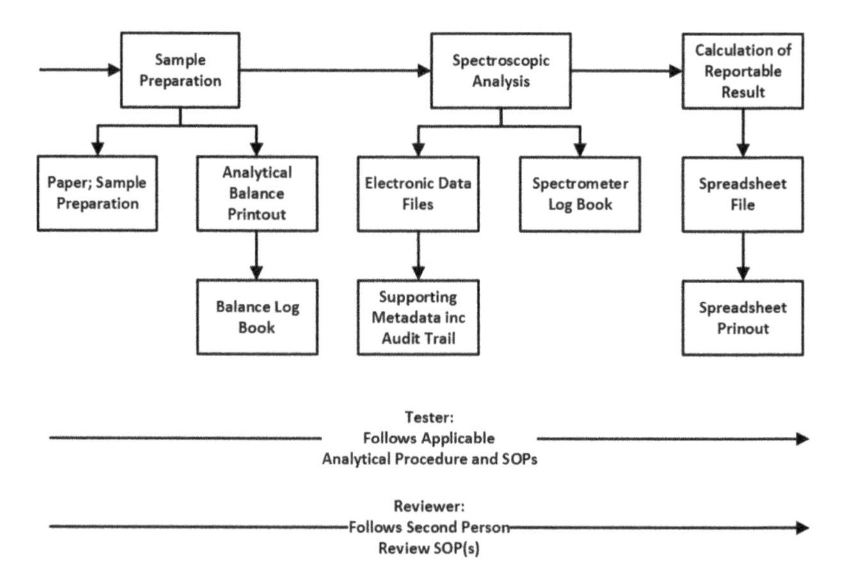

Figure 17.5 Paper and electronic records generated and reviewed by a UV-Vis analytical procedure.

Analysis Dataset Review Form			
Unique Form Number: UV-000297			
Batch Lot Number		Analytical Procedure	
Sample ID		Analyte	

Analytical Task	**Tester: Data/Records Created**	**Reviewer: Data/Records Reviewed**
Sample Preparation Analytical Balance ID		
UV-Vis Spectrophotometer ID. Acquisition File Name Sample Spectra Created Audit Trail Review Date / Time of Work Started Finished:		
Spreadsheet File Name File Location		
Personnel Date and Initials		

Figure 17.6 A data analysis generation and review form.

complete to identify the sample preparation paper records, the instrumental analysis and data sets generated and the name of the spreadsheet file created when calculating the reportable result. To help with the audit trail review, the form could also document the date and time when the analysis was started and completed to help the reviewer search for applicable audit trail entries of modification and, where allowed, deletion. The time and date can be obtained from the bottom right of the workstation screen rather than using the user's wristwatch. This is shown in Figure 17.6 and enables the reviewer to focus specifically on the data generated in a particular analysis rather than hunt through the instrument data files and the network directory where the spreadsheet file is stored.

17.6.7 Original Record or True Copy?

An analytical balance is used in the sample preparation phase of this analytical procedure. A printout from the instrument will contain the following data: weights of the mass(es) used to check that the balance is functioning correctly, the vessel, tare and sample weights taken, tester identity and the date and time that the work was performed. This will be glued to either a controlled form for the analysis or the user's laboratory notebook. If the printout is on thermal paper it is unlikely that this record will last the record retention period, therefore, a true or verified copy of the record is required that will be stuck into the official record by the tester and verified as a true copy. The process for this needs to be under control of a procedure. For the second person review, both the original printout and the true copy must be available so that the reviewer can verify that the copy made by the tester is indeed a true copy.

17.6.8 Have Critical Data Been Entered into the Instrument Data System?

In the analytical procedure shown in Figure 17.5, there are two computerised systems: the spectrophotometer and the spreadsheet. EU GMP Annex 11 clause 6,[3] requires that critical data entered into a computerised system must be checked for accuracy. Therefore, any data entered manually into either the spectrophotometer software and/or the spreadsheet must be checked to ensure that these data are correct and that there are no typographical errors.

This approach also applies to any data entered manually into an electronic system described in Section 17.8.

17.6.9 Review of Electronic Records, Metadata and Audit Trail

There are two computerised systems involved in this workflow: the UV-Vis spectrophotometer data system and the spreadsheet. Both the signed paper printouts must be subject to review and the reviewer must have appropriate user rights to access all the electronic records related to the analysis.

Review of electronic records in the spectrophotometer data system will cover:

- Checks that any data entered manually, *e.g.* sample identity, lot/batch number, *etc.*, are correct.
- Ensuring that all data files for the sample and standards from the analysis have been generated, comply with applicable SOPs and that the number of files matches the number of samples prepared earlier in the analysis.
- Check the time and date stamps on the files generated are consistent, this is especially important for automated analysis. If not restricted by technical controls, check if data files have not been copied from the previous batch analysis (these files will have the same creation date and time stamps) or that data files from the current analysis have not been deleted by checking in the recycle bin or audit trail of the data system.

- Check that the work in the spectrophotometer data system matches the entries in the instrument log book.
- Metadata files used to control the instrument, acquire and process data and store the data are correct and have not been changed without authorisation.
- Check that the file naming convention for the data files generated during the analysis was followed correctly.
- Check that audit trail entries for data modifications and, if allowed, deletions related to the analysis are justified. Review of the audit trail entries should be performed on screen and if there is the functionality, search the audit trail for any data manipulation.
- Check that the file names of the data files and the absorbances on the printouts match the e-records in the data system.

When the review is completed, the evidence that these records have been checked will be documented in the data generation and review form shown in Figure 17.6.

17.6.10 Second Person Review to Ensure Data Have Not Been Falsified

As mentioned by the WHO guidance in Section 17.1.6, the second person reviewer must also look to see if data have been hidden, deleted or falsified. Reiterating my earlier point in Section 17.6.2 and Figure 17.4, if there are technical controls that have been implemented and validated to demonstrate that these options for data integrity violations are greatly limited, the second person review can be reduced and this can be a focus for QA data integrity audits described in Chapter 21.

17.6.11 Do You Really Want to Work This Way?

In Section 9 of the WHO guidance under the attributable section, there is an interesting discussion about the hybrid systems that is reproduced below. All I have done is to bullet point the text and add my comments underneath each sentence and italic text to make it easier to read and understand.

- A hybrid approach may be used to sign electronic records when the system lacks features for electronic signatures.[6]

Regulators will not tell companies that hybrid systems are not suitable as laboratories can still use them, but read on.

- To execute a handwritten signature to an electronic record, a simple means to do so would be to create a single-page controlled form associated with the written procedures for system use and data review, that would list the electronic dataset reviewed and any metadata subject to review, and would provide fields for the author, reviewer and/or approver of the dataset to apply a hand-written signature.[6]

This is different from the way most laboratories review data as the current emphasis is on the paper printouts with little emphasis placed on the electronic records including relevant audit trail entries.

- This paper record with the hand-written signatures should then be securely and traceably linked to the electronic dataset, either through procedural means, such as the use of detailed archives indexes, or technical means, such as embedding a certified true copy scanned image of the signature page into the electronic dataset.[6]

What a waste of time this is – do you really want to work this way?

- *The hybrid approach is likely to be more burdensome than a fully-electronic approach, therefore, utilizing electronic signatures, whenever available, is recommended.*[6]

Here is the killer – get rid of hybrid systems as they waste time and effort, as discussed in Chapter 12. The complexity of some second person reviews may result in the review taking longer than the actual analysis.

17.7 Risk Based Audit Trail Review

As there are now regulations mandating audit trail review as well as a regulatory expectation for this activity, this section will look at risk-based audit trail reviews in more depth.

17.7.1 MHRA GXP Data Integrity Guidance and Definitions

In the MHRA GXP Data Integrity Guidance and Definitions[4] in Section 6.13 on audit trails has the following advice:

> The relevance of data retained in audit trails should be considered by the organisation to permit robust data review/verification. It is not necessary for audit trail review to include every system activity (*e.g.* user log on/off, keystrokes *etc.*).

In plain English, use your brain when reviewing audit trail entries and focus on the requirements of EU GMP Annex 11 Clause 9[3] that are GMP relevant changes and deletions. If the system is configured not to allow deletions, then only focus on the changes to records, as mentioned earlier in this chapter.

17.7.2 Which Audit Trail Should Be Reviewed?

An audit trail is defined by the FDA in the data integrity guidance[5] as:

> A secure, computer-generated, time-stamped electronic record that allows for reconstruction of the course of events relating to the creation, modification, or deletion of an electronic record. An audit trail is a chronology of the "who, what, when, and why" of a record.[5]

Depending on the instrument data system or informatics solution that is installed in your laboratory, the audit trail(s) may vary from the simple to complex depending if the specific application and its architecture. A laboratory software application that has been designed around a database and specifically for use in a regulated GXP environment will have more sophisticated but easier to review audit trails. In addition, there may be utilities or features to help speed the task of the second person reviewer such as search routines or filters.

The FDA also comment that[5]:

> The audit trail for a high performance liquid chromatography (HPLC) run could include the user name, date/time of the run, the integration parameters used, and details of a reprocessing, if any, including change justification for the reprocessing. Electronic audit trails include those that track creation, modification, or deletion of data (such as processing parameters and results) and those that track actions at the record or system level (such as attempts to access the system or rename or delete a file).

The problem comes with a CDS that has been designed for a regulated environment, in that it can have more than one audit trail, as noted above by the FDA guidance on data integrity. The question becomes which audit trail or audit trails needs to be reviewed? Figure 17.7 shows the audit trail structure of a commercial chromatography data system (CDS) designed for a compliant GXP operation. There are system and project audit trails, the latter is sub-divided into four further audit trails together with a results audit reviewer. The functions that each audit trail monitors are also shown in Figure 17.7, therefore from the perspective of a second person reviewer, which audit trails should be reviewed?

EU GMP Annex 11 Clause 9 is concerned with audit trailing GMP-critical data that has been modified or deleted,[3] therefore the focus of second person audit trail review should be on the audit trail containing the details of data modifications/deletions. In short, the second person review focus is on the project audit trails.

17.7.3 How Regular Is a Regular Review of Audit Trail Entries?

As can be seen in Figure 17.7 the audit trails associated with the project contain the data and therefore they should be the focus of the second person

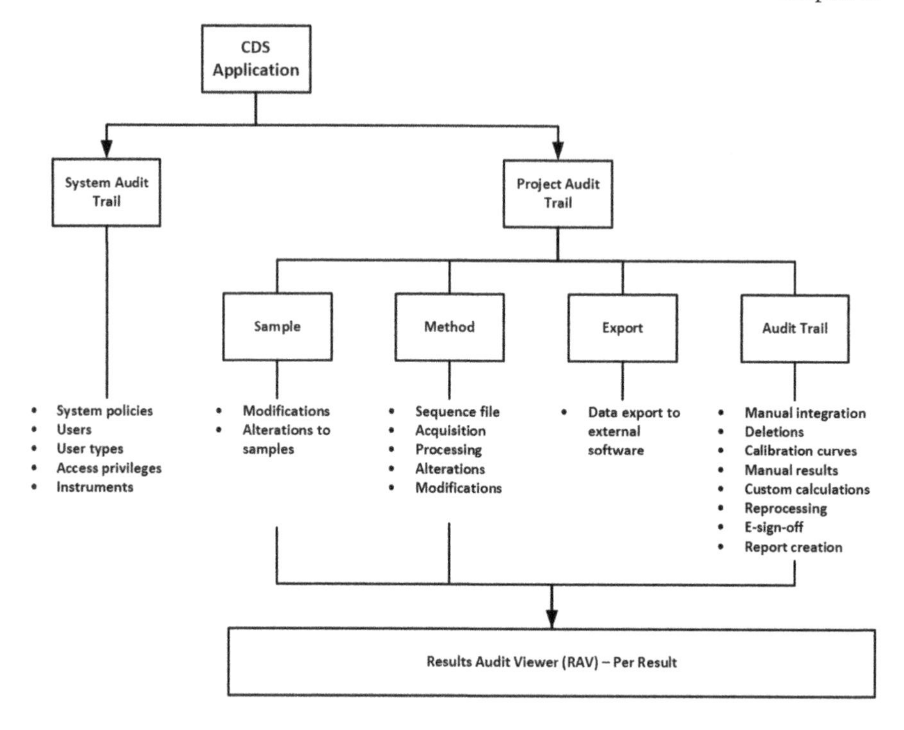

Figure 17.7 Risk based reviews of CDS audit trails.

review. If there are technical controls built into the software these will help identify where data have been modified or (if permitted) deleted by a user rather than simply created in the normal functioning of the CDS.

Of course, the risk based approach advocated by EU GMP Annex 11 [3] is also important. If we look at audit trails from a different perspective as shown in Figure 17.8. Here, the work analysed in a project is identified as their method development, method validation, batch/study analysis and stability testing. We can apply risk-based review of audit trails as follows:

- Method Development: in this situation, it can be argued that the use of a full audit trail with a forced reason for change is not appropriate and a silent audit trail would be better and would not interfere with the work. Audit trail entries would provide information on who changed what and when but not why. However, does a user continually want to input "changed flow rate" or "changed mobile phase composition" in such cases? A method is being developed and is not yet validated. Therefore, I would advocate that a full audit trail is not required and neither is an audit trail review. However, the situation changes when we move to the actual validation of an analytical procedure.
- Method Validation (or Method Transfer): here we move up a regulatory notch and a full audit trail is required here. There are a number

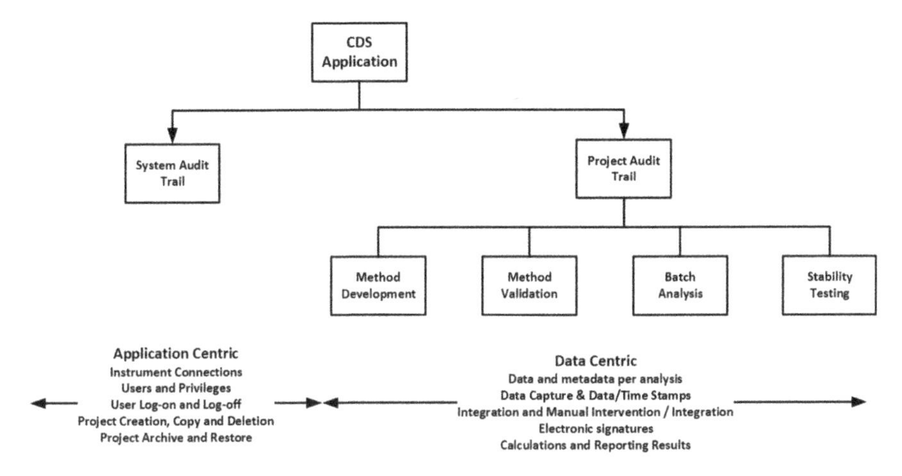

Figure 17.8 Risk based review of audit trails based on the type of work performed.

of experiments to be carried out, depending on the Analytical Target Profile (ATP), before the method is considered validated and can be released for operational use. It depends on the risk profile of the company if the audit trail review is per experiment or per method validation study. However, an audit trail review, as well as other data review, is required before the study report is approved and the method released.

- Batch Analysis: in the GMP arena involves the analysis of raw materials, in-process control samples and finished products. All are involved in the batch release and hence meet the FDA's definition of critical data, as discussed in Section 17.1.5. Therefore, all audit trail entries must be reviewed before the results are approved and before the batch is released.
- Study Analysis: is typically found in clinical and non-clinical studies in the development of potential pharmaceutical products. The choice is either to review the audit trail entries pertaining to each batch of samples analysed or wait until the end of the study to review all pertinent audit trail entries. Given the length of time, some studies could last it is better to perform the audit trail review after each run has been completed.
- Stability Testing: each batch of commercial product is put on long term stability studies for the shelf life of the product. In this case, similar to the study analysis, it is important to review the audit trail entries after each sample pull to ensure that the data are complete and accurate. The main reason for this is that if one-time point is out of specification the company must file a field alert with the FDA within 72 h. Hence the importance of ensuring the integrity and quality of this data

Omitted from the discussion above is the system audit trail which contains information about:

- user log-on and log-off;
- account locking and unlocking;
- user account management;
- instrument and laboratory data server configurations;
- creation, backup and archiving of projects;
- CDS application configuration changes.

Ordinarily, these events would not be subject to a second person review but included as part of a QA data integrity audit, which is discussed in Chapter 21 or as part of a data integrity investigations that will be discussed in Chapter 23 of this book.

17.8 Second Person Review of Electronic Systems and Data

In this section we will consider an electronic solution that uses electronic signatures with a CDS and LIMS interfaced with each other, we will assess how electronic workflows can help our quest for data integrity in a regulated analytical laboratory. This discussion assumes that technical controls have been implemented and validated to ensure that the process is followed and enforced by both the software applications. Configuration settings in both applications are set for the protection of electronic records and electronic signatures, where appropriate.

On a practical note, the date and time stamps of the two systems must be in the same formats, otherwise, data integrity questions can be raised by auditors and inspectors. The two systems may be in different time zones and therefore may need to be synchronised to GMT/UTC/Zulu time for consistency to ensure that both local time and GMT is recorded in each database.

17.8.1 LIMS Interfaced with a CDS

Figure 17.9 shows the outline architecture of the two systems as follows:

- The LIMS is shown on the right-hand side of the diagram. This system is responsible for sample management and collation and reporting of all analyses undertaken on samples for analysis.
- Not shown in Figure 17.9 is an analytical balance interfaced to the LIMS and used to weigh samples and reference standards. The weights can be linked to the correct samples and reference standards already in the LIMS database.
- There is the ability for electronic transfer of data from the LIMS to the CDS to help set up the analytical run that includes download of

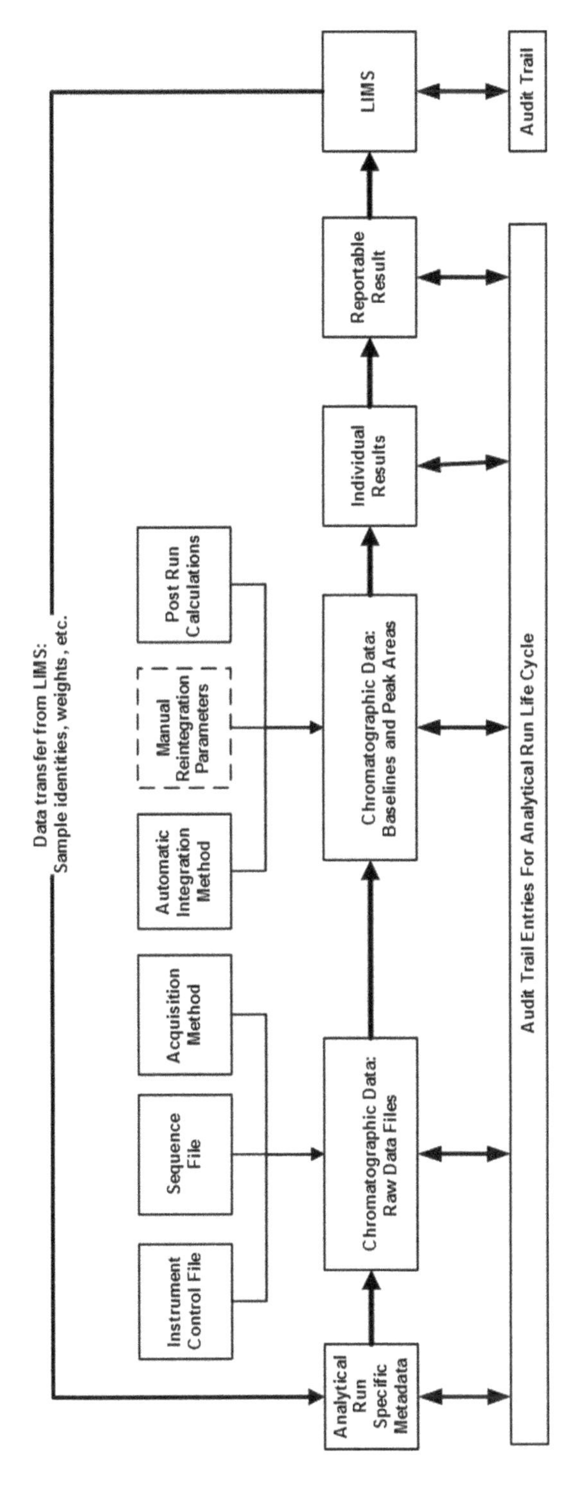

Figure 17.9 A LIMS and CDS interfaced to work electronically.

sample identities and sample weights used for the analysis, reference standard purities and water content depending on the type of analysis performed.
- On the left-hand side of Figure 17.9 is the CDS showing the various data and metadata elements generated during an analysis and interpretation of chromatographic data. In the figure the results of the individual aliquots analyses are calculated along with the reportable result for the sample.
- Any dilutions made during the sample preparation phase of the analysis will be recorded on paper and entered manually into the CDS sequence file. This is critical data and must be checked for accuracy, as discussed in Section 17.6.8.
- In this example, the results are signed electronically in the CDS by the tester and the reviewer after reviewing of the data by the latter.
- Transfer from the CDS to the LIMS can vary from laboratory to laboratory. This transfer can vary from all integrated peak areas per sample identity of all injections so that all subsequent calculations are performed in the LIMS (this is typical for a LIMS for a bioanalytical laboratory) to the calculated reportable results that are compared with the specification in the LIMS. Regardless of the data content, the transfer must be electronic, documented and validated. In Figure 17.9 the sample identities and the reportable result is transferred to the LIMS.
- If a file is transferred between the CDS and a LIMS that is parsed to extract the requisite data, the file must be protected from tampering between transfer and parsing. Ideally, such a mechanism would not be used – direct transfer between the two systems is much better. The transfer file must be retained as part of complete data.

17.8.2 A Second Person Review Is Process Not System Centric

The scope of the second person review should cover the whole of the analytical process from sample storage to reportable result and must not be confined to the boundaries of a single computerised system. If the review is focused on a specific system data integrity issues can fall through the cracks between the applications and not be detected. As the checks for sample preparation have been covered under the hybrid system review in Section 17.6 we will focus here on the second person review between the LIMS and the CDS.

The tester sequence follows these actions:

- The first part of the process is setting up the chromatograph in the CDS and conditioning the column.
- Next, the identities and the weights of samples for analysis are downloaded to the CDS. Depending on the system there should be an export

entry of the data as it leaves LIMS in the LIMS audit trail and a corresponding entry under import into the CDS. This is why it is important to understand the time and date synchronisation between the two systems.
- The analyst should check that the downloaded sample information is in the correct sequence file in the CDS. At this point, the individual may enter manually the sample dilutions from the sample preparation phase. There could be a check to see if there are any transcription errors that could be corrected, resulting in entries in the audit trail of the CDS.
- The analysis proceeds, data are acquired, integrated and the individual results and the reportable result calculated, all of which results in the generation of electronic records and audit trail entries. When complete the tester electronically signs the report of the analysis.

Now we must consider what the second person reviewer must review: both the data and the associated metadata, as shown in Figure 17.9. The problem is that in a risk based world how can we focus on what is critical to meet the requirements of both the FDA data integrity guidance[5] and EU GMP Annex 11?[3] Table 17.1 presents the main questions for consideration in the second person review along with a risk-based approach, the focus is mainly on the electronic records and is not, repeat not, a comprehensive approach as other parts of the analytical process must be reviewed as well, *e.g.* sample preparation.

The areas for review are listed in the left-hand column of Table 17.1 and the areas for consideration for this review are shown in the centre and right-hand columns. The review is divided into primary and secondary areas. Primary is where the focus should be each time a second person review is carried out and secondary as this is where technical controls and validation provide the support that poor data management practices or falsification cannot or are difficult to be carried out. This does not mean that a reviewer ignores these areas but it allows a risk based approach to help focus the review on areas that are not enforced by the electronic workflows in the software or where there is manual input to a system. These are areas where poor data management or errors can occur.

Ideally, the two applications can help with the review. For example, a CDS application has a function where the audit trail entries can be reviewed and by pressing a button the name and time stamp of the reviewer can be entered into the audit trail for the sample to document the review has occurred. Other means of helping the review process are for changes to be highlighted in the audit trail by a colour (*e.g.* green for no changes, yellow for modification and red for deletion) and the reviewer focuses their attention on the modifications and deletions (if allowed).

The initials or signature of a second person show that the original records have been reviewed for accuracy, completeness, and compliance with established standards.

- Batch analysis begins in the LIMS, proceeds to the CDS and reportable results are transferred to the LIMS. Therefore, the second person review covers the overall process and not individual systems.

Table 17.1 Areas and considerations for a second person review of chromatographic data.[11]

Area for second person review	Considerations for primary review	Considerations for secondary review
Correct instrument and column used? Instrument and column logs updated and correct?	• Check the instrument and column logs and the CDS to see that entries are the same and are complete	
Correct reagents, standards and solutions used?	• Check reagents for the mobile phase are correctly made up • All solutions unexpired	
Data input to CDS *via* LIMS (*e.g.* sample IDs, weights, *etc.*)		• Validated data transfer from LIMS
Manual data input to LIMS, *e.g.* dilutions, *etc.*?	• Check original record with CDS • Changes made are acceptable and justified	
Correct order of vials in autosampler?	• Check vial order with run sequence in CDS and any laboratory notebook entries	
Data acquisition: correct acquisition method used?	• Confirm the correct method used • Aborted or shortened runs?	• Technical control may be possible in a CDS to ensure the right method is always used • Rationale for shortened or aborted runs must be scientifically justified
Correct data processing – right processing method used?	• Confirm the correct method used	• Technical control may be possible in a CDS to ensure the right method is always used
Correct data processing – correct integration used?	• Standard and samples integrated in same way? • Can manual intervention or integration be used? • Integration acceptable? • Integrating into compliance?	
All injections reported?	• System evaluation injections reported? • SST injections reported? • Sample injections reported?	
Results calculated in the CDS?		• Validated calculations
CDS audit trail reviewed: are changes acceptable?	• Use application features to target audit trail review • Review by exception	• Audit trail validated • Access privileges enforced
Results transferred to LIMS OK?		• Validated transfer
LIMS audit trail reviewed: data unchanged?	• Use application features to target audit trail review • Review by exception	• Audit trail validated • Access privileges enforced

- Reagents, standards, weights and solutions: have the right ones been used and have they been prepared correctly?
- Transferred data from the LIMS to the CDS (sample IDs, weights, factors, purities, *etc.*) are they in the correct sequence.
- All log books involved in the analysis (depending on the level of automation in the laboratory these can be sample management, reagent and standard preparation, instrument use and column use) are the entries correct and correspond to the work performed?
- Are the vials in the correct order in the autosampler? Do the vials in the autosampler correspond to the vial order in the sequence file?
- Has the correct chromatograph been used for the analysis?
- Have the correct instrument control and data acquisition methods and the correct version of each been used to control the chromatograph and acquire the data?
- If evaluation injections are permitted, are they acceptable and have all been reported?
- Are point of use and SST checks within acceptance limits and have all been reported?
- Has the correct data processing method been used and is the integration acceptable? Where manual integration is permitted, is peak placement scientifically justified? Have standards and sample injections been integrated consistently? Is there any peak enhancement or peak skimming to modify the results?
- Have all errors and any deviations been documented and investigated where appropriate?
- Have the reportable results been calculated correctly?
- Have the audit trail entries in CDS been reviewed and any modifications to data acceptable (this assumes that no user has delete privileges)?
- Have the reportable results been transferred to LIMS correctly?
- Are the time stamps of the CDS import and export and the corresponding LIMS export and import consistent? Comparison of the audit trail data and time stamps for data exchange between the two applications is important. The CDS should have an export audit trail entry and a few seconds or a minute later the LIMS should have an import entry for the data. This again, emphasises the need for time stamp synchronisation.
- Audit trail entries in LIMS reviewed: are the reportable results in the CDS the same as those in the LIMS?

17.9 Recording and Investigating Out of Specification Results

For GMP analyses of samples against a specification, there is the requirement to trend results and investigate out of specification results, as discussed in Section 17.2 and shown in overview in Figure 17.1. When a reportable result fails to meet a product specification for API and product, there needs to be

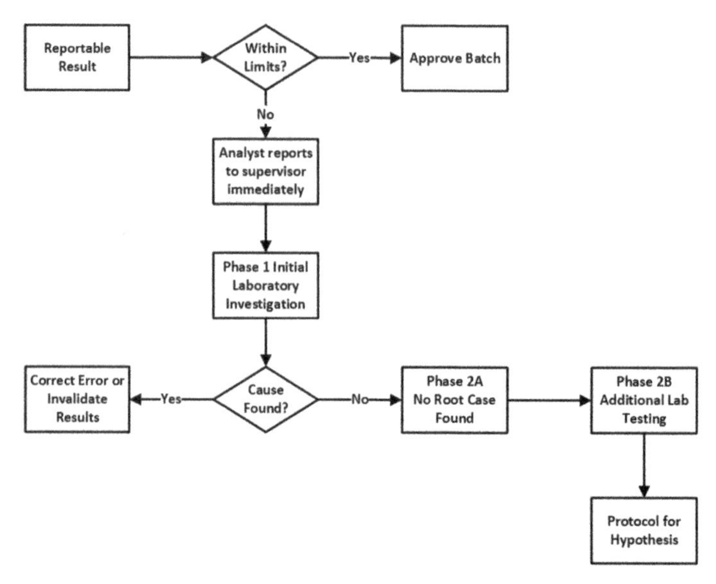

Figure 17.10 Flow chart of OOS Phase 1 and Phase 2B laboratory investigations.

an investigation. A more detailed flow of Phase 1 and Phase 2B laboratory OOS investigations is shown in Figure 17.10. An OOS investigation must be completed within 30 days.

17.9.1 Phase 1: Initial OOS Laboratory Investigation

When an analyst knows that there is a potential out of specification reportable result, they should inform their supervisor immediately and must ensure that no solutions or other materials involved in the impacted analysis are discarded. The reason is that these are required for the investigation.

The analyst and their supervisor should conduct a joint examination of the whole analytical work from the sample, through the sample preparation, instrumental analysis, any data interpretation and calculation of the reportable result. One of the aims is to ensure that the correct analytical procedure was used and that work was correctly performed. Some areas to consider in this process:

- Was the right sample taken for analysis?
- Was the correct reference standard used?
  Is it within expiry date?
  Is there a potential for contamination?
- Was the correct analytical procedure used?
- Were the correct balance, pipettes and other analytical instruments used?
  Were they currently qualified or calibrated?

Were any point of use checks, *e.g.* balance check, SST samples performed and were the results correct?
Visual inspection of the instrumentation to see that there are no leaks or obvious reasons for malfunction.

- Visual inspection of samples and glassware used in the sample preparation work to check for handling and that the correct identification or labelling was used
- Preparation of all reagents, buffers, *etc.*
- Check the records used for the analysis to check for correct entries and that no transcription errors have occurred.
- Verify that the correct instrument files or methods were used for acquiring and processing data and that all data interpretation is correct. This is particularly so for chromatographic integration and for identification of materials by spectroscopy, *e.g.* was the correct library used?
- Check the instrument log books to see what was performed and do they match with the work performed in the analysis.

The objective of this phase of the laboratory investigation is to determine if there is an assignable cause for the OOS result.

- If it is caused by a transcription or calculation error, then this can be corrected and the work passes to the second person review as there is no out of specification result.
- Is there an instrument malfunction or was an instrument used that was out of calibration or qualification?
- Was the analytical procedure not followed correctly during, *e.g.* the wrong sample or standard was used, there was a sample preparation error, instrument problem, data interpretation error, chromatographic column failure, or vials swapped in the autosampler?

It is possible to conduct reanalysis – repeating some portions of the testing using the existing preparations – such as dilutions or selecting preparations to re-extract or dilute, vials to re-inject to check preparation or repeat the measurement stages of the analytical procedure. Where justified this can include the preparation and analysis of a new reference standard. Note that:

- The original OOS sample is not touched.
- All data are retained and nothing is deleted or hidden.

If there is an assignable analytical cause as illustrated in the last two bullet points above, then the result can be invalidated. The analysis can be repeated and only the repeated reportable result and associated data be part of the batch record. The invalidated data must not be deleted as these will be subject to audit and inspection to confirm that the investigation was conducted correctly. Therefore, an investigation report describing the assignable with corrective actions to prevent it occurring again (*e.g.* updating a procedure that caused the problem or instrument maintenance) must be available. The OOS investigation must be scientifically sound to comply with 21 CFR 211.160(b).[1]

If the investigation allows the original reportable result to be invalidated, the sample analysis can be repeated as described in the analytical procedure. This can only happen after the end of the investigation.

A retest – conducting the whole test on a different portion of the same sample – cannot be undertaken at this point. Retesting is only allowed in Phase2B of an OOS investigation and required a scientifically sound and formal test protocol.

17.9.2 Phase 2A Production

If no assignable case can be found in Phase 1, the OOS investigation is elevated to Phase 2A, which is a review of production. This is an activity carried out by Quality Assurance that must generate a report. The whole production batch record is reviewed along with appropriate electronic records and this requires the participation of all departments involved. The aim is to identify a root cause of the OOS. If it is a production failure this will lead to more investigation and CAPA plans that are outside the remit of this book. If a root case cannot be found, the Phase 2B is initiated, which is additional laboratory testing.

17.9.3 Phase 2B Additional Laboratory Testing

This phase of an OOS investigation is only conducted under the remit of a predefined and approved protocol:

- To confirm or reject one or more hypotheses of the root cause of the OOS, *e.g.* an error by the analyst, instrument, analytical variability, *etc.*
- Define the number of retests in the protocol (*i.e.* repeating the complete analytical procedure on a different portion of the same sample) and avoid a claim of testing into compliance. This will depend if the material that is the subject of the OOS is a raw material, starting material, API or finished product and the nature of the analytical procedure (limit test, purity or impurity). For example, raw materials there may require 2 or 3 retests but an impurity analysis involving an API may require up to 12 retests. The design will depend on the potential cause of the OOS and the degree of confidence required coupled with the variability of the analytical procedure.
- The tests may be performed by the same analyst (alone or with a supervisor) or a second analyst or the same or different instrument (if one is available).
- Statistical outlier testing, as defined in USP 1010[12] can be used as an investigational tool but cannot be used to invalidate data and certainly will not identify the cause of the OOS.

- Resampling can only be conducted in Phase 2B if the original sample has been used up, is not representative, compromised or is used to confirm a production failure. However, the rationale for resampling must be justified in the protocol for the Phase 2B investigation.
- The outcome of an investigation may be proof of a root case, *e.g.* evidence of an analytical error and repeating the OOS under certain conditions. However, it may be that a likely cause may be identified in cases where all retests are the same as the OOS such as a handling error or instrument failure if possible supported by other information or a rational and plausible argument. In both cases, the cause should be named so that CAPA plans can be developed for the OOS investigation report.

17.9.4 OOS Investigations: Prevention Is Better than the Cure

As can be seen in this section, an out of specification investigation, especially when a company reaches Phase 2 is very time consuming and requires much detailed and though work. Therefore, prevention is better than cure. To reduce the number of OOS investigations it is essential that:

- Staff are thoroughly trained in the analytical procedures they will execute as well as good documentation practices covering both paper and computerised systems.
- Analytical instruments must be fit for purpose (qualified, calibrated and with point of use checks as necessary). Instruments should also be maintained correctly and in some case minor preventative maintenance performed, *e.g.* replacement of HPLC pump seals with appropriate requalification to ensure reliability.
- Analytical procedures must be fit for purpose, *e.g.* pharmacopoeial methods must be verified for actual conditions of use and in-house developed procedures must be designed and validated and their robustness evaluated to reduce variability. In addition, ensure that the specification limits and the analytical variability are suitable, if the analytical variability is large due to poor method development then there will be many OOS results.
- Define the reportable result carefully and use averaging within the constraints defined by the FDA: the sample is homogeneous, specified in the test methods and there are acceptance limits of the individual results and the SSTs used.[7]
- Trend results to identify and investigate atypical results (*e.g.* out of trend (OOT) or out of expectation (OOE)) before an OOS result arises.

References

1. *21 CFR 211 Current Good Manufacturing Practice for Finished Pharmaceutical Products*, Food and Drug Administration, Sliver Spring, MD, 2008.
2. *EudraLex – Volume 4 Good Manufacturing Practice (GMP) Guidelines, Chapter 6 Quality Control*, European Commission, Brussels, 2014.

3. *EudraLex – Volume 4 Good Manufacturing Practice (GMP) Guidelines, Annex 11 Computerised Systems*, European Commission, Brussels, 2011.
4. *MHRA GXP Data Integrity Guidance and Definitions*, Medicines and Healthcare Products Regulatory Agency, London, 2018.
5. *FDA Draft Guidance for Industry Data Integrity and Compliance with cGMP*, Silver Spring, MD, USA, 2016.
6. *WHO Technical Report Series No. 996 Annex 5 Guidance on Good Data and Records Management Practices*, World Health Organisation, Geneva, 2016.
7. *FDA Guidance for Industry Out of Specification Results*, Food and Drug Administration, Rockville, MD, 2006.
8. *FDA Guidance for Industry Submission of Quality Metrics Data, Revision 1*, Food and Drug Administration, Rockville, MD, 2016.
9. *FDA 483 Observations: Lupin Limited*, Food and Drug Administration, Silver Spring, MD, 2017.
10. *PIC/S PI-041 Draft Good Practices for Data Management and Integrity in Regulated GMP/GDP Environments*, Pharmaceutical Inspection Convention/Pharmaceutical Inspection Co-Operation Scheme, Geneva, 2016.
11. R. D. McDowall, *Validation of Chromatography Data Systems: Ensuring Data Integrity, Meeting Business and Regulatory Requirements*, Royal Society of Chemistry, Cambridge, 2nd edn, 2017.
12. *USP General Chapter <1010> Outlier Testing*, United States Pharmacopoeia Convention Inc., Rockville, MD.

CHAPTER 18

Record Retention

Retention of records generated during regulated GXP laboratory activities, *e.g.* method validation, analysis of non-clinical studies, clinical studies, analysis of starting materials, in process and finished goods and well as stability testing is an important requirement under existing GXP rules (most of which were written when paper was the dominant record format). Retention of paper records is well known and understood. However, electronic records generated by a computerised system will usually outlive the system that generated them. Records retention is the biggest problem facing any electronic or hybrid system as the records are stored in a proprietary format.

18.1 What Do the Regulators Want?

18.1.1 WHO Guidance on Good Data and Record Management Practices

Although this section is focused on outsourcing, the principles apply to all GXP data and records that are required to be maintained for the respective records retention period[1]:

> 7.6 Where data and document retention is contracted to a third party, particular attention should be paid to understanding the ownership and retrieval of data held under this arrangement. The physical location where the data are held, and the impact of any laws applicable to that geographical location, should also be considered. Agreements and contracts should establish mutually agreed consequences if the contract acceptor denies, refuses or limits the contract giver's access to their records held by the

Data Integrity and Data Governance: Practical Implementation in Regulated Laboratories
By R. D. McDowall

Published by the Royal Society of Chemistry, www.rsc.org

contract acceptor. The agreements and contracts should also contain provisions for actions to be taken in the event of business closure or bankruptcy of the third party to ensure that access is maintained and the data can be transferred before the cessation of all business activities.

7.7 When outsourcing databases, the contract giver should ensure that if subcontractors are used, in particular cloud-based service providers, they are included in the quality agreement and are appropriately qualified and trained in GRDP.

Their activities should be monitored on a regular basis at intervals determined through risk assessment. Only validated systems are used for storage of data; however, the media used for the storage of data do not have an indefinite lifespan. Consideration must be given to the longevity of media and the environment in which they are stored. Examples include the fading of microfilm records, the decreasing readability of the coatings of optical media such as compact disks (CDs) and digital versatile/video disks (DVDs), and the fact that these media may become brittle. Similarly, historical data stored on magnetic media will also become unreadable over time as a result of deterioration.

18.1.2 EU GMP Annex 11

Clause 17 of EU GMP Annex 11 is focused on archiving[2]:

Data may be archived. This data should be checked for accessibility, readability and integrity. If relevant changes are to be made to the system (*e.g.* computer equipment or programs), then the ability to retrieve the data should be ensured and tested.

18.1.3 GLP Regulations: 21 CFR 58

In the US GLP regulations there is §58.195(b), which states[3]:

Except as provided in paragraph (c) of this section, documentation records, raw data and specimens pertaining to a non-clinical laboratory study and required to be made by this part shall be retained in the archive(s) for whichever of the following periods is the shortest:

(1) A period of at least 2 years following the date on which an application for a research or marketing permit, in support of which the results of the non-clinical laboratory study were submitted, is approved by the FDA. This requirement does not apply to studies supporting investigational new drug (IND) applications for investigational device exemptions (IDE) or applications for investigational device exemptions (IDE), records of which shall be governed by the provisions of paragraph (b)(2) of this section.

(2) A period of at least 5 years following the date of which the results of the non-clinical laboratory study are submitted to the FDA in support of an application for a research or marketing permit.

(3) In other situations (*e.g.*, where the non-clinical laboratory study does not result in the submission of the study in support of an application for a research or marketing permit), a period of at least 2 years following the date on which the study is completed, terminated, or discontinued.

18.1.4 US GMP Regulations: 21 CFR 211

In the US GMP regulations for finished pharmaceutical products §211.180 (a) there is the following requirement[4]:

> Any production, control, or distribution record that is required to be maintained in compliance with this part and is specifically associated with a batch of a drug product shall be retained for at least 1 year after the expiration date of the batch or, in the case of certain OTC drug products lacking expiration dating because they meet the criteria for exemption under §211.137, 3 years after distribution of the batch.

The key section of the US GMP regulations for laboratory data is §211.194(a)[4]:

> Laboratory records shall include complete data derived for all tests necessary to assure compliance with established specifications and standard, including examinations and assays as follows:
>
> A complete record of all data secured in the course of each test, including all graphs, charts and spectra from laboratory instrumentation, properly identified to show the specific component, drug product container, in process material or drug product and lot tested.
>
> A record of all calculations performed in connection with the test, including units of measure, conversion factors and equivalency factors.
>
> A statement of the results of tests and how the results compare with established standards of identity, strength, quality and purity for the component, drug product container, closure, in-process material or dug product tested.

18.1.5 21 CFR 11 Requirements

In Subpart B of 21 CFR 11, which focuses on electronic records, there is §11.10(c)[5]:

> Protection of records to enable their accurate and ready retrieval throughout the records retention period

As Part 11 is interpreted by the predicate rule the retention period is defined in the applicable US GMP or GLP regulation.

18.1.6 MHRA GXP Data Integrity Guidance and Definitions

The MHRA GXP Data Integrity Guidance and Definitions[6] notes in two sections the following about data retention:

> 6.2: Information that is originally captured in a dynamic state should remain available in that state.
>
> 6.8. Data transfer is the process of transferring data between different data storage types, formats, or computerised systems.
>
> Data migration is the process of moving stored data from one durable storage location to another. This may include changing the format of data, but not the content or meaning.

During the retention period data acquired in a dynamic format should remain as dynamic data. However, during a migration the format of the data can change but if this occurs it should remain in the dynamic state and the content and meaning should not change. This last point is consistent with the FDA's Part 11 Scope and Application Guidance.[7]

18.1.7 FDA Guidance on Data Integrity and cGMP Compliance

In FDA's Guidance for Industry on Data Integrity and cGMP Compliance under question 1 there is the following[8]:

> e. How does FDA use the term "backup" in § 211.68(b)?
>
> FDA uses the term backup in § 211.68(b) to refer to a true copy of the original data that is maintained securely throughout the records retention period (for example, § 211.180). The backup file should contain the data (which includes associated metadata) and should be in the original format or in a format compatible with the original format.
>
> This should not be confused with backup copies that may be created during normal computer use and temporarily maintained for disaster recovery (*e.g.*, in case of a computer crash or other interruption). Such temporary backup copies would not satisfy the requirement in § 211.68(b) to maintain a backup file of data.

This is a prime example of the "c" in cGMP meaning current or an on-going reinterpretation of the GMP regulations by the FDA. When other regulations refer to backup it is to a copy of the data as the first stage of a disaster recovery process – in case of a corrupt file, accidental deletion, malware attack or hardware failure.

18.1.8 EU GMP Chapter 4 Documentation

In Chapter 4 of EU GMP there are the following regulations for retention of records[9]:

> 4.11 Specific requirements apply to batch documentation which must be kept for one year after expiry of the batch to which it relates or at least five years after certification of the batch by the Qualified Person, whichever is the longer. For investigational medicinal products, the batch documentation must be kept for at least five years after the completion or formal discontinuation of the last clinical trial in which the batch was used.
>
> 4.12 For other types of documentation, the retention period will depend on the business activity which the documentation supports. Critical documentation, including raw data (for example relating to validation or stability), which supports information in the Marketing Authorisation should be retained whilst the authorization remains in force. It may be considered acceptable to retire certain documentation (*e.g.* raw data supporting validation reports or stability reports) where the data has been superseded by a full set of new data.
>
> Justification for this should be documented and should take into account the requirements for retention of batch documentation; for example, in the case of process validation data, the accompanying raw data should be retained for a period at least as long as the records for all batches whose release has been supported on the basis of that validation exercise.

18.1.9 FDA Guidance for Industry Part 11 – Scope and Application Guidance

The following is quoted from the August 2003 final guidance for industry with respect to the enforcement discretion for electronic records now being granted under 21 CFR 11.[7]

> The Agency intends to exercise enforcement discretion with regard to the part 11 requirements for the protection of records to enable their accurate and ready retrieval throughout the records retention period (§ 11.10 (c) and any corresponding requirement in §11.30). Persons must still comply with all applicable predicate rule requirements for record retention and availability (*e.g.*, §§ 211.180(c),(d), 108.25(g), and 108.35(h)).
>
> We suggest that your decision on how to maintain records be based on predicate rule requirements and that you base your decision on a justified and documented risk assessment and a determination of the value of the records over time.
>
> FDA does not intend to object if you decide to archive required records in electronic format to non-electronic media such as microfilm, microfiche,

and paper, or to a standard electronic file format (examples of such formats include, but are not limited to, PDF, XML, or SGML). Persons must still comply with all predicate rule requirements, and the records themselves and any copies of the required records should preserve their content and meaning. As long as predicate rule requirements are fully satisfied and the content and meaning of the records are preserved and archived, you can delete the electronic version of the records. In addition, paper and electronic record and signature components can co-exist (*i.e.*, a hybrid situation) as long as predicate rule requirements are met and the content and meaning of those records are preserved.

18.1.10 FDA Inspection of Pharmaceutical Quality Control Laboratories

This FDA guide, which was issued in July 1993, has in Section 13 on laboratory records the following note[10]:

> Expect to see written justification for the deletion of all files.

18.1.11 OECD GLP Regulations

In the OECD GLP regulations, the section on archives has the following requirements[11]:

> The GLP Principles for archiving data must be applied consistently to all data types. It is therefore important that electronic data are stored with the same levels of access control, indexing and expedient retrieval as other types of data. Where electronic data from more than one study are stored on a single storage medium (*e.g.*, disk or tape), a detailed index will be required.
>
> It may be necessary to provide facilities with specific environmental controls appropriate to ensure the integrity of the stored electronic data. If this necessitates additional archive facilities then management should ensure that the personnel responsible for managing the archives are identified and that access is limited to authorised personnel. It will also be necessary to implement procedures to ensure that the long-term integrity of data stored electronically is not compromised.
>
> Where problems with long-term access to data are envisaged or when computerised systems have to be retired, procedures for ensuring that continued readability of the data should be established. This may, for example, include producing hard copy printouts or transferring the data to another system.
>
> No electronically stored data should be destroyed without management authorization and relevant documentation. Other data held in support of computerised systems, such as source code and development, validation, operation, maintenance and monitoring records, should be held for at least as long as study records associated with these systems.

18.1.12 OECD GLP Guidance on Application of GLP to Computerised Systems

Section 3.2 of this guidance document has Clauses 73–78 on Data and Storage of Data and in Clause 77 there are the following requirements[12]:

> 77. Regarding procedures, the test facility management should describe how electronic records are stored, how record integrity is protected and how readability of records is maintained. For any GLP-relevant time period, this includes, but may not be limited to:
>
> a) physical access control to electronic storage media (*e.g.* measures for controlling and monitoring access of personnel to server rooms, *etc.*);
>
> b) logical (electronic) access control to stored records (*e.g.* authorisation concepts for computerised systems as part of computerised system validation which defines roles and privileges in any GLP-relevant computerised system);
>
> c) physical protection of storage media against loss or destruction (*e.g.* fire, humidity, destructive electrical faults or anomalies, theft, *etc.*);
>
> d) protection of stored electronic records against loss and alteration (*e.g.* validation of back-up procedures including the verification of back-up data and proper storage of back-up data; application of audit trail systems); and
>
> e) ensuring accessibility and readability of electronic records by providing an adequate physical environment as well as software environment.

There are further requirements in this section including the possible need to retain software to reconstitute a study report. This is an example, in many cases, of a regulation exceeding the technical possibilities available to the pharmaceutical industry.

18.1.13 Regulatory Requirements Summary

There are many issues to consider with records retention, especially for electronic records from electronic systems but especially for hybrid systems as they must be synchronised with the paper printouts throughout the record retention period:

- Complete data, as required by US GMP, need to be retained as evidence of the work carried out for the duration of the records retention period. This should include all contextual metadata such as autosampler sequence data, instrument control method, acquisition and processing methods, interpretation or integration of the data and calculation of the final results. In addition, all pertinent audit trail entries are required to document who did what to the data and when.
- The records retention period varies with GXP discipline; the minimum is batch expiry plus one year for manufacturing data or two years after

an NDA approval. However, there are potential complications in product liability legislation limits (11 and 20 years in Europe and the US, respectively), the need to support additional license applications with existing data and the electronic Common Technical Document (eCTD) mentioning that product license data may need to be kept for up to 50 years. EU GMP Chapter 4 requires that critical data supporting a Marketing Authorisation must be retained for the time the MA is in force, however, there is an option to retire and replace that data if justified and documented.[9]

- There is the option, based on a documented risk assessment, to covert from the original electronic format to either electronic format or paper provided that the content and meaning of the records is preserved.[7] However, this is converting dynamic to static data and with the current focus on data integrity this option is probably not tenable given the FDA's Level 2 guidance on laboratory records[13] stating that the electronic records are the original records and not paper (or even PDF) printouts. This view is reinforced with the FDA guidance on data integrity.[8] At best the guidance from the FDA is contradictory.
- In contrast, MHRA guidance is must more definitive and very specific: dynamic data must remain dynamic data.[6] This is the better regulatory guidance and implicitly requires that the software be available to view and interpret dynamic data if required.
- If electronic data are archived there needs to be a formal mechanism for this for laboratories operating to GLP.[3,11] However, the WHO data integrity guidance[1] also recommends an electronic archive for computerised system data regardless of regulatory discipline.
- Archived data (either paper or electronic) must be secure and organised for easy retrieval. Note that this is not instant replay. If data are archived off-line replay cannot be instant. However, with the change in the Food Drug and Cosmetic act in 2012, if an FDA inspection is delayed or limited by slow retrieval of GMP data then an organisation's drugs can be deemed to be adulterated.[14]
- When electronic records are deleted there must be a formal documented process and deletion must be authorised by management.

18.2 Laboratory Data File Formats and Standards

Laboratory data currently exists in a proprietary world that means when an instrument data system is purchased, the data typically need the application to interpret the data generated. We will look at some attempts at standardisation from the last century and then look forward to the future.

18.2.1 JCAMP-DX Data Format for Spectroscopy

Spectroscopy has a standard for data files called JCAMP-DX named after the Joint Committee on Atomic and Molecular Physical data extension (JCAMP-DX) that was developed in the 1980s.[15] Put simply, it is a way of formatting spectral data in terms of x, y coordinates. There are viewers that can be used to view data on different scales, expand or shrink spectra and manipulate data files outside of the application that generated them. The problem with this standard is that it only focuses on the data file and there is no consideration of the associated metadata including audit trail data.

18.2.2 Current CDS Data Standards

During the 1990's a standard data file format for CDS files was discussed and selected. The underlying file format is based on the network Common Data Format (net-CDF) which was used as a means of transferring astronomical data between observatories; details of the file format are available in Liscouski's book.[16] In principle, the aim of the standard was to allow data generated on one supplier's CDS to be read and interpreted by a different CDS application.

However, the emphasis of the standards work was on the chromatography data file only and not on the associated contextual meta-data such as instrument control file, acquisition method, processing method, sequence file, report, audit trail, *etc.* Therefore, although the data file can potentially be transferred and read by another data system by a variety of means, it is unlikely that any metadata can be read by another CDS as there is no common standard for these files. The only way this information can be transferred between different data systems is by manual input of the data – a slow, labour intensive and error-prone process. The large volumes of data generated by a CDS over time means that manual transfer is not a practical option. However, it is not a good option for records retention as the file format net-CDF does not allow replay of the data as it requires the original software plus the contextual metadata.

Until we have universal data formats that allow true interoperability between different CDS applications from different vendors, you are effectively locked into a single vendor who has the responsibility of either:

- maintaining stable file formats; or
- providing fully working and documented conversion routines to enable data migration between one version of file to another.

This is an important reason for ensuring that the system selection process is got right first time.

18.2.3 Progress Towards Universal Data File Formats

There has been progress towards universal data file formats based on extensible markup language (XML) that has been modified for the analytical laboratory. Data can be stored in text files rather than binary files using Analytical Information Markup Language (AnIML). This language is being developed under the auspices of the American Society for Testing and Materials (ASTM) E13.15 sub-committee for spectroscopy and separation science over a number of years. Overall, the aim is to permit the portability of data between different applications for the same analytical technique. This standard has been over ten years in development, however, it provides the ability for UV-Vis and chromatographic data to outlive the system that generated the data once incorporated into instrument applications or conversion software added to existing systems.

In parallel, a consortium of pharmaceutical companies has formed the Allotrope Foundation to fund the definition, development and deployment of vendor independent data standards for the major analytical techniques.[17] The objective is to develop standards so that analytical data can be shared within an organisation without the need to use the original software application or hardware, therefore any instrument application that complies with a new data standard could be used to review and reassess the data. Data obtained and stored using the tools being developed by the Allotrope Framework aim to facilitate the compilation of regulatory submissions and their subsequent review by regulatory authorities. Furthermore, these standards covering both the data files and the contextual metadata will facilitate inspections and audits when accessing and retrieving old data. This should save time both for the pharmaceutical companies and the regulatory agencies.

Both AnIML and Allotrope have similar objectives.

BUT

There is a problem. The two standards approaches are going in opposite directions. AnIML is a text based standard. Allotrope are developing digital standards. Therefore, they are incompatible and the market will only support a single standard mainly because the purchasers do not consider what they will do with the data after the instrument is retired. All they currently focus on is the new shiny instrument.

18.3 Options for Electronic Records Retention and Archive

18.3.1 Backup Is Not Archive (Unless You Are the FDA)

It is important to differentiate between backup and recovery and archive and restore (or retrieval); for the purposes of this book the terms are defined as:

- Backup and recovery is used for short to medium term data storage where there is a need for immediate recovery of data (encompassing selected data files to whole disks) in case of file loss, corruption or system failure. This is a pre-requisite for disaster recovery and business continuity planning.
- Archive and restore is concerned with the long-term storage of selected data including the associated contextual metadata. In this context the use of "store" is also used to denote archived electronic records.

As noted in section FDA Guidance on Data Integrity (Section 18.1.7) the current version of the FDA data integrity guidance now interprets backup as archive.[8] However, we will used backup as the first stage of disaster recovery in this book.

Backup is a regular process, specified by the users, that usually occurs daily and is performed by the IT department and is concerned with the whole of the system or the database. In contrast, archive and restore occurs infrequently and is driven by the users (the archiving process may be carried out by the IT department but it is user-specified). The data to be archived are carefully selected and will be by work package rather than all the data contained in the system.

18.3.2 Organising Electronic Records to Retain

You will need to organise the archive of your electronic records to retain around specific packages that will depend on the type of work carried out in your laboratory. Some typical examples could be the complete electronic records including audit trail events from the following types of work packages:

- specific analytical method;
- batch records of the same material;
- stability study;
- pharmacokinetic study or protocol;
- clinical study or protocol.

For example, chromatographic data for a specific development project may be archived if the project is discontinued, a non-clinical or clinical bioanalytical study may be archived once the report has been issued. These are only suggestions for ways to organise the electronic data, as most functions within organisations tend to work slightly differently and you may find other approaches that will fit your specific needs better.

Organisation of the electronic data archive concerns specific rather than general work packages because once you have archived the data you may have to get this back at some time. Spend time in designing a simple way of defining your requirements and get it right first time. Alternatively, you can hope you have retired or the company has merged and it is someone else's problem. The issue is that you could wait a long time before it emerges that you cannot easily restore data, as they are stored in several places or in the worst case, the system does not work at all. That is why the Annex 11 regulations[2] and the WHO, OECD and MHRA guidance documents[1,6,12] mandate checking that the archive works when you first archive the data and at periods thereafter.

When defining the electronic records to archive be sure to include the pertinent audit trail records, if they can be abstracted and are still readable, to comply with the specific requirements of 21 CFR 11 and Annex 11.[2,5]

Some instrument data systems, *e.g.* CDS have databases and organise work packages using projects. The work contained in each project, like the list above, can be archived from the application and this includes all chromatographic data files, contextual metadata and audit trail entries. This is a far better way to go rather than using data files stored in directories created in the operating system.

18.3.3 Options for Electronic Archive

There are two main schools of thought about archive and restore of CDS electronic records: on-line storage or archive off-line to alternative media.

One approach for archive and restore preferred by some people is that we should archive data on-line rather than off-line. As hardware is more resilient and fault tolerant, appropriate servers and disk space can be purchased relatively cheaply. As the electronic records that we generate with a data system grow in number and volume, we must keep pace by purchasing extra disk storage. As we approach the disk capacity of the server, simply purchase a new server and transfer the records. However, there can be problems with this approach as there can be uncertainty with defining the electronic records and associated metadata to transfer to the archive.

System performance may be an issue. Some points to consider as the number and volume of electronic records and data files increase are:

- Can the file management or database system cope with the increasing size of the electronic records over the lifetime of the application use?
- Was this approach tested and supported by the vendor?
- Are the security and integrity of the records maintained? For instance, can records that are approved or final be changed or are they locked?
- Who has access to these records? Is the access read-only or can data be changed?
- What happens if there is a disaster?
  Can all the records be recovered?
  Has this been tested?
  How long will this take? The larger the volume of records the longer it will take and this could be days in extremis.

An alternative to storage *via* a server is to separate data storage from the application and use a fault tolerant storage device such as a Storage Area Network (SAN) or Network Attached Storage (NAS). This should be easier to expand as it is independent of the server running the application.

The length of time before the electronic records are archived will depend on the nature of the work in which you are involved. Some clinical/bioanalytical and product stability studies may last 2–5 years and thus after the report has been prepared and authorised may be a suitable time to consider archiving the data. However, if the data must be accessed later then the decision may need to be delayed.

The records are the organisation's intellectual property and will be the means of meeting regulatory requirements during an inspection. Therefore they must be protected. Unless there are stringent controls on access to the data there may be a risk of corruption of the records. Furthermore, as the size of stored records grows, the performance may suffer if records are kept on line.

18.3.4 Can I Read the Records?

One regulatory requirement is for replay of dynamic data,[6,8] this means that the data are in a format that they can be accessed, viewed and if required reinterpreted. Note that this is NOT instant replay because with an effective off-line archive process, records may be stored off-line or on-line and this will impact the recall time and may have to be restored back to the appropriate computerised system before review. The key issue here is to ensure that whatever archive process you use works and that you can restore and read all the electronic records whenever you wish. The retrieval process needs to be relatively fast to ensure that any FDA inspection is not limited or delayed.[14]

There are a number of questions concerning changes to your data system that will affect the ability to replay data. Look at any differences between when you originally acquired the electronic records and when you want to read, reprocess or replay and consider:

- Are any file formats the same?
- If your electronic records are stored in a database, is the structure the same?
- Have the integration algorithms changed for in the chromatography data system between versions?
- Has the operating system changed and can this make an impact?
- Has the application software changed?
- Has the hardware platform changed?
- Has the archive medium changed if using off-line storage?
- Has the archive software changed as this may impact reading disks created by earlier versions of the application?

As you can see, there are a number of changes that could have an impact on the ability to develop an effective archive. Changes in any one of the above could limit or destroy the ability to restore data from an off-line archive. Technology refresh can have a major impact on an electronic archive throughout the record retention period.

This area presents us with one of the most technically challenging problems for electronic data to be in compliance with GXP regulations over the whole record retention period. The agreement and availability of a standard data format covering all applicable instrument data systems and metadata would help in this area immensely.

18.3.5 Impact of a Changed Data System File Format

What impact does a file format have on archive and restore? There are a number of issues to consider: however, to help you achieve a better understanding of the problems, think back when you were using a word processor. If you have used the word processing application Word for a long time, you will remember the problems with migrating from Word 95 to Word 97. This illustrates what we will face with our electronic records but over a longer period of time. The document file formats were not the same and the migration route originally used conversion to rich text file format. The problems did not end there as the conversion was not perfect and a number of features did not work. For example, a table of contents transferred acceptably, but all the page numbers were migrated as "1". In this case, all you had to do was delete the table of contents and reinsert it, but imagine the audit trail entries for regulated data. Transpose this to your regulatory records: how would you feel if all your data suddenly was changed to "1"?

Changing file format can have a major impact on the ability of a program to read previously acquired data – this can have a direct impact on long running work such as a stability study or a clinical trial. One of the consequences from a business as well as a compliance perspective will be that you will tend to be locked into a specific vendor. You want to look at the method of conversion to ensure that the content and meaning of the records are preserved after the conversion. Recognising this, some instrument suppliers will ensure that old data can still be read by new versions of their software. This is to be encouraged. My book on the Validation of Chromatography Data Systems[18] presents a case study of a migration between CDS applications from the same vendor.

18.3.6 Selection of Off-line Archive Media

There are many problems with selecting the media used for off-line archiving; there are a number of options:

- CD-ROM disks: CD-ROM has a *de facto* standard based on the Sony-Phillips co-development of the technology with a reasonable capacity of 650 MB unless you have large data files to archive, *e.g.* using DAD detector spectra.
- Magneto-optical drives have larger capacity and are stable against magnetic fields but they do not have a universal standard.
- DVD disks: DVD has larger capacity but there are many standards and capacities and it is not known which one will succeed.
- The life time of optical media may limit their long term use.

This is not very encouraging, which is why it is better to use on-line storage.

18.3.7 Changing the Instrument Data System – What Are the Archive Options?

The may be an occasion when you may need to consider alternative approaches to records retention as outlined in the FDA Guidance on Part 11 Scope and Application.[7] Here is the option to transfer electronic records to paper (and other media) providing the content and meaning of the records are preserved. However, with the more recent regulatory guidance about not converting dynamic data, this advice should not be followed.

If faced with changing from one supplier's data system to another, then the options for preserving the records should be reviewed and assessed. What does this really mean in practice? This section should also be read in conjunction with the chapters on data migration (Chapter 36) and system retirement (Chapter 35) of my CDS Validation book[18] to get a fuller picture of the options available.

18.3.8 Overview of Some Options

Some of the options for consideration for the records retention of the replaced data acquisition system are:

- Maintain the old instrument records as paper and formally destroy the corresponding electronic records. This option is not really tenable in light of all regulatory guidance on data integrity that electronic records are the raw data and not paper.[6,8]
- Maintain both the paper records and the corresponding electronic records with a workstation to reprocess the data if there are any requests from laboratory customers or inspectors.
- If possible, virtualise the old system for the storage and replay of data.
- Import the old data files into the new data system *via* a conversion utility and manually enter the other metadata by hand.
- Convert the records to PDF using Adobe Acrobat providing the conversion preserves content and meaning of the records. However, this option does not allow for the replay of data and converts dynamic to static data.[6]
- Convert to electronic format using a Scientific Data Management System (SDMS) but if you cannot reprocess the data this is not a viable option.

Other options may be possible depending on the two data systems involved and technical options available at the time of the work occurring, *e.g.* the two data systems are from the same supplier and a conversion utility for all records is available and is practicable to use.

18.3.9 Assessment of Option Feasibility

As part of the risk assessment process, each of the above options should be reviewed against a series of criteria such as:

- Technical feasibility; each option in the section above should be graded as Easy, Medium or Difficult.
- Preserving record content and meaning: either Yes (preserves content and meaning) or No (does not preserve content and meaning).
- Cost of the approach: evaluated as either Low, Medium and High.
- Regulatory risk: graded as Low, Medium and High.
- Value of records over time can cover a number of issues: *e.g.* how often have archived records been accessed and where the laboratory is in the pharmaceutical value chain (R&D, active pharmaceutical ingredient or secondary production).

Each factor can be scored with a simple method such as technical feasibility options such as low = 1, medium = 2 and high = 3. The same principles can be applied to the other factors for consideration in the risk assessment. To help make a decision, the scores for each factor can be added up and the one with the lowest total is the preferred option. However, it is impossible to give definitive guidance on what to do in an individual instance, as each situation has unique factors to consider.

18.4 OECD Guidance for Developing an Electronic Archive

This guidance provides an approach for creation of an archive in compliance with the principles of GLP and covers both paper and electronic records.[19] However, if the data system is used as a hybrid system there will be both paper and electronic records to consider PLUS the synchronisation between the two media. The focus in this section is on an electronic archive.

18.4.1 Definitions

The archive is a secure facility or room with an archivist specifically responsible for logging material in and out of the archive. Security and access are major issues within a GLP archive: only authorised personnel should have access and visitors are always escorted with a paper archive. An archive is not mentioned in the GMP regulations.

The key definitions used in this document[19] are:

- Archive: A designated area or facility (*e.g.* cabinet, room, building or computerised system) for the secure storage and retention of records and materials.
- Archivist: An individual designated by test facility or test site management to be responsible for the management of the archive, *i.e.* for the operations and procedures for archiving.
- Electronic archives: Facilities and systems provided to maintain electronic records as required by the Principles of GLP.

- Electronic record: All original laboratory records and documentation, including data directly entered into a computer through an instrument interface, which are the results of original observations and activities in a study and that are necessary for the reconstruction and evaluation of the report of that study.
- Metadata: Data that describe the attributes of other data. Most commonly these are data that describe the structure, data elements, inter-relationships and other characteristics of electronic records.
- Migration: The transfer of electronic records from one format, media or computerised system to another.
- Process Owner: The manager, or designee, of the department that is most impacted by, or is the primary user of, the system.

18.4.2 Roles and Responsibilities

OK, so who is involved and what do they do in this archiving process? The guidance document outlines several roles and responsibilities such as test facility management, study director, QA, IT and the archivist. I want to focus on just two roles for the purposes of this discussion: the archivist and IT. If you work in a GMP regulated laboratory you can interpret "archivist" as the person in either QA or the laboratory who will be assigned the task of records manager.

Archivist: The archivist is the gatekeeper for allowing material into and out of the archive. If looking in a GMP facility, it will be the staff member(s) responsible for deciding when to move electronic records from the data system to the archive. Inevitably there must be procedures for controlling and recording this process. Therefore, get the structure and the organisation of the records right: remember the old saying Garbage In – Garbage Out (GIGO). Therefore, file naming conventions for projects and, if required data files, including the associated contextual metadata are important. File naming tends to be automated in an instrument data system with a database and the only naming convention required would be for each individual project.

Information Technology (IT) Personnel: IT personnel involved in archiving operations (such as ensuring integrity of electronic records) should be adequately trained and understand GXP regulations as it relates to their work. They must work closely with the archivist and have either SOPs and/or written agreements to cover this work under Clause 3.1.[2] Establishment of physical security of the server and the room where it is housed and the logical security of who can access the electronic records are vital, as we will see next.

18.4.3 Archive Facilities

The guidance discusses construction of the archive at some length as this, however, let us focus on the facilities for storage of electronic records.

- The components that provide storage of unique electronic records should be physically secure. These would typically be housed in a cage that should be locked in a data centre or computer room.
- The computerised archive facility should have procedures to prevent unauthorised access and virus protection. This should include both logical and physical security elements.
- The security controls necessary to restrict access to electronic records will usually be different from those applied to other record types. Since many electronic storage media can be re-used (*e.g.* overwritten), measures should be implemented to ensure that records cannot be altered or deleted, *e.g.* write protected disks.
- *For electronic archives .... as a minimum deletion or alteration of electronic records in electronic archives should be avoided.*[19] This is too tactfully worded in my view; I would interpret this as where it says "should" read "must". Any deletion or purging of electronic data from an archive must be a documented process authorised my management at the end of the record retention period.

In addition, the environmental conditions of the room where the archive servers are housed should be monitored and have an uninterruptible power supply (UPS) to protect the system from power loss. Not mentioned explicitly in the guidance, the computer hardware should be sufficiently resilient to overcome common problems such as disk or processor failures. In addition, disaster recovery including fires, floods should have contingency measures that are documented and work to ensure that electronic records are not lost in the event of failure or disaster. Depending on the nature of the work done, the archive could be storing electronic records supporting licence applications or batch release of products quite literally worth hundreds of millions or billions in sales. Therefore, great care in designing the platform of the archive is essential to ensure security, reliability and resilience. A separate facility should be established to have a second electronic archive in a separate location. The two archives should not be mirrored in case of a malware attack that would destroy both. Thus, there should be a reconciliation procedure periodically (*e.g.* monthly or quarterly) to ensure that the two instances hold identical data sets.

18.4.4 Archiving Electronic Records

The requirements for archiving of electronic records have additional needs compared with paper which the guidance presents in five main subject areas and I will discuss below.

Decision to Archive Electronic Records: There needs to be a conscious decision to work with electronic records and also retain them. This may require the need for file naming conventions if the current application software cannot do this. However, as the FDA has stated in their level 2 guidance on their web site electronic records are the raw data not paper, then this decision has been made for you.[13]

Storage Media: The decision to retain electronic records has major ramifications for any laboratory and will influence the choice of storage media. Avoid optical off-line storage as there will be the need for checking that the disks are still readable. Will more than one copy be needed? Do I use different optical disk suppliers? One key issue is that off-line storage media change relatively rapidly. From my perspective, the best way to store records is on line to avoid the media problems.

This now brings the next issue – can we read the data in the same format for the length of the retention period? We may need the original software or we may need to migrate records to a new file format to ensure readability. Computer technology is developing rapidly and devices capable of reading storage media in common use today probably will not be available in the future. If records need to be migrated, procedures must be in place to ensure that integrity of data as the records move from one format to another. Do not be surprised if data elements are lost during the migration but this needs to be documented and the impact assessed. This is reflected in the guidance as follows: Electronic media may be discarded when the media itself no longer permits evaluation (due to hardware or software issues) provided the disposal is authorised, documented, and electronic records are migrated and any record losses documented.[19]

Defined Archive Area on a Computerised System: Electronic records may be moved from the production part of a computerised system to a discrete, secure archive area on the same computer system (physically separated, *e.g.* file record systems), or explicitly marked as archived (logically separated, *e.g.*, database record systems). Archived records should be locked so that they cannot be altered or deleted without detection.

Again, in a pragmatic way, the OECD guidance[19] offers the following advice; viewing electronic records without the possibility of alteration or deletion of the archived electronic record or replicating within another computerized system does not constitute retrieval of a record. This offers a major advantage over a paper archive – you need to be inside the archive to see the physical paper record. In the electronic world, virtually anyone could have read only access to the electronic record.

Dedicated Electronic Archive System: Complete electronic records must be entered into the archive system. This includes all the data and metadata including electronic signatures required for reconstruction of a study, otherwise the migration is useless. When electronic records are being transferred into the archive the person doing this needs to ensure that there is agreement between what is documented to be transferred and what has actually been transferred. Ideally, the archivist should be the system owner and have an agreement with IT to manage the system.

Maintenance and Preservation of Electronic Records: This is the difficult bit – we have about 2000 years' experience with handling and storing paper records but less than 50 years with computerised systems, less if you consider that it was only in the 1980s when computers became widely available in the laboratory. Procedures should be in place to ensure that essential information remains complete and retrievable throughout the specified retention period this also includes the mechanisms for backing up the systems.

What may be problematic is the statement: If the record medium requires processing in order to render the retained records into a readable format, then the continued availability of appropriate equipment should be ensured. This is the problem, how many vendors provide a future proof method of maintaining electronic records? Standards to help migration are emerging but need to be adopted widely.

If electronic record migration is necessary, the process of migration should be fully documented, and validated to ensure complete and accurate migration of the original records before they are lost or destroyed. If it is impossible to migrate the records to new electronic media it may be necessary to migrate to paper records.

References

1. *WHO Technical Report Series No. 996 Annex 5 Guidance on Good Data and Records Management Practices*, World Health Organisation, Geneva, 2016.
2. *EudraLex – Volume 4 Good Manufacturing Practice (GMP) Guidelines, Annex 11 Computerised Systems*, European Commission, Brussels, 2011.
3. *21 CFR 58 Good Laboratory Practice for Non-clinical Laboratory Studies*, Food and Drug Administration, Washington, DC, 1978.
4. *21 CFR 211 Current Good Manufacturing Practice for Finished Pharmaceutical Products*, Food and Drug Administration, Sliver Spring, MD, 2008.
5. *21 CFR 11 Electronic records; electronic signatures, final rule*, in *Title 21*, Food and Drug Administration, Washington, DC, 1997.
6. *MHRA GXP Data Integrity Guidance and Definitions*, Medicines and Healthcare Products Regulatory Agency, London, 2018.
7. *FDA Guidance for Industry, Part 11 Scope and Application*, Food and Drug Administration, Rockville, MD, 2003.
8. *FDA Draft Guidance for Industry Data Integrity and Compliance with cGMP*, Silver Spring, MD, USA, 2016.
9. *EudraLex – Volume 4 Good Manufacturing Practice (GMP) Guidelines, Chapter 4 Documentation*, E. Commission, Brussels, 2011.
10. *Inspection of Pharmaceutical Quality Control Laboratories*, Food and Drug Administration, Rockville, MD, 1993.
11. *OECD Series on Principles of Good Laboratory Practice and Compliance Monitoring Number 1, OECD Principles on Good Laboratory Practice*, Organisation for Economic Co-Operation and Development, Paris, 1998.
12. *OECD Series on Principles of Good Laboratory Practice and Compliance Monitoring Number 17 on Good Laboratory Practice Application of GLP Principles to Computerised Systems*, Organisation for Economics Co-Operation and Development, Paris, 2016.

13. *FDA Questions and Answers on Current Good Manufacturing Practices, Good Guidance Practices, Level 2 Guidance – Records and Reports*, 2010 [27 May 2016], Available from: http://www.fda.gov/Drugs/GuidanceCompliance-RegulatoryInformation/Guidances/ucm124787.htm.
14. *FDA Guidance for Industry Circumstances that Constitute Delaying, Denying, Limiting, or Refusing a Drug Inspection*, Food and Drug Administration, Rockville, MD, 2014.
15. R. S. McDonald and P. A. Wilks Jr., JCAMP-DX: a standard form for exchange of infrared spectra in computer readable form, *Appl. Spectrosc.*, 1988, **42**(1), 151–162.
16. J. Liscouski, *Laboratory and Scientific Computing – A Strategic Approach*, Wiley, New York, 1995.
17. J. M. Vergis, *et al.*, Unlocking the power of data, *LCGC North Am.*, 2015, **33**(4), 270–281.
18. R. D. McDowall, *Validation of Chromatography Data Systems: Ensuring Data Integrity, Meeting Business and Regulatory Requirements*, Royal Society of Chemistry, Cambridge, 2nd edn, 2017.
19. *OECD Series on Principles of Good Laboratory Practice and Compliance Monitoring Number 15: Establishment and Control of Archives that Operate in Compliance with the Principles of GLP*, Organisation for Economic Co-Operation and Development, Paris, 2007.

CHAPTER 19

Quality Metrics for Data Integrity

Quality metrics are objective measures used by management and other interested parties to monitor the overall state of quality of a GXP organisation, activity or process or study conduct, as applicable. In the WHO and PIC/S guidance documents quality metrics are also advocated for data integrity, *e.g.* Measures to assess the effective functioning of pharmaceutical quality system controls and of the performance, quality and safety of medicinal products and reliability of data.[1]

However, it is important that metrics or key performance indicators (KPIs) do not drive activities that are counter to a data integrity culture, *e.g.* managers performance is based on a low number of deviations and investigations.[2] Similarly, aggressive laboratory turn-around times can influence analyst behaviour and compromise data integrity.

19.1 What Do the Regulators Want?

19.1.1 EU GMP Chapter 6 Quality Control

As analytical procedures are subject to variation, it is an expectation to see not only OOS but also out of expectation (OOE) and out of trend (OOT) results. There is a specific EU GMP regulation requiring trending in EU GMP Chapter 6[3]:

> 6.9 Some kinds of data (*e.g.* tests results, yields, environmental controls) should be recorded in a manner permitting trend evaluation.
>
> Any out of trend or out of specification data should be addressed and subject to investigation.

Data Integrity and Data Governance: Practical Implementation in Regulated Laboratories
By R. D. McDowall

Published by the Royal Society of Chemistry, www.rsc.org

6.16 The results obtained should be recorded. Results of parameters identified as quality attribute or as critical should be trended and checked to make sure that they are consistent with each other......

What would be an OOS rate in a regulated laboratory? If you do not know – there might be some issues with the next inspection.

19.1.2 FDA Quality Metrics Guidance for Industry

The FDA's draft guidance for industry on quality metrics, published in November 2016,[4] proposed three quality metrics for companies to submit to the agency, one of which was focused on the laboratory:

Invalidated Out-of-Specification (OOS) Rate (IOOSR) as an indicator of the operation of a laboratory which was defined as:

IOOSR = the number of OOS test results for lot release and long-term stability testing invalidated by the covered establishment due to an aberration of the measurement process divided by the total number of lot release and long-term stability OOS test results in the current reporting timeframe.[4]

Invalidated OOS results rate is an important data integrity criterion, as shown in the 483 Observations issued by the FDA in April 2017 to Lupin Limited.[5]

Observation 1. There is a failure to thoroughly review any unexplained discrepancy and the failure of a batch or any of its components to meet any of its specifications whether or not the batch has been already distributed. Specifically, from January 2016 to March 2017, your firm invalidated several OOS results as summarised in Table 19.1:

19.1.3 WHO Guidance on Good Data and Record Management Practices

The WHO guidance document[1] has several references to quality metrics:

Table 19.1 Invalidated out of specification results from the Lupin Limited 483 observations 2017.[5]

Product	No. OOS investigations	No. OOS investigations deemed invalid	% OOS invalidated
Finished product	89	67	75
Stability	31	30	97
Raw material	48	34	71

1.4: Modernization of quality assurance inspection techniques and gathering of quality metrics to efficiently and effectively identify risks and opportunities to improve data processes.

4.6: Management governance. To establish a robust and sustainable good data management system it is important that senior management ensure that appropriate data management governance programmes are in place.

Elements of effective management governance should include: ….

- application of appropriate quality metrics;

6.4: Management reviews and regular reporting of quality metrics facilitate meeting these objectives. This requires designation of a quality manager who has direct access to the highest level of management and can directly communicate risks, so that senior management is made aware of any issues and can allocate resources to address them. To fulfil this role the quality unit should conduct and report to management formal, documented risk reviews of the key performance indicators of the quality management system. These should include metrics related to data integrity that will help identify opportunities for improvement. For example:

- Tracking and trending of invalid and aberrant data may reveal unforeseen variability in processes and procedures previously believed to be robust, opportunities to enhance analytical procedures and their validation, validation of processes, training of personnel or sourcing of raw materials and components;
- Adequate review of audit trails, including those reviewed as part of key decision-making steps (*e.g.* GMP batch release, issuance of a GLP study report or approval of case report forms), may reveal incorrect processing of data, help prevent incorrect results from being reported and identify the need for additional training of personnel;
- Routine audits and/or self-inspections of computerized systems may reveal gaps in security controls that inadvertently allow personnel to access and potentially alter time/date stamps. Such findings help raise awareness among management of the need to allocate resources to improve validation controls for computerized systems;
- Monitoring of contract acceptors and tracking and trending of associated quality metrics for these sites help to identify risks that may indicate the need for more active engagement and allocation of additional resources by the contract giver to ensure quality standards are met.

19.1.4 PIC/S PI-041 Good Practices for Data Management and Integrity in Regulated GMP/GDP Environments

There is a section on quality metrics in the PIC/S guidance[1] that is listed below with both requirements with a warning:

6.4.2 The company's Quality Management System should be able to prevent, detect and correct weaknesses in the system or their processes that may lead to data integrity lapses.

The company should know their data life cycle and integrate the appropriate controls and procedures such that the data generated will be valid, complete and reliable.

Specifically, such control & procedural changes may be in the following areas: ......

Quality metrics and reporting to senior management

6.5.1 There should be regular management reviews of quality metrics, including those related to data integrity, such that significant issues are identified, escalated and addressed in a timely manner.

Caution should be taken when key performance indicators are selected so as not to inadvertently result in a culture in which data integrity is lower in priority.

19.1.5 MHRA GXP Data Integrity Guidance and Definitions

The final version of the MHRA GXP data Integrity Guidance and Definitions has in Section 3.3 the following statement[6]:

3.3 The impact of organisational culture, the behaviour driven by performance indicators, objectives and senior management behaviour on the success of data governance measures should not be underestimated.

The data governance policy (or equivalent) should be endorsed at the highest levels of the organisation.

19.1.6 Regulatory Guidance Summary

There is a regulatory requirement in EU GMP Chapter 6 to trend analytical data,[3] however, with the publication of data integrity guidance documents there is a regulatory expectation to improve data integrity, QA and QMS processes through metrics. There is a warning from the PIC/S guidance that collecting metrics can have the opposite effect from that intended.

19.2 KPIs and Metrics for the Laboratory

Key Performance Indicators (KPIs) or metrics are used by many laboratories to measure the operational and quality performance of processes and the laboratory itself. As Peter Drucker said – if you cannot measure it; you cannot control or manage it. Quality metrics are also a requirement of the ISO 9001 quality standard and six sigma initiatives for continuous improvement of processes. In this chapter, we look at the regulatory requirements for data integrity metrics so that laboratory managers can understand how their laboratories are performing with respect to data integrity and to identify where they may have potential problems.

19.2.1 Understanding Laboratory Metrics

Consider a laboratory analysis: a supervisor or manager needs to know how many samples are being analysed and how long it takes to do each one. They need to know if there are any bottlenecks and if they have enough resources to do the job. Metrics can provide the information to measure a process or activity. If used correctly, metrics help staff understand the performance measures that a laboratory is judged against and some common laboratory metrics are shown in Table 19.2.

19.2.2 Metrics Must Be Generated Automatically

One of the key requirements for collection of metrics is that the process must be automatic and not manual. Why you may ask? The simple reason is that if humans are used to collect and collate the data manually it is error prone, slow, tedious and labour-intensive process. Collection could also subject to falsification (*e.g.* I am late to give in my metrics, so I will make them up or if they do not make me/us look good so I will alter them to make us look better than we are). This is not the best option when generating data integrity metrics.

Automatic metric generation is the only way to create metrics that are objective and reliable in a timely and repetitive manner. This is where

Table 19.2 Some common laboratory key performance indicators (KPIs) or metrics.

Metric	Measurement
Turn Around Time (TAT)	• Time from sample receipt to delivery of final report/ Certificate of Analysis • If required, further sub-division is possible, *e.g.*: Time between receipt and start of analysis Time for the actual analysis Time for review of results Time for report writing
Analytical runs	• Number of runs per time period • Number of aborted runs • Number of short runs • Percentage of aborted runs
Instrument utilisation	• Percentage utilisation per instrument (weekends included or excluded) • Machine breakdowns: time off line • Samples analysed between maintenance periods
Injections	• Number of injections per chromatograph per run • Peaks analysed with automated *versus* manual integration • Percentage of peaks manually integrated

suppliers of laboratory informatics solutions such as instrument data systems and Laboratory Information Management System could help greatly in having database search functions that generate both general laboratory and data integrity metrics automatically. The search routines can be either provided as standard or user defined.

19.2.3 Why Metrics for Data Integrity?

We now need to ask why data integrity metrics are important. Put simply it is now a regulatory expectation as we can from the PIC/S guidance PI-041 [2] outlined in Section 19.1.4. As can be seen, there is an expectation that data integrity metrics of processes and systems are collected for management review. Of course, there is the implicit expectation to act if the metrics indicate an activity or trend that compromises data integrity.

19.2.4 Do Quality Metrics Lead Behaviour?

Note the last part of Clause 6.5.1 of the PIC/S guidance presented in Section 19.1.4, it says be careful when selecting a KPI or metric so *as not to result in a culture in which data integrity is lower in priority*. Let us consider this statement further.

Take the example of Turn Around Time (TAT) shown in Table 19.2. Imagine that the laboratory target for TAT is set for 15 days and over the past few months the target has been missed and the laboratory manager is not pleased. The sample you are starting to analyse has been in the laboratory for thirteen days and your analysis will take three days to perform. You will miss the TAT target unless......

If the target TAT is too challenging or work is too close to the target, will analytical staff be tempted to cut corners to ensure that the metric is met and thus compromise data integrity? This is where quality culture, management and data integrity collide and it is where the PIC/S guidance caution about metrics comes into play – are some metrics intended to monitor activities capable of inducing behaviours that compromise data integrity? As noted by Churchward in an MHRA blog[7]:

> Performance metrics can also contribute to this pressure. Instead of assessing performance based on a 'zero defects' or 'right first time' indicator, perhaps measures relating to 'released when correct' may be more appropriate?

Figure 19.1 shows the turnaround time for samples in a GMP regulated analytical laboratory that consists of three sections: raw materials, finished products and stability testing. Ninety analysis reports, thirty from each section, were selected at random to determine the turnaround time for analysis. What is interesting is the wide range of times – these are real data generated before data integrity was an industry issue. The bulk of the

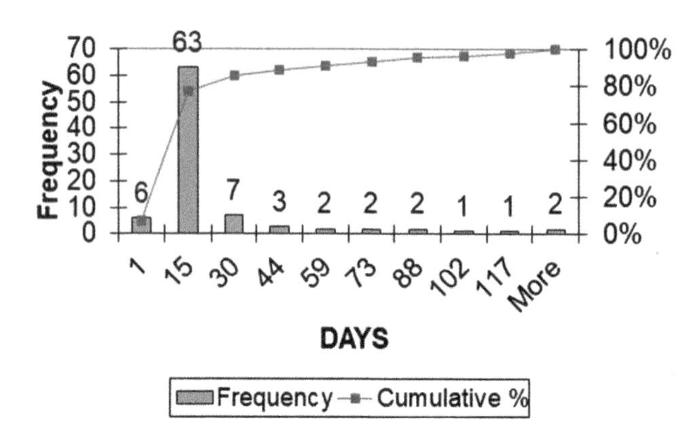

Figure 19.1 Turn Around Time for a regulated laboratory (n = 90).

samples, *circa* 77%, were analysed within 15 working days. However, nearly a quarter of the samples took between 15 and 150 days to complete analysis. There were several reasons for this, but one of the main reasons is the time taken to adequately investigate out of specification results and to find the root cause.

The market value of batch(es) of finished product impacted by an OOS result can be of the order of millions of pounds/dollars/euros. Therefore, there can be terrific organisational pressure on the laboratory to invalidate an OOS result, reanalyse and release the batch. If the OOS is produced in stability testing, then regulatory agencies must be informed, *e.g.* FDA requires a field alert to be sent within 72 h with the potential for a product recall. As can be seen, there can be enormous pressure within an organisation, pharmaceutical company, CRO or CMO to gloss over a scientifically sound OOS investigation and put the error down to one of the two classic laboratory excuses:

- analyst error – resulting in retraining of an individual;
- instrument malfunction – "fix" the problem.

The problem is as you can see from the Lupin 483 inspectional observations[5] and presented in Table 19.1, the root cause is never identified, the OOS results were invalidated and product released to the market or in the case of the stability studies avoiding informing the regulatory authorities.

However, "instrument problems" as a reason to invalidate laboratory results can give rise to warning letters. Elan Pharmaceuticals had a facility in Georgia, USA that when inspected by the FDA in the mid-1990s, had multiple "instrument failures" in 24 hour extended release dissolution testing. This was stated to be due to air bubbles in the sampling tubes between the vessels in the dissolution bath to the UV spectrophotometer used for automated analysis. Examination of the data found that most of the failures occurred at about 16 h. Typically, the automated dissolution test was started in the late afternoon and when the analyst came in the next morning, *circa* 16 h later, they could see that the test would fail, resulting in the "observation" of air bubbles in the sample tube and the excuse to invalidate the run. Hence, the inevitable warning letter and consent decree.

19.2.5 Are Incidents Hidden Metrics?

Deviations from any procedure requires documenting, investigation to identify the root cause and may result in CAPA actions for major or critical deviations. They should also be subject to metrics such as time to identify the root cause and effectiveness of the CAPA plans. However, as described in Chapter 16, are deviations disguised as "incidents" and hence off the deviation radar? There is no regulatory requirement for an incident (apart from IT incident in EU GMP Annex 11[8]), but there is GMP requirement for deviations to be investigated.[3,9]

19.3 Data Integrity Metrics in an Organisation

19.3.1 Overview: Start Small and Expand

It is tempting to set up automatic collection of metrics for as much as possible and then sit at a table at regular intervals and gaze at them wondering what they all mean. This is the start of analysis paralysis. The best situation is to start small and gather a relatively small number of meaningful data integrity metrics and adjust the criteria by which you judge them by, *e.g.* acceptance, warning or alarm limits.

Another factor to consider here is as you monitor some activities, they will improve because they get attention of the laboratory and the organisation. However, this approach can be at the expense of other activities that are not monitored. Be aware of what you do not monitor as well, for example, measuring metrics on quality control production turn-around time could cause stability tests to be neglected.

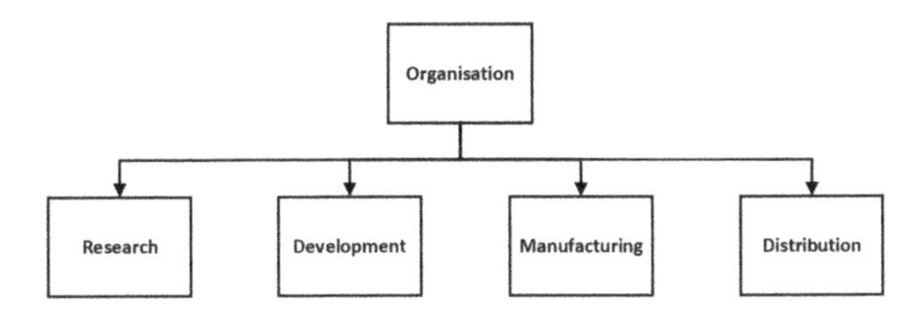

Figure 19.2 Scope of a pharmaceutical organisation.

19.3.2 Scope of the Organisation

The scope of a typical pharmaceutical organisation is shown in Figure 19.2 and ranges from research through development to manufacturing and distribution of finished products. The first question for the data integrity committee is what is the scope of the metrics collection? Given that people move around organisations and that research records can be included in pharmaceutical applications to regulatory authorities, the best approach would be to extend the collection of metrics to the whole organisation.

19.3.3 Some Suggested Data Integrity Metrics

This is an evolving subject but the aim in this chapter is a discussion of data integrity metrics that could be generated as part of a data integrity programme of work as well as for routine analytical work. Let us look where in an organisation, metrics for data integrity could be generated in Figure 19.3. Obviously in a book written for regulated laboratories, we will not consider manufacturing or distribution data integrity metrics. Although focused on a Quality Control laboratory, the same principles will apply to laboratories in pharmaceutical R&D as well as laboratories in contract research and manufacturing organisations.

The scope of data integrity metrics can cover four main areas within an organisation:

- data governance such as data integrity policy(ies) and the associated training;
- assessment and remediation of laboratory manual processes and computerised systems;
- development and production activities including outsourcing of laboratory analysis;
- quality assurance oversight.

Each area will be discussed in the following sections.

19.4 DI Policies, Assessment and Remediation of Processes and Systems

The first two areas from Figure 19.3 to consider for data integrity metrics are data integrity policies with associated procedures and assessment and remediation of processes and systems.

19.4.1 Data Integrity Policy and Associated Procedures

A data integrity policy and the associated procedures, covered earlier in Chapter 7, should include the following:

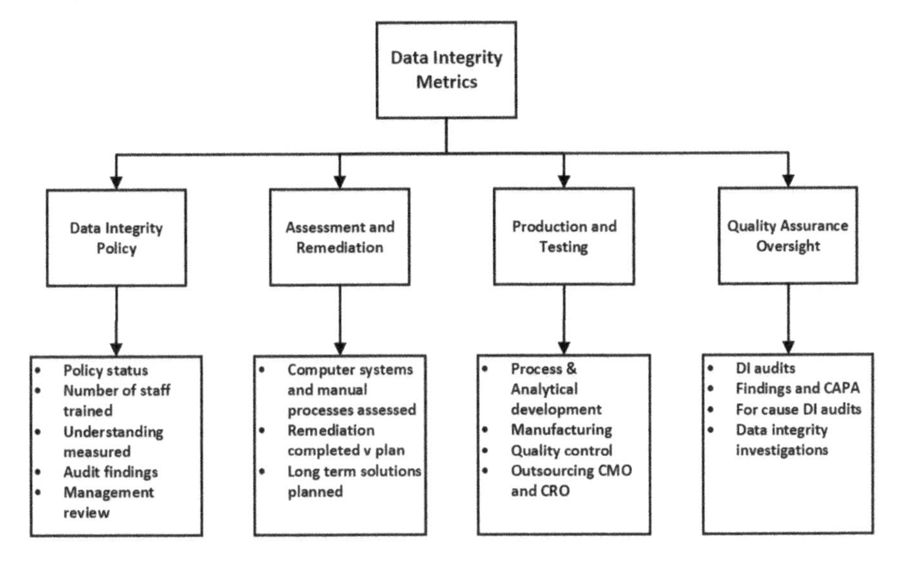

Figure 19.3 Scope of data integrity metrics within an organisation.

- DI Policy outlining company expectations, behaviours and culture;
- Good Documentation Practices SOP covering paper, hybrid and electronic processes and systems;
- data acquisition followed by data interpretation, processing and calculation;
- second person review SOP including audit trail & e-records review;
- investigating out of specification results;
- updating current procedures to include validation processes that ensure the integrity of records and detection of improper actions.

Metrics for the policy and procedures can be seen in Table 19.3 and focus on the training for the policies and procedures and its effectiveness. This is a subject for which a read and understand approach is not tenable and demonstrable evidence of understanding is required, as discussed in Chapter 7.

19.4.2 Assessment of Processes and Systems

Assessment of processes and systems should cover the following activities:

- Assessment of new systems for integrity gaps prior to purchase.
- Identification of existing computerised systems and paper processes for assessment including databases and spreadsheets and their listing in an inventory.
- Processes and systems are prioritised by risk and record impact.
- Assessment of systems and paper process;
  Data process mapping to highlight records generated and vulnerabilities.

Table 19.3 Metrics for data integrity policies, assessment and remediation of paper processes and computerised systems.

DI subject area	Suggested metrics
Data governance integrity policy and associated procedures	• DI Policy Status: Drafted, Issued, In-Revision *versus* schedule • Staff training in DI policies and procedures Staff training: number completed *versus* plan Number of staff passing first time • Metrics on HR issues related to DI
Assessment of processes and systems	• Number of systems and processes, risk priority • Percentage of high, medium and low risk systems and processes assessments completed *versus* plan
Remediation of processes and systems	• Percentage of short term quick fixes completed *versus* plan: high, medium and low risk processes and systems • Longer term improvement or solution projects completed *versus* plan: high, medium and low risk processes and systems • Manual processes: percentage of uncontrolled blank forms replaced by a computerised process or control over master templates and controlled blank forms

For each process and system there should be short and long-term remediation plans.

- Proposed remediation of some spreadsheets, maybe their elimination and incorporation of the calculations in an instrument data system or LIMS.

The metrics shown in Table 19.3 are focused on the completion of assessments for high, medium and low risk systems.

19.4.3 Executed Remediation Plans

Metrics should be generated for execution of remediation plans processes and computerised systems. The implementation order of both short-term remediation and long-term solution should be based on the risk posed by an individual process or system.

Some suggested remediation metrics are listed below and also presented in Table 19.3:

- Short term quick fixes to remediate critical risks to records, *e.g.* eliminate shared user identities or restrict access to data directories to meet ALCOA+ criteria.
- Longer term solutions (*e.g.* update the system to have a database or effective audit trail, or replacing hybrid systems with electronic systems and the elimination of paper and spreadsheets).
- Replacement of manual processes utilising uncontrolled blank forms by either a computerised process, or controlled master templates and blank forms with reconciliation.

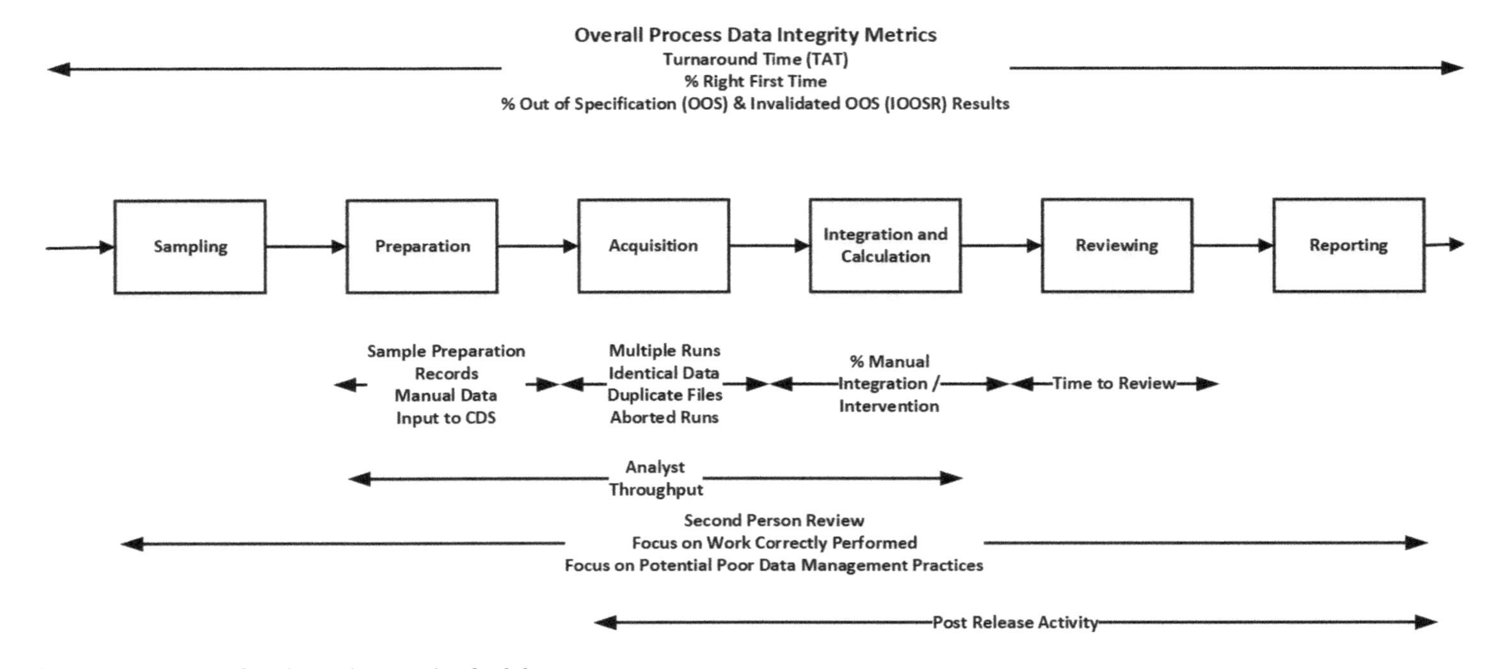

Figure 19.4 Some data integrity metrics for laboratory processes.

19.5 Laboratory Data Integrity Metrics

Looking at the regulated laboratory and the analysis carried out therein we can see some of the data integrity metrics that could be generated in Figure 19.4.

19.5.1 Some Preliminary Considerations for Laboratory Data Integrity Metrics

It is nearly impossible to generate metrics automatically for manual activities or hybrid computerised systems. For example, you could detect missing files in a system by comparing the instrument use log (paper) against a report of sample identities from all runs stored on a computer system. However, this approach is too laborious and tedious to be done on a regular or reliable basis. Such searches are best left for either a data integrity audit or data integrity investigation. In contrast, sample IDs in a LIMS or ELN could be compared against the data files on laboratory data systems using an automated script.

Owing to the time involved in their preparation, manual reports should be limited to specific scenarios over a specific time period (*e.g.* detecting fraud from a single suspected individual as part of an investigation). Most data integrity reports can only identify points of concern that would suggest further investigation as it is rare where a single electronic record equates to one data integrity breach. You should think which metrics in which areas are required as too much data can bury you in an attempt to help you. As noted in Section 19.3.1, start small with data integrity metrics, evaluate the data and refine as necessary. Consider the following items:

- Not all metrics are equal, therefore neither should their review be equal.
- There are some activities that are just hard to catch with an automated report. For example, once a user changes the naming convention for repeat injections, your search will probably not detect it in a routine report.
- Some things cannot be reported. For example, analysts can delete data on a standalone balance or other simple instrument as demonstrated in the Sun Pharmaceuticals warning letter: "*He said that he recalibrated the balance and prepared new documentation, and subsequently discarded the original record. Furthermore, we learned that additional original calibration records of other balances had similarly been discarded.*"[10] The data were not saved to create a GXP record. This is where the investment in data capture to a central informatics system can improve the laboratory operations and avoid regulatory citations.
- Metrics could also include assessment of the performance of individuals, *e.g.* measuring the time to complete a method for several different people and looking for someone working too quickly. This is where regulatory compliance clashes with, in some countries, the works council who are concerned with protecting the rights of workers. The balance

between ensuring data integrity of the organisation producing and marketing pharmaceutical products needs to be balanced with the rights of individuals. However, the regulatory expectation is that processes must ensure the integrity of records generated.

Some metrics for laboratory processes are shown in Table 19.4 and cover the main areas of chromatographic analysis.

19.5.2 Outsourced Laboratory Testing

The assessment of a CMO or CRO laboratory is presented in detail in Chapter 23 and can be summarised as does the CRO have a similar approach to data integrity as your organisation? Some of the same metrics can be used to monitor the outsourcing laboratory's work as outlined in the preceding sections. The metrics of most interest to outsourced work are:

- number of OOS results;
- number of invalidated OOS results with a copy of the investigation report and the root cause analysis.

19.6 Quality Assurance DI Metrics

The main areas for quality assurance oversight in data integrity are DI audits and investigations and the resulting CAPAs that are raised following them. Typically, there will be:

- DI audits following a schedule of both announced and unannounced visits for all processes, systems and areas of an organisation. The reason

Table 19.4 Some data integrity metrics for laboratory processes.

DI subject area	Suggested metrics
Re-processed sample analysis	• Sample results that were released, withdrawn and released again. (This is usually done for errors discovered after release.)
Injections	• Run times longer than average. • Run times shorter than average.
Exported runs	• A missing metric from some chromatography data systems is a list of all runs exported to LIMS/ELN systems. This missing piece can permit users to run a sample multiple times and release their "favourite". A report of runs with 1–2 injections per vial can help detect some of this behaviour, but in general, accounting for all chromatography injections is an issue.
Runs stopped/aborted	• This metric could indicate behaviour where an analyst does not like what they see as they look at the run in-progress, so they stop data collection.

for unannounced audits is to see a better picture of work carried out, although there is a risk of disrupting work.

- Data integrity investigations raised either by an audit or inspection finding or a staff concern.
- CAPAs arising from audits or investigations:
  Close out date *versus* planned completion date
  Effectiveness of each action plan after a follow-up assessment.

QA data integrity metrics are shown in Table 19.5.

19.7 Management Review of DI Metrics

19.7.1 Management Are Responsible for Data Integrity and the PQS

Data integrity metrics need to be reviewed regularly by senior management as they are responsible for the whole of the Pharmaceutical Quality System. The review should be formal with minutes kept of the meeting and action items should be raised and their progress monitored. This is especially true of high risk/impact systems along with rapid implementation of short term fixes to ensure any major data integrity gaps are remediated. Demonstrable progress is important as management activity in this area is best evidenced by their actions and not words. Management review and follow-up emphasises the importance of data integrity to the organisation, and assures that process/resource bottlenecks are exposed and removed.

Management must be aware that metrics can be used to monitor for some potential integrity issues but not all. Management must take care to ensure that metrics do not drive behaviours that compromise data integrity. To be sustainable and timely, collection of metrics must be automated.

Table 19.5 Data integrity metrics for quality assurance oversight.

DI subject area	Suggested metrics
QA oversight metrics	• DI audits progress against plan for scheduled and unscheduled audits • Number of DI investigations • Number of data integrity concerns raised by staff • CAPAs raised from audits and investigations Number and classification of findings Number of CAPAs closed *versus* planned completion CAPA effectiveness on follow-up • Data integrity findings from inspections (metric for the effectiveness of QA data integrity audits) • Consider some time metrics here: *e.g.* Average, maximum and minimum time to release test results, sorted by analytical procedure.

It is essential that metrics enable management monitoring of data integrity as this is a key part of success for an overall data governance and data integrity programme in an organisation. However, not all metrics are equal, they need to be chosen carefully. Use long term data integrity remediation as an opportunity to also improve productivity in laboratory processes.

19.7.2 How Regular Is Regular Review?

To demonstrate that management are engaged with both the PQS and data integrity metrics, management review needs to be both meaningful, as discussed in Chapter 6, and regular. The question now arises, how regular is regular? At the start of a data integrity programme of work, regular may mean monthly as management come to grips with the programme strands and the quality metrics for data integrity. As mentioned earlier, quality metrics for data integrity is an evolving subject and therefore for the first few months data need to be captured and analysed to see if the metric needs refinement or if it is a proper metric to gather.

Once the programme of assessment, remediation and training is underway, then the management reviews may be reduced in frequency to every two months and then quarterly. However, it is important that the communication programme discussed in Chapters 6 and 8 explains that this is not due to management indifference but due to programme progress.

Acknowledgement

I would like to thank Mark Newton for input and review in the writing of this chapter.

References

1. *WHO Technical Report Series No. 996 Annex 5 Guidance on Good Data and Records Management Practices*, World Health Organisation, Geneva, 2016.
2. *PIC/S PI-041 Draft Good Practices for Data Management and Integrity in Regulated GMP/GDP Environments*, Pharmaceutical Inspection Convention/Pharmaceutical Inspection Co-Operation Scheme, Geneva, 2016.
3. *EudraLex – Volume 4 Good Manufacturing Practice (GMP) Guidelines, Chapter 6 Quality Control*, European Commission, Brussels, 2014.
4. *FDA Guidance for Industry Submission of Quality Metrics Data, Revision 1*, Food and Drug Administration, Rockville, MD, 2016.
5. *FDA 483 Observations: Lupin Limited*, Food and Drug Administration, Silver Spring, MD, 2017.

6. *MHRA GXP Data Integrity Guidance and Definitions*, Medicines and Healthcare Products Regulatory Agency, London, 2018.
7. D. Churchward, *MHRA Inspectorate Blog: Too Much Pressure: A Behavioural Approach to Data Integrity (Part 2)*, 2017, Available from: https://mhrainspectorate.blog.gov.uk/2017/03/30/too-much-pressure-a-behavioural-approach-to-data-integrity-part-2/.
8. *EudraLex – Volume 4 Good Manufacturing Practice (GMP) Guidelines, Annex 11 Computerised Systems*, European Commission, Brussels, 2011.
9. *21 CFR 211 Current Good Manufacturing Practice for Finished Pharmaceutical Products*, Food and Drug Administration, Silver Spring, MD, 2008.
10. *FDA Warning Letter Sun Pharmaceuticals*, Food and Drug Administration, Rockville, MD, 2014.

CHAPTER 20

Raising Data Integrity Concerns

Analytical staff are often the first people to realise that there are data integrity problems in a laboratory. In Chapter 8 when discussing an open culture, there is the iceberg of ignorance where individuals close to the work know about the problems but only 4% of those are known by senior management. There needs to be in place a policy for staff to raise their concerns in confidence and without fear of retaliation either by the organisation or other employees. These concerns must be investigated and, if found to be true, further action can be initiated. Raising data integrity concerns is also known as whistleblowing.

20.1 What Do the Regulators Want?

20.1.1 WHO Guidance on Good Data and Record Management Practices

The WHO guidance has one clause on raising data integrity concerns by staff[1]:

> 5.1 All organizations performing work subject to GXP are required by applicable existing WHO guidance to establish, implement and maintain an appropriate quality management system, the elements of which should be documented in their prescribed format, such as a quality manual or

Data Integrity and Data Governance: Practical Implementation in Regulated Laboratories
By R. D. McDowall

Published by the Royal Society of Chemistry, www.rsc.org

other appropriate documentation. The quality manual, or equivalent documentation, should include a quality policy statement of management's commitment to an effective quality management system and to good professional practice.
These policies should include a code of ethics and code of proper conduct to assure the reliability and completeness of data, including mechanisms for staff to report any quality and compliance questions or concerns to management.

20.1.2 PIC/S PI-041 Good Practices for Data Management and Integrity in Regulated GMP/GDP Environments

This PIC/S good data management and integrity guidance offers more detail about what is expected in an organisation for raising data integrity concerns[2]:

> 5.2.3 An effective data governance system will demonstrate Senior Management's understanding and commitment to effective data governance practices including the necessity for a combination of appropriate organisational culture and behaviours (Section 6) and an understanding of data criticality, data risk and data lifecycle.
>
> There should also be evidence of communication of expectations to personnel at all levels within the organisation in a manner which ensures empowerment to report failures and opportunities for improvement. This reduces the incentive to falsify, alter or delete data.
>
> 6.1.2 Inspectors should be sensitive to the influence of culture on organisational behaviour, and apply the principles described in this section of the guidance in an appropriate way. An effective 'quality culture' and data governance may be different in its implementation from one location to another.
>
> Depending on culture, an organisation's control measures may be:
>
> - 'open' (where hierarchy can be challenged by subordinates, and full reporting of a systemic or individual failure is a business expectation)
> - 'closed' (where reporting failure or challenging a hierarchy is culturally more difficult)
>
> 6.2.5 There should be a confidential escalation program supported by company policy and procedures whereby it encourages personnel to bring instances of possible breaches to the Code of Conduct to the attention of senior management without consequence. The potential for breaches which may be directed by senior management should be recognised and a suitable reporting mechanism for those cases should be available.

20.1.3 NELAC Quality Standard

Although not a pharmaceutical regulations or guidance, a US environmental laboratory accreditation standard (NELAC) has the following requirements for data integrity[3]:

> 5.4.2.6.1. Management will provide a confidential data integrity reporting mechanism. This will assure confidentiality and a receptive environment for raising ethical issues.
>
> 5.4.2.6.2. The mechanism will inform laboratory management of the need to investigate data integrity further.

20.1.4 Regulatory Guidance Summary

As part of an overall data integrity and data governance programme, there needs to be a policy on data integrity and one part of this document is to describe the means of raising data integrity concerns to management without fear of any retribution. The PIC/S guidance also is sensitive to the issue of culture that means in some countries challenging management or company policy is difficult.

20.2 Data Integrity Problem or Concern?

It is important to understand the difference between a data integrity problem and a concern.

- A problem with data integrity may be the wording in a procedure, how a laboratory computerised system is configured or an issue with a working practice. These problems can be dealt with openly as part of the overall quality management system by discussing with a laboratory supervisor, manager or a member of Quality Assurance. Problems such as these can be dealt with relatively quickly.
- A concern may involve the working practices of an individual or a group that could be peer pressure as described in an MHRA blog.[4] Here, a new analyst is pressured by more experienced staff not to report an OOS result and to doubt her own skills and training as the attitude of the current laboratory staff is "we have never had results like that here before". Alternatively, a concern is if an analyst sees another individual or group who are committing falsification of results, deleting data, *etc.* In both cases what should these people do?

There is a clear regulatory expectation, outlined in Section 20.1, that within the pharmaceutical quality system there is a mechanism to raise data integrity concerns to management without fear of retaliation.

20.3 What Is Needed to Raise a Data Integrity Concern?

20.3.1 A Section in the Corporate Data Integrity Policy

The data integrity policy of any organisation must have a section dedicated to presenting the way that staff can raise concerns as they are often the first people to realise that there may be something seriously wrong within a laboratory. Some elements of this section should include the following:

- The policy must make a statement that raising a concern or whistleblowing is viewed by the company as a positive act that can make a valuable contribution to the organisation's ethical reputation with regulatory agencies.
- It is a business expectation of senior management for staff to raise data integrity concerns.
- It is not disloyal to colleagues or the organisation to speak up as it demonstrates to regulators that the company is committed to achieving high ethical standards to ensure data integrity.
- The policy should also emphasise that the person raising a concern will be protected from victimisation, harassment, discrimination and dismissal if you have made any disclosure in good faith.
- It should also state that if a data integrity concern has been raised, there will be formal feedback to the person who initiated the concern.

The aim of this section in the corporate data integrity policy is to encourage staff who have serious data integrity concerns to come forward and voice those concerns. Chapter 7 covers the data integrity policy in more detail along with the associated training. In addition to the data integrity policy, there will be an associated SOP within the QMS for this process and how a concern and the individual raising it will remain confidential.

20.3.2 Communicate and Train How to Raise Data Integrity Concerns

We discussed an open culture in Chapter 8 and here all levels of management have a responsibility to communicate and inform all staff about the data integrity policy including how to raise concerns. During training for the data integrity policy and as part of annual GMP refresher training how to raise concerns must be discussed. In addition, if Gemba walks are conducted by management this is an ideal opportunity for a senior manager to mention the process and encourage staff to use it if they have a data integrity concern. An alternative communication method may be to use posters in corridors or laboratories detailing how a concern can be raised.

20.3.3 Raising a Concern or Airing a Grievance?

It is important for all staff to realise that their focus must be on raising a data integrity concern under the umbrella of the data integrity policy.

It is not and must not be seen as a mechanism for raising personal grievances or harassment, which should be handled through normal Personnel or Human Resources procedures. If, however, a data integrity concern is raised frivolously, maliciously or for personal gain, appropriate action that could include disciplinary action, may be taken against the individual making the claim. These are important points when writing this section of the data integrity policy or procedure as well as the associated training for the implementation.

20.3.4 What Should Be Reported?

Any serious data integrity concerns that an individual has should be raised with sufficient information to support the concern, where possible. Where the concern involves actions of laboratory staff there may not be documented evidence available and the individual may only be able to voice their concern.

20.3.5 Protecting the Whistleblower

Any company should recognise that a decision to report a concern can be a difficult one to make. If the person honestly and reasonably believes what they say is true, the individual should have nothing to fear because they will be doing their duty to the company and to his or her colleagues, as well as protecting patients who take the company's medicines. Harassment or victimisation of a whistleblower (including informal pressures) must result in appropriate action when the individual has raised a concern in good faith and the company must treat this as a serious offence that will be dealt with through the company's disciplinary rules and procedures.

Support of the individual raising a concern should come from senior management who must take the concerns raised seriously. In some cases, it could be necessary to temporarily re-deploy the person for the period of the investigation.

If a concern is raised by an individual in good faith and reasonably believing it to be true, but it is not confirmed by an investigation, the company will recognise the concern and the individual should have nothing to fear.

20.3.6 Confidentiality

All concerns must be treated in confidence and every effort will be made not to reveal your identity if that is your wish. If disciplinary or other proceedings follow the investigation, it may not be possible to act as a result of the disclosure without the help of the individual, so they may be asked to come forward as a witness. If this occurs, then the company must offer advice and support to the individual.

20.3.7 Raising Concerns Anonymously

The data integrity policy or related procedure should encourage staff to put their name to their concern whenever possible. If a concern is raised anonymously it is much more difficult for the organisation to protect an individual or to give feedback regarding the concern.

20.4 Raising a Concern

20.4.1 Who Should You Raise Your Concern with?

This will depend on several things such as the seriousness and sensitivity of the issues involved and who is suspected of the data integrity issue. Concerns should normally be raised with one of the following:

- Line or Department Manager;
- Division Head;
- Qualified Person;
- Quality Assurance;
- Site Head or Senior Management.

20.4.2 How to Raise a Concern

There are several ways that a concern may be raised: telephone, in person, e-mail or in writing. The earlier the concern is raised, the easier it is to investigate and, if required, take action. You will need to provide the following information:

- the nature of your concern;
- why you believe it to be true;
- any background and history of the concern (with relevant dates if possible).

Although the person is not expected to prove beyond doubt the truth of their suspicion, they will need to demonstrate to the person contacted that they have a genuine concern relating to suspected malpractice and that there are reasonable grounds for the concern. It may be appropriate to consider discussing your concern with a colleague first and it may be easier to raise the matter if there are two (or more) of you who have had the same concerns.

20.4.3 Raise an Issue *via* Management or Quality Assurance?

Providing a single route to raise a data integrity concern to management may be difficult as the problem may be about a manager. An option that a company must consider is the need to have an alternative route for raising concerns *via* quality assurance that will report to a higher management level than an immediate supervisor or laboratory head.

20.4.4 What the Organisation Must Do

The company must acknowledge the concern and begin to investigate it. At this stage the organisation has not accepted or rejected the concern but is verifying that the information provided is accurate. This needs to be done within a reasonable time frame (*circa.* 10 days in some cases). The overriding principle for the organisation must be to ensure that data integrity and data quality of the work are sufficient to protect patients. To be fair to all employees, including those who may be wrongly or mistakenly placed in the spotlight, initial enquiries will be made to decide whether an investigation is appropriate and, if so, what form it should take.

The investigation may need to be carried out under terms of strict confidentiality, *i.e.* by not informing the subject of the complaint until (or if) it becomes necessary to do so.

Where appropriate, the matters raised may investigated by:

- management;
- quality assurance;
- internal audit;
- third party consultant.

Within the specified time of a concern being raised, the person investigating the concern will communicate back to the person who raised the concern:

- acknowledging that the concern has been received;
- indicating how the company proposes to deal with the matter;
- supplying the individual with information on staff support mechanisms;
- informing the individual whether further investigations will take place and if not, why not.

The amount of contact between the individual and the organisation's staff considering the issues will depend on the nature of the matters raised, the potential difficulties involved and the clarity of the original information. It is likely that the concern raiser will be interviewed to ensure that the concern is fully understood. This interview and any meeting can be arranged away from the company, if required.

The person raising the concern needs to be assured that it has been properly addressed and should be informed of the progress and outcome of any investigation. In some cases, a concern may require a formal data integrity investigation that is covered in Chapter 23.

20.4.5 What If the Company Is the Problem?

If the company and the management are the problem, there is not much point in an individual raising the issue internally. Where does that leave the concern and the person? Fortunately, there is at least one case study that points the way. We turn full circle and welcome back Able Laboratories. As discussed in Chapter 2, Able Laboratories had undergone seven FDA pre-approval inspections with no adverse findings, however, it was an internal person who contacted the local FDA field office with their concern and the rest is history.

Therefore, if there are issues that management or the company cannot or will not address, there is always the option to raise the issue with a regulatory agency.

References

1. *WHO Technical Report Series No. 996 Annex 5 Guidance on Good Data and Records Management Practices*, World Health Organisation, Geneva, 2016.
2. *PIC/S PI-041 Draft Good Practices for Data Management and Integrity in Regulated GMP/GDP Environments*, Pharmaceutical Inspection Convention/ Pharmaceutical Inspection Co-Operation Scheme, Geneva, 2016.
3. *NELAC Quality Standard*, National Environmental Laboratory Accreditation Conference (NELAC), Weatherford, TX, 2003.
4. D. Churchward, *MHRA Inspectorate Blog: Too Much Pressure: A Behavioural Approach to Data Integrity (Part 2)*, 2017, Available from: https://mhrainspectorate.blog.gov.uk/2017/03/30/too-much-pressure-a-behavioural-approach-to-data-integrity-part-2/.

CHAPTER 21

Quality Assurance Oversight for Data Integrity

In Chapter 5 we discussed a data integrity model and one of the three areas emerging from the Foundation level was Quality Assurance oversight that was responsible for checks of current work, data integrity audits and data integrity investigations. This chapter will look at QA oversight in general and data integrity audits and Chapter 22 will focus on data integrity investigations.

Quality Assurance oversight, can be defined as the process of keeping a process or initiative under observation to assure that it complies with applicable regulations and company procedures. The aim of QA oversight with respect to data integrity is independent assurance that procedures, processes and computerised systems are working compliantly and correctly for the generation of complete and accurate data and records. This chapter should be read in conjunction with Chapter 24 on a Data Integrity Audit Aide Memoire.

21.1 What Do the Regulators Want?

21.1.1 EU GMP Chapter 9 Self-inspections

The shortest chapter in EU GMP is Chapter 9 on Self-Inspections,[1] the aim of which is outlined in the Principle:

Data Integrity and Data Governance: Practical Implementation in Regulated Laboratories
By R. D. McDowall

Published by the Royal Society of Chemistry, www.rsc.org

Self-inspections should be conducted in order to monitor the implementation and compliance with Good Manufacturing Practice principles and to propose necessary corrective measures.

21.1.2 US GMP 21 CFR 211 Current Good Manufacturing Practice for Finished Pharmaceutical Products

In Section 211.192 on production record review,[2] there is the following regulatory requirement:

> All drug product production and control records, including those for packaging and labeling, shall be reviewed and approved by the quality control unit to determine compliance with all established, approved written procedures before a batch is released or distributed. Any unexplained discrepancy .... or the failure of a batch or any of its components to meet any of its specifications shall be thoroughly investigated, whether or not the batch has already been distributed.
>
> The investigation shall extend to other batches of the same drug product and other drug products that may have been associated with the specific failure or discrepancy.
>
> A written record of the investigation shall be made and shall include the conclusions and follow up.

The problem with the FDA regulations is the term quality control unit that is interpreted flexibly by the agency. Sometimes, it will refer to the quality control laboratory and, more often, in recent data integrity warning letters the term "quality unit" appears meaning quality assurance.

21.1.3 FDA Compliance Program Guide 7346.832 for Pre-approval Inspections

This CPG was rewritten after the Able Laboratories fraud case and became effective in May 2012,[3] and Objective 3 is the Data Integrity Audit:

> Audit the raw data, hardcopy or electronic, to authenticate the data submitted in the CMC section of the application. Verify that all relevant data (*e.g.*, stability, biobatch data) were submitted in the CMC section such that CDER product reviewers can rely on the submitted data as complete and accurate.

If this is the approach taken by the FDA when conducting pre-approval inspections, then a similar approach needs to be taken internally to ensure that all data are complete, consistent and accurate.

21.1.4 21 CFR 58 Good Laboratory Practice for Non-clinical Laboratory Studies

In US GLP regulations 21 CFR 58.35 from 1978,[4] there is the requirement for the Quality Assurance Unit (QAU) to:

> (a) A testing facility shall have a quality assurance unit which shall be responsible for monitoring each study to assure management that the facilities, equipment, personnel, methods, practices, records, and controls are in conformance with the regulations in this part. ....
>
> (b) The quality assurance unit shall: ....
>
> (3) Inspect each nonclinical laboratory study at intervals adequate to assure the integrity of the study and maintain written and properly signed records of each periodic inspection showing the date of the inspection, the study inspected, the phase or segment of the study inspected, the person performing the inspection, findings and problems, action recommended and taken to resolve existing problems, and any scheduled date for reinspection.
>
> Any problems found during the course of an inspection which are likely to affect study integrity shall be brought to the attention of the study director and management immediately. ....

21.1.5 MHRA GXP Data Integrity Guidance and Definitions

The 2018 GXP Data Integrity Guidance and Definitions guidance document[5] has the following requirements:

> 3.5 Organisations are not expected to implement a forensic approach to data checking on a routine basis. Systems should maintain appropriate levels of control whilst wider data governance measures should ensure that periodic audits can detect opportunities for data integrity failures within the organisation's systems.
>
> 6.4 Data Integrity (part) .... Assuring data integrity requires appropriate quality and risk management systems, including adherence to sound scientific principles and good documentation practices.
>
> 6.15 Data Review (part) .... Periodic audit of the data generated (encompassing both a review of electronically generated data and the broader organisational review) might verify the effectiveness of existing control measures and consider the possibility of unauthorised activity at all interfaces, *e.g.* have there been IT requests to amend any data post review? Have there been any system maintenance activities and has the impact of that activity been assessed?

A data integrity audit (or periodic review as above) assesses data but also looks wider at a computerised system but should also consider paper based processes as well.

21.1.6 WHO Guidance on Good Data and Record Management Practices

In Section 6 on Data Governance and Quality Audits there are the following clauses[6]:

> 6.4. .... adequate review of audit trails, including those reviewed as part of key decision-making steps (*e.g.* GMP batch release, issuance of a GLP study report or approval of case report forms), may reveal incorrect processing of data, help prevent incorrect results from being reported and identify the need for additional training of personnel;
>
> Routine audits and/or self-inspections of computerized systems may reveal gaps in security controls that inadvertently allow personnel to access and potentially alter time/date stamps. Such findings help raise awareness among management of the need to allocate resources to improve validation controls for computerized systems;
>
> 6.5 Quality audits of suppliers, self-inspections and risk reviews should identify and inform management of opportunities to improve foundational systems and processes that have an impact on data reliability. Allocation of resources by management to these improvements of systems and processes may efficiently reduce data integrity risks. For example, identifying and addressing technical difficulties with the equipment used to perform multiple GXP operations may greatly improve the reliability of data for all of these operations. Another example relates to identifying conflicts of interests affecting security. Allocating independent technical support personnel to perform system administration for computerized systems, including managing security, backup and archival, reduces potential conflicts of interest and may greatly streamline and improve data management efficiency.

Other items pertaining to quality oversight are found in the following sections:

> 8.2 This may require specific training in evaluating the configuration settings and reviewing electronic data and metadata, such as audit trails, for individual computerized systems used in the generation, processing and reporting of data. For example, the quality unit should learn how to evaluate configuration settings that may intentionally or unintentionally allow data to be overwritten or obscured through the use of hidden fields or data annotation tools. Supervisors responsible for reviewing electronic

data should learn which audit trails in the system track significant data changes and how these might be most efficiently accessed as part of their review.

8.3 Management should also ensure that, at the time of hire and periodically afterwards, as needed, all personnel are trained in procedures to ensure GDocP for both paper and electronic records. The quality unit should include checks for adherence to GDocP for both paper records and electronic records in their day-to-day work, system and facility audits and self-inspections and report any opportunities for improvement to management.

11.10 For example, during self-inspection, some key questions to ask are: Am I collecting all my data? Am I considering all my data? If I have excluded some data from my decision-making process, what is the justification for doing so, and are all the data retained, including both rejected and reported data?

21.1.7 PIC/S-PI-041 Good Practices for Data Management and Integrity in Regulated GMP/GDP Environments

The PIC/S guidance has the following requirements for quality oversight[7] in two sections of the document:

5.6 Data Governance System Review

5.6.1 The effectiveness of data integrity control measures should be assessed periodically as part of self-inspection (internal audit) or other periodic review processes. This should ensure that controls over the data lifecycle are operating as intended.

5.6.2 In addition to routine data verification checks, self-inspection activities should be extended to a wider review of control measures, including:

- A review for consistency of reported data/outcomes against raw data entries.
- In situations where routine computerised system data is reviewed by a validated 'exception report', a risk-based sample of computerised system logs/audit trails to ensure that information of relevance to GMP activity is reported as expected

6.7 Dealing with data integrity issues found internally

6.7.1 In the event that data integrity lapses are found, they should be handled as any deviation would be according to the pharmaceutical quality management system. It is important to determine the extent of the problem as well as its root cause, then correcting the issue to its full extent and implement preventative measures. This may include the use of a third party for additional expertise or perspective, which may involve a gap assessment to identify weaknesses in the system.

21.1.8 Regulatory Compliance Summary

Summarising the regulations and regulatory guidance documents there are three areas of quality oversight that need to be considered in any organisation for ensuring data integrity:

1. QA oversight of any regulated analytical work such as method validation, batch and study analysis and stability testing to ensure the integrity of data is both a regulatory requirement and a regulatory expectation. The oversight is to ensure that work performed has followed the correct procedures, is complete, consistent and accurate and that good documentation practices have been followed. The process of this type of quality oversight is similar to a second person review, except that it may be a wider sampling exercise depending on the individuals who generated and reviewed the data. Of analytical interest would be QA oversight of deviations and OOS investigations. For a better understanding and discussion of a second person review please refer to Chapter 17.
2. Review and audit of processes, systems and procedures of the quality management system associated with data integrity and data governance. These audits can be either internal or external (*e.g.* supplier assessment, covered in Chapter 23) to assure management that all work is undertaken as required, risks are managed and that any poor data management practices and data integrity violations are identified. If this is a repeat audit of a process where CAPA actions have been implemented, the audit should assess how effective they have been. Where appropriate, management should be informed and in some cases a data integrity audit may trigger a data integrity investigation.
3. Investigations of data integrity violations. This is the subject of Chapter 23 and will not be covered further in this chapter.

In all three items if there are any findings this will trigger corrective and preventative actions to resolve the immediate problem as well as prevent the problem from occurring again.

The MHRA GXP guidance[5] states in Section 21.1.5 that, *Organisations are not expected to implement a forensic approach to data checking on a routine basis.* We will return to this topic later in Section 21.4.

21.1.9 Role of the Laboratory in Ensuring Data Integrity

The sole purpose of Quality Assurance is not to identify and correct the mistakes made in a regulated laboratory. The responsibility for ensuring the integrity and quality of the data generated falls mainly to two people in the laboratory. The performer of the test as described in Chapter 16 and the second person reviewer, as discussed in Chapter 17. Both individuals must ensure that the work performed is correct, accurate and that data are complete. This work must be supported by management so that analytical staff have the time and resources to perform the work correctly and completely. This is a reiteration of Section 1.7.2 in the introduction to this book.

The bottom line is that data quality and data integrity are everybody's job.

The function of QA is to have oversight of the work performed and to assure that analytical work is performed in compliance with the applicable regulations and company procedures including those for data integrity. As part of the change in culture, management must ensure that all staff take responsibility for their work and do not leave it to QA to clean up the mess.

21.2 Data Integrity Audits: Planning and Execution

21.2.1 Rationale for Data Integrity Audits

Data integrity audits are now a regulatory expectation as we can see from the various quotations in Section 21.1. Under EU GMP, data integrity audits would fall under the remit of Chapter 9, Self-Inspections.[1] From this, data integrity audits need to be planned, there should be evidence that they took place and any resulting CAPA actions have been implemented.

The reports of any data integrity audit would be confidential but inspectors will want to see that there is a schedule of planned audits, that they took place as scheduled with evidence of the audit and the CAPA outcomes. One of the problems with a data integrity audit is that there could be potential overlap with periodic reviews of computerised systems, and this topic will be discussed later in this chapter.

21.2.2 What Are the Objectives of a Laboratory Data Integrity Audit?

There are two objectives of any data integrity audit:

1. To assure that the process or computerised system being audited complies with applicable regulations and company policies and procedures for data integrity and good documentation practice.
2. To identify if any poor data management practices or data integrity violations have occurred. This will result in a confidential QA report and CAPAs, the latter will be accessible by inspectors unless there is a for cause inspection.

21.2.3 What Will We Audit? The Data Integrity Inventory and Data Criticality

The first stage in planning data integrity audits is to understand the processes and computerised systems to be included within scope. The first port of call will be the inventory of computerised systems kept for inspections and audits, however, this will not contain the manual processes that must also be included in a data integrity programme. As the assessment of manual processes progresses, each one will be added to the data integrity inventory.

We now have a problem. Will there be two inventories or just one? Two inventories will comprise all the computerised systems, spreadsheets and databases within the organisation plus a separate inventory of manual processes. Or should the two be integrated together? The choice is yours.

Regardless of the number, an inventory should list each system or process, the scope of the system or process, the records generated by it, the risk rating (*e.g.* high, medium or low) plus a reference to the assessment and remediation plan as shown in Figure 21.1. The criticality of the system coupled with a risk rating should identify the spectrum of processes and systems from high risk to low risk for audit.

21.2.4 What Is the Order and Frequency of Audit?

It is obvious that the order of data integrity audit will be based on the criticality of the system or process and the current risk to the records generated by each one. This will provide an initial sorting of the inventory. This can be refined further with the number of users, frequency of use and impact either on a regulatory submission or product release to give a list of the most to least important.

The next stage is more difficult as there are two factors to consider:

- How often to audit?
- How many trained staff do you have to conduct these audits?

Audits of processes and computerised systems can be disruptive, so these need to be planned carefully. What needs to be balanced is the time taken to perform the audit *versus* the risk of undiscovered poor practices or data integrity violations. The frequency will be decided upon the criticality of the system, the data it generates and the type of controls in place to ensure data integrity and detect falsification. For example:

- Validated technical controls are better than procedural controls.
- Segregation of duties (IT administrators) *versus* local administrators in the laboratory with two user roles.
- Networked *versus* standalone systems.

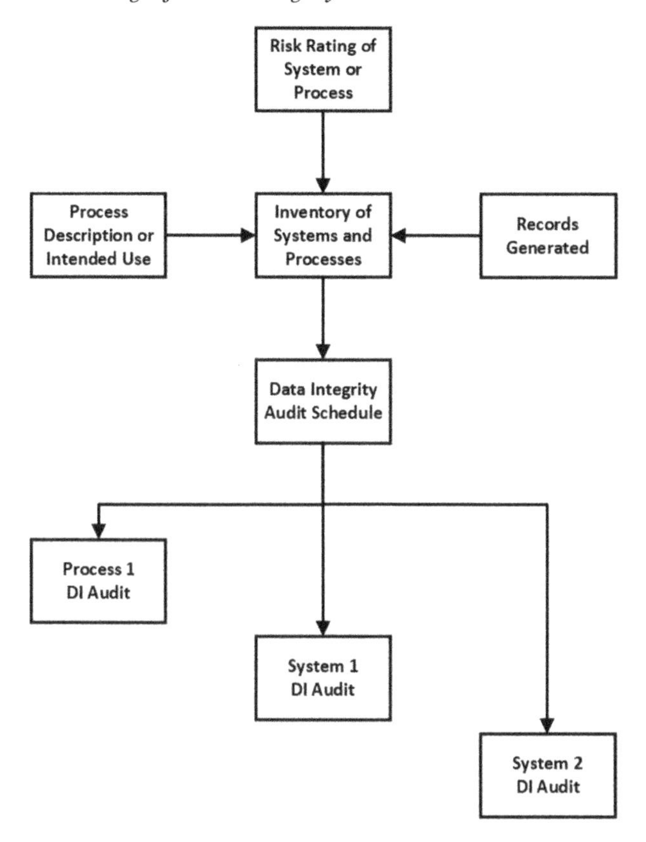

Figure 21.1 Planning data integrity audits of processes and systems.

Therefore, critical hybrid standalone systems that use procedural controls and have laboratory administrators represent a potentially higher risk than a large multi-user networked system working electronically with validated technical controls that limit user actions and with segregated application administration.

The frequency of audit is problematic as it depends on the size of the organisation, willingness of management to conduct these audits and the availability of trained auditors to conduct them. In education courses that I have participated in as a trainer many attendees have suggested that a 2 or 3 year cycle is acceptable as a frequency for even critical systems. To these and their management I would offer the following argument: if there is a major data integrity violation in the laboratory that is not identified for up to three years after it has occurred, what is the cost to the organisation? Consider items such as inspectional reputation, costs to remedy, data integrity investigation(s), batch recalls or drug applications refused or revoked? Strangely, when money is tight in any organisation, adverse inspection findings suddenly brings a surfeit of financial largesse to resolve the problems. Always remember when making such decisions that the cost of compliance is always cheaper than the cost of non-compliance, as discussed in Chapter 2 and shown in Figure 2.1. Therefore, I would suggest a much shorter audit timeframe be considered.

In any case, there must be a planned schedule of data integrity audits for processes and systems available for audits (if a supplier assessment) or regulatory inspections.

21.2.5 Who Will Conduct the Audit?

As noted in EU GMP Chapter 9,[1] audits or self-inspections must be independent and therefore the auditors will be members of Quality Assurance or a different department. All will be trained auditors with data integrity training and knowledge. Auditing paper processes and simple instruments such as pH meters and analytical balances is a relatively straightforward task. The problem comes when computerised systems are audited. Each software application, performing the same function work or be operated differently. For example, all chromatography data system software applications have the same basic functions, such as controlling an instrument, acquiring and processing data, *etc.*, but the way each one operates is different depending if the application is standalone or networked and flat files or database. Therefore, either the auditor must be an expert in all instrument data system software or they require expert assistance. It is the latter option that will probably be the most practical approach to take. When a computerised system is audited, the laboratory will provide an experienced person to operate the application under the direction of the auditor.

Could the auditor be misled by such a person wishing to hide any data integrity violations? Quite possibly. To overcome this, the auditor will have sufficient training on auditing computerised systems and will know general ways to audit to find data falsification and must be reliant on determination to go into detail and not be deflected or distracted by the laboratory administrator.

An alternative approach could be the use of a lead auditor trained in general data integrity principles with a subject matter expert or experts for the computerised systems being audited from different parts of the organisation. The subject matter expert also needs to be independent to comply with Chapter 9 requirements.

21.2.6 Data Integrity Audits and Periodic Reviews of Computerised Systems

For computerised system already subject to periodic reviews as required by EU GMP Annex 11,[8] data integrity audits add another layer of quality oversight. There is the potential for overlap between the two so that work is duplicated or if the two are separated, there is the possibility of items falling between the cracks and being missed by either the audit or the review. Therefore, we need to understand what are the objectives of each one:

- Data integrity audits, as outlined in Section 21.2.2, aim to assure the data integrity generated by the system and to identify any poor data integrity practices or violations.
- Periodic reviews aim to assure that the computerised system remains in a state of control and is validated. For more detail on conducting a periodic review see Chapter 33 in Validation of Chromatography Data Systems.[9]

Although the two have different objectives, there are potential overlaps and the duplication possible can be seen in Table 21.1. By knowing the overlap, the work can be divided between the more frequent data integrity audit and the less frequent periodic review. Many of the common items in Table 21.1 could, and in many cases should, be covered in a data integrity audit. Owing to the different frequencies of audit and review the temptation to combine both into one review, say every two years, should be resisted strongly.

21.2.7 Procedure and Checklist for a Data Integrity Audit

Of course, there must be a procedure for conducting a data integrity audit that covers the whole process from planning, conducting the audit, reporting the results and following up any corrective actions. A key issue is should there be a checklist for the audit?

Table 21.1 Comparison of areas to be reviewed in laboratory data integrity audits and periodic reviews.

Data integrity audit	Periodic review
• Trace data from sample to report and *vice versa* • Correct procedures followed and complete data generated during analysis • Searches of audit trail entries for possible violations • Check technical controls to restrict user working practices are still in place and are working • Protection of electronic records from unauthorised changes • Instruments used are qualified and calibrated • Identification of any potential data falsification	• Assess if the validation documentation is still current: URS, traceability matrix, test suite, *etc.* • Check change control records for the period under review: application and IT infrastructure levels • Review extent of revalidation after approved changes • Review any emergency change control requests • Handling and resolution of incidents and problems • Platform support procedures and records • User training
Areas common to both the audit and review	
• Application configuration: the same in the application and documentation • Configuration of user types and access privileges • Allocation of user type to each authorised user • Authorised user list: current and historical users • Data and system backup and recovery • Data archive and restore	

No.

There are several reasons for this answer but they can be summarised as checklists do not always work.

- Checklists restrain an auditor and stop creative auditing: follow the data, the analytical process and the documented evidence (paper and electronic as appropriate) and not the checklist.
- There is the great temptation to complete the checklist and miss some data integrity issues.
- Following a checklist question by question is boring for both the auditee and the auditor.

Instead, use an aide memoire that gives general direction for an audit rather that proscriptive checklist, this will be covered in Chapter 24.

21.3 Conducting a Laboratory Data Integrity Audit

21.3.1 Relationship Between the Data Integrity Model and a Data Integrity Audit

Before conducting and during a laboratory data integrity audit, it is important to keep in mind the levels of the Data Integrity Model described in Chapter 5 and consider the following:

- Foundation:
  Data integrity policies, procedures and training.
  Management involvement, open culture.
- Level 1:
  Qualified analytical instruments.
  Validated computerised systems including spreadsheets.
- Level 2:
  Validated analytical procedures and verified pharmacopoeial procedures.
  Analytical procedures are verified under actual conditions of use especially methods transferred to a site or laboratory.
- Level 3:
  Analysis of the sample from sample plans, sample management, sample preparation, analysis, interpretation of data, calculation of results and investigation of out of specification results.

Although the data integrity audit will be focused on Level 3 initially, it is important to remember that this layer relies on those underneath being performed correctly if it is to succeed, see Figure 16.1 to see the interaction of the Model levels. Therefore, when conducting a data integrity audit, it is important that when looking at analytical data consider if:

- The analytical procedure used to acquire and process the data was validated.
- Analytical instruments and computerised systems have been qualified and validated, respectively.
- Analytical staff are appropriately trained and there are records to demonstrate this.

The data integrity auditing aide memoire in Chapter 24 provides more information about this approach and should be used in conjunction with this chapter to conduct a data integrity audit.

21.3.2 Overview of the Analytical Process for a Laboratory Data Integrity Audit

The overall analytical process at Level 3 of the Data Integrity Model is represented in Figure 21.2 and shows the main stages for a chromatographic analysis. Other instrumental analysis technique such as spectroscopy can be accommodated by changing some of the items such as column ID and means of interpreting data.

It is important to realise that a laboratory data integrity audit is not simply having a look at a computerised system. There are many places in the analytical process that falsification can be hidden and that does not include a computerised system. Sampling and sample preparation are two main areas where changes can be made without any record as these are manual processes and therefore very difficult to identify any repeat work.

The basic premise of a data integrity audit is that the auditor should be able to trace from the sample to a reportable result easily and transparently if all records are available, complete and accurate. The same applies in reverse, an auditor should be able to pick a reportable result and be able to trace backwards to the sample.

As the FDA states in the Inspection of Pharmaceutical Quality Control Laboratories[10]:

> Laboratory records and logs represent a vital source of information that allows a complete overview of the technical ability of the staff and of overall quality control procedures.

As the audit progresses and an instrument, computerised system or analytical method is found, the auditor can delve in the qualification, computer validation or calibration checks to find out further information. If required, training records of analytical staff can be requested. When the applicable procedures and specifications are identified, they can be read to see how they have been followed by the analyst and the second person reviewer if that the individual results and reportable result comply, respectively.

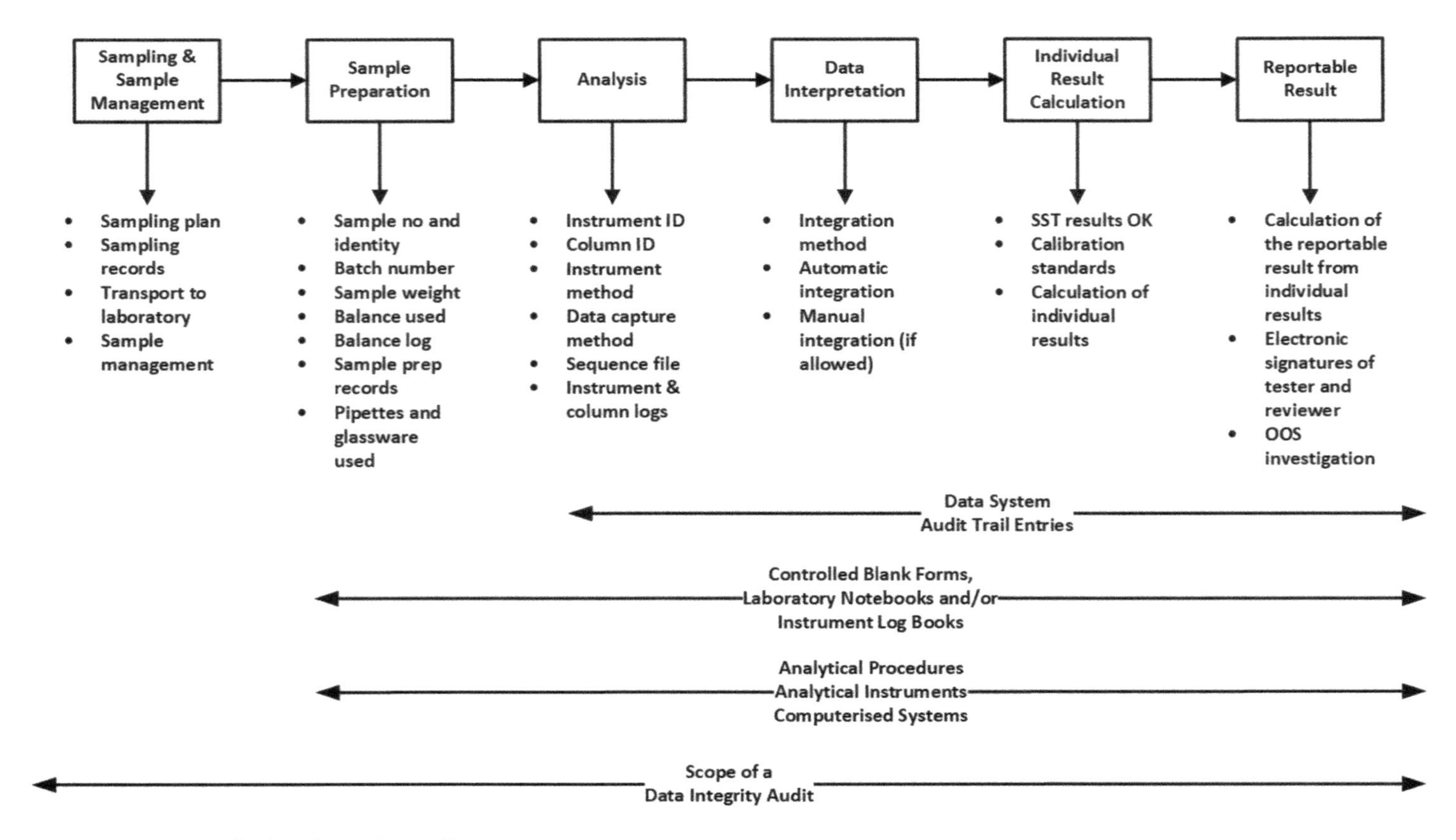

Figure 21.2 Scope of a data integrity audit.

21.3.3 Expectations for Laboratory Records

From the perspective of a data integrity audit laboratory records must be recorded as follows:

- Bound laboratory notebooks.
- Controlled and numbered blank forms for which there is accountability and reconciliation.
- All records are kept regardless if they are paper or electronic.
- True copies of records must be verified by the person making the copy and checked by a second person.
- Secure storage of all electronic records, and where used electronic signatures.
- Identification of all individuals creating or modifying records either paper or electronic.

This means that paper records do not include data written on scrap paper, post it notes, envelopes, cigarette packets, sleeves of laboratory coats or body parts. Electronic records need to be recorded on suitable and secure media – this does not include USB sticks or CD disks.

21.3.4 Auditing Records and Data from Sampling to Report

Depending on the nature of the process or system being audited the stages, shown in Figure 21.2 and the Analytical Data Life Cycle described in Chapter 9, will be the same or reduced in scope. There needs to be clear traceability of data from the sample to the result or result back to the sample. An approach to the data integrity audit is based upon Table 21.2.

Ensure that the integrity of the sample(s) is acceptable in that the sample was taken correctly according to the protocol or sampling plan, the sample was transported to the laboratory and stored under the correct conditions. The documentation, log or computerised system records are correct the dates and, where appropriate, time is consistent. If required refer to the manufacturing or study records to ensure congruence with sampling records.

During sample preparation check to see that uncontrolled blank forms are not being used, otherwise it will be impossible to check if the work is original, or second, third or fourth attempt at analysis. Check the logbooks, associated forms, laboratory notebooks and compare these records with the electronic records and audit trail entries in the computerised system associated with the analysis. If necessary, be prepared to examine all records and worksheets for accuracy and authenticity and to verify that electronic records correspond to and support the reported results.

Check to see if any data have been changed and why this is so, check by cross referencing the data that have been corrected to authenticate them. Conform that the sequences are correct and no sample data are missing. If results are missing there must be a written explanation for this.

Table 21.2 Phases of analysis with items to consider in a data integrity audit.

Analysis phase	Items to consider when auditing
Sampling	• Identification of the sample, *e.g.* study number, study, time or product name and lot number • Check the protocol or sampling plan to check if the sample was taken correctly and the correct sample containers were used
Sample transport	• How was the sample transported to the laboratory? • Were storage conditions correct throughout transport to the laboratory? • When was the sample received in the laboratory? • Any sample condition issues noted on receipt? • Test dates should be after the date of sample receipt
Sample storage and retrieval	• Correct storage conditions used? Evidence of this? • If storage is not ambient, how many times has the sample been removed from and replaced in storage? • Sample storage stability data available?
Reagent, buffer and mobile phase preparation	• Records of reagents, *etc.* available and follow procedures? • If solutions still available, are they correctly labelled as per EU GMP Chapter 6.21? If not check current solutions and see if they comply • Follow the reagent through the analysis to see correctly documented
Reference standard solutions	• If a primary or pharmacopoeial reference standard check the certificate of analysis to confirm use is within the expiry date • If a secondary standard is used check cross check to a primary standard • Reference standard correctly prepared and stored? • Use of solution within expiry date?
Sample preparation	• Are records recorded in laboratory notebooks or on controlled blank forms • Suspect perfect records with no errors and associated corrections • Identification of all apparatus used, *e.g.* glassware, pipettes, *etc.* used • Check balance and pH meter calibration and instrument log books with preparation records Be wary of the ability to weigh consistently to the nearest milligram • Instruments should have printers attached to record the actual values and not rely on observation or acquire data directly to an electronic system, *e.g.* LIMS or ELN • Ensure that true copies and the original are both available
Instrumental analysis	• Instrument set up and point of use or SST checks are correctly performed • Examine the data system electronic records to ensure that correct acquisition and processing parameters used • Data interpretation follows procedures and has not been tested into compliance • Integration of chromatograms treats standards and samples the same • Manual integration of chromatograms is allowed and is scientifically justified

Table 21.2 (*continued*)

Calculation of results	• Calculation of results including rounding is correct • Calculations in a validated data system, LIMS or spreadsheet • Spreadsheet files are saved and linked to the printout • Results within specification or OOS? • File based systems are suspect if a user can access the operating system, clock and recycle bin • Results hidden and retested? • Consistency of electronic record data and time • Are there any other data with the same date and time stamps? • Audit trail entries support the work performed without suspicion?
Reporting results	• Results in Certificate of Analysis or Study Report are correct • Results can be traced back to the sample from which they were analysed • Has the laboratory carried out scientifically sound OOS investigations? • Has the laboratory improperly invalidate OOS results?

In the event that a data integrity violation is identified, thoroughly document the unreliable data and identify the personnel associated in creating and supervising such data.

21.3.5 Checking the Configuration Settings of Computerised Systems

As noted in Table 21.1, there are many items to check with the configuration settings of the application software used in a regulated laboratory. Although the configuration specification is a key validation document, it is also a key part of a data integrity audit. The audit needs to assure senior and laboratory management that the system is being operated correctly and the technical controls in place to ensure the protection of electronic records remain in place and have not been changed by unauthorised individuals. Any changes to configuration settings of a live and validated system must be linked to an approved change control request along with update of validation documents and any regression testing required.

From a risk perspective, networked systems with IT administrators could be a lower risk than standalone laboratory systems where a laboratory user has two roles: administrator and a user. If this is the case, then the checks for the configuration could be left to a periodic review. However, each organisation must carry out their own risk assessment and determine where to perform this check and how often.

It is not just a check of the documentation *versus* the application settings (static check), the auditor must also look at the audit trail to see if any changes to the configuration settings have been made. If they have, are they supported by an authorised change control request? One area in particular to look out for are audit trail entries for turning the audit trail off and then back on a few minutes later. This will raise many deep and searching questions as to why and what activities, most probably unauthorised, occurred when the audit trail was turned off.

21.3.6 Identification and Investigation of Laboratory Out of Specification Results

One of the most critical areas for a data integrity audit to review are out of specification (OOS) results and the subsequent investigations. Remember, that one of the FDA's quality metrics is the invalidated out of specification result rate[11] and the subject was discussed in Section 17.8 of this book. Questions to consider as part of a data integrity audit are:

- What is the OOS rate for the laboratory across all tests?
- Are OOS results being hidden and documented as incidents?
- Are laboratory investigations being conducted correctly and are they scientifically sound?

OOS results identified as laboratory errors will occur when an analyst makes a mistake in following an analytical procedure, using the wrong standard or standard solution strength, makes a dilution error or a miscalculation of a result. Be suspicious if all OOS results are invalidated due to "analyst error" and that the only outcome is retraining of an individual. Here, the root cause(s) has/have not been identified and further OOS results will be recorded. As the FDA note[12]:

> The exact cause of analyst error or mistake can be difficult to determine specifically and it is unrealistic to expect that analyst error will always be determined and documented. Nevertheless, a laboratory investigation consists of more than a retest. The inability to identify an error's cause with confidence affects retesting procedures, not the investigation inquiry required for the initial OOS result.

The laboratory investigation should be completed within 20 working days – check to see how often this has occurred.

21.3.7 Photographs to Support Audit Observations and Findings

In certain situations, photographs will aid a data integrity audit to show the report readers of circumstances when a finding or recommendation is made. When performing an internal data integrity audit there will not be a problem as this will be a self-inspection carried out under the auspices of Quality Assurance and the right to take photographs will be in company procedures.

However, if carrying out a supplier assessment or data integrity audit of a supplier, taking photographs may be an issue unless agreed in advance with the supplier or contract acceptor.

Regardless of an internal or external audit, if a photograph is taken it should be date and time stamped and the GPS co-ordinates should be added if undertaking a supplier audit for the first time.

Similarly, if an inspection takes place the laboratory or organisation needs a procedure that an inspector cannot take photographs. Instead the company will provide a photographer who will work under the direction of the inspector taking photographs as required. Then, both the inspector and the company have copies of the same photographs.

21.3.8 Reporting the Audit

There should be a report of each data integrity audit to serve as a record over time for each specific process and computerised system. The report should identify who was involved and what was observed, *e.g.* manual processes, record and data sets and software applications. Any findings should be documented and graded as to their severity.

If any data violations are identified, more detail should be provided and the names of the analysts involved should be given along with the dates of the violation and the studies and batches impacted by them. This will probably be a trigger for a data integrity investigation, as discussed in Section 21.5 and Chapter 22.

21.4 What Is a Forensic Approach to Data Checking?

As stated in Section 21.1.5, the MHRA GXP data integrity guidance[5] notes that, *Organisations are not expected to implement a forensic approach to data checking on a routine basis*. This is not the best phrasing as there is no definition of *forensic approach* to help companies interpret what is meant by it. In the normal course of events forensic data checking will not be used, but in cases of data integrity investigation it may be of use and as the subject is on the borderline of audits and investigations, it is discussed here.

21.4.1 Forensic Data Analysis

It is ironic that this subject has the initials FDA and for this reason these initials will not be used in this section. The aim of forensic data analysis is to search for patterns in data that identify potential fraudulent activity. The current focus of forensic data analysis is on financial systems to identify cases of money laundering, financial and e-commerce fraud, however, the word forensic means that this is generating evidence that could be presented in a court of law. I do not think that this is the aim of the MHRA guidance, rather the guidance is aimed at using the techniques of forensic data analysis to identify regulated data patterns that might indicate or show fraudulent acts or recover deleted records post falsification.

21.4.2 Recovery of Deleted Files

The first thing to do is to look in the recycle bin to see if there are any deleted regulatory files. These can be identified and recovered to the location from which they were deleted from. It is important to take a note of the file names or take a screen shot if there are a large number so that you can search for them after restoration. An example of the contents of a recycle bin can be seen in Figure 21.3, here you can see that there are several sample related files in the recycle bin as well as the criminal deletion of the David Bowie album. Interestingly, there were more music files on this QC workstation than scientific data.

If the recycle bin has been emptied then we can move to the second option. If the computer is networked and backed up, then it may be a matter of restoring the last full backup followed by all incremental backups available or the last differential backup. This recovery would be to a separate computer so as not to disrupt an operational system. Sometimes, the software application would also be recovered from backup to allow searching of data.

There are software utilities available that can be used to recover deleted files, even if the recycle bin has been emptied. This approach can be used

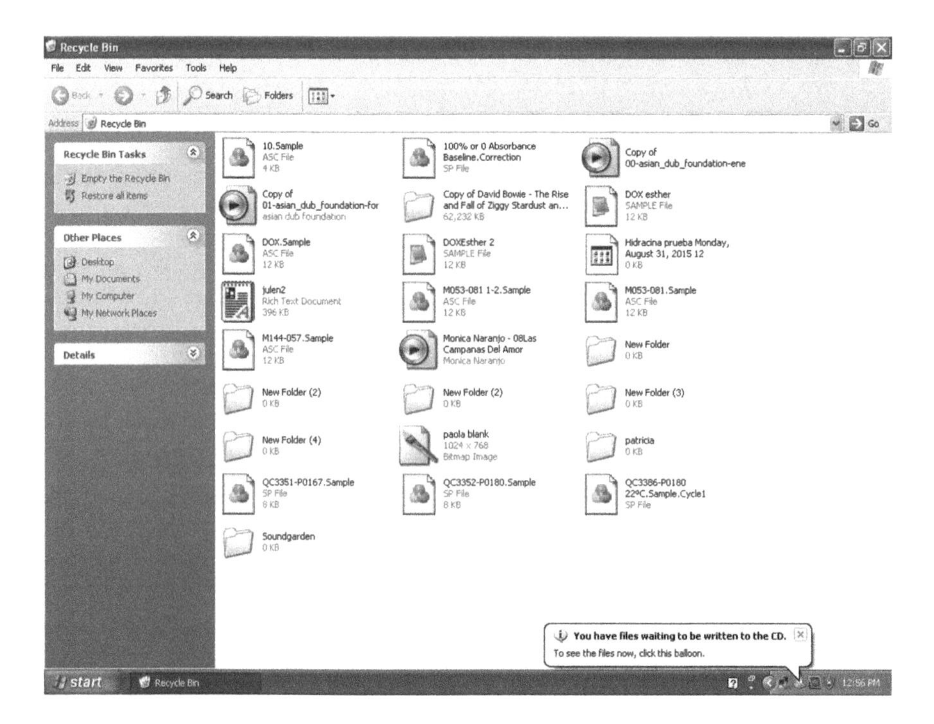

Figure 21.3 Contents of a recycle bin from a workstation operating a spectrometer.

especially with file based laboratory data systems to see if any regulatory records have been deleted to mask attempts at falsification. Such utilities can scan a computer hard drive and prepare a list of all deleted files. This list needs to be searched to see if any files have regulatory significance and recover them if required.

However, there are limitations to this approach and you need to understand how the Windows operating system deletes a file. The pointer to the file location on the disk is deleted but the file is still on the disk. Windows also marks the areas of the disk occupied by the file as available for storage so over time the location will be overwritten and when this occurs the data are gone for good. Thus, this approach is better sooner after the event rather than later where the file could be overwritten.

21.4.3 Forensic Data Analysis Techniques

Think of the pain of reviewing on screen audit trail entries that were programmed by a nerd who does not know one end of a test tube from another, has no clue what a GXP regulation is and has written the audit functionality from a specification drafted by a supplier's business analyst. The audit trail has multiple entries in one large protected file with limited ability to search, perhaps deciphering Egyptian hieroglyphics is an easier task?

Here is where data analytics can help internal auditors – and by extension inspectors – by using computer utilities and tools to identify trends in data. This is a complex area and is outside the scope of this book as the sampling process can be judgemental or statistical depending on the experience of the auditor. Large volumes of data are scanned and therefore forensic data analytics requires an expert auditor and the techniques are more likely to be used during a data investigation. The book by Richard Cascarino on Data Analytics for Internal Auditors is recommended for further reading on this subject.[13]

However, there are some simple alternatives that could be used with laboratory data. If the audit trail entries can be exported from the application as a text string or in a comma separated value (.csv) file the entries can be imported into a spreadsheet or analytics software such as Tableau. The data can then be searched for trends. This may be easier than reading entries on the screen. Care needs to be taken with the data as the text or csv file is unprotected and could be inadvertently altered.

21.5 Triggers for a Data Integrity Investigation

One of the outcomes from a data integrity audit could be a finding that is a trigger for a larger and more extensive data integrity investigation. Triggers for such investigations may be the discovery of a single case where passing data has been copied to save performing an analysis or deletion and reanalysis to pass a batch of material. Data integrity investigations are more involved, extensive and may involve informing regulatory authorities of their impact on a regulatory submission or product safety. Data integrity investigations are covered in more detail in Chapter 22.

References

1. *EudraLex – Volume 4 Good Manufacturing Practice (GMP) Guidelines, Chapter 9 Self Inspection*, European Commission, Brussels, 2001.
2. *21 CFR 211 Current Good Manufacturing Practice for Finished Pharmaceutical Products*, Food and Drug Administration, Silver Spring, MD, 2008.
3. *FDA Compliance Program Guide CPG 7346.832 Pre-approval Inspections*, Food and Drug Administration, Silver Spring, MD, 2010.
4. *21 CFR 58 Good Laboratory Practice for Non-clinical Laboratory Studies*, Food and Drug Administration, Washington, DC, 1978.
5. *MHRA GXP Data Integrity Guidance and Definitions*, Medicines and Healthcare Products Regulatory Agency, London, 2018.
6. *WHO Technical Report Series No. 996 Annex 5 Guidance on Good Data and Records Management Practices*, World Health Organisation, Geneva, 2016.
7. *PIC/S PI-041 Draft Good Practices for Data Management and Integrity in Regulated GMP/GDP Environments*, Pharmaceutical Inspection Convention/Pharmaceutical Inspection Co-Operation Scheme, Geneva, 2016.
8. *EudraLex – Volume 4 Good Manufacturing Practice (GMP) Guidelines, Annex 11 Computerised Systems*, European Commission, Brussels, 2011.
9. R. D. McDowall, *Validation of Chromatography Data Systems: Ensuring Data Integrity, Meeting Business and Regulatory Requirements*, Royal Society of Chemistry, Cambridge, 2nd edn, 2017.
10. *Inspection of Pharmaceutical Quality Control Laboratories*, Food and Drug Administration, Rockville, MD, 1993.
11. *FDA Guidance for Industry Submission of Quality Metrics Data, Revision 1*, Food and Drug Administration, Rockville, MD, 2016.
12. *FDA Guidance for Industry Out of Specification Results*, Food and Drug Administration, Rockville, MD, 2006.
13. R. E. Cascarino, *Data Analytics for Internal Auditors*, CRC Press, Boca Raton, FL, 2017.

CHAPTER 22

How to Conduct a Data Integrity Investigation

One of the Quality Assurance tasks in the current regulatory environment is to conduct data integrity investigations. The trigger for a DI investigation would occur when a data integrity violation has been identified and it aims to document the extent of the problem, the impact of it, the root causes and the subsequent corrective and preventative action plans. Following the presentation and examination of the regulatory guidance on the subject, two case studies are presented to outline the process and how to conduct and report a data integrity investigation. It is important to understand that data integrity investigations are not only due to human actions but also to technology issues. Therefore, the case studies described in this chapter cover both situations.

22.1 What the Regulators Require

There are no specific GXP regulations covering data integrity investigations, however, deviations from procedures must be reported and investigated from which a data integrity investigation may be triggered, but these regulations are not specifically listed here. Instead, there are four regulatory data integrity guidance documents including a policy that go into much more detail about what is required in such an investigation, as can be seen in Figure 22.1.

Data Integrity and Data Governance: Practical Implementation in Regulated Laboratories
By R. D. McDowall

Published by the Royal Society of Chemistry, www.rsc.org

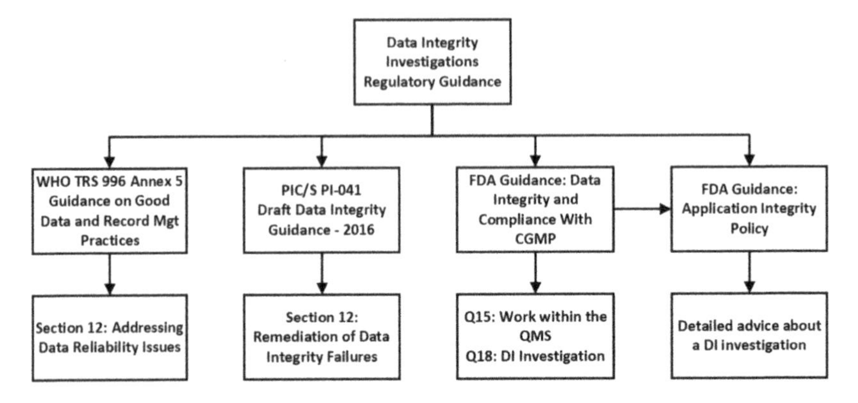

Figure 22.1 Summary of data integrity investigation regulations and guidance.

22.1.1 WHO Guidance on Good Data and Record Management Practices

Section 12 of this guidance addresses data reliability issues[1]:

> 12.1 When issues with data validity and reliability are discovered, it is important that their potential impact on patient safety and product quality and on the reliability of information used for decision-making and applications is examined as a top priority. Health authorities should be notified if the investigation identifies material impact on patients, products, reported information or on application dossiers.
>
> 12.2 The investigation should ensure that copies of all data are secured in a timely manner to permit a thorough review of the event and all potentially related processes.
>
> 12.3 The people involved should be interviewed to better understand the nature of the failure and how it occurred and what might have been done to prevent and detect the issue sooner. This should include discussions with the people involved in data integrity issues, as well as supervisory personnel, quality assurance and management staff.
>
> 12.4 The investigation should not be limited to the specific issue identified but should also consider potential impact on previous decisions based upon the data and systems now found to be unreliable. In addition, it is vital that the deeper, underlying root cause(s) of the issue be considered, including potential management pressures and incentives, for example, a lack of adequate resources.
>
> 12.5 Corrective and preventive actions taken should not only address the identified issue, but also previous decisions and datasets that are impacted, as well as deeper, underlying root causes, including the need for realignment of management expectations and allocation of additional resources to prevent risks from recurring in the future.

22.1.2 FDA Guidance on Data Integrity and Compliance with CGMP

Question 18 of this guidance[2] is *How does FDA recommend data integrity problems identified during inspections, in warning letters, or in other regulatory actions be addressed?*

> FDA encourages you to demonstrate that you have effectively remedied your problems by: hiring a third-party auditor, determining the scope of the problem, implementing a corrective action plan (globally), and removing at all levels individuals responsible for problems from CGMP positions. FDA may conduct an inspection to decide whether CGMP violations involving data integrity have been remedied.

The guidance then references the FDA's Application Integrity Policy that is presented in the next section.

22.1.3 FDA Application Integrity Policy

The FDA have an Application Investigation Policy that was published in 1991 following the generic drug scandal in the 1980s where applications containing falsified data were submitted to the agency. This policy can be invoked if the FDA find that data in a drug application have been falsified. There are four key requirements in CPG 120.100 (previously CPG 7150.09) for remedial action[3]:

1. Cooperate fully with FDA and other Federal investigations to determine the cause and scope of any wrongful acts and to assess the effects of the acts on the safety, effectiveness, or quality of products
2. Identify all individuals who were or may have been associated with or involved in the wrongful acts and ensure that they are removed from any substantive authority on matters under the jurisdiction of FDA
3. Conduct a credible internal review designed to identify all instances of wrongful acts associated with applications submitted to FDA, including any discrepancies between manufacturing conditions identified in approved applications and manufacturing conditions during actual production.
   The internal review is intended to supplement FDA's ongoing, comprehensive investigation to identify all instances of wrongful acts.
   The internal review should involve an outside consultant or a team of consultants who are qualified by training and experience to conduct such a review.
   All oral or written reports related to the review that are provided by the consultant to the applicant should be made available simultaneously to FDA for independent verification;

4. Commit, in writing, to developing and implementing a corrective action operating plan to assure the safety, effectiveness, and quality of their products.

   This commitment ordinarily will be in the form of a consent decree or agreement, signed by the president, chief executive officer, or other official most responsible for the applicant's operations, and submitted to FDA.

   The corrective action operating plan will, as appropriate, address procedures and controls to preclude future instances of wrongful acts and noncompliance with regulatory requirements for approved applications, as well as procedures and controls to preclude any recurrences of other violations which may have been found (*e.g.*, a comprehensive ethics program).

22.1.4 PIC/S PI-041 Good Practices for Data Management and Integrity in Regulated GMP/GDP Environments

Section 12 of this guidance focuses on Remediation of Data Integrity Failures,[4] this is a long section and only key paragraphs are quoted here and the reader is encouraged to see the full text.

12.1 Responding to Significant Data Integrity issues

12.1.1 Consideration should be primarily given to resolving the immediate issues identified and assessing the risks associated with the data integrity issues. The response by the company in question should outline the actions taken. Responses should include:

12.1.1.1 A comprehensive investigation into the extent of the inaccuracies in data records and reporting, to include:

- A detailed investigation protocol and methodology; a summary of all laboratories, manufacturing operations, and systems to be covered by the assessment; and a justification for any part of the operation that the regulated user proposes to exclude;
- Interviews of current and former employees to identify the nature, scope, and root cause of data inaccuracies. These interviews may be conducted by a qualified third party;
- An assessment of the extent of data integrity deficiencies at the facility. Identify omissions, alterations, deletions, record destruction, non-contemporaneous record completion, and other deficiencies;
- Determination of the scope and extent and timeframe for the incident, with justification for the time-boundaries applied;
- Data, products, processes and specific batches implicated in any investigations;

- A description of all parts of the operations in which data integrity lapses occur, additional consideration should be given to global corrective actions for multinational companies or those that operate across multiple differing sites;
- A comprehensive retrospective evaluation of the nature of the testing and manufacturing data integrity deficiencies, and the potential root cause(s). The services of a qualified third-party consultant with specific expertise in the areas where potential breaches were identified may be necessary;
- A risk assessment of the potential effects of the observed failures on the quality of the drugs involved. The assessment should include analyses of the risks to patients caused by the release of drugs affected by a lapse of data integrity, risks posed by ongoing operations, and any impact on the veracity of data submitted to regulatory agencies, including data related to product registration dossiers;

12.1.1.2. Corrective and preventative actions taken to address the data integrity vulnerabilities and timeframe for implementation, and including:

- Interim measures describing the actions to protect patients and to ensure the quality of the medicinal products, such as notifying customers, recalling product, conducting additional testing, adding lots to the stability program to assure stability, drug application actions, and enhanced complaint monitoring.
- Long-term measures describing any remediation efforts and enhancements to procedures, processes, methods, controls, systems, management oversight, and human resources (*e.g.*, training, staffing improvements) designed to ensure the data integrity.

22.1.5 Summary of Data Investigation Regulations and Guidance

As we can see, the regulatory requirements of a data integrity investigation are potentially complex and the four main regulations and guidance documents are shown in Figure 22.1. Most of the guidance documents standalone: however, there is an interaction between the FDA draft guidance for industry on Data Integrity and cGMP Compliance[2] that invokes the Application Integrity Policy.[3]

The next step is to abstract the requirements from the documents and meld them together into a unified approach to ensure that any data integrity investigation is acceptable to any regulatory agency in the world. This is shown in Figure 22.2 and this should be modified in light of an individual data integrity investigation. The main stages are:

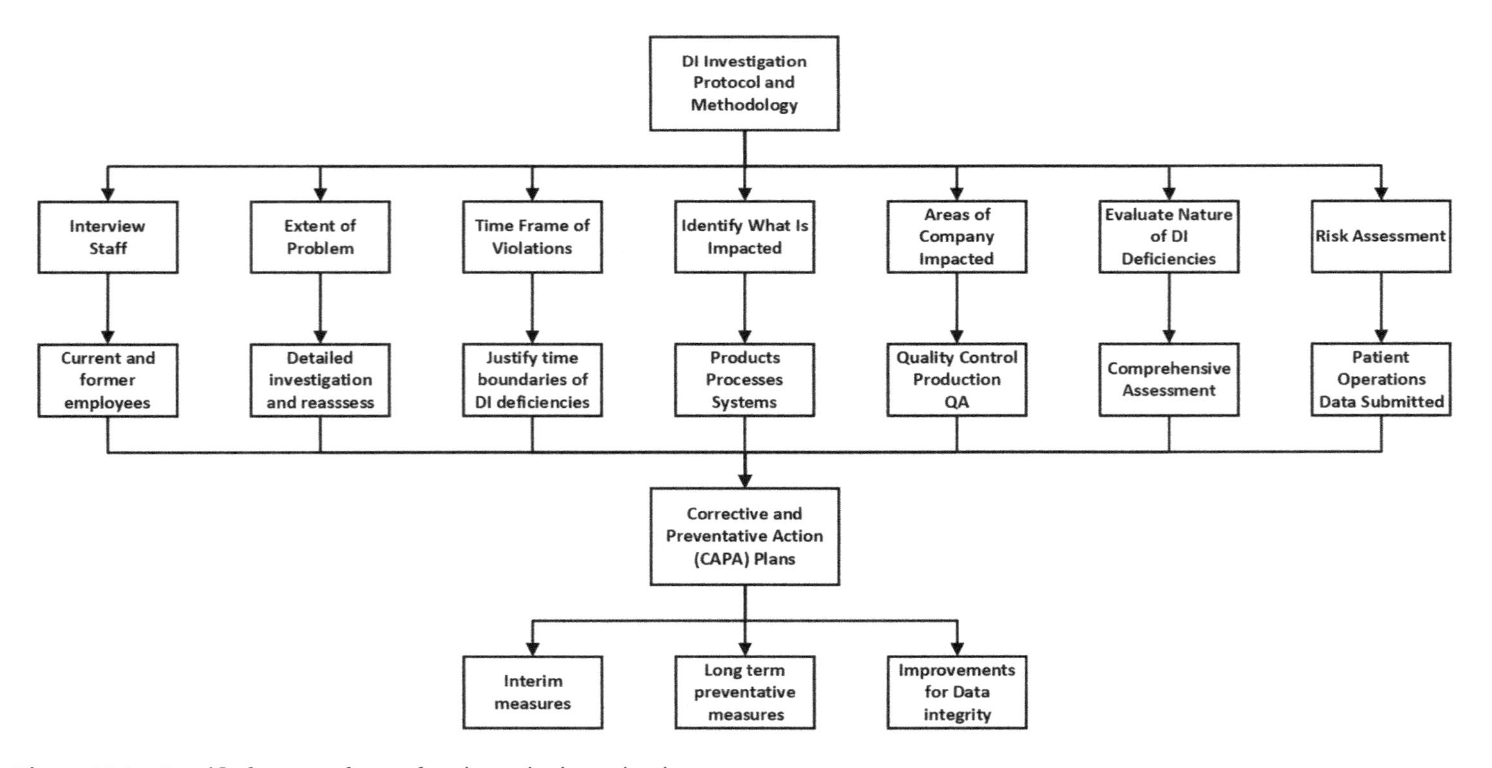

Figure 22.2 A unified approach to a data integrity investigation.

- Sequester a copy of the impacted data/data sets/database.
- Interview staff including current and former employees to understand why the data integrity violations occurred.
- Determine the extent of the problem and this is followed by a detailed investigation of processes and systems. With detailed information the scope of the problem can be reassessed.
- Define and justify the time frame of the data integrity violations.
- Identify the processes, systems and products impacted by the data integrity violations, *e.g.* production, quality control and/or quality assurance.
- A comprehensive assessment of the impact of the violations including a risk assessment on the products, the patient and/or regulatory submission(s).

We will see how this evolves when we look at each of the case studies involving data integrity investigations.

22.2 Case Study 1: Software Error Investigation

22.2.1 Case Study 1 Background

It is important to understand that not all data integrity investigations are predicated on evil analytical staff falsifying data huddled in a dark, smoke-filled room behind a laboratory (assuming, of course, local health laws have not banned smoke-filled rooms). In this case study, a data integrity investigation was triggered by a technology issue not a human one.

The investigation was triggered by a software error found during a periodic review of a critical GXP application that held records that formed part of global regulatory submissions. The issue was only seen in one version of a database. It is only seen when user types and associated access privileges are cascaded from a master list, all users, regardless of their assigned access privileges, were granted write access to any database record. Although the problem was confined to a specific version of the database but there was no patch available to resolve the error. The database either had to be updated to the next version or the cascade of user roles with the current database version turned off to resolve the issue.

In this situation, there are no staff to interview as the data integrity violation is caused by a technology issue. The database supplier knew of the problem but this was not communicated to customers and there was no entry of the problem in the US National Vulnerability Database run by the National Institute of Standards and Technology (http://www.nvd.nist.org).

22.2.2 Sequester a Copy of the System and the Data

The first stage in the investigation was to sequester a copy of the whole system as found: application, configuration settings, users and access privileges, database and data. Several copies were made, either by disk to disk copying as well as tape backups to ensure that were copies available for the investigation.

The general principles here are:

- Do not rely on a single electronic copy because if there is an error, the problem cannot be investigated accurately. This applies to any data integrity investigation. If data cannot be copied, then the system must be isolated so that the problem can be investigated.
- Verify that each copy, regardless of type, is complete. Many normal data backups are not verified due to the extra time that the verification process takes when the backup is read and compared to this original. A data investigation is an exception, all copies must be verified as there is no room for error in these cases.

22.2.3 Temporary Resolution of the Problem

As a temporary measure, the cascading of user roles and access privileges in both the production and validation instances of this application was turned off and this was achieved after a change control request was approved. This enabled the application to keep operating without the possibility of unauthorised data manipulation when a user accesses the database *via* the software error.

However, this brought another problem. The rationale for cascading roles was to save time by copying roles and accompanying access privileges and when this function was turned off, the large number of user roles had to be organised manually.

22.2.4 Systems Approach to the Issue

The database was the organisation's standard; therefore, the first stage of the investigation was to understand the scope of the problem. There were over 1000 applications running in the organisation globally using this database but less than 50 instances that used the specific version of the database that had the software error. Of the latter, only a single application was found to cascade user roles and hence was impacted by the software error. This simplified the data integrity investigation to a single application where the problem was originally observed.

A copy of the application and the database was sequestered and backed up twice to allow investigations of possible data manipulation and determine the impact of the problem without impacting the production system and interrupting normal work.

22.2.5 Time Frame of the Potential Data Integrity Vulnerability

It was quickly determined from the validation documentation and change control records for the computerised system that the user cascade had been implemented some 27 months prior to discovery of the security issue.

Therefore, there was a substantial amount of data in the database that could have been changed without anybody knowing. In addition, the data was high impact as it was used for regulatory submissions, including some applications already lodged with regulatory authorities globally.

22.2.6 Investigating the Impacted Database

Using copies of the sequestered database, the main issue for investigating the impact of the software error is that an unauthorised user accessing the database avoids the application and hence the audit trail. Any unauthorised change made using this route will not generate an audit trail entry and will be undetectable.

It is important to understand that to change any data in the database it is necessary for a user to:

- know about the database vulnerability;
- know that they can access the database, this is not immediately obvious as users are only trained to use the system *via* the application software;
- know the structure of the database to find specific records to change;
- have the technical expertise to change data within the database, *e.g.* know SQL coding.

From this list, there will be a very small proportion of the user base that could make any changes within the database. However, regardless of this, it is important to undertake due diligence to ensure as far as is technically possible that an unauthorised user could not have made any changes. If the investigation finds that this was done, identify the impacted data and correct the records.

To identify if any changes have been made directly *via* the database and not the application it is necessary to access the transaction logs for the database. Normally these are backed up as part of the data and can be recovered from the backups going back 27 months. The transaction logs can be analysed not manually but by a script to automatically identify changes. Then, once changes are identified it is possible to see who made the changes and if they did not have write or modification privileges. Fortunately, after extensive analysis there was no apparent evidence of unauthorised changes identified.

Note that in some countries, transaction logs are not backed up due to local data protection laws for individuals or there is an agreement with the Works Council. If this was the case, then the data integrity investigation would have been far more complex and expensive.

22.2.7 Informing Regulatory Authorities

As the data in the system contained GXP regulated data that was submitted as part of registration dossiers, it was necessary to inform regulatory authorities. A report of the data integrity investigation was generated, authorised by the company and then passed to the various regulatory authorities. The report did not trigger any for cause inspection as the company had disclosed the issue proactively, however, it did raise questions from some authorities after the disclosure that were answered satisfactorily. During routine regulatory inspections that followed, the system was included in the inspection programme and additional questions asked.

22.3 Case Study 2: Data Falsification Investigation

This case study of a data integrity investigation involves falsification of GMP laboratory data by a single operator in an API manufacturer.

22.3.1 Case Study Background

During the preparation for a pre-approval inspection (PAI), the data generated from three batches of product submitted in a regulatory submission application were reviewed to ensure that the data were complete, consistent and accurate. During this review, some discrepancies were found between entries in the instrument log books and chromatography data system records (both electronic records and paper printouts) used to release the three batches. The initial investigation revealed the following problems and inconsistencies with data and records:

- In some chromatographic analyses, more SST injections had been made than defined by USP <621>[5] and the laboratory SOPs.
- Some injections were not documented in the instrument log book but existed in an instrument data system.
- Some injections in the data systems have duplicate times indicating that they were injected at the same time but the log book only shows a single set of injections.
- Data have been copied from earlier analysis and used for later work in validating an analytical method.

The discrepancies appeared to be due to a single individual who left the company when these issues were raised. Thus, a data integrity investigation was started to determine the extent and impact of these data integrity breaches.

It is important to note that although we discuss here a data integrity investigation into data falsification, it also uncovered far more instances of poor data management practices than instances of falsification. This is compounded by the poor data management practices that contribute to an overall impression that the company involved has a poor approach to GMP compliance. However, the actual situation is not that simple as the staff are well motivated to do a good job, it is the lack of knowledge of GMP regulations that contributed to the issues that we will discuss here.

22.3.2 Meeting the Intent of the Application Integrity Policy

As presented in Section 22.1, to meet the intent of the Application Integrity Policy, FDA encourages companies to demonstrate that they have effectively remedied discovered problems, see Figure 22.3. There are four essential requirements to demonstrate this, as follows:

- Hire a third-party auditor.

  External and independent resource(s) are essential in a data integrity investigation as they are objective and emotionally detached from the immediate data integrity issues and internal politics of the organisation. They also bring expertise that the company may not have internally to investigate the issues.

- Determine the scope of the problem.

  Defining the scope and boundaries of the problem is essential to know what must be investigated but also to have a reasonable chance of completing the work in a timely manner. Data integrity investigations take time to perform and involve the analysis and correlation of data and records from several sources. In contrast, too narrow a scope may miss some key issues that if left unresolved may impact the data integrity of future work.

- Implementing a corrective action plan.

  A final report will provide the conclusions of the investigation but will also spawn multiple CAPA plans outside the report for short-term remediation and long-term solutions. The report should contain root cause analysis of the issues to determine comprehensive corrective and preventative action plans. During the investigation, additional issues typically will have been discovered, unrelated to the initial investigation, and these should be included in both the report and the associated CAPA plans for the laboratory.

- Removing at all levels individuals responsible for problems from cGMP positions.

  The investigation needs to include interviews with current and past staff where appropriate, in the case study the individual responsible for the falsification had already left the company and did not respond to requests to be interviewed. However, no evidence of collusion with other members of staff was identified when data from all batches and all tests was evaluated.

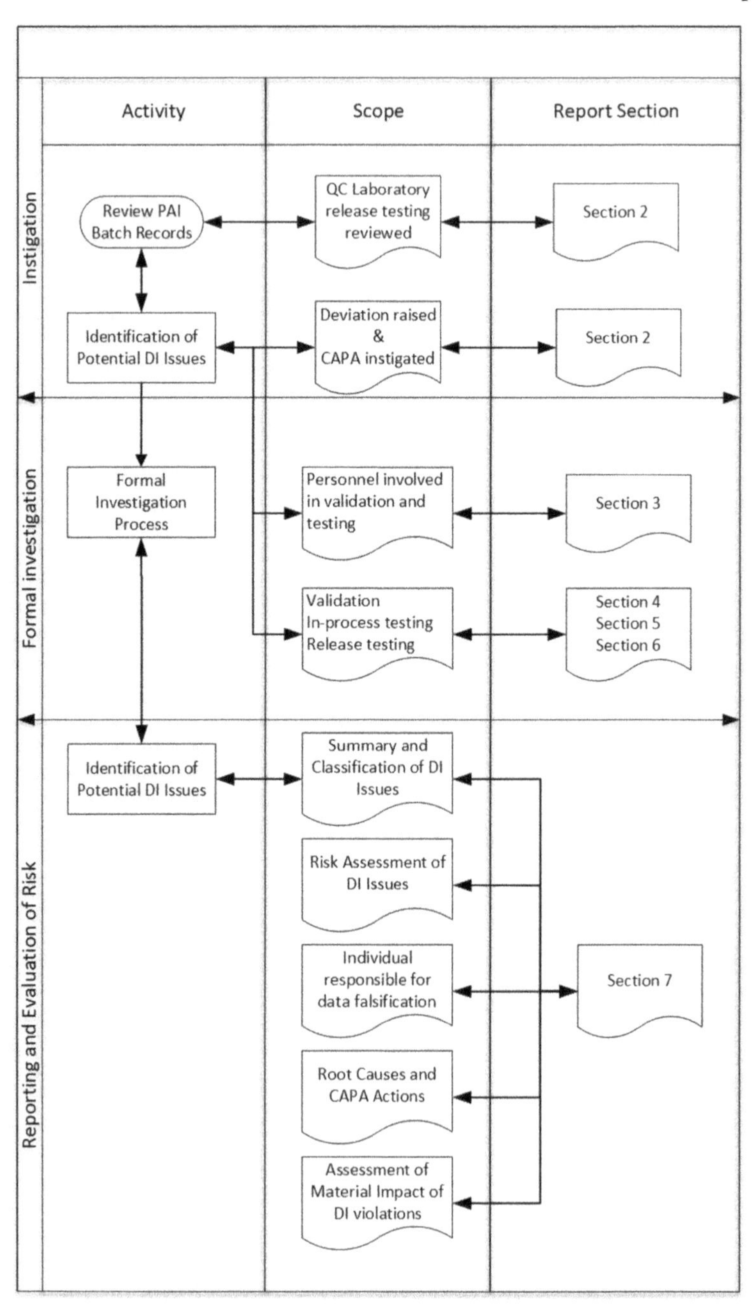

Figure 22.3 Process flow from a data investigation case study.

22.3.3 Scope of the Data Integrity Investigation

The scope of a GMP data integrity investigation is shown in outline in Figure 22.4, here the following areas were investigated:

- The scope was limited to investigating the data integrity issues associated with the application dossier for the FDA.
- Validation of all methods used for raw materials, in-process and release testing.
- All raw material and in-process testing for the three batches included in a regulatory submission.
- Release testing for three batches submitted to regulatory agencies.

It should be noted that this is a relatively narrow scope – one product and all associated method validations and the complete test suite. However, do not underestimate the amount of work and time involved in investigating data integrity issues.

If considering a GLP/GCP bioanalytical data integrity investigation, then the work could still involve validation of the bioanalytical methods but the work packages would be based on protocols from either non-clinical or clinical studies.

22.3.4 Approaches to the Investigation of Laboratory Data Integrity Issues

In this case study, the Quality Control laboratory has many different analyses, however, they can be classified into the following types:

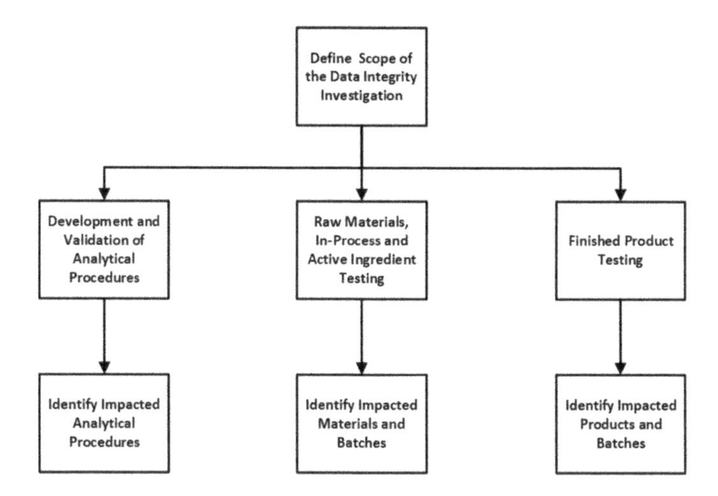

Figure 22.4 Defining the scope of a GMP data integrity investigation (Case Study 2).

- Simple tests with observation as no analytical instrumentation is involved, *e.g.* colour, odour or microbiological colony counting.
- Simple analytical tests using USP <1058> Group B analytical instruments, *e.g.* pH measurement, loss on drying.
- Analytical tests involving chromatography and chromatography data systems installed on standalone workstations, *e.g.* raw materials, in process, active ingredient and impurity testing. The CDS systems are used in hybrid mode with electronic records and signed paper printouts. However, there is not a single CDS application but six different ones.
- Method validation data generated in each analytical procedure validation or verification was also investigated.

22.3.5 Do Not Just Focus on Data Integrity Violations – Look Also for Poor Practices

Any organisation undertaking a data integrity investigation will expend a large amount of time and resources; it is also important that the maximum benefit is obtained from this investment in resources. You may disagree about the use of the word benefit in the last sentence but look at the investigation as an opportunity to clean up not just the violation but also other poor data management practices that may exist in a laboratory. This is to ensure that not only the current issue is remediated and resolved but that poor practices are eliminated that could cause violations in the future.

Due diligence is the name of the game when conducting a data integrity investigation. Find not just the root causes of the violation but also identify and include in the CAPA plan those poor data management practices that could contribute both directly and indirectly to the violations. We shall see this as we progress through this case study.

22.3.6 Investigation of Tests Using Observation

The records of the simple tests using observation such as solubility, appearance, odour and colony counting were reviewed as part of the investigation. One of the surprising findings of the investigation was that there was no effective second person review in that only the records of the test were examined but not the test itself. For example, the colony counting was performed by a single person who wrote the number of colonies down on the analysis sheet associated with the batch. The second person review was only focused on correct completion of the form and the reviewer did not recount the colonies to verify that the recorded number was correct. Furthermore, there was no evidence of the plate (*e.g.* in the form of a photograph) for the batch record, so after the plate was discarded there was no independent means of verifying the colony count.

Similar findings were found for the solubility, appearance and odour tests: the form was reviewed but there was no check that the observation itself was correct. As discussed in Chapter 17 under the second person review, in these instances the reviewer should repeat (rather than witness) such a test.

There was no evidence of falsification of these tests but here are examples of poor practices that would be ignored if an investigation only focused on finding data integrity violations.

22.3.7 Investigation of Simple Analytical Testing

The tests involving pH measurement of in-process testing had similar findings to the tests by observation in that readings were recorded by a tester but not verified as there was no printer attached to the instrument. There was an analytical balance in a sterile area without a printer attached to record the data, results were recorded by observation without verification by a second person and there was no instrument log book. As above there was no effective second person review.

Karl Fischer results were also recorded by observation and verified by a second person but, yet again, there was no objective evidence available. There was also a failure to weigh samples within the range required by one analytical procedure and this was not identified as a failure to follow an analytical procedure by a second person reviewer. Finally, some test data were invalidated without documented justification or supervisor or QA review.

22.3.8 Investigation of Analytical Testing by Chromatography

Many chromatographic tests were used in raw materials, in-process and release testing for purity, residual solvents, impurities, *etc.* One problem is that the laboratory had six different chromatography data systems from different suppliers installed on standalone workstations that were not connected to the company network. The various CDS applications were mostly file based and there was access to the data files, the system clock and the recycle bin outside of the CDS software. On some systems there was no segregation of duties and all users shared the same account. Use of these CDS applications is relatively crude in that the peak areas are manually entered into spreadsheets for calculation of SST and reportable results, making the overall process very slow, error prone and tedious. Finally, the laboratory defines raw data for these systems as paper–electronic records are not considered.

The approach to investigation of the chromatography records is shown in Figure 22.5. The aim is to check consistencies between the instrument log book, the dates and times on the CDS files and metadata, audit trail entries (where this function has been activated) and the paper printouts. This is essentially a second person review.

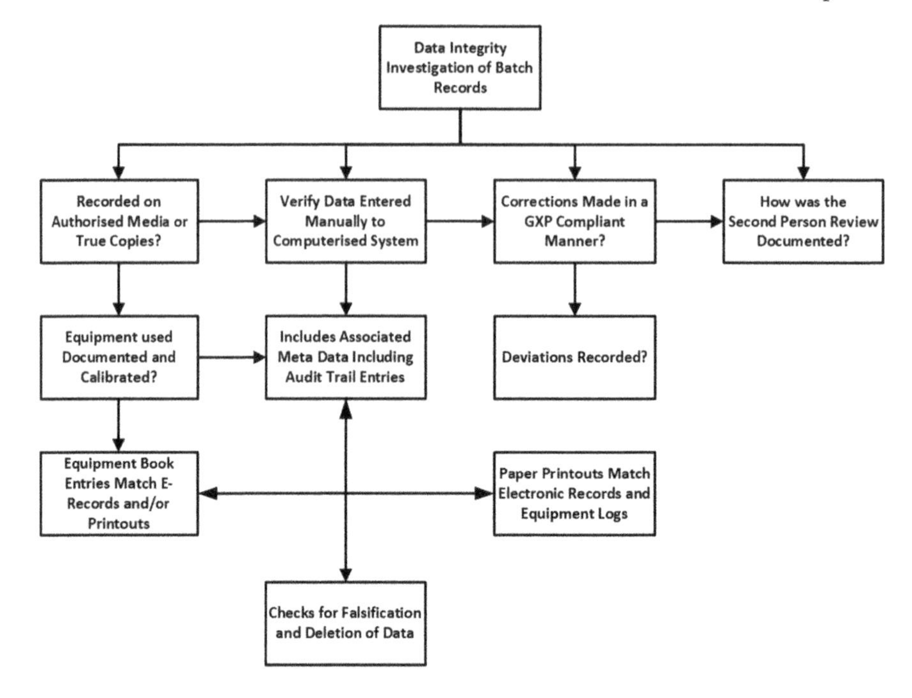

Figure 22.5 Cross referencing chromatography records in a data integrity investigation.

22.3.9 Staff Interviews

One element of a data integrity investigation are staff interviews. Here, there needs to be care taken with the approach – an over-zealous and aggressive questioning is not the best way to elicit information about falsification of data. Many staff may not be involved in any falsification and will take an accusation of collusion wrongly that will impact their future motivation in the laboratory. Before starting the interview ensure that all the facts are available and are substantiated, then base the questions to staff on these facts. Try to empathise with the staff that are questioned and try to understand what were the contributory factors to any falsification.

In this case study, the evidence was that only one person was involved, who had left the company when the falsification was discovered and never responded to requests for interview. Other staff members did not have any knowledge of the falsification. This is similar to the MHRA falsification court case in the UK.[6]

It should be noted that if an effective second person review process had been in place that correlated entries in the instrument log book, paper records and electronic records including audit trails, as shown in Figure 22.5, the first time falsification occurred it would have been picked up and the issues resolved quickly rather than a full-scale data integrity investigation having to take place several months after the events.

22.3.10 Findings and Their Classification

The findings from the data integrity investigation were classified into two areas:

- deliberate falsification/data integrity violations with the intent to deceive;
- poor data management practices caused by lack of GMP knowledge.

These are presented in Tables 22.1 and 22.2, respectively. It is interesting to note that although the data integrity investigation focused on falsification there was over double the number of poor data management practices identified. These were also included in the report along with the CAPA actions to resolve them. An overview of the falsification and poor data management findings is shown in Figure 22.6.

Table 22.1 Data integrity violations from a data integrity investigation.

Data integrity violation	Violation detail
Incomplete data reporting	Many instances involved addition of one more SST injection than required by the analytical instruction but not reported. However, there are more severe cases where whole runs were analysed and not reported. Contravenes GMP regulations.
Data file deletion	Deleting data files contravenes GMP regulations as complete data is not captured and maintained, *e.g.* §211.194(a).[8]
Unjustified analysis	There are several instances across multiple methods and validations where additional runs were performed and not used. This is linked with time travelling, incomplete reporting and data file deletion. Testing into compliance and is a GMP violation.
Failure to follow method or validation protocol	1. Failure to follow the method or protocol to falsify data. This is intent to deceive and is a serious and high risk GMP violation. 2. Manually stopping a run after the last peak has eluted to save time. No potential impact on the resulting data generated. However, a method has not been followed without any documented justification and is a GMP violation as early termination could hide impurities.
Work not recorded in instrument log book	Work is undocumented and therefore is a GMP failure.
Time travelling	A major problem as all laboratory systems are standalone workstations with access to the system clock. • Easy to perform and can be difficult to detect. • The intention is to falsify data and deceive.
Copying data from previous work	Deliberate data falsification with an unacceptably high impact
Testing to pass	Work is typically undocumented or at best partially documented. Does not meet the requirement for complete data.

Table 22.2 Identification of some poor data management practices from a data integrity investigation.

Poor data practice	Poor practice detail
No effective second person review	• Several tests do not have a second person review. A key requirement for GMP is a second person review as this is the one item that would have picked up most data integrity issues highlighted here, if staff were aware and understood the scope and importance of this review. • When dealing with a computerised system, the second person review must include checks between electronic records (including audit trail entries and associated metadata), paper printouts, instrument log books and analysis sheets. This was not performed. • There is no SOP or training for second person review. • The second person review must be documented.
No SOP for chromatographic integration	• A key data integrity requirement is the control of integration in a CDS. The document should define what are the key integration parameters and what they do. What are the acceptable and unacceptable practices for chromatographic integration?
No SOP for invalidating results	• A chromatographic run was invalidated due to "particulate matter" but there was no second person verification of this fact. An SOP is required for this activity. • Invalidating data without a formal investigation or documentation was not scientifically sound, the current OOS SOP only focuses on final results.
Results recorded by observation with no objective evidence for review	• Colony counting is a manual process but there are no photographs taken of the plate to allow effective second person review. Unable to provide objective evidence of testing for second person review.
No instrument log book	• An incubator does not have an instrument log book.
Maintenance log book contains loose and uncontrolled blank sheets	• Uncontrolled blank sheets must not be used as they are high risk and subject to guidance from FDA, PIC/S and EMA. • Check across all log books to ensure that all instances of this practice are identified and resolved.
Shared user identities and passwords	• Several CDS applications used a single user identity and password combination. • Access privileges include all operators having administration privileges and the conflict of interest needs to be removed.
Analysis and method validation sheets are not uniquely numbered	• Although controlled and issued by QA, analysis and method validation sheets are not uniquely numbered and there does not appear to be a reconciliation mechanism as required by FDA, WHO, PIC/S and EMA guidance documents.
Failure to identify OOS results	• There is a lack of understanding as to when an OOS result occurs. • For example, if the SST samples fail their acceptance criteria then there cannot be an OOS result. • Analysts performing a test should be trained to identify OOS results and inform their supervisor. • Reviewers should be trained to identify and act upon OOS results.

Table 22.2 (*continued*)

Poor data practice	Poor practice detail
Open access to data, system clock and recycle bin	• Open access to the operating system, clock and recycle bin means that some of the data falsification activities can be easily performed. • Windows security needs to be enabled on all workstations to prevent laboratory users from accessing these features provided it does not restrict usability of the software application. If this occurs there needs to be a documented rationale and description of the procedural controls to be implemented.
Software applications are not validated	• Although all software used in the Quality Control laboratories has undergone IQ and OQ using supplier qualification materials. These systems have not been formally validated, *e.g.* validation plan, user requirements specification, traceability matrix, user acceptance testing, *etc.*

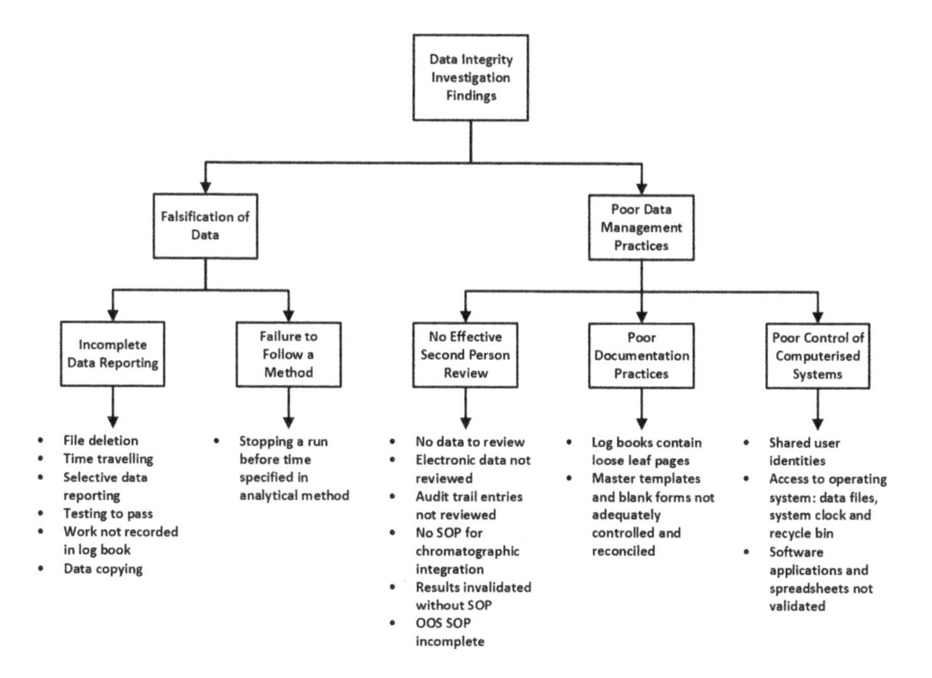

Figure 22.6 Overview of the data integrity findings.

The data integrity violations/falsification identified in the investigation were classified in the range of Routine Violation to Intentionally Misleading based on Figure 8.2 of the GAMP Guide on Records and Data Integrity.[7] The detail of the data integrity violations identified during the investigation are listed in Table 22.1.

Table 22.2 presents the main poor data management practices identified, these are not directly related to the data integrity violations listed in Table 22.1 but were discovered during the data integrity investigation. Some of these poor data management practices are major failings of the overall quality management system.

22.3.11 Root Cause of Data Integrity and Poor Data Management Practices

In this section, the root cause of the data integrity falsification as well as contributory factors will be discussed. The focus in this analysis is on the controls that should have been in place in the overall quality management system of the company and would have prevented the falsification rather than analysing the list of data integrity violations themselves.

The root cause of the data integrity violation was:

- Failure of Management to Set Company Expectations for Data Integrity

  The company does not have a data integrity policy covering paper, hybrid (paper printouts from underlying electronic records) and electronic systems, along with associated training and evaluation of employee understanding. Senior management are responsible, under GMP, for the quality management system and within it for ensuring data integrity. Without this policy and training in place, did the employee realise that what they did was not ethical or compliant with the regulations? Furthermore, there are other examples of poor data management practices documented in this report that could be minimised or eliminated if a data integrity policy were available and training in its contents had been given.

It should be noted that the root cause is not the individual who committed the falsification of data but with the lack of controls for ensuring data integrity in the overall quality management system.

Closely associated with the root cause is:

- Lack of Effective Good Documentation and Record Management Practices Procedures

  There is a lack of one or more procedures for good documentation practices (GDocP) and the associated record management practices that cover both paper and electronic records. These procedures sit underneath the company data integrity policy and have some common elements across the company, *e.g.* good documentation practices for paper and electronic systems. Some ancillary procedures will be production or QC specific or even for a single computerised system, *e.g.* procedure for chromatographic integration for a CDS. Contained in one or more of these procedures will be the definition of electronic records including the associated contextual metadata as the original records that must

be collected and not the signed paper printouts when a computerised system is involved. This includes spreadsheets.
In addition, selective reporting of data, time travelling and unjustified invalidation of a run should be eliminated with an effective second person review of electronic records together with the paper printouts and properly completed instrument log book entries (GDocP).

Subsidiary, but contributory, causes to the data integrity problem are the following:

- Lack of Effective Second Person Review

  As the falsification was easily identified when reviewing the instrument log book *versus* the signed paper printouts and the corresponding electronic records (mainly chromatograms), it indicates the lack of an effective second person review. If such a review process were in place when these events occurred, the falsification would have been identified immediately the first time it occurred and resolved quickly, obviating the need for an investigation. Any second person review can be conducted by an appropriately trained peer with the combination of training, education and experience so that the burden is shared between QC staff and allows for absences.
  Some analytical procedures do not even have a second person review, *e.g.* appearance and solubility, respectively, in contravention of both US and EU GMP regulations. All current methods lacking a second person review must be revised to include a second person check at the time of testing or after.

- Lack of Objective Evidence for Second Person Review

  Some analytical procedures do not generate objective evidence that would allow for a proper second person review, *e.g.* microbiological assay colony counting. These methods need to be reviewed and updated to ensure that there is objective evidence available for review. For example, colony counting should have an automated colony counter that generates a digital photograph that allows for second person review and later audit or inspection. For non-destructive testing, such as odour or appearance, the second person review could include a visual check of the sample used for the test as well as the associated documentation.

- Inappropriate Access Privileges to Computerised Systems

  If the analyst who caused the data integrity violations could delete data from a CDS, it implies that inappropriate access privileges were granted to the individual. As there is a single shared administrator account and password for the ion chromatograph controlled by one CDS, this may not be an isolated case. The recycle bin of this instrument contained multiple deleted directories, reports and data files over a period of 2 years, indicative of a consistent, widespread but undocumented practice. No laboratory user should have data deletion privileges on any laboratory system, data can only be deleted with written management approval coupled with a procedure. This must be applied throughout the company with no exceptions to avoid dual standards and issues with any Research data that could be submitted to regulatory agencies.

- Access to Operating System Functions

  All current computerised systems in the QC laboratories consist of standalone workstations that are not controlled adequately in that any user can access the system clock, recycle bin and the electronic records that usually, but not exclusively, reside in directories in the operating system. Access to these functions must be restricted to IT administrators only.

- Computerised Systems are not Validated

  Each computerised system must be validated with application user types and access privileges controlled and documented. As part of this process, the vulnerability of the electronic records created by the system must be assessed and risk based controls put in place to ensure that they are protected.

- Focus on Paper Records

  During the data integrity investigation, the focus of the second person review, when it is carried out, is on paper records. It does not appear that electronic records were reviewed before method validation reports were approved or batches released before this data integrity investigation. Furthermore, discrepancies between the paper records and entries in the instrument log book were not identified prior to this investigation. Each analytical procedure needs to be reviewed to ensure that all records are identified, the tester is trained to ensure that they are collected and the second person review trained to check that they have been collected as required and have been reviewed, including date and time stamps on any electronic records. In addition, the reviewer also needs to be trained in detecting poor data management practices and falsification.

- Failure to Following Analytical Instructions

  In many cases in this report, staff have failed to follow instructions in analytical procedures. Stopping chromatographic runs after peaks of interest have eluted but before the specified run time is a very common practice that is a gross failure to comply with GMP. The rationale to save time is inadequate to justify the failure to follow instructions especially when there is no written documented rationale for this practice.

- Blank Forms and Master Templates

  Analysis and method sheets are issued by QA but they are not uniquely numbered. There does not appear to be a reconciliation mechanism as required by FDA, WHO, PIC/S and EMA data integrity guidance documents.

- Sample Weights for Water Content Determination

  The issue with the sample weights taken for water content determination is that the analyte is hygroscopic and the weights need to be taken quickly before the sample absorbs moisture from the air. This is the main reason for the differences in sample weight taken *versus* target weight in the analytical procedure. A new Karl Fischer titrator instrument that has an instrument data system that produced records is being qualified and validated. The instrument will be sited in a glove box with a nitrogen atmosphere and will be qualified and validated.

- No Chromatographic Integration SOP

  As there is not a chromatographic integration SOP there does not appear to be any control on the number of times a peak can be integrated. This leaves the company open to the accusation that they are integrating into compliance.

- Equipment Maintenance Logs are Inadequate

  Examination of the maintenance log book for one instrument found that uncontrolled loose-leaf sheets were used to document maintenance of the instrument. This is unacceptable as blank forms are the subject of comments in the FDA, WHO, PIC/S and EMA data integrity guidance documents. They should be replaced with bound notebooks with sequentially numbered pages that are issued and reconciled by QA.

22.3.12 Assessment of Material Impact

As discussed by the FDA and PIC/S,[3,4] there should be an assessment of material impact of the data integrity violations on the PAI batches on the following factors:

- Impact on product quality
  Is under- or overstrength product being released on the market?
  Should any method be revalidated or batches reanalysed to see if a product recall is required?
- Impact on patient safety
  Related to the above point what would be the impact on the patient taking this product?

Are there higher impurity levels in the product released that could adversely impact a patient?

- Impact on data integrity

  What is the impact of the falsification on the results in the specific application dossier discussed here or on released batches in general?

For reasons of confidentiality, the case study discussions and conclusions cannot be presented here but the three areas must be discussed and the outcome presented in the data integrity investigation report. From this, a decision needs to be made to inform the regulatory authorities or not.

22.3.13 CAPA Plans: Short-term Remediation and Long-term Solutions

It is important when developing any CAPA plans as outcomes from a data integrity investigation that a systems approach be taken to resolving the problems identified. Table 22.3 outlines the main CAPA actions from this case study. These CAPA plans contain both short-term remediation (*e.g.* shared

Table 22.3 Corrective and preventative action plans arising from the data integrity investigation.

CAPA plan	Detail
Implement Company Data Governance & Data Integrity approach	• Management Leadership and communication • Executive Sponsor for the DG/DI programme • Data Governance plans • Data Integrity Policy and training • Good Documentation Practices procedures and training • Quality oversight of data integrity • Data integrity audits and investigations
Implement IT Controls for Standalone Systems	• Short term: Remove shared user access and implement access control restrictions for all systems • Long term: global account management system with new qualified IT infrastructure
Implement Effective Second Person Review	• Write an SOP for second person review that includes check for data falsification • Training in the SOP for reviewers
Extend scope of OOS SOP	• Extend OOS SOP to include how to recognise when an OOS is present and how to investigate and, where appropriate, invalidate results correctly • Training for staff
Control of Master Templates & Blank Forms	• Implement a process and/or system for control of master templates and uniquely numbered blank forms • Implement a track and trace system to account for blank form use
Improved log book management	• Review all current logbooks to determine how many use loose-leaf pages • Devise and implement a standard equipment maintenance and use log that includes review of entries

Table 22.3 (*continued*)

CAPA plan	Detail
Implement an SOP for chromatographic integration	• Write a scientifically sound SOP for chromatographic integration including where and when it is possible to manually integrate peaks • Training in the procedure
Implement an SOP for Computerised System Validation	• The approach will be based upon an integrated AIQ-CSV as outlined in the new version of USP <1058> • Write an SOP for flexible risk based instrument qualification and computer validation • Training in the procedure
Consolidation of different CDS applications	• Business improvement and compliance project • Implement and validate a single networked CDS application • Electronic working with electronic signatures • Eliminate spreadsheets and paper from the chromatographic process

user accounts, user access to data files, data integrity policy and associated training) and long-term solutions (*e.g.* implementation of a networked CDS).

22.4 Summary

Data integrity investigations require much detailed and painstaking work and take up much internal and external resource. It is much better to prevent the situation from occurring if possible. Undisclosed software errors can make an investigation unavoidable. However, human instigated investigations can be avoided if there is the correct management direction, culture, policies, procedures and training in place to prevent the data integrity violation from taking place.

References

1. *WHO Technical Report Series No. 996 Annex 5 Guidance on Good Data and Records Management Practices*, World Health Organisation, Geneva, 2016.
2. *FDA Draft Guidance for Industry Data Integrity and Compliance with cGMP*, Silver Spring, MD, USA, 2016.
3. *FDA Application Integrity Policy: Fraud, Untrue Statements of Material Facts, Bribery, and Illegal Gratuities (Compliance Policy Guide Section 120.100)*, Food and Drug Administration, Rockville, MD, 1991.
4. *PIC/S PI-041 Draft Good Practices for Data Management and Integrity in Regulated GMP/GDP Environments*, Pharmaceutical Inspection Convention/Pharmaceutical Inspection Co-Operation Scheme, Geneva, 2016.
5. *USP General Chapter <621> Chromatography*, United States Pharmacopoeia Commission Inc, Rockville, MD.

6. *Scientist Steven Eaton Jailed for Falsifying Drug Test Results*, 2013, Available from: http://www.bbc.co.uk/news/uk-scotland-edinburgh-east-fife-22186220.
7. *GAMP Guide Records and Data Integrity*, International Society for Pharmaceutical Engineering, Tampa, FL, 2017.
8. *21 CFR 211 Current Good Manufacturing Practice for Finished Pharmaceutical Products*, Food and Drug Administration, Silver Spring, MD, 2008.

CHAPTER 23

Data Integrity and Outsourcing

Outsourcing analytical testing to Contract Manufacturing Organisations or Contact Research Organisations is pervasive throughout the pharmaceutical industry. Indeed, some pharmaceutical companies are virtual with a small number of permanent staff where all work from development, through regulatory submission and manufacturing of the product is outsourced to contract organisations. As part of supplier assessment and on-going monitoring of the work, the CMO/CRO's approach to data integrity is an important consideration as inspection of the contracted work will be undertaken there and not at the sponsor or contract giver's site.

A pharmaceutical company (contract giver) must now include requirements for data integrity for the outsourcing organisation (contract acceptor) in quality and/or technical agreements. Moreover, with the emphasis on electronic records generated from the work, where will these be stored and how easily can they be retrieved in case of an audit or inspection? Should a pharmaceutical company just accept a summary report or does due diligence require that the electronic records be reviewed by the contract giver as well? It all depends on the sponsor or contract giver's approach to the management of regulatory and business risk.

23.1 What the Regulators Want

23.1.1 WHO Guidance on Good Data and Record Management Practices

This WHO guidance has several clauses that discuss the approaches to outsourcing work from a data integrity perspective[1]:

Data Integrity and Data Governance: Practical Implementation in Regulated Laboratories
By R. D. McDowall

Published by the Royal Society of Chemistry, www.rsc.org

6.5 Quality audits of suppliers, self-inspections and risk reviews should identify and inform management of opportunities to improve foundational systems and processes that have an impact on data reliability. ...

7.1 The increasing outsourcing of GXP work to contracted organizations, *e.g.* contract research organizations, suppliers and other service providers, emphasizes the need to establish and robustly maintain defined roles and responsibilities to assure complete and accurate data and records throughout these relationships. The responsibilities of the contract giver and acceptor, should comprehensively address the processes of both parties that should be followed to ensure data integrity. These details should be included in the contract described in the WHO GXPs relevant to the outsourced work performed or the services provided.

7.2 The organization that outsources work has the responsibility for the integrity of all results reported, including those furnished by any subcontracting organization or service provider. These responsibilities extend to any providers of relevant computing services. When outsourcing databases and software provision, the contract giver should ensure that any subcontractors have been agreed upon and are included in the quality agreement with the contract accepter, and are appropriately qualified and trained in GRDP. Their activities should be monitored on a regular basis at intervals determined through risk assessment. This also applies to cloud-based service providers.

7.3 To fulfil this responsibility, in addition to having their own governance systems, outsourcing organizations should verify the adequacy of the governance systems of the contract acceptor, through an audit or other suitable means. This should include the adequacy of the contract acceptor's controls over suppliers and a list of significant authorized third parties working for the contract acceptor.

7.4 The personnel who evaluate and periodically assess the competence of a contracted organization or service provider should have the appropriate background, qualifications, experience and training to assess data integrity governance systems and to detect validity issues. The nature and frequency of the evaluation of the contract acceptor and the approach to ongoing monitoring of their work should be based upon documented assessment of risk. This assessment should include an assessment of relevant data processes and their risks.

7.5 The expected data integrity control strategies should be included in quality agreements and in written contract and technical arrangements, as appropriate and applicable, between the contract giver and the contract acceptor. These should include provisions for the contract giver to have access to all data held by the contracted organization that are relevant to the contract giver's product or service as well as all relevant quality systems records. This should include ensuring access by the contract giver to electronic records, including audit trails, held in the contracted organization's computerized systems as well as any printed reports and other relevant paper or electronic records.

7.6 Where data and document retention is contracted to a third party, particular attention should be paid to understanding the ownership and retrieval of data held under this arrangement. The physical location where the data are held, and the impact of any laws applicable to that geographical location, should also be considered. Agreements and contracts should establish mutually agreed consequences if the contract acceptor denies, refuses or limits the contract giver's access to their records held by the contract acceptor. The agreements and contracts should also contain provisions for actions to be taken in the event of business closure or bankruptcy of the third party to ensure that access is maintained and the data can be transferred before the cessation of all business activities.

This is an extensive and comprehensive approach to the data integrity issues when outsourcing laboratory work to a third party. The buck stops with the contract giver or sponsor.

23.1.2 PIC/S PI-041 Good Practices for Data Management and Integrity in Regulated GMP/GDP Environments

In Section 10 of this guidance document about data integrity considerations for outsourced activities there are the following clauses[2]:

10.1.1 Data integrity plays a key part in ensuring the security and integrity of supply chains. Data governance measures by a contract giver may be significantly weakened by unreliable or falsified data or materials provided by supply chain partners. This principle applies to all outsourced activities, including suppliers of raw materials or, contract manufacturers, analytical services, wholesalers and contracted consultation service providers.

10.1.2 Initial and periodic re-qualification of supply chain partners and outsourced activities should include consideration of data integrity risks and appropriate control measures.

10.1.3 It is important for an organisation to understand the data integrity limitations of information obtained from the supply chain (*e.g.* summary records and copies/printouts), and the challenges of remote supervision. These limitations are similar to those discussed in Section 8.11 of this guidance This will help to focus resources towards data integrity verification and supervision using a quality risk management approach.

Under a section on routine document versification there is the following:

The supply chain relies upon the use of documentation and data passed from one organisation to another. It is often not practical for the contract giver to review all raw data relating to reported results. Emphasis should be placed upon robust supplier and contractor qualification, using the principles of quality risk management.

10.3 Strategies for assessing data integrity in the supply chain

10.3.1 Companies should conduct regular risk reviews of supply chains and outsourced activity that evaluate the extent of data integrity controls required. Information considered during risk reviews may include:

- The outcome of site audits, with focus on data governance measures
- Review of data submitted in routine reports, for example:

Areas for review:

- Comparison of analytical data reported by the contractor or supplier *vs.* in-house data from analysis of the same material
- To look for discrepant data which may be an indicator of falsification

10.3.2 Quality agreements should be in place in place between manufacturers and suppliers/contract manufacturing organisations (CMOs) with specific provisions for ensuring data integrity across the supply chain. This may be achieved by setting out expectations for data governance, and transparent error/deviation reporting by the contract acceptor to the contract giver. There should also be a requirement to notify the contract giver of any data integrity failures identified at the contract acceptor site.

10.3.3 Audits of suppliers and manufacturers of APIs, critical intermediate suppliers, primary and printed packaging materials suppliers, contract manufacturers and service providers conducted by the manufacturer (or by a third party on their behalf) should include a verification of data integrity measures at the contract organisation.

10.3.4 Audits and routine surveillance should include adequate verification of the source electronic data and metadata by the Quality Unit of the contract giver using a quality risk management approach. This may be achieved by measures such as:

Site audit:
Review the contract acceptors organisational behaviour, and understanding of data governance, data lifecycle, risk and criticality.

Material testing *vs.* CoA:
Compare the results of analytical testing *vs.* suppliers reported CoA. Examine discrepancies in accuracy, precision or purity results. This may

be performed on a routine basis, periodically, or unannounced, depending on material and supplier risks.

Remote data review:
The contract giver may consider offering the Contracted Facility/Supplier use of their own hardware and software system (deployed over a Wide Area Network to use in batch manufacture and testing). The contract giver may monitor the quality and integrity of the data generated by the Contracted Facility personnel in real time. In this situation, there should be segregation of duties to ensure that contract giver monitoring of data does not give provision for amendment of data generated by the contract acceptor.

Quality monitoring:
Quality and performance monitoring may indicate incentive for data falsification (*e.g.* raw materials which marginally comply with specification on a frequent basis).

10.3.6 Care should be taken to ensure the authenticity and accuracy of supplied documentation (refer to Section 8.11). The difference in data integrity and traceability risks between 'true copy' and 'summary report' data should be considered when making contractor and supply chain qualification decisions.

Although this is an extensive quotation from the PIC/S document, the inclusion of these statements is entirely justified by the extensive contracting prevalent throughout the pharmaceutical industry.

23.1.3 Regulatory Guidance Summary

This section is very extensive and this is quite deliberate as it illustrates the major concerns of regulatory authorities with the problems associated with data integrity when outsourcing R&D projects, bioanalytical analysis, stability studies, analytical development, analytical procedure validation, routine analysis and pharmaceutical manufacturing to third parties.

In Chapter 4 on Data Governance, three elements for outsourcing laboratory analytical work were identified as:

- supplier data governance;
- technical and quality agreements;
- supplier audits.

Abstracting from the regulatory guidance above, these three main themes regarding the outsourcing of any analysis work can be expanded to cover:

- Before placing any work at a CRO or CMO sufficient due diligence must be done to ensure that adequate data governance and approaches to data integrity are present in the contract organisation. This is not a box ticking exercise but the process is a due diligence assessment. This is especially so as many data falsification cases have occurred in facilities where work has been sub-contracted. The assessment for data integrity needs to examine if data have been falsified or not.

- The technical and quality agreements or contracts must have data integrity clauses in them as well as the roles and responsibilities of both contract giver and contract acceptor. There must also be an on-going communication programme as well as the right of the contract giver to audit.
- Performance of the work needs to be reviewed on an on-going basis both remotely as well as with on-site audits. This is vital as the contract giver is accountable for the work carried out on their behalf by the CMO/CRO. Proactive monitoring is essential and not sign the contract and think that all is well. As can be seen from a multitude of warning letters and 483 observations to API manufactures and contract laboratories, ignore this aspect at your peril!

The data integrity requirements for outsourcing analytical work have been ramped up by the regulators. Therefore, the due diligence assessment and audit of suppliers must be increased by sponsors and contract givers in response.

23.2 Cetero Research Laboratories Data Falsification Case

Imagine your company has submitted a New Drug Application to the FDA containing bioequivalence data generated by a CRO called Cetero Research. The study compares the absorption and distribution of your generic version of a drug and the ethical version. Your application has been accepted by the FDA but later you receive a letter from the Agency containing the following paragraphs[3]:

> The pervasiveness and egregious nature of the violative practices by Cetero has led FDA to have significant concerns that the bioanalytical data generated at Cetero from April 1, 2005 to June 15, 2010, as part of studies submitted to FDA in New Drug Applications (NDA) and Supplemental New Drug Applications (sNDA) are unreliable.
>
> FDA has reached this conclusion for three reasons:
>
> (1) the widespread falsification of dates and times in laboratory records for subject sample extractions,
> (2) the apparent manipulation of equilibration or "prep" run samples to meet pre-determined acceptance criteria, and
> (3) lack of documentation regarding equilibration or "prep" runs that prevented Cetero and the Agency from determining the extent and impact of these violations.

> Serious questions remain about the validity of any data generated in studies by Cetero Research in Houston, Texas during this time period. In view of these findings, FDA is informing holders of approved and pending NDAs of these issues.[3]

One of data integrity issues at Cetero was caused by staff who wanted to claim additional money for overtime and falsified data to show that they were present outside of normal working hours and at weekends. Cetero received an untitled letter in 2011.[4] Many sponsors have had to repeat bioequivalence and bioavailability studies to ensure that their regulatory dossiers are acceptable to the agency. If Cetero Research was a GMP laboratory, then a product recall, warning letter or an import ban may be the possible regulatory outcomes.

Outsourcing is supposed to save money – however, in this and many other instances what is perceived to be a financial panacea turns into a financial disaster.

You will recall that in Section 2.4 of Chapter 2, a CRO called Semler Research Private Limited also falsified bioequivalence data that resulted an untitled letter and also sponsor companies having to repeat bioanalytical studies.[5,6] This is not an isolated instance.

Repeating an earlier statement in this chapter, the issue here is clear that the selection of a contract laboratory is not a mere box ticking exercise but one where adequate due diligence must be performed. This is to ensure that a selected facility has an equivalent approach to data integrity as the sponsor's or contract giver's organisation. Outsourcing analytical work is not a passive contract and forget approach but requires an active approach of trust but verify. Due diligence including data integrity also applies when acquiring a company. Baxter acquired Claris Injectables Ltd on the same day that the FDA started an inspection that resulted in a warning letter for data integrity violations issued in July 2018 (https://www.fda.gov/ICECI/EnforcementActions/WarningLetters/ucm613538.htm). It appears that due diligence during the acquisition of the company did not extend to a data integrity assessment of manufacturing and testing operations.

23.3 Include Data Integrity in Supplier Assessment/Audit

In this chapter, the terms assessment and audit are used interchangeably, the aim is to determine that the organisation selected for outsourcing work is fit and able to perform the work and that this includes data integrity assessment. It does not matter if it is called an assessment or an audit, what does matter is that you do sufficient due diligence for data integrity to ensure that your organisation has the confidence to place the work with an external supplier.

23.3.1 Current Approaches to Laboratory Audit

Supplier assessment is already a key requirement of GXP regulations and regulatory expectations, *e.g.* EU GMP Chapter 7 on outsourcing.[7] If a contract laboratory is required to undertake routine or specialist analysis it is important to undertake sufficient due diligence to ensure that the work will be carried out to your specifications and analytical procedures and that this results in an acceptable analytical report. Some typical areas for assessment are:

- Facilities;
- quality management system;
- scientific knowledge and training of the staff;
- analytical instruments used and their qualification;
- application software used and the quality of validation;
- procedures and record keeping practices.

Of these in the list above only the record keeping practices comes close to assessing an organisation's or a laboratory's understanding of data integrity, as we have seen throughout this book. However, with the emergence of data integrity issues, it is essential to extend supplier assessment to include how the outsourcing organisation approaches data integrity and data governance. It is important to ensure that an expanded assessment does not just focus on data integrity but also includes the factors in the bullet list above.

23.3.2 Extending Assessment to Include Data Integrity

As you can see from the regulatory guidance in Section 23.1 for outsourcing regulated work, there are extensive requirements to consider from a data integrity perspective. At a high level, Figure 23.1 shows three main stages in the process:

- initial selection and Quality Management System;
- client technical and quality agreements;
- monitoring and auditing the on-going work packages.

This section will give you an overview of the process and cover key points but there will be additional information found in the other chapters of this book as well as other ref. 1,2 and 8.

Figure 23.2 shows further detailed considerations when outsourcing regulated laboratory work. This covers the initial assessment of the laboratory, what should be included in quality and technical agreements between the CRO and the sponsor/contract giver as well as some questions for monitoring the on-going work for data integrity.

23.4 Initial Data Integrity Assessment of a Facility

In this section, the focus is on assessing the approach to data integrity and data governance within the selected contract laboratory. In addition to this section, the reader is referred to Chapter 21 where data integrity audits are

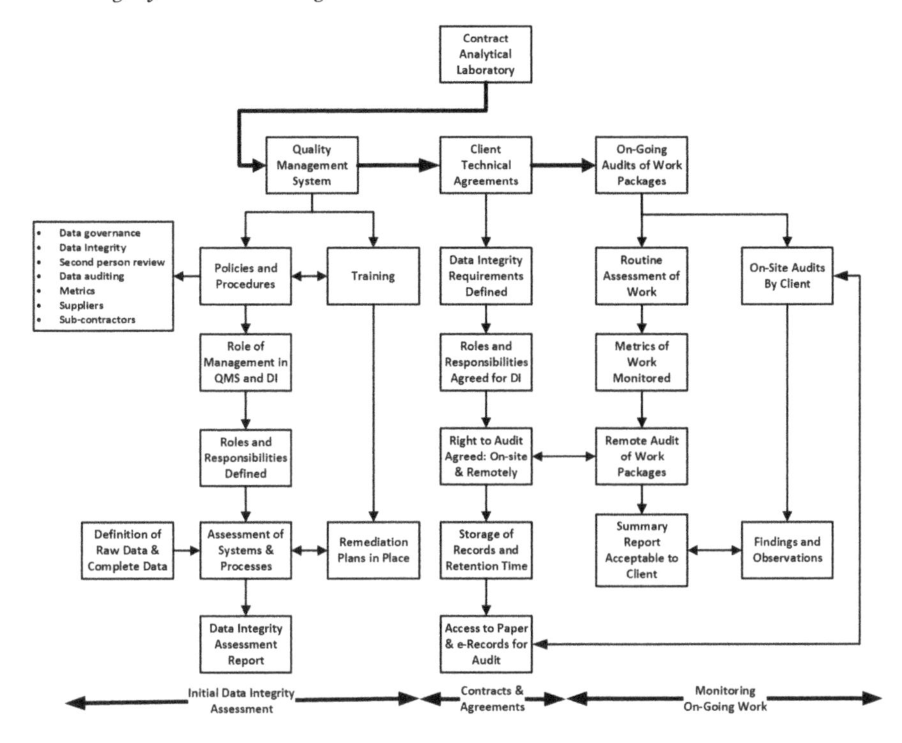

Figure 23.1 Overview of data integrity issues when outsourcing analytical work.

covered and Chapter 24 that is an audit aide memoire based on the data integrity model discussed in Chapter 5.

This section is not intended to be a discussion on how to audit a laboratory but more about a data integrity focus when conducting an audit. Therefore, we will not be discussing the process of auditing, *e.g.* conducting the opening or closing meetings or how to ask questions, *etc.*

23.4.1 Initial Selection of the Contract Laboratory

When outsourcing work, there will be a specification of the analytical procedure(s) to be applied (or developed and validated), the analytical technique and analytical instrumentation to use and the skills required to perform the work. These will be the primary selection criteria to be used in an initial screening process. Secondary criteria may also be applicable such as location, price, recommendations from colleagues or placement of previous work with the facility.

From the available laboratories, a selection will be made of one or two facilities that are worth investigating further. The order will be prioritised typically with a primary candidate and a secondary one. There may be remote discussions before a decision is made to audit the primary candidate.

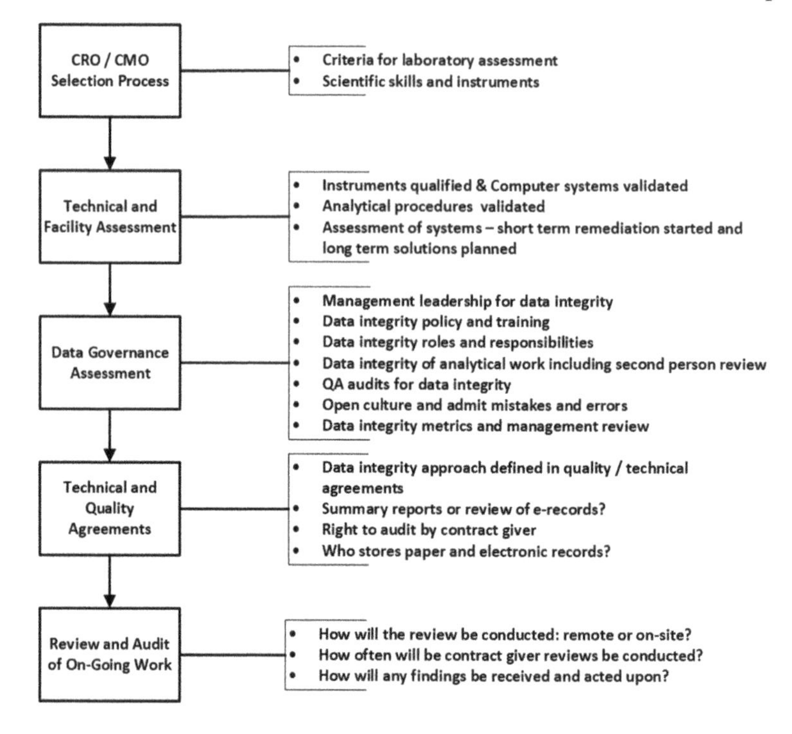

Figure 23.2 Considerations when outsourcing laboratory work: assessment, technical agreements and monitoring.

23.4.2 Do Not Forget the Scientific and Technical Competence of the Supplier

Although this chapter is focused on data integrity, there is not much use in having a laboratory that has a great approach to data integrity but has poor scientific and technical competence. Consideration of how the analytical instrumentation and computerised systems are selected, qualified or validated, calibrated and operated is also a key factor in selecting a contract laboratory or assessing a quality control laboratory as part of contract manufacturing organisation. Additionally, the scientific knowledge such as how to validate or transfer an analytical procedure is just as important as the laboratory's approach to data integrity.

For many CRO and CMO organisations, the extension of any assessment to include data integrity represents a major change in how they are assessed.

23.4.3 Request for Pre-audit Information

On-site audits are frequently time limited, especially if the contract facility has many clients or customers. To ensure the best time is made on site it is essential to prepare for the audit. You will need to get information about the following items:

- quality management system;
- list of policies and procedures;
- data integrity and data governance approaches;
- site Master File (if a GMP CRO/CMO facility).

However, you cannot afford to select another Cetero Research and if insufficient time is allowed (say one day) then you have a few choices to consider:

- request extra time for the audit;
- add an additional auditor;
- walk away and find another contract laboratory.

Factors that can influence the contract laboratory will be who the contract giver is, the type of work and the potential value of the work.

23.4.4 Planning the Audit

On-site you need to be make the best use of time and avoid some of the common mistakes such as being led by the laboratory. Ensure that you are in charge and have a timetable that you need to stick to and have a working lunch to cover more areas. Remember that audits sample and you will not have time to go into too much detail, unless a problem is found that needs to be understood in more detail.

Plans for the audit should be simple and flexible and ensure that the auditee does not run the show. For example, you need an introduction to a new company and their work but not 50 marketing slides extolling their virtue and excellence, similarly a three-hour lunch break with copious volumes of alcohol does not find any data integrity problems. Therefore, manage the auditee's expectations carefully and specify an overview presentation focused on your topics of interest with a short working lunch to make the most of your time. The timetable needs to be planned prior to the audit covering the agreed timeframe between the CRO and you. We will now look at some of the topics we need to cover from a data integrity perspective.

23.4.5 Data Governance and Data Integrity in the Context of a PQS

As part of the pre-audit data gathering, the contract facility should have provided a site master file or equivalent containing a high level list of QMS procedures and methods. From this an overview of the QMS can be gathered: what procedures are there for data integrity and data governance? However, from a simple listing it is unlikely that you can come to a view on how data integrity is covered at the facility and the role that senior management have in the leading role of management in the Pharmaceutical Quality System, *e.g.* leadership for data integrity as well as management reviews of the QMS and data integrity. Similarly, the culture of the organisation is extremely difficult to assess from a list of PQS procedures.

However, where there are written data integrity policies and procedures this at least provides a lever from which the organisation's PQS approach to data integrity can be started and then branching out to the associated training materials with records of training effectiveness, as presented and discussed in Chapter 7.

Other elements of data integrity and data governance that should be present in a QMS are:

- How are master templates and blank forms managed? Typically, the way this is performed should be found in a records management or document control procedure.
- How have computerised systems been assessed for data integrity and how are the identified issues being remediated? Assess the overall level of automation in the laboratory and technical controls are or will be used to enforce data integrity.
- Assessment of the data integrity of paper processes. This is obviously an important area but many organisations only focus on hybrid systems and electronic processes for assessment. As we have seen in Chapters 11 and 13 it is paper processes where poor data management practices or even falsification can occur without any or little traceability. What remediation is planned for paper processes?
- Understand how the roles and responsibilities for data integrity have been defined, *e.g.* data owners, data stewards, *etc.* as presented in Chapter 6. The role of the audit should be to assess how effective these roles have been in relation to system assessment and remediation as well as operational use of these systems.
- How are raw data defined for paper, hybrid and electronic processes? Where is the SOP that defines this?
- Second person review procedure is important. As we have seen in Chapter 22, lack of an effective second person review procedure can lead to some serious data integrity issues that could have been stopped quickly with such a procedure.
- Control of the potential CMO/CRO's suppliers: how do they assure that their supplier's data integrity is acceptable?
- The role of senior management and open culture should also be assessed. Thin profit margins in contract organisations can result in pressure being placed on individuals to get results finished and cut corners.

23.4.6 Investigate Electronic Record Controls

One important area to look at are controls for the generation, interpretation, reporting and storage of electronic records. This will not be a static audit but will require the auditor to talk around the laboratory and view the systems themselves. In particular:

- What are the controls to prevent users going around the back of the application to delete records and time travel to repeat the work? These must be investigated and verified by seeing them in operation.
- Procedural controls are not unreliable and inconsistent. What the auditor needs to see is that the available technical controls are specified, implemented and validated with whatever procedural controls are required.
- However, there should also be an on-going plan to replace older and sub-standard data systems with networked data systems, together with progress against the plan.

Reviewing the actual systems in the laboratory is a key audit requirement that should have been notified in advance to the auditee.

- If access to systems is refused.
  The option?
  Walk away.
- If client confidentiality is cited as refusing to show these systems, this is a clear indicator that the CMO/CRO does not have an effective system for handling confidential information and hiding the sponsor's identity.
  The option?
  Walk away.
- If the CRO does not show you the technical controls for ensuring data integrity.
  The option?
  Walk away.

Walking away is cheap compared with the cost of non-compliance if there are data integrity issues that are found later by inspectors. What is the cost of finding a new supplier compared with repeating a bioequivalence study or a product recall due to data integrity issues? Minimal.

23.4.7 Conclusion of the Audit

At the end there will be a written report with the overall assessment of the facilities, infrastructure, scientific and technical competence and the data integrity status of the organisation. Together with the report, there may be a list of findings (against regulations or procedures) and recommendations (not non-compliances but opportunities for improvement).

Depending on the overall findings and CAPA responses to the findings, a decision will be made to use the contract facility or not. Let us assume that the decision is to use the organisation and we can proceed to the next stage of the process outlined in Figure 23.1, which is negotiating the agreements and contracts for the work and how data integrity must be included in them.

23.5 Agreements and Contracts for Data Integrity

The next stage of the process is to ensure that work is agreed jointly in a technical or quality agreement or contract. Apart from the usual terms and conditions of such documents, which are out of the scope of this book, there is the need to ensure that data integrity is included in these documents and that roles and responsibilities are defined adequately between the two parties.

23.5.1 Main Data Integrity Contractual Responsibilities

Adapting EU GMP Chapter 7 on outsourcing,[7] here are the roles and responsibilities for outsourcing analytical work at a CRO/CMO that need to be included in agreements between the two parties. The two parties are the contract giver or sponsor and the contract acceptor or CRO/CMO.

The minimum contract giver/sponsor needs and responsibilities are:

- Provide to the laboratory robust and accurately written analytical procedures and technical information, with records.
- Provide troubleshooting advice if required when transferring, establishing or running a procedure.
- For routine analytical work, the sponsor needs access to paper and especially electronic records to verify the work was performed correctly.
- The right to audit remotely and on-site (including for cause where necessary) to assess the analytical work including the integrity of any records generated.

Contract Acceptor/Contract Laboratory needs and responsibilities:

- Defining original records and raw data as electronic where a computerised system is involved.
- Accurate, complete and reliable documentation throughout the analysis following ALCOA+ principles.
- Effective second person review of all work from sampling to reportable result (as applicable to the work).
- Internal QA oversight – assessment of QA procedures and activities and schedule of internal audits. The latter must include the schedule of data integrity audits and evidence they have been carried out *e.g.* report front page and CAPA plans.
- Monitoring and trending results with identification of out of expectation and out of trend results and communication to the contract giver/sponsor when there are issues such as OOS results, *etc.*
- Permit remote and on-site audits of the work performed including electronic records.

These needs and responsibilities must be included in the agreements and contracts between the two parties.

23.5.2 Using the Same Chromatography Data System

As many analytical procedures use chromatography as the analytical technique, one of the requirements from the contract giver/sponsor may be for the contract laboratory to use the same CDS. If feasible, this can offer several advantages for both organisations as CDS methods can be transferred to the CRO and electronic records transferred back to the sponsor where appropriate for review. The advantages for monitoring the quality and integrity of on-going work are very high. It also has the means of ensuring data retention by the contract giver, independent of the contract acceptor that generated the data.

This approach may not be applicable for all analytical techniques but only for those techniques that are common and where there is a high sample throughput.

23.5.3 Storage of the Records Generated

In the past, when paper records were the norm it was easy to determine where the records would be stored either short-term or long-term as it was a simple matter of moving cellulose from one location to another. Now there is more of a problem as there are both paper and electronic records. Paper from the manual stages of the analytical process is simple but far more problematic are the electronic records as typically these need the application that generated them to open and view them. When the contract giver and contract accepter have the same system, this is relatively straightforward as the data can be transferred on an optical disc or *via* a secure FTP site.

When there are different systems at the two sites then it would make sense to keep the records with the laboratory where they were generated for the record retention period. However, depending on the time of retention, there may be one or two upgrades of software involved. What does the agreement say about this? Probably nothing. What needs to be included in the agreement is the due diligence of the CRO to inform the sponsor that software will be upgraded and that the results from a standard data set come to the same decision before and after the upgrade.[9]

In all cases, storage of the records must be part of any agreement or contract and include what happens in case of inspection at the contract giver's site, how will access to the electronic records be achieved?

23.6 On-going Monitoring of Work and Audits

Once the agreements and contracts are signed and work has commenced, the contract giver now must monitor the on-going work, especially the data integrity aspects to assure themselves that these aspects of the quality or technical agreement are being followed. If this is not undertaken, it may be because of a sponsor's attitude that they have confidence in the CRO. However, be prepared for nasty surprises when an inspector calls to say that there are problems in the facility you outsourced your work to, *e.g.* Cetero and Semler.[4,6] Therefore, it is vitally important to realise that you cannot outsource work and forget about it – the work must be monitored and this takes time and effort within the sponsor's organisation.

There are three types of monitoring and audit that will be considered in this section:

- remote monitoring of results and associated data;
- remote audit;
- on-site audit including for cause audits.

23.6.1 Risk Based Approaches to Monitoring

It important to understand where the greatest analytical and data integrity risks are in the work that is being outsourced so that the monitoring can be matched with the risk posed by the analysis. For example, which of the following QC analyses are the most critical and high risk?

- appearance;
- colour;
- label claim;
- impurity profile.

Typically, it will be label claim and impurity profile analysis by chromatography that will pose the highest data integrity risk. From the pre-contract audit, the chromatography data system used by the laboratory will be known, *e.g.* a networked system with a database or standalone systems with directories for storing data files. This knowledge will help to focus the monitoring and auditing efforts of the sponsoring organisation.

23.6.2 Monitoring the Results

The type of monitoring undertaken by the sponsor will depend on the nature of analytical work undertaken by the contract laboratory. For bioanalytical work it may be checks of the standard curves and back calculated concentration values, monitoring the results from the three concentrations of quality control samples, viewing blank injections, plasma concentration *versus* time profiles and incurred sample reanalysis results. In contrast, quality control monitoring may be looking at the chromatograms, individual values from aliquots as well as the reportable results and trending these over time and batch, impurity profiles between batches, *etc.*

Actively monitoring outsourced analytical work means that trends or issues could be identified, discussed with the contract laboratory and rectified before they become major problems. Equally so, the contract facility should also be monitoring and discussing with the sponsor the same issues if they are monitoring the work correctly.

23.6.3 Remote Assessment of Work Packages

Following on from the monitoring of results is the remote assessment of work packages, *e.g.* review of the whole of the analytical batch or study records. This can be a combination of video conferencing and direct access *via* a secure internet link between the CRO and the sponsor. Electronic records including the metadata and audit trails can be reviewed remotely. Access can be *via* a link where a member of the contract laboratory staff operates the data system following the requests of the sponsor's representative. An alternative could be where restricted access is granted to a sponsor's reviewer or auditor for read only access to a networked computer system when a batch is complete. Paper records can be reviewed remotely *via* video link but, depending on the resolution of the link, may only give an impression and it may be difficult to ascertain if the record is original or not. However, it is the approach to sponsor's due diligence that is the basis for ensuring work is carried out correctly. Any findings identified remotely can be documented and CAPAs generated by the contract laboratory.

The main issue is that the sponsor needs conformation that the data integrity of the batch or study records are good and that they can rely on the results to take decisions. There would typically be more audits early in a contractual relationship between a sponsor and CRO as the number and frequency may be reduced over time as the two organisations become confident of working together.

23.6.4 On-site Audits

On site audits come in two forms: regular scheduled audits or for cause audits. The latter is the audit of a specific batch or batches due to a problem such as an increase in serious adverse events associated with a product batch. In either case, the approach is usually the same:

- Access to the electronic records and paper records for all systems to check the work has been done correctly and data integrity is assured.
- Have procedures been followed and have there been any attempts at short cuts, copying data from one batch to another or other poor data management practices?
- Audit of a specific or several work packages.
- Checking the effectiveness of previous CAPAs. This may not be applicable to a for cause audit.

Out of these audits may come findings and recommendations for the contract laboratory to act upon and generate CAPAs.

23.6.5 Contract Analytical Work with Your Eyes Open

You may be thinking after reading this chapter that contracting analytical work is a major problem. Far from it, contracting analytical work has been a fact of life in the pharmaceutical industry for over 50 years. If handled correctly, the process should be considered as an extension of your own laboratory and a collaborative process. If handled wrongly, it can be a millstone hanging around an organisation's neck for a decade or so. Approach using contract facilities with your eyes open and trust but verify, especially when it comes to data integrity. To help in assessment of internal and external audits and assessments of laboratories, there is an auditing aide memoire in Chapter 24.

References

1. *WHO Technical Report Series No. 996 Annex 5 Guidance on Good Data and Records Management Practices*, World Health Organisation, Geneva, 2016.
2. *PIC/S PI-041 Draft Good Practices for Data Management and Integrity in Regulated GMP/GDP Environments*, Pharmaceutical Inspection Convention/Pharmaceutical Inspection Co-Operation Scheme, Geneva, 2016.
3. *FDA Letter to ANDA Sponsors Conducting Bioequivalence Studies at Cetero Research*, 2011, Available from: http://www.fda.gov/downloads/drugs/drugsafety/ucm267907.pdf.
4. *Cetero Research Untitled Letter (11-HFD-45-07-02)*, Food and Drug Administration, Silver Spring, MD, 2011.
5. *Notification to Pharmaceutical Companies: Clinical and Bioanalytical Studies Conducted by Semler Research Are Unacceptable*, 2016, Available from: https://www.fda.gov/Drugs/DrugSafety/ucm495778.htm.
6. *FDA Untitled Letter: Semler Research Center Private Limited*, Food and Drug Administration, Silver Spring, MD, 2016.
7. *EudraLex – Volume 4 Good Manufacturing Practice (GMP) Guidelines, Chapter 7 Outsourced Activities*, European Commission, Brussels, 2013.
8. *GAMP Guide Records and Data Integrity*, International Society for Pharmaceutical Engineering, Tampa, FL, 2017.
9. R. D. McDowall, *Validation of Chromatography Data Systems: Ensuring Data Integrity, Meeting Business and Regulatory Requirements*, Royal Society of Chemistry, Cambridge, 2nd edn, 2017.

CHAPTER 24

Data Integrity Audit Aide Memoire

This chapter contains an auditing aide memoire that is based on the Data Integrity Model described in Chapter 5 and should be used in conjunction with Chapter 21 (Quality Assurance Oversight), Chapter 22 (Data Integrity Investigations) and Chapter 23 (Data Integrity and Outsourcing).

Note that the questions in the sections in this chapter and annotations in the figures are intended for experienced auditors to stimulate further questions and are not intended to be comprehensive; they are intended to be aide memoires rather than definitive and detailed checklists. The main reason is that checklists typically do not work and they constrain an auditor. In many parts of this chapter, the aide memoire is unusual in that the questions are presented as figures as this is easier to refer to during an audit.

The bullet points and questions presented here do not always ask for documented evidence as an audit is assumed to be conducted by an experienced auditor.

24.1 What the Regulators Want

24.1.1 EU GMP Chapter 9 Self-inspections

The shortest chapter in EU GMP is Chapter 9 on Self-Inspections,[1] the aim of which is outlined in the Principle:

> Self-inspections should be conducted in order to monitor the implementation and compliance with Good Manufacturing Practice principles and to propose necessary corrective measures.

Data Integrity and Data Governance: Practical Implementation in Regulated Laboratories
By R. D. McDowall

Published by the Royal Society of Chemistry, www.rsc.org

The remainder of the chapter is presented below:

> 9.1 Personnel matters, premises, equipment, documentation, production, quality control, distribution of the medicinal products, arrangements for dealing with complaints and recalls, and self-inspection, should be examined at intervals following a pre-arranged programme in order to verify their conformity with the principles of Quality Assurance.
>
> 9.2 Self-inspections should be conducted in an independent and detailed way by designated competent person(s) from the company. Independent audits by external experts may also be useful.
>
> 9.3 All self-inspections should be recorded. Reports should contain all the observations made during the inspections and, where applicable, proposals for corrective measures. Statements on the actions subsequently taken should also be recorded.

24.1.2 Data Integrity Guidances for Audits

Presented in Chapters 21 and 23 are additional regulatory requirements that will not be repeated here due to the number and size.

24.1.3 Regulatory Requirements Summary

Self-inspections, including data integrity audits, must be:

- Planned with a schedule of processes and systems to be audited.
- Conducted by independent auditors (individuals from outside of the functional area and with the necessary expertise or using subject matter experts, where appropriate).
- A report of each data integrity audit with observations seen together with any findings and associated CAPA plans will be written.
- As reports are QA audits they will not be shown to inspectors but the audit plans, a certificate for each data integrity audit and the resulting CAPAs will be subject to regulatory inspection.

24.2 Audit Aide Memoire for the Foundation Layer: Data Governance

The Foundation Layer of the Data Integrity Model, presented in Chapter 5, focuses on the data governance with right culture and ethos for data integrity and this involves the following main topics:

- management leadership;
- writing and training of a corporate data integrity policy;

- further data integrity procedures and training, *e.g.* Good Documentation Practices;
- allocation of data owners for computerised systems and manual processes;
- establishment and maintenance of an open culture.

These areas are shown diagrammatically in Figure 24.1 and some questions to ask are described in the following sections.

24.2.1 Management Leadership for Data Integrity

Demonstrable evidence of senior management involvement in a data integrity programme is required, such as:

- Is there Management involvement for data integrity, *e.g.* Executive Sponsor, Data Integrity Committee or equivalent?
- Is there an approved corporate data integrity and data governance policy? This document may include a code of conduct.
- Is there an on-going communication programme from management to employees about data integrity initiatives within an organisation?
- Is there support for the initial and on-going data integrity training programme and integration into the annual GMP update training programme?
- Does management introduce the corporate data integrity training session?
- How are resources and budget for the assessment and remediation of processes and systems allocated?
- Is there identification of high risk processes and systems and management of risk to records?
- What metrics are generated for monitoring data integrity in the Pharmaceutical Quality System?
- Is there management review of data integrity programme: progress against plans?
- How often does management review the Pharmaceutical Quality System?
- Do Gemba walks by management occur and, if so, how frequently?

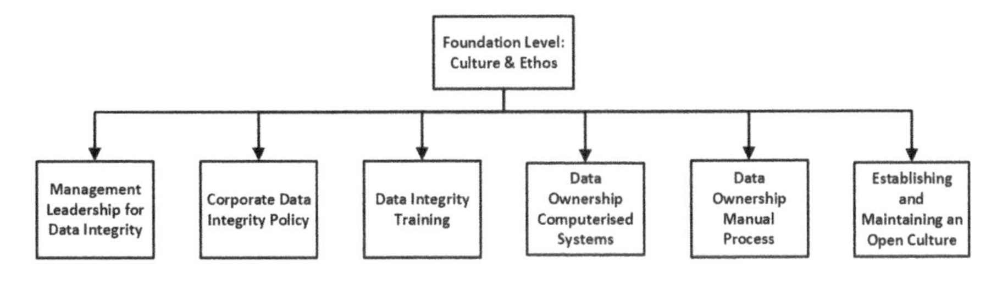

Figure 24.1 Main areas to audit in the foundation level of the data integrity model.

24.2.2 Corporate Data Integrity and Ethics Policy

Some of the questions to consider for a corporate data integrity and ethics policy are shown below and in Figure 24.1:

- Is there a corporate data integrity policy?
- What is the scope of the policy? Does it cover all staff?
- Is there a policy statement and expected behaviour of staff?
- Are roles and responsibilities from executive management to all employees outlined?
- Are there data Integrity procedures, *e.g.* written procedures and employee training, involvement of quality assurance, monitoring, internal data integrity audits, management review?
- Does the policy cover raising compliance concerns, prohibited actions, open door policy, reporting a suspected violation?
- Does the data integrity investigation include: non-retaliation for reporting employees, confidentiality of the investigation, consequences of a violation of the policy for an employee?
- How have the staff been trained in the policy? Is there demonstrable understanding of each employee in the content of the policy?
- Has each member of staff signed a statement that they understand and will follow the data integrity policy?

It may be that instead of a single document, the topics in Figure 24.2 have been split over several policy, guideline or procedures depending on a company's terminology for controlled documents.

24.2.3 Data Integrity Training

Data integrity must cover the following areas:

- Data integrity training has been or will be given to all existing employees and all new starts?
- Data integrity is or will be part of annual GMP refresher training all current employees from now?
- The training must include organisational data integrity policy, the critical need for honesty and full disclosure in all analytical reporting, plus how and when to report data integrity issues and record keeping?
- Training will include discussion of all data integrity SOPs and training documentation including how to document analytical records?
- Employees must understand that failure to follow the laboratory data procedures will result in a detailed investigation that could lead to very serious consequences?
- A copy of the corporate data integrity policy and all training materials is given to each trainee?
- All data integrity training has attendance sheets or equivalent?

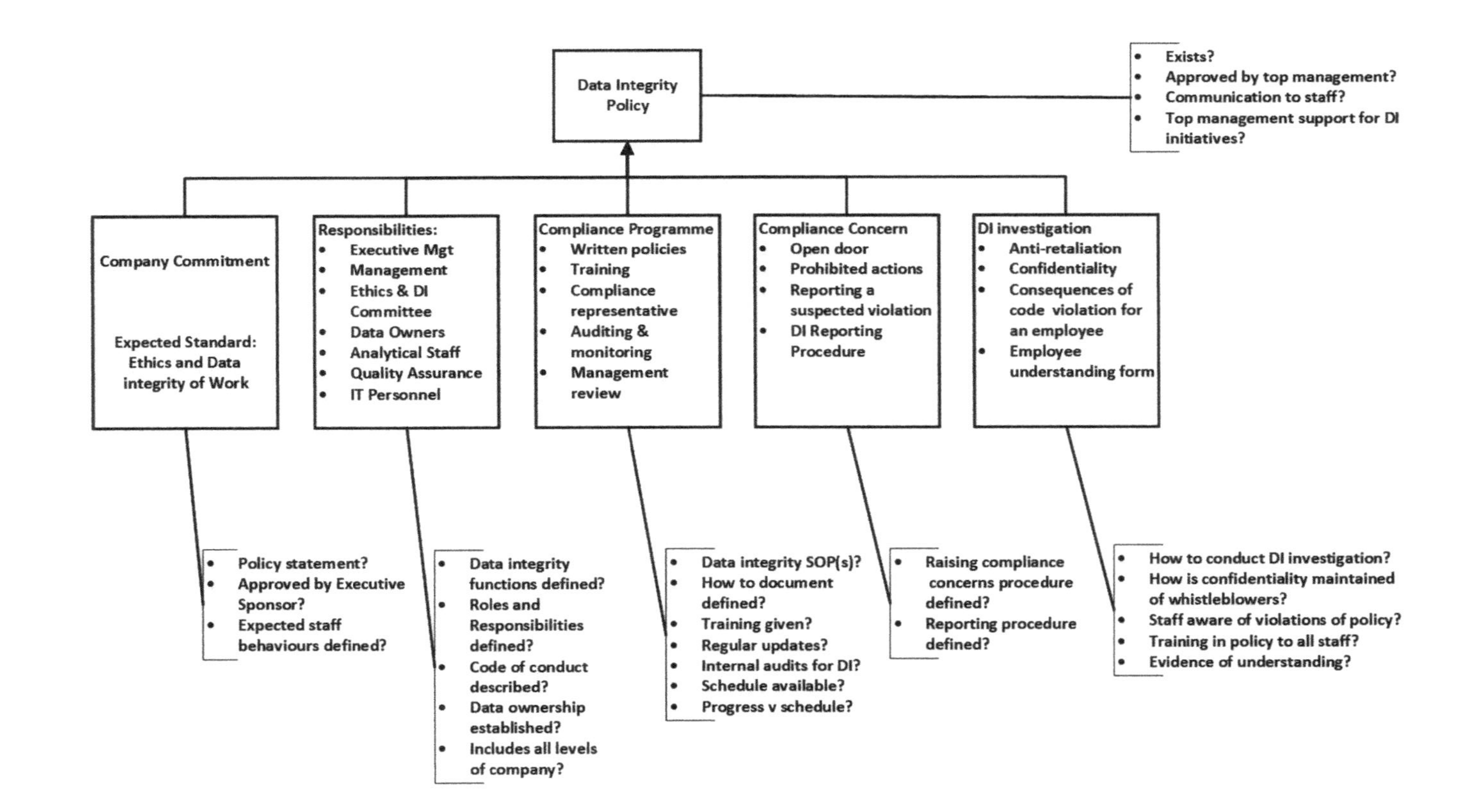

Figure 24.2 Audit questions for a corporate data integrity policy.

- Attendees must demonstrate that they understand their obligations related to data integrity, *e.g.* assessment questionnaire or equivalent.
- Senior managers actively support and implement the data integrity procedures (see Section 24.2.1).

This training is in addition to any specific training on laboratory procedures such as interpretation of spectra, chromatographic integration or second person review.

24.2.4 Data Ownership for Computerised Systems

Computerised systems require an individual to be identified as a data owner:

- Is there an inventory of computerised systems used in regulated laboratories?
- Is each system and the records generated classified for criticality?
- For each system is there a data owner identified who is responsible for the data generated and stored within the system (typically this would be the same person who is the process owner)?
- When data are transferred to another system, is the ownership of the data transferred to the data owner of the receiving system?

24.2.5 Data Ownership for Manual Processes

Manual processes should be included in a data integrity programme:

- Is there an inventory of manual processes (*e.g.* sampling, sample preparation, manual tests) in the laboratory?
- Has a data owner been allocated to each process?
- Is there a risk assessment of data integrity risk for the records generated in each process?
- Are there uncontrolled blank forms used in any manual process?
- Is there a list of manual processes for mitigation documented in a plan?
- Is there mitigation of the data integrity risks for each manual process?

24.2.6 Establishment and Maintenance of an Open Culture

An open culture, described in Chapter 8, is required to ensure that all effort put into data integrity is of value:

- How was an open culture established by senior management?
- How is the open culture maintained and encouraged?
- How are staff encouraged to report data integrity problems and issued with processes and systems?

- How is staff honesty in reporting data integrity problems rewarded by the company?
- Are Gemba walks used to gain feedback about problems in laboratory areas and to mentor staff to work in an open and honest way?

24.3 Audit Aide Memoire for Level 1: AIQ and CSV

24.3.1 Overview

At Level 1 of the data integrity model we need to consider the following interrelated topics:

- Analytical Instrument Qualification (AIQ);
- Computerised System Validation (CSV);
- Interfacing between two computerised systems.

24.3.2 Analytical Instrument Qualification

The qualification approach described in the 2017 version of USP <1058> on AIQ[2] has been interpreted into a simplified V model in Figure 24.3 to show the relationship and traceability between the user requirements specification (URS) and the operational qualification (OQ).

24.3.3 Computerised System Validation

A simplified GAMP software category 4 life cycle is shown in Figure 24.4 and is copied from Validation of Chromatography Data Systems, 2nd Edition[3] along with suggested questions for a computerised system validation that could be asked. If there is any category 5 software (custom or bespoke) then further details about specification, coding, code reviews, unit and integration testing should be requested and reviewed.

24.3.4 Validating Interfaces Between Computerised Systems

Figure 24.5 shows some of the questions to consider when reviewing the validation of a computer to computer interface. This figure can also be used in Level 3 if auditing a process where two computerised systems are interfaced.

24.4 Audit Aide Memoire for Level 2: Analytical Procedure Validation Life Cycle

Rather than base the audit aide memoire on ICH Q2(R1),[4] this section, as well as Chapter 15, are based on the more recent draft USP <1220> on analytical procedure validation lifecycle[5] and the updated ICH Q2(R1) as a life

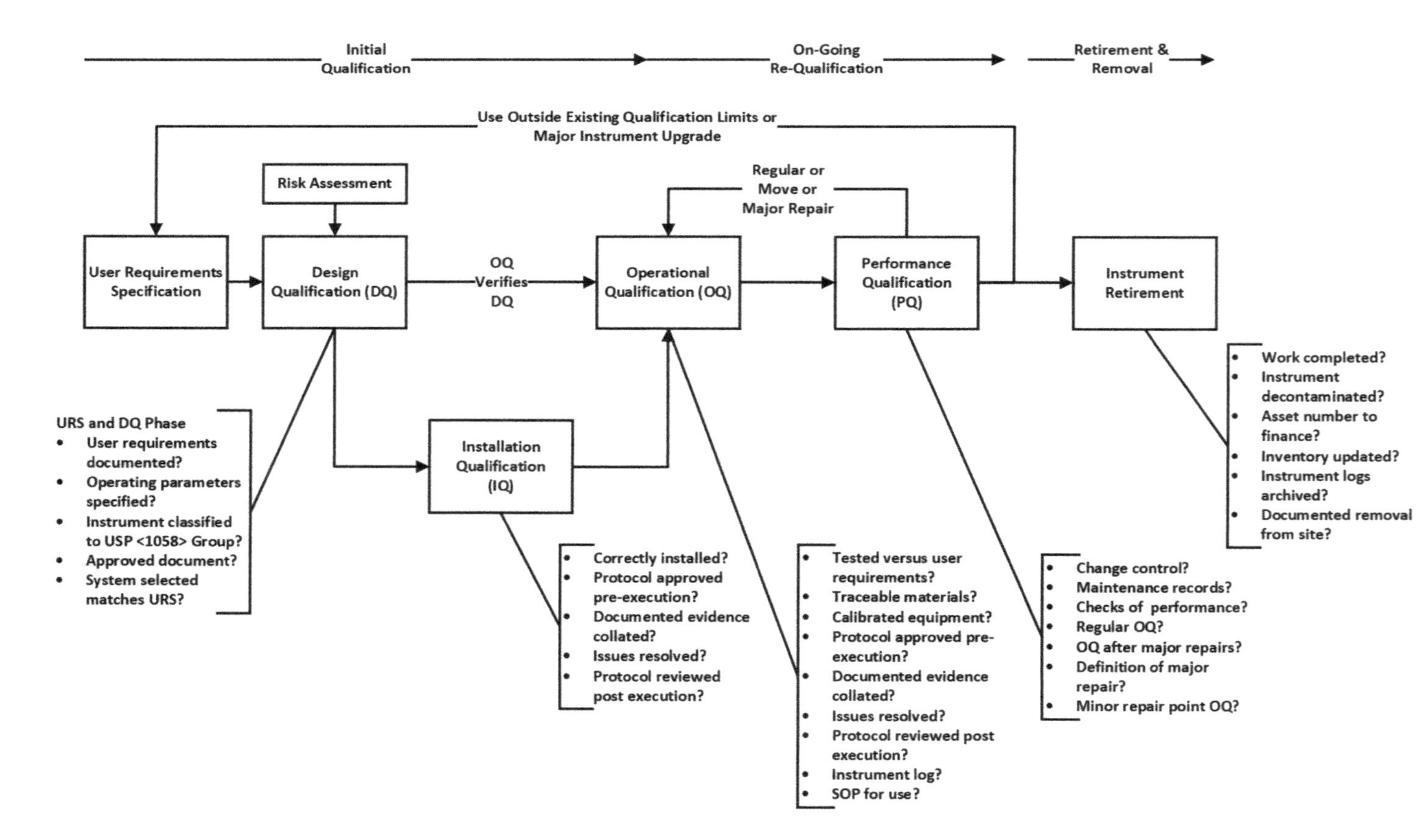

Figure 24.3 Questions to consider about analytical instrument qualification.

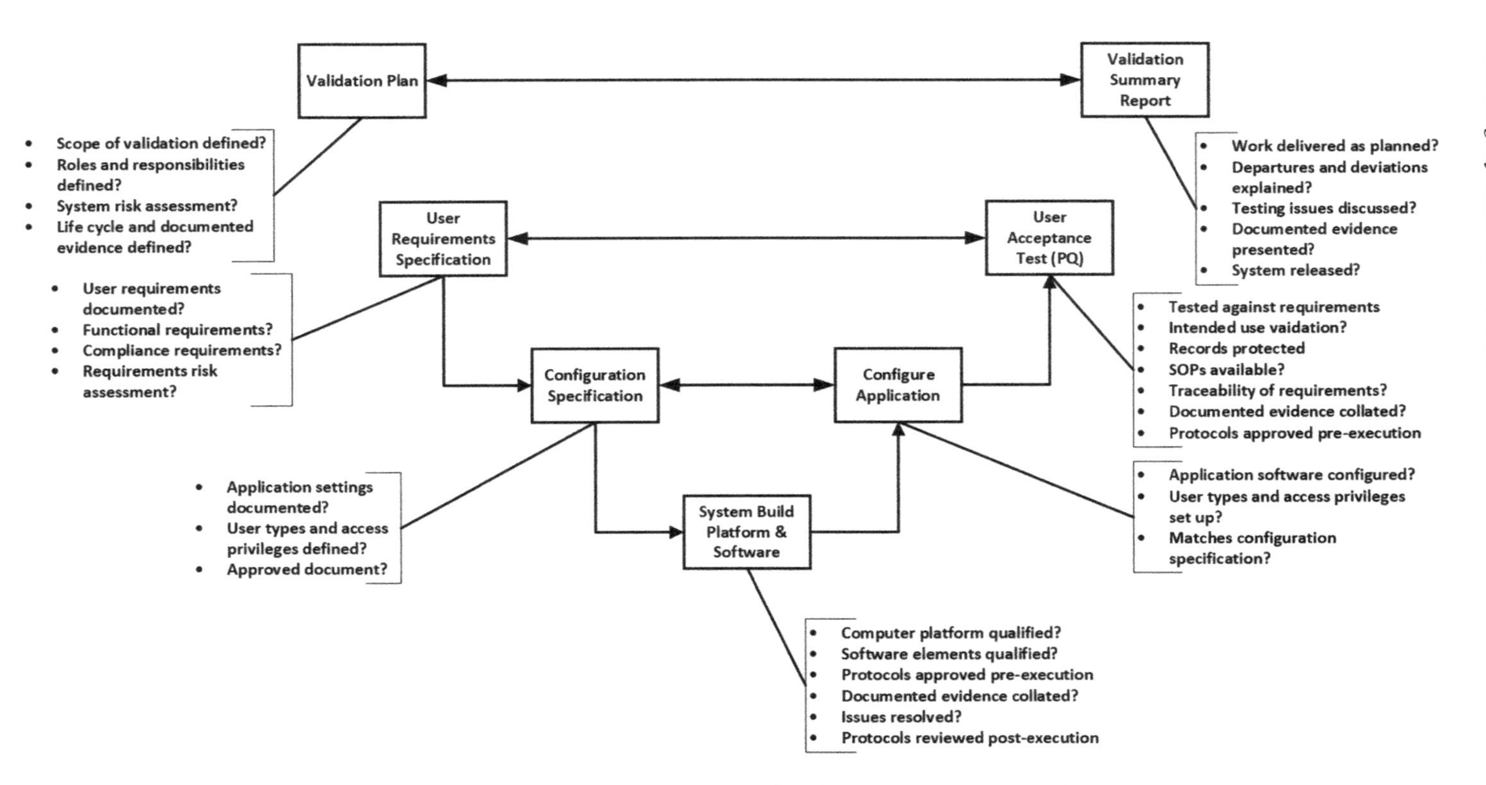

Figure 24.4 Some questions to consider when auditing the validation of a laboratory computerised system.

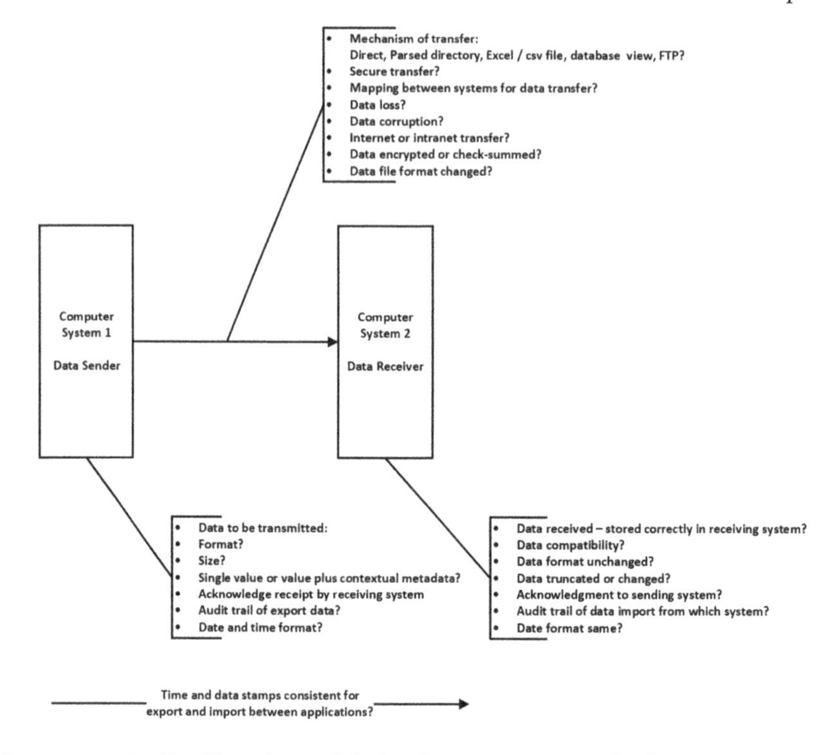

Figure 24.5 Audit of interface validation between computerised systems.

cycle approach to analytical procedures[6] will become the norm for method validation over time.

The life cycle approach consists of defining the ATP or analytical objectives and the three stages outlined below and presented in Figure 24.6:

- defining the analytical objectives or analytical target profile;
- Stage 1: Procedure design (method development);
- Stage 2: Procedure performance qualification (method validation);
- Stage 3: Continued performance verification.

The three subsections that follow give further detail about what to audit at Level 2 for analytical procedures throughout their life cycle.

24.4.1 Procedure Design (Method Development)

Development of an analytical procedure begins with defining the Analytical Target Profile and the measurement uncertainty (precision and accuracy) required of the method,[7] from this the analytical technique and analytical principles can be derived. Method development is conducted through the design of experiments that identify critical process parameters and understand their influence on the analytical method. At the end of the development there is an analytical procedure with a defined analytical control strategy, as shown in Figure 24.7.

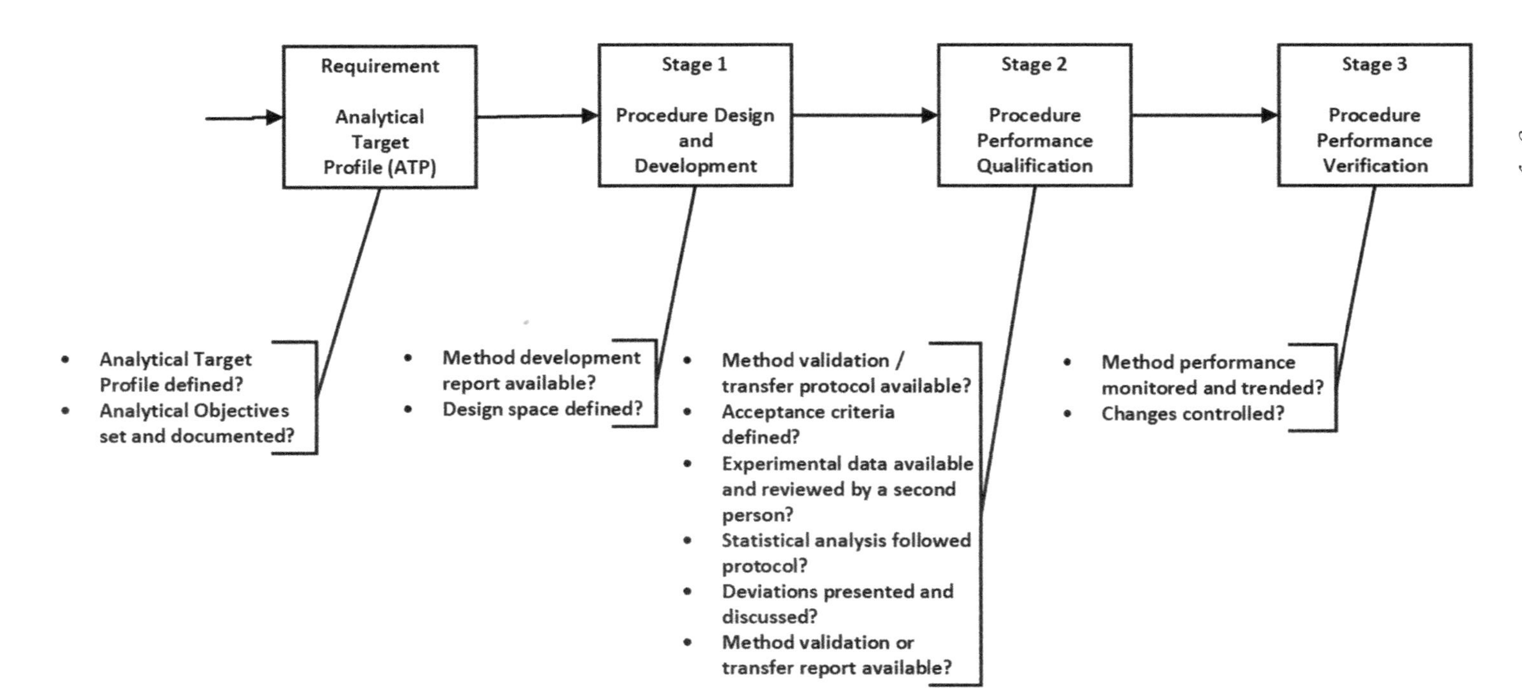

Figure 24.6 Overview of an analytical procedure validation lifecycle.

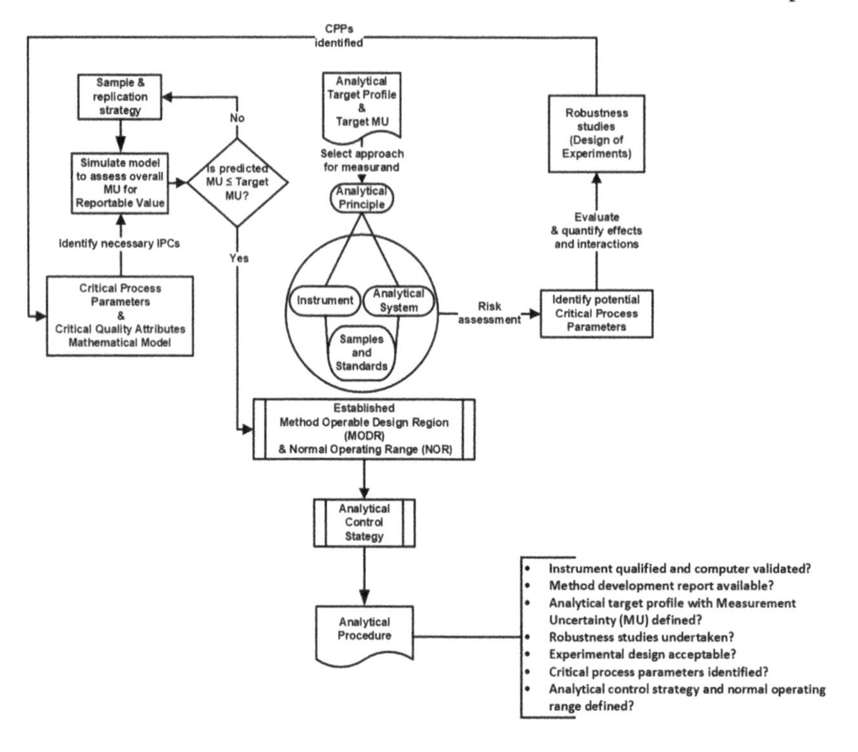

Figure 24.7 Development of an analytical procedure using quality by design.

24.4.2 Analytical Procedure Performance Qualification (Method Validation)

After the analytical procedure has been developed it undergoes procedure performance qualification to demonstrate that the procedure meets the Analytical Target Profile. Figure 24.8 shows the process and some of the questions to ask to determine if the procedure has been validated correctly.

24.4.3 Method Application: Control and Monitoring

During routine use of the method there should be trending of the results and where appropriate the system suitability test or point of use checks performed on the instrument to ensure that the procedure and results are in control, see Figure 24.9.

24.5 Level 3: Study and Batch Analysis Data Integrity Aide Memoire

At Level 3 of the data integrity model, all elements of the Foundation and Levels 1 and 2 should be present to ensure the integrity of data generated at this level. In this section, there are audit aide memoires for audits of study or batch records:

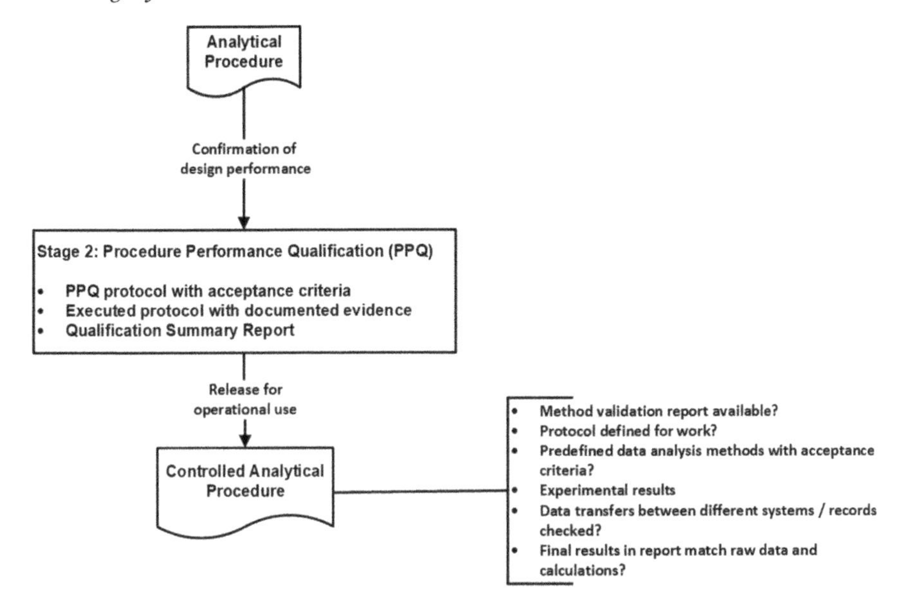

Figure 24.8 Analytical procedure performance qualification (method validation).

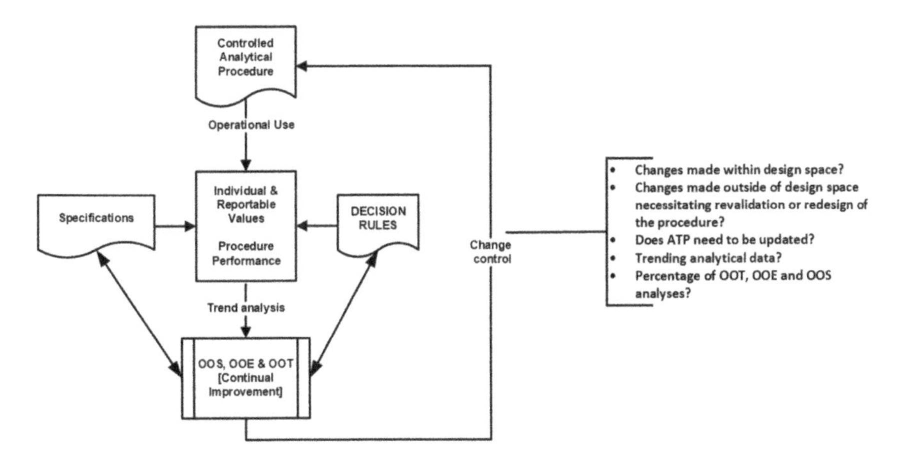

Figure 24.9 On-going analytical procedure performance verification.

- general audit aide memoires;
- paper records;
- spreadsheets;
- Laboratory Computerised System;
- Chromatography Data Systems.

24.5.1 Routine Analysis Data Integrity Aide Memoire

Some of the areas to consider for the assessment of analytical data are shown in Table 24.1:

Table 24.1 Analysis data integrity aide memoire.

Topic	Points to consider
Laboratory management	• Management emphasis on laboratory throughput • No mechanism for reporting problems • Analysts not allowed to use their judgement
Access to facilities and records	• Denial of access or conditional access to areas • Distraction of the audit • Records not readily available – delaying the audit
Procedures and training	• Analysts not able to describe the data generation or review process • Managers clarifying or translating analyst's responses to questions
Good documentation practices	• Correct documentation of work • Error corrections conform to GXP requirements • True or exact copies documented appropriately • No use of pencils and typewriter correction fluid • Uncontrolled records • All notebooks and logbook entries signed, dated and reviewed • Extremely clean, neat lab notebooks and instrument logs – have they been used?
Analytical procedures and SOPs	• Methods current or out of date? • Discrepancies between the method and SOP • Discrepancies between SOP and practice
Reagents and standards	• Preparation records for reference standards, *etc.* • Check log books of balances, pH meters for work performed • Storage of reference standards, *etc.* • Check for expired standards and reagents • Check for unlabelled reagents, containers, *etc.*
Sample management and sample preparation	• Sample inventory log does not match physical samples • Sample volume discrepancies • Samples or blanks diluted without explanation • Blank forms used in these processes?
Instrument records	• Lab note books with no errors • Missing analyst and reviewer signatures • Discrepancies between data and report • Time sequence of e-records do not match printouts • Unexpected gaps or changes in records • Instrument calibration records missing • Use of print screens for reports • Sequential chromatograms not displaying suitable background noise, matrix effects • SST failures ignored • Samples used as SST injections • Deleted data files • Check manual integration – consistent for both standards and samples? • Integration enhances or shaves peaks?
Study and batch records	• Incomplete data package for the analysis • Unexpected sample results not investigated • Too perfect results • Documentation without errors
Out of specification results	• Number of OOS reported in a specified time period • Number of OOS invalidated • Review OOS investigations • Assignment of the cause of an OOS scientifically sound?

24.5.2 Audit of Paper Analytical Records

An aide memoire for auditing a typical manual or paper process is shown in Figure 24.10. The arrow at the bottom of the figure referring to good documentation practices implies that the GXP requirements are applied throughout the whole process.

24.5.3 Audit of Hybrid Laboratory Computerised Systems

When auditing a laboratory computerised system used as a hybrid system with electronic records and signed paper printouts, consider the following items shown in Figure 24.11.

In addition to the items in Figure 24.11 consider asking questions in the following areas:

- review the instrument log and compare with the paper printouts and electronic records;
- procedures and logs for backup and recovery of the electronic records;
- changes and approved and rejected change control records;
- number of emergency changes to the system;
- revalidation following an approved change.

Most of these questions can also be applied to audits of electronic systems covered in Section 24.5.5.

24.5.4 Validation and Use of a Spreadsheet

When spreadsheets are used for regulated activities, the aide memoire in Figure 24.12 should be used. This figure covers both the validation of the spreadsheet and its use for regulated activities.

24.5.5 Chromatography Data System Aide Memoire

Chromatographic analysis is the major analytical technique used throughout the pharmaceutical industry and chromatography data systems have been involved in poor data management and falsification cases in both GMP and GMP regulated environments. Figure 24.13 shows some of the areas to consider in an audit of such systems.

24.6 Quality Assurance Oversight Aide Memoire

QA oversight is essential to see that the data integrity and data governance of an organisation are working and if not to identify any gaps and to oversee any corrective and preventative measures.

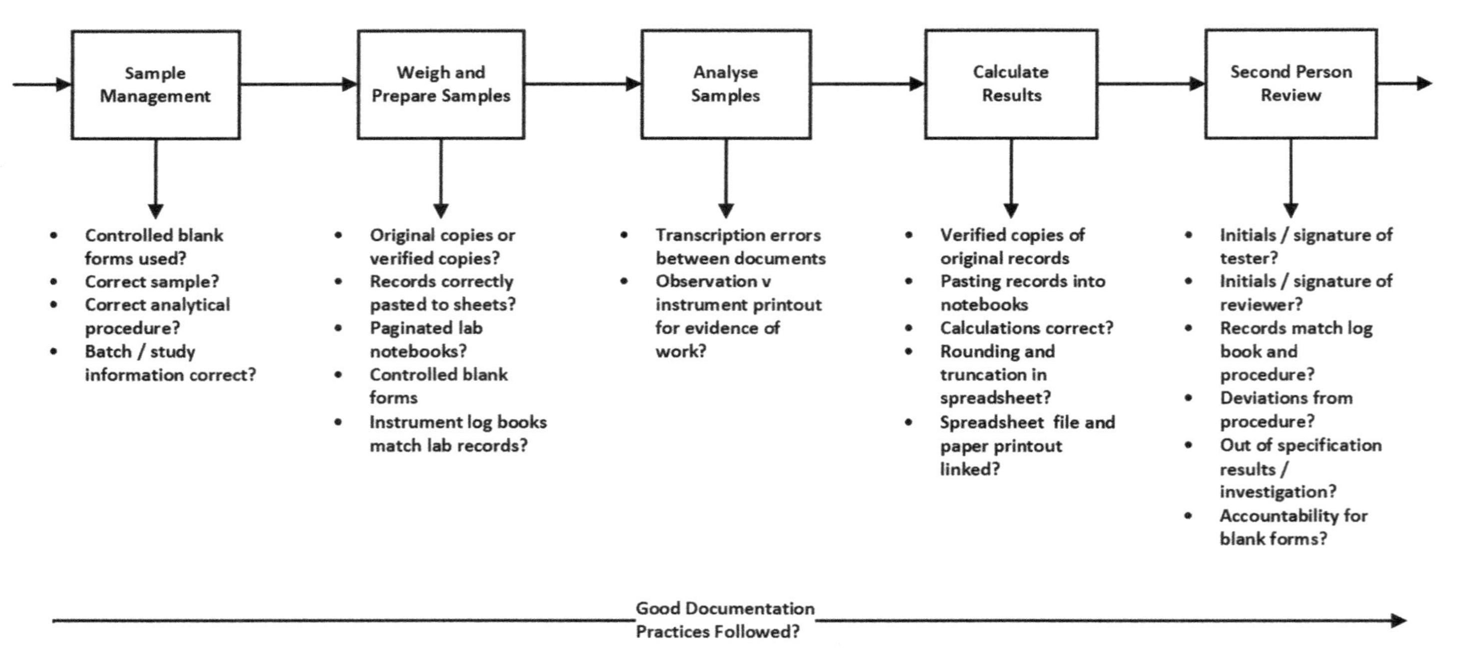

Figure 24.10 Aide memoire for audit of a paper process.

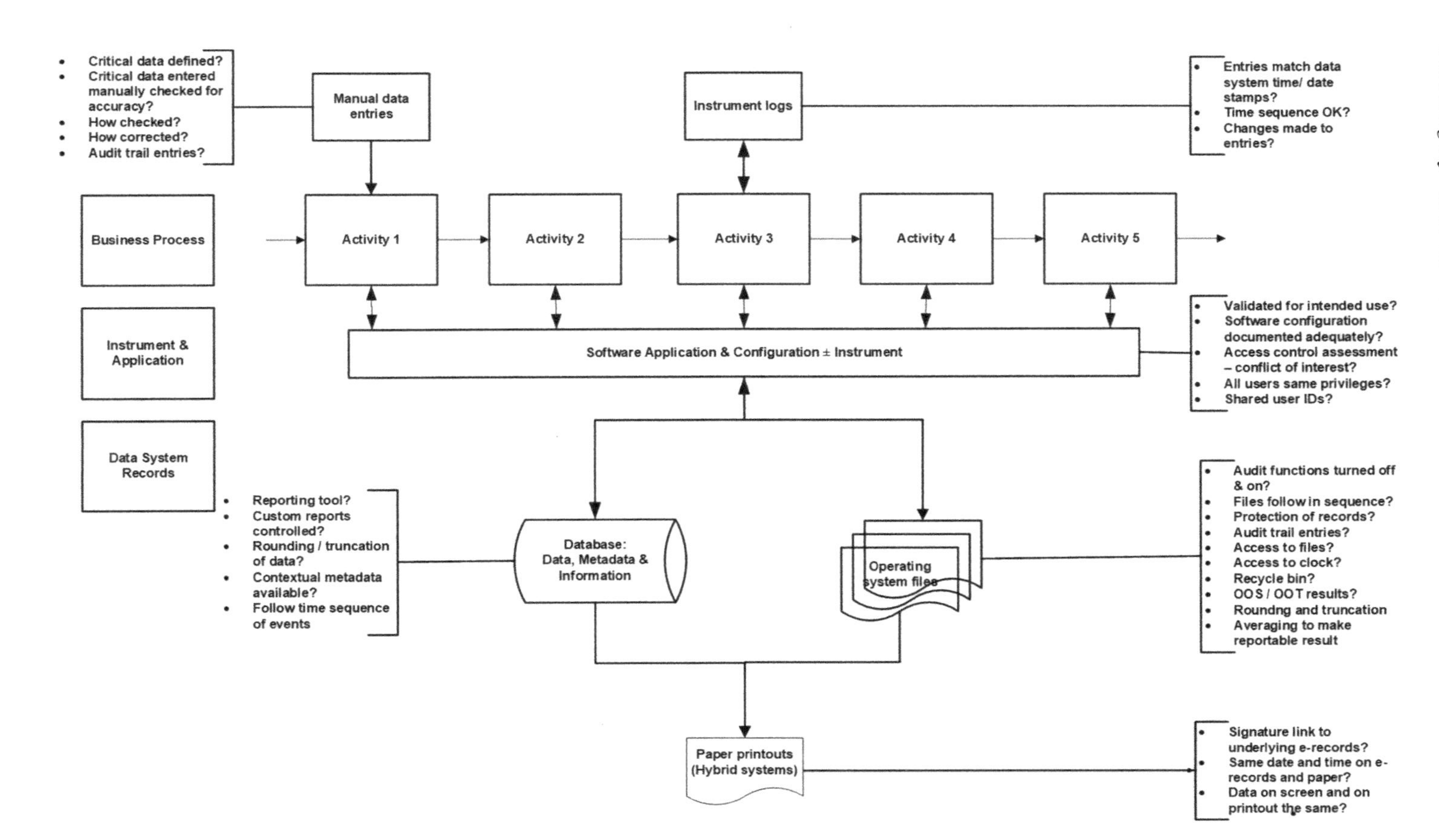

Figure 24.11 Audit of a laboratory computerised system.

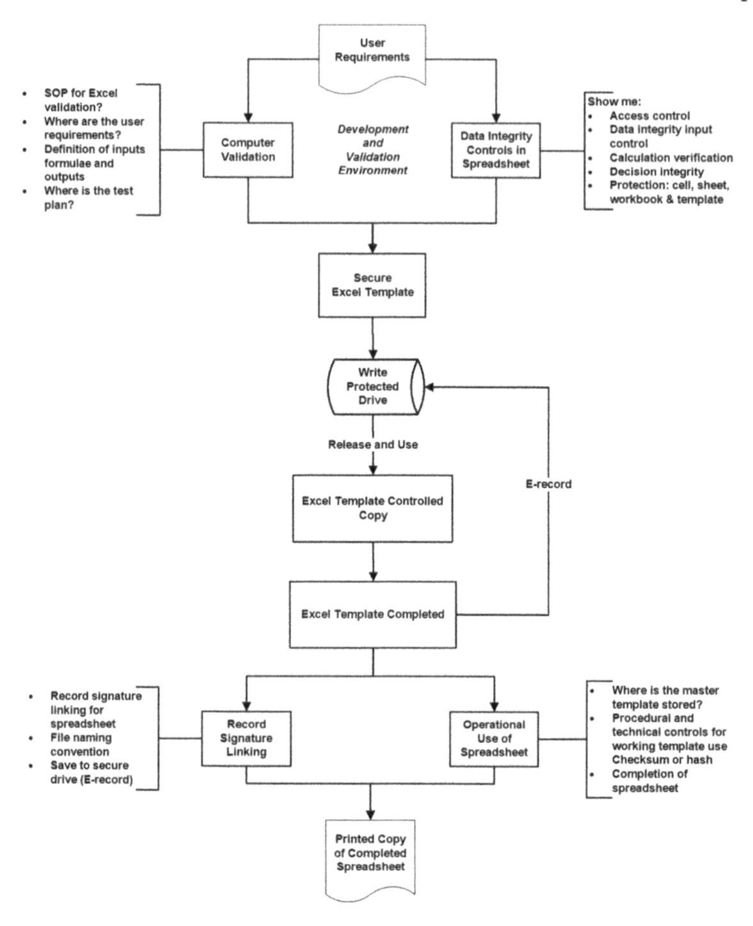

Figure 24.12 Aide memoire for an audit of a spreadsheet.

24.6.1 Routine Checks of Study or Batch Records

QA review of study and batch records must occur and ensure that good documentation practices are followed for paper, hybrid and electronic system data that comprise the record set. The QA oversight should check electronic records in hybrid and electronic records match paper printouts and include the associated contextual metadata including the appropriate audit trail entries.

24.6.2 Data Integrity Audits

Either have separate data integrity audits or integrate them with the audit schedule (but in the latter case ensure sufficient time and resources are available):

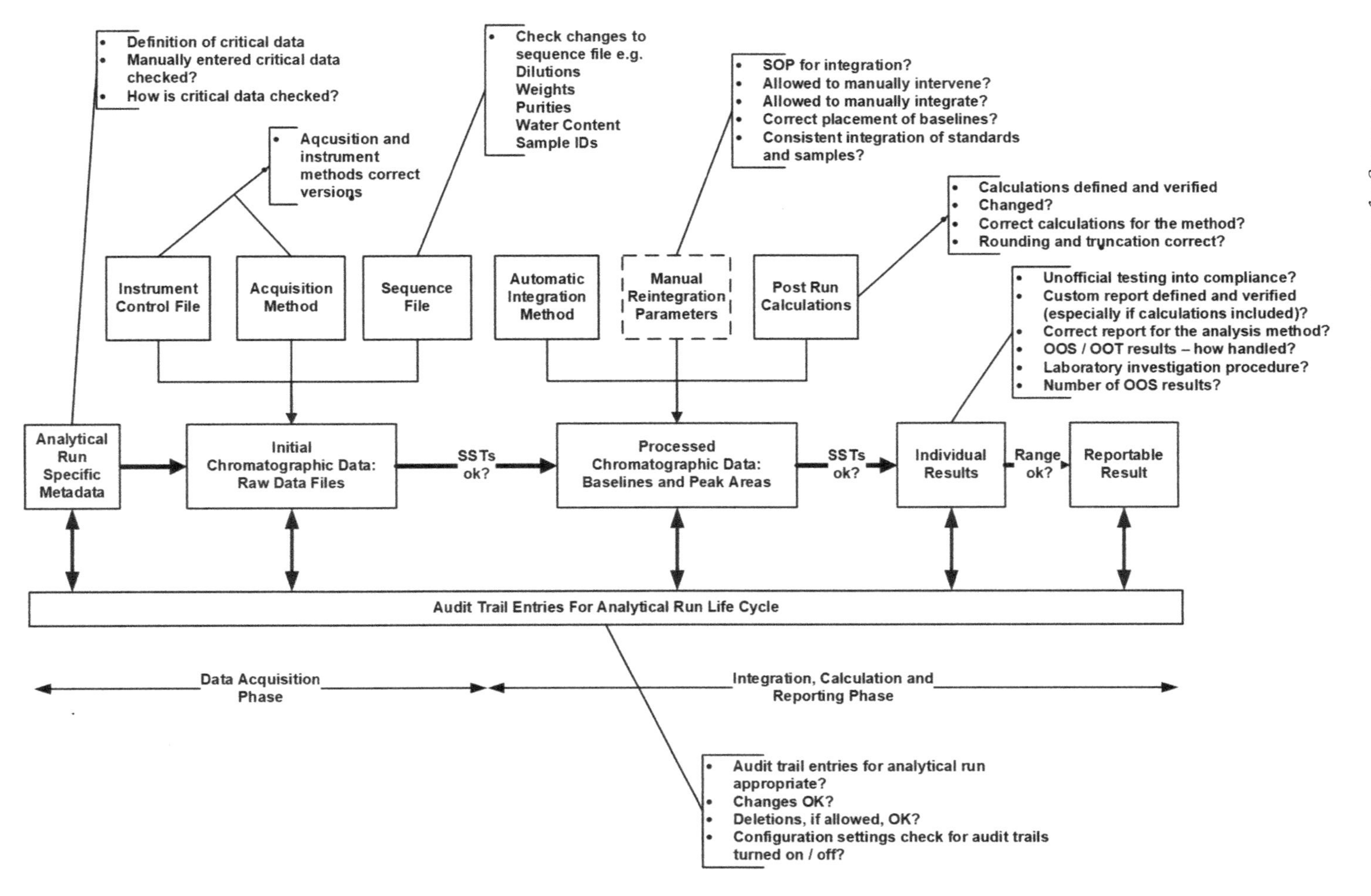

Figure 24.13 Chromatography data system audit aide memoire.

- Procedure for data integrity audits internally and suppliers available?
- Auditors are trained for data integrity audits?
- Data integrity audits are scheduled for the year?
- Audits scheduled have taken place?
- CAPAs have been raised in a timely manner from audits already performed?
- Input for management review?
- Reports for each data integrity audit available?

It is important that data integrity audits should not overlap with periodic review coverage of computerised systems.

24.6.3 Data Integrity Investigations

When data validity and reliability issues are discovered, the investigation should:

- Ensure that copies of all data are secured in a timely manner to permit a thorough review of the event and all potentially related processes.
- Interview staff (lab staff, supervisors, QA and management) to understand the nature of the failure and how it occurred.
- Identify what might have been done to prevent and detect the issue sooner.
- Investigate not just the specific immediate issue identified but also consider potential impact on historical events.
- Identify the root cause(s) of the issue to be considered including potential management pressures and incentives, including lack of adequate resources that may have led to the issue.
- Any corrective and preventative actions taken should not only address the identified issue, but also historical events and datasets.
- CAPA should address root causes, including the need for realignment of management expectations and allocation of additional resources to prevent risks from recurring in the future.
- Regulatory agencies should be informed if the investigation identifies material impact on patients, products or reported information or application dossiers.

Acknowledgements

I would like to thank Chris Burgess for the use of Figures 24.7–24.9 in this chapter.

References

1. *EudraLex – Volume 4 Good Manufacturing Practice (GMP) Guidelines, Chapter 9 Self Inspection*, European Commission, Brussels, 2001.

2. *USP 41 General Chapter <1058> Analytical Instrument Qualification*, United States Pharmacopoeia Convention, Rockville, MD, 2018.
3. R. D. McDowall, *Validation of Chromatography Data Systems: Ensuring Data Integrity, Meeting Business and Regulatory Requirements*, Royal Society of Chemistry, Cambridge, 2nd edn, 2017.
4. *ICH Q2(R1) Validation of Analytical Procedures: Text and Methodology*, International Conference on Harmonisation, Geneva, 2005.
5. G. P. Martin, *et al.*, Stimuli to the Revision Process: Proposed New USP General Chapter: The Analytical Procedure Lifecycle ⟨1220⟩, *Pharmacop. Forum*, 2017, **43**(1).
6. *Concept Paper: Analytical Procedure Development and Revision of ICH Q2(R1) Analytical Validation*, International Council on Harmonisation, Geneva, 2018.
7. M. L. J. Weitzel, *et al.*, Stimuli to the Revision Process: Measurement Uncertainty for the Pharmaceutical Industry, *Pharmacop. Forum*, 2018, **44**(1).

Subject Index